OBJECTIVE BIOMETRIC METHODS FOR THE DIAGNOSIS AND TREATMENT OF NERVOUS SYSTEM DISORDERS

ELSEVIER
science & technology books

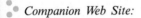

Companion Web Site:

https://www.elsevier.com/books-and-journals/book-companion/9780128040829

Objective Biometric Methods for the Diagnosis and Treatment of Nervous System Disorders
Elizabeth Torres

Available Resources:

- Selected figures from chapters
- Sample data from selected chapters
- Sample Matlab scripts to generate figure panels from selected chapters
- A bonus general methods folder with Matlab and python code
- Videos illustrating the experimental set up for scenarios and methods described in selected chapters

ELSEVIER

ACADEMIC PRESS

OBJECTIVE BIOMETRIC METHODS FOR THE DIAGNOSIS AND TREATMENT OF NERVOUS SYSTEM DISORDERS

ELIZABETH B. TORRES

Rutgers, The State University of New Jersey, New Brunswick, NJ, United States

ACADEMIC PRESS

An imprint of Elsevier

Academic Press is an imprint of Elsevier
125 London Wall, London EC2Y 5AS, United Kingdom
525 B Street, Suite 1650, San Diego, CA 92101, United States
50 Hampshire Street, 5th Floor, Cambridge, MA 02139, United States
The Boulevard, Langford Lane, Kidlington, Oxford OX5 1GB, United Kingdom

Notices
Knowledge and best practice in this field are constantly changing. As new research and experience broaden our
understanding, changes in research methods, professional practices, or medical treatment may become necessary.

Practitioners and researchers must always rely on their own experience and knowledge in evaluating and using
any information, methods, compounds, or experiments described herein. In using such information or methods
they should be mindful of their own safety and the safety of others, including parties for whom they have a
professional responsibility.

To the fullest extent of the law, neither the Publisher nor the authors, contributors, or editors, assume any liability
for any injury and/or damage to persons or property as a matter of products liability, negligence or otherwise,
or from any use or operation of any methods, products, instructions, or ideas contained in the material herein.

British Library Cataloguing-in-Publication Data
A catalogue record for this book is available from the British Library

Library of Congress Cataloging-in-Publication Data
A catalog record for this book is available from the Library of Congress

ISBN: 978-0-12-804082-9

For Information on all Academic Press publications
visit our website at https://www.elsevier.com/books-and-journals

Working together
to grow libraries in
developing countries

www.elsevier.com • www.bookaid.org

Publisher: Nikki P. Levy
Acquisition Editor: Natalie Farra
Editorial Project Manager: Kristi L. Anderson
Production Project Manager: Mohanambal Natarajan
Designer: Maria Ines Cruz

Typeset by MPS Limited, Chennai, India

Contents

Preface   ix

1. The Closed Feedback Loops Between the Peripheral and the Central Nervous Systems, the Principle of Reafference and Its Contribution to the Definition of the Self

Part I: Searching for Volition While in a Comma State   1

Part II: Physical Growth and Neurodevelopment in Neonates   18

Conclusions   41

Appendix—First-Order Stochastic Rule   42

References   43

2. Critical Ingredients for Proper Social Interactions: Rethinking the Mirror Neuron System Theory

Part I   45

The Human Mirror Neuron Systems   52

Rethinking Simulation Theory   59

How Do I Know It's Me Moving My Own Body With Agency?   66

Blindly Predicting the Task From the Performance Variability   83

Some Additional Thoughts   86

Part II: A Small Pedagogical Parenthesis About Taking Motor Control to the Classroom in a Do-It-Yourself Setting   89

Physical Movement Kinematics as a Window Into Mental Decisions   90

Where to Go With This?   91

Self-Supervision: Distinguishing My Self-Generated Kinesthetic Noise From Extraneous Noise   91

Self-Discovering Cause and Effect: Autonomy From the Bottom-Up   92

The Bridge Between Mental Intent and Physical Volition   100

Habilitation and Enhancement of Volition to Evoke Vocalization   104

References   107

3. The Case of Autism Spectrum Disorders: When One Cannot Properly Feel the Body and Its Motions From the Start of Life

Part I: How Do You Measure Autism?   111

Measuring ASD Objectively, That Is   113

One Plain Random Process in Nature vs Many Running in Tandem Within Interconnected Systems Poised to Develop Intelligence   118

An AHA! Moment   126

Mind the Gap: Behavior is Much More Than Mouse Clicks   135

Science Meets Technology Transfer and Commercialization   140

Parameterizing Changes in the Nervous Systems in Near Real Time   142

Part II: Physiological Signals Underlying Cognitive Decisions   145

Motor Research With Children is More Challenging so it Requires New Methods   150

Sensing Through Movement: When not all Noise is Created Equal   162

The Autistic Somatic-Motor Phenotype   168

Forty Subjects are not Enough   173

When Open Access Came to the Rescue   176

References   181

4. The Case of Schizophrenia: Is that My Arm Moving on Purpose or Spontaneously Passing by?

Part I: Predictive Reach: From Brownies
    to a PDE   183
Finding a Problem Worthy of a PhD Thesis in
    Cognitive Science   186
Rethinking Some Questions to Ask in Neural Motor
    Control   190
Two Vantage Points: The External Observer *vs.* the
    Internal Nervous System   193
Justifying a Geometric Approach to Movement
    Modeling   196
The Geometrization of the DoF Problem to Move
    along the Shortest Paths in Space and
    Time   197
Generating Unique Geometric Solutions   201
Locally Linear Isometric Embedding of X
    into Q   206
DoF Decomposition, Recruitment, and Release
    According to Task Demands   209
Empirical Evidence for Speed Invariance   211
Invariants of the Geodesic-Generating PDE   219
My Time at CALTECH: In Search for Evidence
    within the Neural Code   221
Geometric Invariants of Unconstrained
    Actions   231
Part II: Switching Research Paths for Clinical
    Applications   235
The Experimental Paradigm   237
Parsing DoF Reveals Excess Deliberateness in
    Automatic Movement Segments of PD
    Patients   243
Schizophrenia Patients: Did I Mean to Do What I
    Just Did?   248
Appendix 1   259
Coordinate Charts and the Inner
    Product   259
Appendix 2   260
The Gauss Map and Its Differential   260
Appendix 3   261
Testing Geodesic Property and Obtaining Curvature
    Measures with the Model   261
Appendix 4   263
Switching between Tasks   263
Appendix 5   265

Coordinate Transformations and New Distance
    Metrics   266
Appendix 6 Numerical Estimation of Positive
    Definite Coordinate Transformation Matrix   266
References   267

5. Learning to Detect Expertise in Sports Aided by the Gift of Our Students

Uncertain Times   271
Spotting Genius   280
Measuring Motor Learning in Sports   293
Designing a Motor Control Experiment using
    Sports   297
Bernstein's DoF Problem in the Boxing
    Routines   305
Sensing the Future Speed in Different
    Contexts   317
References   322

6. Rethinking Diagnoses and Treatments of Disorders: The Third (Objective) Neutral Observer Assessing the Interactions between the Examiner and the Examinee or the Therapist and the Client

From Ballet to Personalized Precision
    Psychiatry   325
Some Thoughts on Our Social Mental Spaces   327
Nobody Prints!   329
But First, Mind the Gap!   332
Ballet Partnering: The Dance of Coupled
    Biorhythms   339
Dynamic Diagnostics and Outcome Measures
    Using Phylogenetically Orderly
    Taxonomy   353
Closing the Feedback Loops in Parametric
    Form   372
Toward Dyadic Interactions between Children
    and Avatars   377
The ADOS Dyadic Exchange: Another Form
    of Social Dance   380
Two Wings of the Same Bird   381
References   389

7. Different Biometrics for Clinical Trials That Measure Volitional Control

Part I Opening Pandora's Box   391
The Hope Keepers of SMIL   392
From Deliberate Autonomy to a Measure of Quality
   of Life   393
From Involuntary Head Motion in FMRI to Faulty
   Diagnosis of ASD   396
Dual Diagnosis and Recommended Treatments in
   Neurodevelopmental Disorders Under
   Pre-Imposed Assumptions   399
The DSM   403
Pandora's Box Let Out Something Awful   415
Excess Noise in Autism   425
Significance and Potential Consequences
   of These Results   428
The Age-Dependent Shifts in Probability
   Distribution Functions   433
The ADOS   439
The 5:1 Females to Males (Statistically Impossible)
   Ratio of ASD   440
Part II Biomarkers for Clinical Trials   454
SHANK3 Deletion Syndrome   456
Tracking a Clinical Trial Using Gait
   Biometrics   457
Longitudinal Tracking of Somatic-Motor Change
   Across the Group   472

Individualized Tracking of Trial Effects   479
References   495
Appendix   499

8. Adding Dynamics to the Principle of Reafference: Recursive Stochastic Feedback Closed Control Loops to Evoke Autonomy

Part I Everything Is Sound   503
A Wink and a Blink Contain a Spontaneous
   Segment—And so Does a Reach, a Walk,
   and Every Other Complex Bodily
   Movement   508
Let's Play Back the Sounds We Make When We
   Move   516
Exploring Sound Preferences: Entrainment Beneath
   Awareness   525
Objective Outcome Measures of Treatments:
   Drug Trial Revisited   535
Objective Outcome Measures of Treatments:
   Sensory-Based OT   536
Closing Remarks   551
References   551

Index   553

Preface

This book is a personal journey through the science-making of my lab and the process of discovery that brought me here today. Originally, I intended to write a different book, one more formal to serve as a textbook or a recipe of sorts to perform behavioral experiments and do behavioral analyses in a new, objective and personalized manner. But as I started to write the book, the stories flowed more naturally when I narrated them as they took place over the years. Then, it became much easier and more fun to write about the process of discovery, as it happens from day to day in the collective of a laboratory. There, amidst instruments, computers, manuals, peer reviewed papers, books, and participants of all ages with different clinical conditions, is where one realizes how beautiful the scientific enquiry is and how surprising and at times serendipitous, the path to discovery can be.

Most of our studies started with a question in mind that evolved over time. The line of enquiry we followed set us on a path of detective work, deeper into hidden aspects of the problem at hand, without any preconceived agenda. We let that path of enquiry unfold and take us to unexplored places. Discovery self-emerged from these travels and in a synergistic cooperative effort, we all contributed to these discoveries. It has been so much fun to work with the people that have crossed paths with the lab on their way through life. Together, in those transient periods at my lab, we have built a harmonious environment where thoughts flow freely and creativity is nurtured. From undergraduate to doctorate students, from postdoctoral scholars to well established professors and clinicians, we have learned to respect each other's skill sets and contributed to the accumulation of new knowledge -unprecedented at times and complementary in nature. We have learned to build our own vision of the brain-body functionality and have adopted new philosophies to help unravel hidden mechanisms leading to the self-emergence of autonomy and agency, as fundamental properties of our human existence. It is my hope that you find the book interesting and useful.

There is no prescribed order to follow when reading the book. One can go in sequential order, from Chapter 1 to Chapter 8; or open the book at random in any chapter and start reading. The figures will help the reader follow the story and many of them can be reproduced using scripts in Matlab and Python, which I will place in a companion website to that end.

Although the stories are written in a somewhat informal way, the work is rigorous, and the material could be difficult at times. Because of that, I will place much of the material to reproduce figures and results in the companion site with further explanations, sample data and heavily commented sample code. This will help the reader recreate some of the key figures and use these analytics in their own work. It is my hope that the new personalized methods are adopted as a starting point for a radical departure from current subjective methods of behavioral analyses. Whether studying sports, the performing arts, or assessing clinical cases, whether tracking a student-teacher interaction, a parent-child exchange or a clinician-patient communication, the methods in the book offer a new way to track social cohesiveness and the emergence or absence of rapport between two interlocutors, in an objective, data-driven way.

This is a starting point to create a new behavioral science. As such, the book will greatly benefit from the contributions of students and instructors who will hopefully adopt and teach the new methods. These methods are aimed at the development of a truly open objective way to do science in the brain- and health-related disciplines.

I submit to you the work of 20 years, that started when I was a graduate student. This work was done in collaboration with many colleagues, and enriched by the creative thinking and the young spirit of folks who think out of the box. Keep it moving, onwards and forward, to transform and innovate!

Elizabeth B Torres, PhD

(Dedicated to my parents)

1

The Closed Feedback Loops Between the Peripheral and the Central Nervous Systems, the Principle of Reafference and Its Contribution to the Definition of the Self

Voluntary movements show themselves to be dependent on the returning stream of afference which they themselves cause. **Erich Von Holst and Horst Mittelstaedt**

PART I: SEARCHING FOR VOLITION WHILE IN A COMMA STATE

Serendipitous Encounters

Sometime in the fall of 2016 I visited the neonatal intensive care unit (the NICU) of the Robert Wood Johnson Medical Hospital. The Director of the NICU, Dr. David Sorrentino, hosted me on occasion of setting up collaborative work between my lab and his unit. Earlier that year, I had finished the analyses of data from neonates and had discovered a way to detect stunting in the development of neuromotor control.[1] The work with the babies sparked my interest on the question concerning the emergence of volition in the nascent nervous systems of the neonate.

As I toured the unit and saw six premature babies fighting for their lives, I marveled at the miracle of life winning small battles day by day. Amidst tubes and probes recording the vital signs of the baby, autonomy was gradually emerging and gaining stability. First, the heartbeat stabilizes, and then the respiration gains autonomy. Finally, the digestion

1

and gut providing nourishment and cleansing the body through autonomic processes supported by the neuroimmune systems[2] begins the path of shifting from peripheral to central control. The central nervous systems (CNS) will stabilize and find its way to the volitional control of the body in motion. The evolution and critical milestones of this precognitive state of being, even prior to the development of cortical and subcortical bodily maps of actions and their consequences, can be captured by statistically tracking the adaptive capacity of biophysical rhythms and their stochastic signatures.

The work on neonates involved collaborating with an infant-development lab from the University of Southern California (USC) directed by Dr. Beth Smith. Beth and I had met at an Autism conference hosted by the Profectum Foundation. This conference took place on March 21, 2014, in Pasadena, CA (https://profectum.org/awakening-potential-through-brain-science-conference-agenda/). I was invited by Dr. Serena Wieder and Dr. Ricki Robinson to deliver a keynote lecture. At the end of the lecture, Beth came to the podium and showed me a wearable sensor developed by a company in Portland Oregon, APDM (Fig. 1.1); with a number of features amenable to expand the type of research I was doing using high-grade tethered sensors. The size, weight, battery functioning, and memory storage capacity of the sensor made it ideal for the type of "on the go" research concept I was starting to develop in the lab. Indeed, we had created a new statistical platform for individualized behavioral analyses (we called it SPIBA). I was using SPIBA to develop new data types and methods for Precision Medicine[3] and mobile Health (m-Health). These new research programs in the lab required the type of portability such sensors offered. During that time and as a result of funding granted by the National Science Foundation (NSF), my PhD student Jillian Nguyen and I had taken the path of Innovation Corps (I-Corps) aimed at translating scientific discoveries into societal innovations. Our patent pending technology (https://www.google.com/patents/US20140336539) was disruptive as it called for a radical change in the ways we conduct scientific work and measure the outcome of interventions in clinical practices.[4] As part of the NSF program, we needed to find the market fit for our technology. To that end, we interviewed over 100 stakeholders in the autism ecosystem. This interview process revealed a critical need for objective outcome

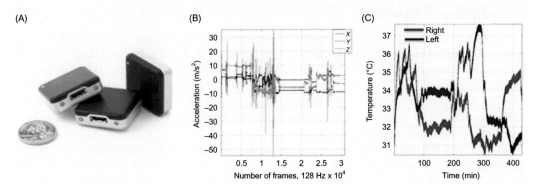

FIGURE 1.1 **Sample wearables and output traces.** (A) Wearable sensors from the APDM company shown at scale next to a dime. APDM sensors 128 Hz, (Portland, OR) at the time of our study in 2014 could register data from triaxial accelerometers shown in (B), gyroscopes, temperature shown in (C) and a magnetometer. They could store 8 Gb of data and had a battery that lasted at least 12 h of continuous recordings.

measures of the types we had developed using high-grade sensors in the lab. As it turned out, a huge problem in autism is the lack of insurance coverage to enable the diversification of therapeutic interventions leading to improvements on *sensory-motor patterns* and better motor control. Most such interventions involve occupational therapists (OTs) in the pediatrics field. Our SPIBA was something OTs really needed and desperately wanted.

To provide a complete platform for m-Health (also known as smart and connected health), we needed to add the portability of wearables with the ability to harness biophysical rhythms from the nervous systems. Such biorhythms could be collected as the person *naturally* interacted with the environment, performed activities of daily living, and rested during sleep. The market was beginning to be flooded with such sensors, ranging from Fitbits to smart watches from major companies like Google, Samsung, and Apple. However, using off-the-shelf sensors we could only access already-filtered data and outcome measures that lacked the reliability we needed for our research program. Our clinical research in particular required access to the actual raw data because the filtered data that off-the-shelf sensors offered may miss or mask frequencies with physiological relevance, particularly in nascent nervous systems or in nervous systems with pathology of unknown origins.

The activity trackers or the wellness and fitness bracelets in the market did not offer the features we needed for our clinical research. The APDM wearables that Beth showed me at the conference did offer the possibility of accessing raw signals of broader bandwidth than other commercial sensors (Fig. 1.1).

Beth and I kept in touch after the meeting and developed a collaborative link that led us to the discovery of patterns of stunting in the neonate[1] that I will describe in the second part of the chapter. Her lab coordinated the study—a major feat—and collected the data to track the development of walking patterns in preterm versus full-term babies.[5] My lab designed biometrics to track the longitudinal evolution of such patterns across many hours per day, during multiple visits. I used this Big Data set to design indexes of stunting in the neurodevelopment of their nascent nervous systems.

The data type and analytics that I originally developed to study the emergence of neuromotor control in the neonates did not come from newborn babies though. They came from work I did concerning the biophysical rhythms that the wearables harnessed from the nervous systems of a pregnant lady who had slipped into a coma because of a debilitating seizure linked to a large brain tumor in her frontal lobe.[6]

The work with the coma patient gave me the ability to study a nervous system where most likely the bridge between the CNS and the peripheral nervous systems (PNS) had been disrupted. The work with the neonates provided me with access to a nascent nervous system where that bridge was developing. Both data sets could give me a window into the inner workings of the PNSσCNS closed-loop connection as the nervous systems gained deliberate autonomous control, control at will of the brain over the body in motion. This form of control is called *volition* in our movement neuroscience field.

From Spontaneous Random Noise to Well-Structured Signals

Looking back at that Profectum Conference, a series of serendipitous events took place the day of my lecture that led me to the development of the methods I describe in this chapter. But before I delve into the methods, let me first take a detour into how I came to derive the analyses of the data from Melissa Carleton, the pregnant patient undergoing a coma state at the time (https://www.facebook.com/supportmelissacarleton/).

The day of the Profectum Conference where I met Beth, as I was approaching the podium to deliver the keynote address, I received a phone call from Brian Lande. At the time, Brian was a consultant for one of the Defense Advanced Projects Agency (DARPA) programs on strategic social interactions. I had met him while attending a DARPA meeting in Washington, DC, on December 2013. At that meeting I delivered a lecture that Brian recalled during the phone call that day in Pasadena, a few months later, on March 2014.

I barely knew Brian and had no name on the caller ID, but answered the call anyways because I recognized the area code was from the Bay Area, where I have family. To my surprise, Brian re-introduced himself and explained what had happened to his pregnant wife Melissa. At the DARPA meeting, he had talked about their recent wedding. I remembered the picture from her as she looked strikingly beautiful (Fig. 1.2). I was frankly a bit shaken by the news that she had slipped into a coma. Brian sounded quite desperate. He mentioned my research on intentional motions and asked me if there was anything that I could think of, to help him proof that his wife was still there (i.e., that she had will.) He was worried that with a few months left of her pregnancy and the sudden coma diagnosis it was hard to know the full prognosis of her case and the possibilities for the birth of their baby. Clearly, between the shocking news and the talk I had to immediately give, I was a bit perturbed. I asked Brian to give me a moment to think about it and told him that I would call him back soon. I somehow put that question on hold in the back of my head for the next hour or so. At the end of my talk, when Beth brought me the APDM sensor, and I saw the portability of it, I got my answer.

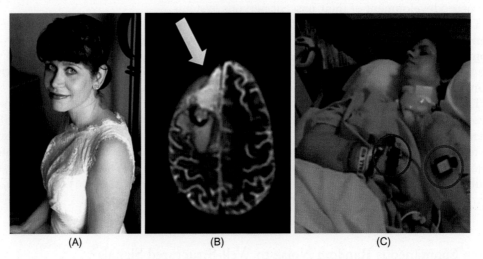

| (A) | (B) | (C) |

FIGURE 1.2   **The comma patient.** (A) Melissa Carleton photographed at her wedding day. (B) Brain scan showing the frontal lobe after the removal of a tumor that induced a debilitating seizure preceding her coma state. (C) The use of wearables at the hospital.

All I needed to do was to monitor her motions continuously and detect volition. But volitional control was precisely what I had been working on since 1998, the year when I finished the first draft of a computational model of volition in systems with redundant degrees of freedom (such as the arms and their end effectors). The preliminary work toward my PhD Thesis at the University of California, San Diego, performed under the supervision of Prof. David Zipser,[7] was a model that embodied the guiding principles of the research program I later developed in my lab, a program that I used to help detect volition in Melissa's motion. I will discuss that model in great detail in Chapter 4.

At the heart of this problem are spontaneous fluctuations inherently present in the biorhythms that we can harness from the nervous systems. At a first glance, these moment-by-moment fluctuations are random and have high noise-to-signal ratio (NSR). They are subtle in nature and tend to go unnoticed by the naked eye of an observer. One may generally need proper instrumentation to capture their evolving signatures. Nevertheless, the ever-changing peaks and troughs of the nervous systems biorhythms contain information of relevance to our quest on volition. But to extract such information, one needs to employ techniques that *empirically* characterize the inherent variability of such data, i.e., without enforcing theoretical assumptions that do not fit the data well.

Up until that point, the bulk of the research from my field (motor neuroscience) had a main focus on the neuromotor control of overt goal-directed movements.[8] However, research on the types of spontaneous motions that I was interested in was nonexistent. I had to create a new notion of different and interlayered classes of movements[9] with taxonomy of controllability and phylogenetic order of emergence and maturation.

This architecture of the neuromotor control problem afforded the measurement of biorhythms harnessed from the nervous systems output to profile their stochastic signatures and find their typical ranges (e.g., those in Fig. 1.3). In this way we could begin the path of identifying deviations from typical neuromotor control trajectories. Yet, to study such biophysical signals, we required a new statistical platform and new data types to accomplish two tasks: (1) integrate data from different interconnected layers of the nervous systems (e.g., the systems for autonomic and voluntary control) and (2) map the different levels of variability inherently present in these different systems onto different levels of neuromotor control (Fig. 1.3).

The first step toward gaining a better understanding of the various nervous systems pathologies was to begin an exhaustive statistical characterization of the nervous systems biorhythms as they naturally evolved and matured across the human life span. In this sense, we had to break away from the current *"one-size-fits-all"* statistical model of the brain and health sciences. Such a model did not leave room for the emerging concept of personalized medicine[3] that my lab was already pursuing. The current statistical approach (Fig. 1.4) simply assumes "ideal" populations' mean and averages out as noise the very signals we needed to understand: the subtle moment-by-moment fluctuations in the amplitude and timing of these biorhythms peaks and valleys. To accomplish these two tasks we created *the micro-movements* and paired them with the SPIBA framework (Fig. 1.5).

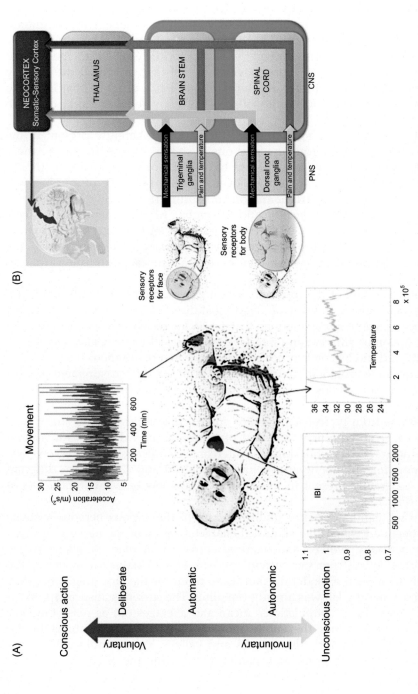

FIGURE 1.3 Taxonomy of neuromotor control and ways to measure variability in the nervous systems biorhythms from interlayered levels spanning from autonomic to voluntary control. (A) Phylogenetically orderly appearance and maturation of control levels in the PNS and CNS and their composition of involuntary motions contributing to or interfering with volition (voluntary control at will). Examples of involuntary motions with "good" variability facilitating the emergence of voluntary control manifest in the evolving reflexes of the neonate, leading to the awakening of the body and the mapping of peripheral motions, their sensations and consequences onto the CNS. Examples of involuntary motions with "bad" variability interfering with volition are the various types of resting and intentional tremors in Parkinson's disease. All these biorhythms can be harnessed with wearables today. They can also be integrated in combined signals from multiple layers and their combined contribution to neuromotor control statistically characterized as such. (B) Different maps in the PNS contribute to different maps in the CNS. They involve the probabilistic sampling of self-generated and self-sensed biorhythms' fluctuations and their statistics. *Source: Adapted from Purves Neuroscience 2008.*

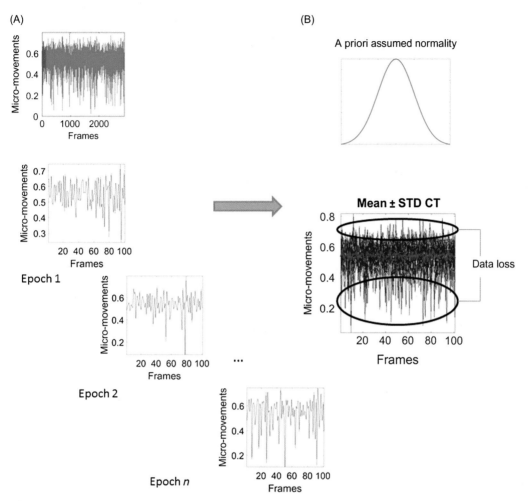

FIGURE 1.4 **Data waste with current statistical assumptions.** (A) The standardized micro-movements extracted from some nervous system signal are scaled between 0 and 1. These spike trains represent a continuous random process modeled here as a Gamma process. The NSR of these spike events are of interest in the statistical analyses of the SPIBA framework. Traditional models assume a Gaussian random process with additive statistics. As such, the assumed theoretical (population) Gaussian moments (the mean and the variance) are used to process the data using the average of the waveform's peaks across a preset number of frames. (B) Typically, preselected epochs of the data are averaged under this Gaussian population mean assumption. This grand averaging method smooths out the fluctuations in the signal, thus incurring in gross data loss. This example shows the mean $+/-$ the standard deviations comprising the data that gets analyzed. The circled 'excess' is thrown away as noise, but it actually contains the very information we need. This is the traditional "*one-size-fits-all*" approach to data analyses in the health and brain sciences today.

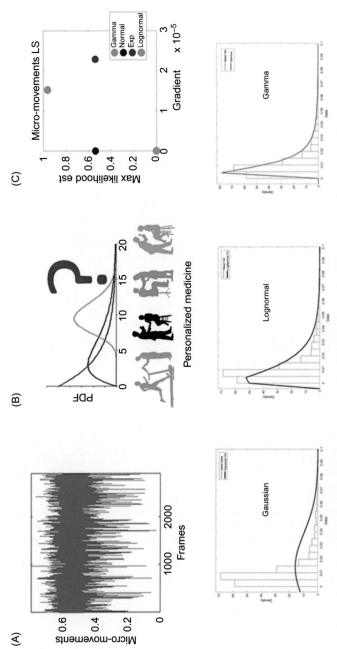

FIGURE 1.5 **SPIBA as a new step toward personalized approaches and the analyses of natural human behaviors over time.** (A) The micro-movements waveform of fluctuations in amplitude of a nervous system biorhythm in Fig. 1.4A is analyzed using SPIBA. This step provides a data transform that scales the original waveform and standardizes it to avoid allometric effects that arise from disparities in anatomical lengths across the population. (B) SPIBA does not assume a priori any theoretical distribution. Instead, it accumulates events till the estimation process yields tight confidence intervals for the fitting of various families of PDFs. (C) In this case, maximum likelihood estimation (MLE) is used to evaluate the goodness of the fitting of various sample probability families shown at the bottom panels: the Gaussian distribution, the lognormal distribution, and the Gamma distribution. Panel C shows the results of running MLE. The horizontal axis shows the values of the gradient output from the optimization function (they should be very small) and the vertical axis shows the likelihood value, which is clearly highest for the Gamma family. One should test different families of probability distributions before settling on one to characterize the random processes our data represent. In this way we do not waste as much data as it is currently done and have a chance to use better statistical inference and interpretation of our data's inherent variability.

The Micro-Movements Perspective

How could we study Melissa's subtle movements, i.e., those movements seemingly spontaneously generated by her nervous systems? Were they volitional in nature? Averaging those fluctuations, as it was commonly done, would smooth out the very signal I was interested in. It would throw away as noise the information that I needed to determine if these motions had in any way a nonrandom signature. In other words, was there anything anticipatory or predictable about these seemingly spontaneous random motions? The micro-movements that we had invented to study volition in autism[10] could hold the answer to our questions here.

But, what are the micro-movements that the SPIBA framework uses? The raw biophysical data continuously registered from physiological sensors (i.e., sensors registering *physiological rhythms* such as electroencephalography, electrocardiogram, respiration patterns linked to muscles electromyographic activity, kinematics from bodily, head and eye movements, tremor data, etc.) give rise to continuously changing peaks and valleys of various amplitudes occurring at variable times within a time series. These peaks understood as spikes of variable fluctuations in amplitude and timing can be construed as a continuous point process, i.e., following a continuous random process. In this process, events in the past may (or may not) accumulate evidence toward prediction of future events. The spike trains derived from such peaks and valleys in the continuous analogue data from high sampling resolution sensors could serve as input to different classes of (stochastic) random processes. These *"micro-movements"* data type that I invented are described at length in the Supplementary Materials of various papers from my lab, e.g., 1,11,12. To create them, I was first inspired by electrophysiology research on neocortical neurons,[13] but then realized their utility in the context of other bodily biorhythms from the periphery.[9]

The micro-movement waveforms derived from the time series of multiple parameters harnessed from the nervous systems biorhythms (e.g., kinematics signals in Fig. 1.3) are used to represent a continuous random process under the general rubric of Poison random process. To be more precise, we treat the spikes in the first rate of change in various signals' motions as spikes of random amplitudes and random times. To model them, we build on our original work[10] whereby the amplitudes and inter-spike interval times are modeled as independent and identically distributed (iid) random variables following a Gamma process where the continuous Gamma family of distributions is used to model the process. This iid assumption is one that we relax later in Chapter 8, to be in tune with the time dependencies of nervous systems processes and their accumulation over time. In this chapter, however, I will focus primarily on iid assumptions because even with the limitations they may impose on data harnessed from a self-supervising and self-correcting biological system—such as the nervous system—these new methods pose nontrivial improvements over old ones in the field. As such, this is a first step of many iterations until we get closer to the true nature of an intelligent system that learns to heal itself. Sample pipeline of data processing are shown in (Fig. 1.6)

Under iid assumption and upon empirical estimation of the Gamma parameters, we track their values on the Gamma parameter plane, compute the empirical probability distribution functions (PDFs), obtain the empirical summary statistics (the moments), and integrate various such signals from multiple layers of the nervous systems (Fig. 1.7).

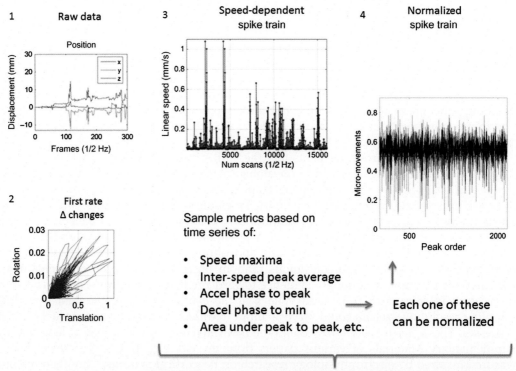

FIGURE 1.6   **Example of micro-movements extraction from kinematics data** (Step 1) Obtain raw positional data (these can be from linear displacements or angular rotations). (Step 2) Examine excursions of various types, e.g., first rate of change in displacements and rotations (plotted here), cumulative excursions, etc. (Step 3) Obtain the scalar (speed) magnitude of the first-order rate of change time series (i.e., commonly termed velocity-dependent data). In this case the velocity obtained from positional data is computed and the scalar value (speed) obtained. The peaks are used to study their fluctuations in amplitude and timing (spike trains). Sample kinematic metrics derived from the velocity-dependent data are (among others) spike trains of speed maxima, of inter-speed peak average, inter-minima speed average, acceleration (rising phase) to the peak, deceleration (decay phase) to the minima, area under peak to peak, area under minima to minima, etc. (Step 4) Normalize the micro-movements to create unit-less quantities that provide a standard scale and account for allometric effects due to disparity in anatomical features in cross sections of the population, age disparity, among others.

Amidst this process of gathering data in tandem from multiple layers of the nervous systems and performing stochastic analyzes to empirically characterize their inherent variability, we profiled the micro-movements from multiple layers of the nervous systems across the human population.[11] Part of the goals of our research program was to provide a standardized measure of all these levels of control and their neurodevelopment, independent of anatomical differences and chronological age. Another long-term goal was to provide a scale-invariant metric amenable to integrate *discrete* (ordinal) clinical scores, currently used

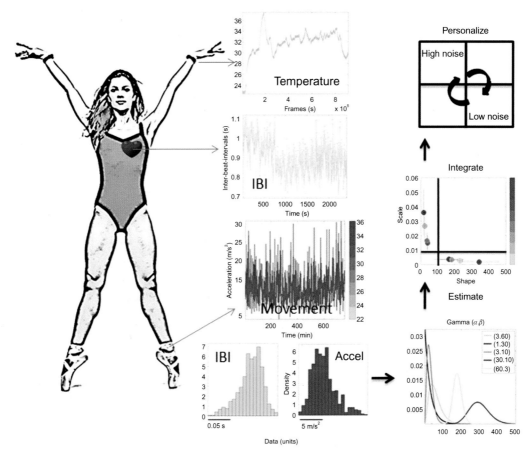

FIGURE 1.7 **Integrated micro-movements from different nervous systems' biorhythms to provide an empirical characterization of volitional control or lack thereof.** Wearables provide different signals. The micro-movements waveforms are extracted and their statistical signatures estimated from the Gamma process to derive the shape and dispersion (signal-to-noise ratio) of the micro-movements. They determine transitions from spontaneous random noise to well-structured signal with systematically predictive power indicative of adaptive control and the emergence of volition.

to classify pathologies of the nervous systems, with *continuous* physiological signals obtained noninvasively from the naturally functioning nervous systems.

In this sense, our research program aimed at creating new ways to track and visualize performance outcome in near real time, involving various levels of enquiry. These levels of enquiry range from a macro-level of observational description of behavior to a functional micro-level, considering as well within these disparate scales, different states of controllability (from spontaneous to intentional). Ultimately, our goal has been to produce objective outcome measures that enable evaluation of the person's quality of life, through the assessment of his/her volitional control and self-agency over the body in action. Arguably, without such autonomy it becomes difficult to have high quality of life.

The medical field is filled with observational inventories and self-reports. This is their data. As such, somehow basic science registering continuous physiological data is constantly forced to correlate "apples and oranges." The medical literature has abundant cases where the continuous biophysical data has been (inappropriately) correlated with discrete ordinal data without any justification about their underlying statistical assumptions. For example, observational scores that a neurologist or trained clinician may take from a patient with Parkinson's disease using the Universal Parkinson's Disease Rating Scale are often correlated with kinematics data (speed, acceleration, end-point kinematic errors in a target-pointing task, etc.) However, while the clinical scores assume normality to build their scales, such data from kinematics parameters are not normally distributed. Furthermore, the kinematics of human movements arise from highly nonlinear interactions across systems in motion. Such motions are generated with a high number of degrees of freedom causing highly nonlinear interactions. Yet, the basic research literature tends to apply multivariate linear regression analyses and other parametric tests based on assumed statistical features that such data violate.

The same type of problem exists in the literature of basic research on developmental disorders like autism spectrum disorders (ASD). There, the Autism Diagnosis Observational Schedule,[14,15] built under assumptions of normality and linearity, generates discrete scores with no proper metric. Further, there is no neurotypical data to build a relative scale of departure from normality. Researchers are forced to correlate these ordinal discrete scores with continuous physiological data in the real or complex domain. It is all very puzzling to me. The kinematics of bodily biorhythms also arises from highly nonlinear processes. The developmental data obtainable from trajectories describing voluntary movements show evolving families of distributions unique to each child (Fig. 1.8). These distributions are empirically well characterized by the continuous Gamma family spanning skewed distributions, inclusive of the (memoryless) exponential. As such, the Gaussian assumption of a population ideal mean tends to mask the maturation process that typically takes place in a growing child with a developing nervous system.[11] The methods currently in use by most of the developmental research of the brain and health sciences mask the very information we need to detect the risk of developing a problem during neurodevelopment. This methodological flaw of the static statistics enforced on the dynamically changing data also prevents us from intervening early, before the neurodevelopmental problem becomes obvious to the naked eye. In Part II, we will revisit these issues in the context of neonates and detection of risk for neurodevelopmental stunting. This will soon be important to help us develop a proper understanding on the differences between staged actions and spontaneous motions.

Distinguishing Deliberateness From Spontaneity in Kinematics Signals

The spontaneous covert motion segments I discovered in an earlier work involving athletes[9] were far more sensitive to the external environmental and the internal bodily-driven influences than the overt deliberate motion segments coexisting in complex boxing routines. In that sense, the mental intent to move in a certain way could be captured in the physical realization of the actions carrying high certainty in the prediction of their possible consequences. Indeed, using our SPIBA I could extract and differentiate the

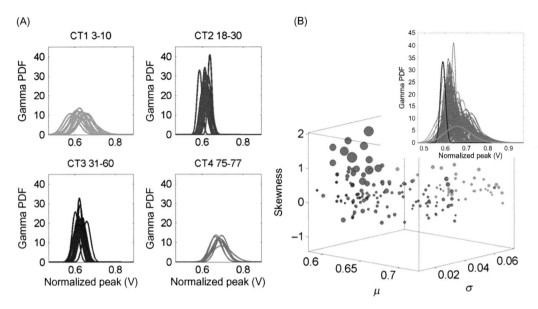

FIGURE 1.8 **Developmental evolution throughout the human life span of PDFs characterizing the peak speed in our voluntary motions.** (A) From 3 years of age to 77 years of age, the stochastic signatures of the hand trajectory speed in route to a spatial target change dramatically. This simple pointing experiment was performed in 178 participants of multiple ages and conditions. This figure focuses on a subset of neurotypical individuals (with no diagnosis of a nervous systems disorder). Each curve represents one participant who performed over 100 of the pointing task in different intervals (to avoid fatigue). Then the micro-movements were extracted from the speed profiles of the goal-directed hand motions to the target and a Gamma process used to empirically estimate the signatures of the PDF best fitting the data in an maximum likelihood estimation sense (as in Fig. 1.5). From 3 to 10 years old the distributions are skewed and have large dispersion, then they turn more symmetric and with less dispersion (low NSR). By middle age they start to regress toward more dispersion and skewness and by the 70s they begin to look a lot like those of a 3-year-old child. (B) The estimation of the Gamma moments in the population at large are represented in a four-dimensional plot. The x-axis is the mean, the y-axis is the variance, the z-axis is the skewness, and the size of the circle is the kurtosis. The plot includes patient data as well, from participants with Schizophrenia (bigger circles with large kurtosis); Parkinson's disease of various levels of severity (middle of the plot) and the college (red) students contrasting with the 3-year-old (green) participants. Even without knowing the age labels, we can blindly identify two clusters far apart in this Gamma moments' space. The health and brain sciences assume Gaussian distributions regardless of age.

signatures of intentionally performed motions from motions spontaneously performed and largely beneath awareness. The key feature of the kinematics was that the variability patterns of the deliberate segments of the continuous trajectories from the body in motion were far more robust to changes in the body dynamics (e.g., changes in speed) than those which were spontaneously co-occurring. I will show these features in detail in Chapter 4, but for the purposes of this chapter, the important piece of information is that the stochastic signatures of the Gamma process used to describe the transitions from high to low noise, or those from random to predictable micro-movements, were well defined and consistently different for each type of motion. Volition had distinct signatures captured in the rates of change and transitions across the left upper (LUQ) and the right lower quadrants (RLQ) of the Gamma plane (denoted by LUQ and the RLQ in Fig. 1.9).

In the case of Melissa, her coma state was preventing her from performing overt movements and timely controlling her limbs in tandem with her mental commands. However, the micro-fluctuations of her seemingly spontaneous and small motions could also carry the stochastic signature of deliberate actions signaling the degree of intent her brain had to control her body. Our methods are scalable. They harness the micro-movements from both large overt movements and invisible minute motions. The stochastic nature of the fluctuations in amplitude and timing that these continuous biorhythms have could be tracked from day to day, over 12 hours of continuous recordings each day. These stochastic

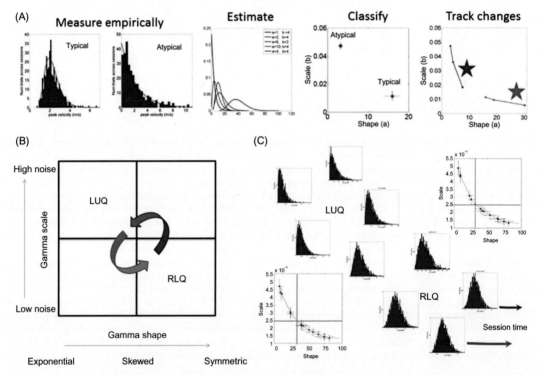

FIGURE 1.9 **SPIBA for independent identically distributed (IDD) events assumed in the spike trains (micro-movements) derived from biophysical rhythms harnessed from the nervous systems using non-intrusive wearable sensors.** (A) Gather the micro-movement signals into a frequency histogram. Estimate the probability density function, e.g., using the maximum likelihood estimation (MLE) methods. In this case the continuous Gamma family of probability distributions is a good fit. As such the Gamma process is used to estimate for each minute-long block of data the shape and scale parameters, plotted here on the Gamma parameter plane with 95% confidence intervals for two representative cases (TD vs ASD). The coordinates of the Gamma estimates on the Gamma parameter plane are tracked over time, as they shift along a stochastic trajectory across the 13 h of recordings. (B) The shifts jump from the LUQ of the Gamma plane (higher NSR and more skewed distribution shapes) to the RLQ with higher signal content (lower NSR) and more symmetric shapes. The amplitude and frequency of the shifts can be measured and tallied to quantify stationary from nonstationary transitions (see Fig. 1.16). (C) Representative evolution of probability distributions and their empirically estimated PDFs representing the correspondence to the LUQ and RLQ scenarios.

rhythms could then be construed as an amalgamate of random events coming from different levels of the nervous systems, ranging from autonomic to automatic to voluntary (as in the taxonomy of Fig. 1.3) and nonetheless tell us about the level of mental intent the brain was exerting over the physical body.

From Volition to the Path of Rehabilitation

Immediately after my talk ended at the conference, I called Brian and texted him a picture of the APDM sensor that Beth had introduced me to. Through the magic of the internet I sent the picture along with the company information. Brain called APDM and miraculously fast, he secured two sensors that he placed on his wife's wrists (Fig. 1.2). We set up a data-transfer system whereby he could send me the wearable sensors' data regularly (via the cloud) for 4 consecutive months. From April to July 2014 I regularly checked for patterns of wrist acceleration, rotation, and temperature in search of a sign of self-organizing patterns and changes in somatic-motor physiology suggestive of deliberate control. I reasoned that through the continuous tracking of the PNS activity—as we were routinely doing in the lab while using overt movements, I could gain a window into the amount of control Melissa's brain was likely exerting over her bodily motions. Indeed, the peripheral activity at the end effectors (the wrists in this case), as subtle as they may be, could serve as a proxy of her internal self-control and agency; perhaps she was trying to show others around her that she was still there, despite her clinically declared comma state.

Minute by minute, the methods integrated the micro-movements from the wrist acceleration and those harnessed from sensor's temperature combining the readings from her skin temperature, the battery's energy consumption and the ambient temperature (Fig. 1.10). The more actively she moved, the higher the fluctuations in temperature were on average. But it was the NSR derived from the Gamma process that told us something about her volition. The rates of change in this quantity served to distinguish, over the course of four months, those motions that were spontaneous and random in nature from those which were systematically predictive and high in signal content (Fig. 1.10C).

On the week of May 19–25 I detected a flurry of self-organized activity and a spike of change in the temperature output by the sensors. Tracking the patterns on the Gamma plane quadrants alerted me of a dramatic change in the levels of re-organization and predictability in her self-generated motions (Fig. 1.11A and B). Such systematic patterns flagged a level of volition that emerged during those days, i.e., the type of body control that we know from the literature on intentional actions that the brain exerts at will, with a form of consistent deliberateness that contrasts with spontaneous random fluctuations in motion patterns.[9,16]

The emergence of increasingly organized signals showed a clear transition from spontaneous random noise on the Gamma plane LUQ to well-structured signal on the Gamma plane RLQ. The statistical meaning of the Gamma plane that we had empirically mapped across a large cross section of the general population (represented schematically in Fig. 1.9B) paired with our empirical evaluation of Melissa's nervous system—as it longitudinally evolved, helped me statistically infer the potential significance of these stochastic shifts in the biorhythms these sensors were continuously outputting at the periphery. Indeed the precise prediction of the week when her baby was born (shown months later with his mom Melissa in Fig. 1.11C) was very encouraging for us.

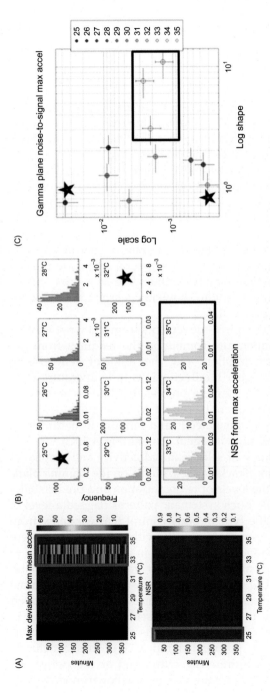

FIGURE 1.10 **Integrated micro-movements waveform from kinematics and temperature.** (A) The Gamma scale parameter is the NSR of the micro-movements. Here we can integrate acceleration and temperature by selecting for each temperature degree those acceleration peaks above the mean overall acceleration. This data type is provided on the top panel, where the highest range of acceleration peaks (minute by minute) is highlighted by a square and the color bar provides the actual range of motion (m/s²). The micro-movements from this integrated waveform (temperature-dependent motion) are then obtained and input to a Gamma process. The resulting color map matrix reveals the temperature range with the highest and the lowest NSR. The color bar shows the overall range of the data from 360 min. During those 6 h of continuous data registration we can track the evolution of the stochastic signatures in (B). Here we highlight the temperature-dependent motions with the least skewed shapes corresponding in (C) to the points on the Gamma plane enclosed in a rectangle. These are the data for 1 day across 6 h. By itself such data provides limited information, but the evolution of these stochastic signatures over 4 months revealed very important trends regarding volition.

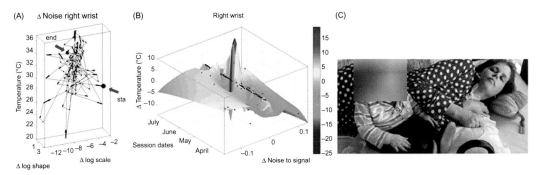

FIGURE 1.11   **Detection of adaptive change and the re-emergence of volitional control over time.** (A) As the micro-movements transition from spontaneous random noise to well-structured signal over 4 months of continuous recordings from Melissa's nervous systems, the fluctuations in temperature-dependent motion turn high in signal content and (B) pinpoint the birth date of her baby. (C) Melissa and her baby some months later.

FIGURE 1.12   **Volition.** Melissa reached for a bottle and moved the hand away from it using a controlled retraction motion.

At their end, Melissa and her family had initiated the path toward rehabilitation with the ups and downs the recovery period from such a traumatic brain injury entails. At some point, I was able to see Melissa reaching for a bottle in a short video that I decomposed frame by frame in Fig. 1.12. Executing such a motion is the hallmark of volitional control. She was without a doubt trying to do so very hard. Her resilience and strong will are extraordinary. Her family knows that, but as a researcher who has spent over 20 years studying volition and modeling the mathematics of reaching actions I know what performing that motion entails. This figure told me much more than meets the eyes. What an incredible journey this has been for her!

Revisiting the Principle of Reafference

From moment to moment and from day to day, the data that I analyzed predicted with ever-growing certainty the extent to which future events could be statistically linked to past events. In a way, I was stochastically characterizing von Holtz and Mittlestaedt principle of reafference[17–19] stating that "*Voluntary movements show themselves to be dependent on the returning stream of afference which they themselves cause.*" The powerful notion of this principle lies in the system's self-recognition of cause and effect, assumed to be directly derived from the expected consequences of its self-generated motions. What the research work with Melissa suggested was that spontaneous fluctuations (up to now neglected in the literature of motor control) provide a window into the *unexpected consequences* of self-generated motions, i.e., the sort of stuff that made visible to the nervous system these otherwise invisible motions.

In this sense, contrary to common belief, the type of variability the field rendered useless and averaged out as noise or a nuisance contained the signal I was looking for to detect volitional control in Melissa's case. Once discoverable, e.g., by "the surprise factor" with an outcome that was unexpected to her awakening nervous system, the spontaneous micro-movements could turn controllable and then transition from "invisible" to "visible" and meaningful. Through their evolution from totally spontaneous to discoverable fluctuations, the CNS could deploy trial and error states, whereby initially nonobvious goals would become interesting to the system. Something emerging as a detectable feature in the environment could serve as an anchor for trial and error (e.g., to the sense of touch, something as simple as a skin ripple could be detectable by the touch sensors on the hand surface). In this sense, trial and error could evolve from seemingly random to systematic. The signatures of fluctuations in the amplitude and timing of wrist acceleration could provide such information. We will see later that in the neonate system such evolution also manifests as the newborn infant transitions from flailing of the arms to goal-directed pointing. The theoretical model I had derived earlier on in my scientific career[7,20] to study sensory-motor integration, action ownership (agency), and volition within redundant, highly nonlinear systems once again provided me with a road map to pose my questions and formalize the problems I needed to solve.

Our work with Melissa suggested that our SPIBA framework and micro-movements data type integrating these multiple biorhythms across the developing layers of controllable signals were appropriate to capture change in their statistical patterns. Armed with these tools we proceeded to examine the longitudinal changes that neonates undergo as their bodies move and their brains begin building maps to help form a bridge between the PNS and the CNS.

PART II: PHYSICAL GROWTH AND NEURODEVELOPMENT IN NEONATES

Some Detective Work

How does volition emerge in neurodevelopment? How is neurodevelopment linked to physical growth? And how does physical growth is measured? These were some of the questions my lab started to ask after we completed the research work with the comma patient.

Then, I learned that according to the Food and Agriculture Organization (FAO) of the United Nations, as of 2013, there are 161 million children under 5 years of age estimated to be *stunted* on their growth, with half of all stunted children located in Asia and over a third in Africa. Wasting from starvation and malnutrition are considered among the causes of stunting with a global prevalence of stunting at almost 8% of the reported total and 3% of that number accounting for severe stunting (see **Notes**[1-3]). Malnutrition has also been studied using *weight* as one of the parameters to measure it, with criteria that include both underweight and overweight conditions, whereby according to these 2013 reports, the World Health Organization (WHO) estimated worldwide 99 million children less than 5 years of age are underweight, while 42 million of that same age are overweight (see the URLs given earlier).

I was not exactly aware of these problems before Sejal Mistry, an undergraduate student from Biomathematics at Rutgers University, came to my lab to volunteer as a researcher. As a premed student, Sejal had shadowed several pediatricians in the low-income districts of Central New Jersey. She brought to the lab the concept of *the growth charts* and their use to track the infant's wellness and growth rates in the first 5 years of life. The charts are part of an effort by both the Center for Disease Control, CDC[21] and the WHO[22] to track developmental progress of infants and young children in standardized ways, i.e., tracking the physical growth of breast-fed babies of all ethnic and racial backgrounds.

The charts popped up in the lab around the time we had completed the coma study and predicted the birth of Melissa's baby (in Part I). Sejal kept mentioning that the longitudinal measurements we had designed to detect the birth date of the baby would be ideal to build something like the growth charts, but instead do the longitudinal tracking for pregnant women in general. However, I had no idea what she meant by that, because I had never come across "*the growth charts*" she referred to.

It wasn't until December 2015, when Beth provided us with the neonates' data from her lab, that I actually made the connection to Sejal's comments and then took a deep interest in these growth charts. Part of the longitudinal data Beth shared with us consisted of measures of body length, head circumference, and body *weight*. When I realized that these measures that came with the data and the charts Sejal had mentioned were related, I asked her and Caroline Whyatt, the postdoctoral research associate in the lab, to go "hunting." Let's do some detective work—I said—on how exactly these growth charts are built. I wanted to know everything about them: actual data if there were any available, what type of statistical techniques they used to derive them and more generally, what statistical assumptions they had made. I also wanted to know if there was any basic science backing up any of the assumptions that went into the making of such charts.

While Sejal and Caroline did their independent search, I decided (in parallel) to educate myself on the matters related to the basic science of physical growth, particularly in relation to anatomic measurements. The idea was for each of us to do our detective work independently and then report back to brainstorm together on possible next steps. First, we had to reproduce in the blind what each one independently found and then see if we were all on the same page. This is how we do research in my lab. It is always an adventure with not specific hypothesis to test at first, but rather a data-driven exercise that we can all reproduce in the blind if we are indeed on the right track. Any discrepancies on

[1]http://www.fao.org/docrep/015/i2490e/i2490e02b.pdf

[2]http://www.fao.org/fileadmin/user_upload/raf/uploads/files/129654.pdf

[3]http://www.fao.org/news/story/en/item/176888/icode/

our outcomes are usually "food for thought" that helps us reframe the question or reformulate the problem at hand.

I went on to read everything related to the statistical methods and graphs used in the making of these charts. They seemed very relevant to our general quest of neurodevelopmental disorders and the possibility of detecting a problem early to intervene. It occurred to me that if every parent (willingly) regularly took their newborn baby to a pediatrician to check on the progression of physical growth, they may be interested too in regularly checking neurodevelopmental progression. I thought that in particular, they may be interested in learning about the progression of neuromotor control and the emergence of volition in the bodily motions of their baby. Indeed, I pondered if we could build a dynamic index of neuromotor control and express it as a function of physical growth? But to do it, I needed to understand the rates of change of various related quantities during neurodevelopment. That required some further research.

Physical growth tracked in isolation would not be revealing of the underlying neural development. But using biometrics that could longitudinally track the development of neuromotor control, and in particular identify volition, would be a step in the direction of early detection of *risk* for stunting in neurodevelopment. Such a concept did not exactly exist in the pediatric literature or in the developmental literature at large. Early detection of a problem really meant the problem had become obvious and was reliably detectable, i.e., detected with high certainty that grew in reliability as the child continued to age.

By 3 years of age in the United States we have early intervention programs (EIP) in place because by then, neurodevelopmental issues are already visible. Indeed, the rates of autism prevalence have dramatically increased and continue to shift, depending on the inventories that parents fill in.[23] All pediatricians I had met had the general consensus that even 3 years of age was much too late for a diagnosis, particularly if that diagnosis was reliable. Something about the nervous systems' development seemed already stunted, even in the presence of physical growth.

Based on my experience in neuromotor control research I knew that by 3 years of age, the nervous systems have reached a point of maturation.[24–26] Indeed around 3 years of age the nervous systems are poised to transition into a new statistical landscape with different probability distributions characterizing its voluntary biorhythms (Fig. 1.13).[10,11] In this sense, the EIP based on expectations from typical development could not possibly provide the supports and accommodations that a coping nervous system gone awry would need. This is particularly the case when each nervous system coping with a neurodevelopmental issue does so differently, in a rather uniquely individualized way. Once again, the *"one-size-fits-all"* approach prevalent in the health and brain sciences would work against us because time was of the essence. What I did not know was how much "of the essence" time was in the first few months of life. That information came from our detective work pointing at an accelerated rate of change in all parameters of growth that we examined: head circumference, body length, and body weight.[1]

The first thing I wanted to do about growth charts, even before considering how they were built, was to understand the nature of anatomic increment data. When parents visit the pediatricians' office for the wellness check, the doctor takes the baby's physical measurements and localizes the values on the WHO−CDC growth chart. To that end, the clinician matches the baby's values and the chart values at the corresponding age of the baby

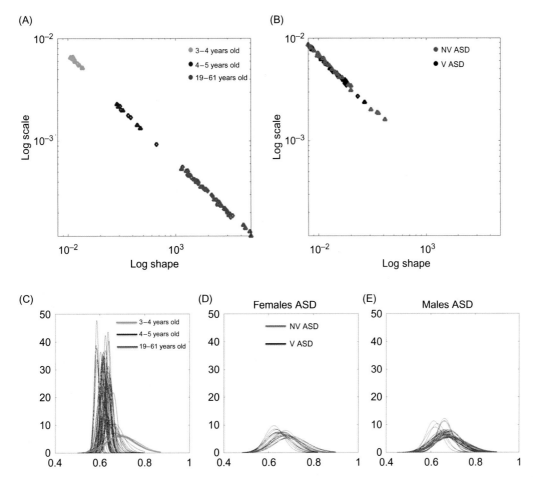

FIGURE 1.13 **Poised for maturation at 3 years of age: Typical and atypical developmental trajectory of empirically estimated PDFs characterizing the variability in speed peaks-dependent micro-movements.** (A) Gamma process outcome from different cross sections of the population ranging from 3 to 61 years old. Each point represents the Gamma PDF shape and scale where a power law describes typical maturation patterns whereby the shape of the distribution transitions from exponential at 3 years of age to skewed at to 4–5 years of age and from skewed to nearly symmetric by college age. As the shape tends toward the symmetric range, the NSR (the scale parameter) decreases. In the typical group a broad range of distributions is found across the human life span. (B) The case of ASD is very different. Here an absence of transition into a symmetric case is evident and the distributions remain with highly skewed, toward exponential shapes and much higher high NSR than their peers of similar age. Panels C–E contrast the broad bandwidth of PDFs in the typical development with the narrow range of PDFs in ASD at large.

to assess deviations from the expected value of the growth parameters on the chart. This reading gives the doctor a sense of how the baby compares to some percentile of the population of babies that age. That percentile is based on the Gaussian distribution, which as it turned out was the distribution of choice *pre-imposed* on the growth data used to build the charts.[27,28] We will get back to this in more detail, but it suffices to say that this form of tracking growth leaves out the individual variability of the baby's own trajectory. As such, the historicity of the baby's incremental growth trajectory (i.e., relative to its own life span evolution) cannot be compared to the trajectory of the expected absolute values the pediatrician measures. The very concepts of *change* and the rates at which change occurs, so important when the baby is rapidly growing, are masked by the methods the pediatricians use. In other words, *how is the baby's system changing with respect to itself, rather than with respect to an ideal population mean?*

The "ideal" population mean is just that, i.e., a theoretical mean from a PDF that has been assumed to represent the data a priori. As a matter of fact, the true empirical distribution does not agree with the pre-imposed theoretical one for a given (presumed steady state) time period in development, let along with the irregular nature of the timeline of growth from 0 to 5 years of age that the growth charts cover. Indeed, Fig. 1.14 delves into this point in schematic form (A vs B), suggesting different types of relations that we could find using different timescales within a given time window of development (from 0 to 5 years of age, for instance). Fig. 1.15 illustrates the same point in the schematic of Fig. 1.14 using actual data from physical body weights obtained from the breast-fed babies whose data went into the making of the CDC−WHO charts. Here these babies were longitudinally tracked during the first 2 years of their lives and then cross-sectional data including data from different babies were obtained until the children reached the age of 5 years old and the chart's timeline was completed. Throughout the 5-year period, different time windows showed very different profiles in weight evolution and its rate of change. For example, in the initial weeks the babies lost weight but later recovered it and then went on to rather increase weight at highly accelerated rates. By 4 or 5 years of age, these rates slowed down. How would these complex rates of physical growth impact their neurodevelopment of motor control? First, I needed to understand the actual nature of anatomical increment since the data from the WHO−CDC growth charts strongly pointed at a nonlinear dynamical process.

The answer to my questions did not come directly from the growth charts per se. It came from a combination of factors including: (1) understanding the methods used to

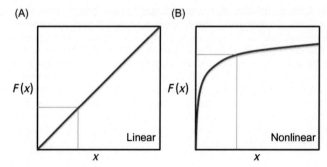

FIGURE 1.14 **Contrasting linear versus nonlinear change.** (A) In the linear profile the function of x ($f(x)$) whose values are on the vertical axis, change proportionately to those of x on the horizontal axis. This change is such that a unit increment in x leads to a unit increment in $f(x)$, e.g., doubling one results in doubling the other. The quantities scale additively. (B) The nonlinear profile is such that a wide range of dependencies between x and $f(x)$ are possible. They do not have to change proportionately to each other, e.g., a change on x at some rate may lead to much faster rate in the change on $f(x)$, thus producing much higher values that scale multiplicatively.

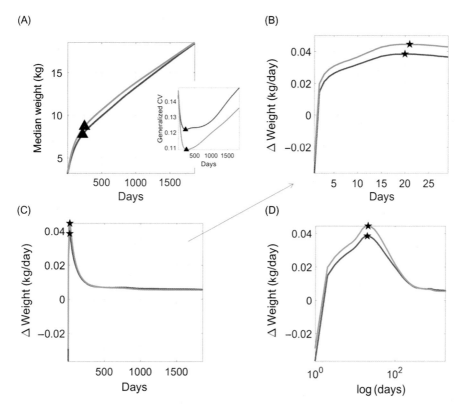

FIGURE 1.15 **Nonlinear and stochastic nature of the rates of change in body weight.** (A) Progression of the change in weight day by day in male and female newborn babies according to the median weight summary drawn from 26,985 babies per summary point (13,623 females in magenta, 13,362 males in cyan). Babies were longitudinally tracked for 24 months upon which cross-sectional data was used to build the charts up to 5 years of age.[27,37] Inset reflects inflection point in the curve tracking the generalized coefficient of variation from the weight data. Female babies reach the significant minimum at 224 days, almost a month earlier than male babies at 252 days. (B) Zooming in the data for the first month shows that the two groups separate in the first week after birth. (C) The change in weight per day is early on characterized by a considerable weight loss. Points mark the days when the generalized coefficient of variation reached inflection points, thus marking critical significant departures in variability in males versus females. (D) Log plot of (C) tracking the median weight per day over the first 5 years of life. Points mark the change in the underlying variability according to the inflection point in the generalized coefficient of variation inset in (A).

build the charts; (2) examining the actual data used to build these charts (since the data are publicly available); and (3) studying the scientific literature on human anatomic data, dating all the way back to the 1930s, particularly the literature on increment anatomic data assessing the statistical features of the first rate of change in parameters of physical growth across different time windows of development.

The assumption of the literature up to now was a linear, uniform process along the lines of that represented in Fig. 1.14A. The actual picture from human data yielded a curve along the lines of that in Fig. 1.14B. (I use schematic graphs to illustrate our point on the disparity between assumed and actual shapes of the curves representing these developmental processes.)

The actual data from weight parameters and their incremental trajectory came from the publicly available data that Sejal downloaded and that I plotted in a format that made the differences obvious. Instead of using absolute values of the weight, I used incremental values per day. The outcome of my inquiry on incremental data was useful to analyze the neonate USC data later. However, what the analyses revealed to me was also troublesome because the empirical data had statistical signatures that were far from what pediatricians were assuming in their practices, i.e., they were not normal. This meant that many of the diagnoses they were producing had a chance of being problematic. Likewise, researchers trusting these charts would be reporting inaccurate results. Both areas of the health and developmental arenas were working with babies under the assumption that the anatomic increment data followed a Gaussian distribution. But these anatomic increment data were highly nonlinear (Fig. 1.15), non-Gaussian (Fig. 1.16), and very irregularly changing over time, e.g., as revealed in the patterns of data for preterm babies developing with a coping nervous system (Fig. 1.17).

I found a brief communication paper from 1962 by H.V. Meredith[29] discussing another contemporary paper at the time by Garn, Rohmann, and Robinov, 1961.[30,31] The commentary by Meredith was on the nature of frequency distributions of incremental physical

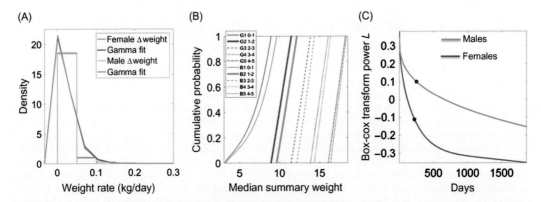

FIGURE 1.16 **Non-Gaussian nature of growth data characterized by skewed probability distributions underlying the anatomic increment weight parameter used to track physical growth.** (A) Here weight increment (kg/day) is used as the parameter of interest from Fig. 1.15 to plot the frequency histogram best fit by a Gamma distribution than by the Gaussian distribution. (B) The longitudinal changes in the median summary weight parameter across the population of male and female infants. Note the early differences in the shapes of the cumulative distributions, which later on attenuate. Note as well the evolution of the PDFs with developmental age period (in the legend G stands for girls and B for boys). Each curve reflects the year-by-year male and female trajectories. (C) Tracking the L parameter (the Box-Cox transformation power value reported in the methods to build the WHO–CDC charts to enforce symmetry on the skewed probability distributions from the actual empirical data[28] reveal the non-normality of the incremental growth data. Quoting the Methods paper[27] *"The assumption is that, after the appropriate power transformation, the data are closely approximated by a normal distribution"*. Notice that as in all other parameters the required transformation power L is different for male and female babies, denoting different families of probability distributions underlying their physical growth over different time windows.

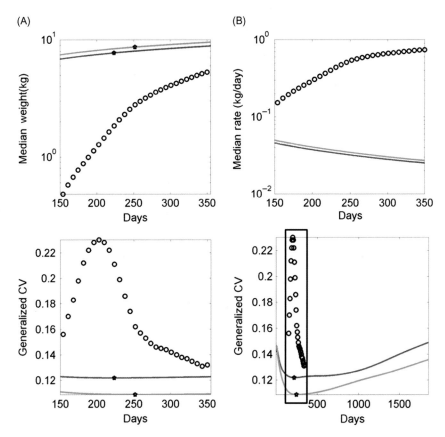

FIGURE 1.17 **Irregular growth profile of preterm babies.** Data from premature babies superimposed on the data from typical babies obtained from the WHO growth charts (weight parameter in Fig. 1.15A). Each circle represents the summary median data for a week taken across 25,000 premature babies (no sex is reported). For clarity, the log of the median (A) and its rate of change per week days (B) are used along the vertical axis.

growth parameters. Meredith had a history of papers dating as far back as 1935 claiming that their distributions were symmetric, i.e., with zero skewness. In contrast, the 1961 paper by Garn et al.[31–33] challenged that notion and referred to the *"natural skewness characteristic of incremental data."* Their findings were confirming *"skewness as a general property of growth increments in the age range considered"* (emphasis added). The note by Meredith was rather harsh, clearly defending his earlier findings. Who was right or wrong? Or were they all correct in some form?

I did my homework and tried to understand the discrepancy between the two reported results. I reasoned that in this day and age testing such issues should not be hard with so much publicly available data lying around. In fact, Sejal had just downloaded the longitudinal data that went into building the curves for the first 2 years of the WHO–CDC growth charts along with the rest of the cross-sectional data that went into building the curves from that point onward until the 5 years of age, when the chart resumes the tracking. The data comprised of 26,985 babies per summary point, 13,623 females and 13,362 males.

The Fig. 1.15 that I built using the CDC–WHO data and examining the anatomic increment values since birth until 5 years of age clarified the argument between Meredith and Garn et al. Meredith had made measurements of growth parameters after 4 years of age and until 7 years of age; but zooming out of that time window and examining the full range since birth gave us the answer. As we can see in Fig. 1.15, around 4 years (1460 days) of age the rate of growth has slowed down and reached some stability. Very likely, the underlying processes can be modeled as *locally linear,* even though, zooming out of that portion of the curve; one can appreciate the nonlinear nature of the process that takes place along the first 5 years of human development. The timescale and sliding time window used to examine physical growth were therefore important to make other inferences about hidden processes of neuronal growth and nervous systems' maturation, i.e., neurodevelopment.

I underscore that in terms of the underlying statistics of the variability inherent in that local window after 4 years of age, it is likely that an additive Gaussian process captures well the frequencies that Meredith measured in the anatomic incremental parameters. In contrast, Garn et al. referred to data inclusive of younger infants whereby the distributions of the anatomic incremental data were rather skewed. The age range they used across development yielded rates of change that were highly nonlinear in nature, likely comprising processes with multiplicative rates of change. The fundamental difference between the two studies was given by *when* in neurodevelopment they took their longitudinal measurements.

Because of the nonlinear, highly accelerated early rates of change in the time window right after birth, those first 3 years would be critical to build toward a major phase transition later on (Fig. 1.13), i.e., setting up the system for an important developmental milestone. Garn et al. claims made total sense in light of the longitudinal anatomical increment data from the WHO–CDC that we used to learn about the making of the growth charts. This enquiry told us that earlier in infancy the baby is changing at a highly accelerated rate that later on slows down.

Fig. 1.15 shows the trajectory of the rate of change in body weight increments (kg/day). We found the same outcome for the rates of change of body length (cm/day) and head circumference (cm/day), which we later examined in the 36 babies we studied from the USC Smith lab. In the end, both studies were correct, but they were examining different epochs of neurodevelopment and implicitly assuming uniformity across the full time length of neurodevelopment. Clearly the message from the WHO–CDC data spanning the full period is that one should not make such simplifying assumptions in developmental phenomena.

The authors did not consider the irregular and nonlinear nature of development in their argument (Fig. 1.16). Owing to the irregular changes across different time windows (Fig. 1.16B and C), they should have considered both a small sliding time window within a time period and a larger sliding time window to capture the differences, particularly after the first year and ½ of life. Otherwise, seemingly conflicting outcomes emerge and even when both parties were locally correct, their global inference and interpretation were at odds. Indeed, without access to the full picture on the longitudinal rates of physical

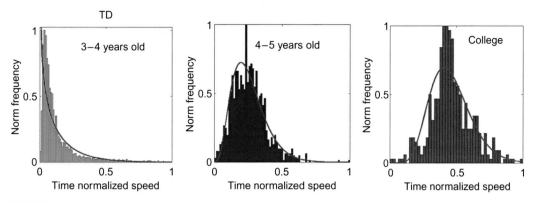

FIGURE 1.18 **Evolution of empirically estimated PDFs in typical development of voluntary control in pointing behavior.**

growth, from infancy to 5 years of age, these researchers were butting heads and bruising egos, yet missing the main point on the highly nonuniform early human development, particularly within a coping nervous system (Fig. 1.17).

At my end, when I saw the table of anatomic parameters and their incremental values listed on the Meredith paper[29] and the age column listing ages from 4 to 7 years of age, I was suspicious of the argument. Even without knowing much about anatomical growth, I had done research about the underlying variability of neuromotor control. I already knew that from 3 to 4–5 years of age there is a dramatic statistical shift, a transition in the shapes and dispersions of the PDFs characterizing voluntary pointing motions (Fig. 1.18).

Pointing motions require stabilization of the physical arm (weight and length) for controllability of the motion dynamics. If a 4-year-old child were to change weight at the rate of a 1- to 3-month-old infant, pointing would not be possible. The brain of the child would have to recalibrate all maps of peripheral body sensations (including kinesthetic sensations from movements that would be affected by the change in the arm mass) to build the model of internal dynamics leading to successful (accurate) pointing at adequate speeds. The stability of these maps ought to be important to develop proper neuromotor control. Thus, the rapidly changing physical body in the early life of an infant requires recalibration of the nerves' networks and of the emerging maps of the embedded nervous systems. These systems better be growing and evolving at comparably fast rates as those of the body, if neuromotor control is to take place. The data from the USC Smith lab would hold the answer to my question on the types of relationships between the variability on physical growth and that in the emergence of neuromotor control.

The detective exercise taught us several lessons: (1) do not believe the literature but rather think critically about everything that gets published and more importantly, dissect the Methods section of every paper you read because the "how to" is there and with it, lays the actual result; (2) go out of your way to re-examine the questions that have been previously asked or assumed as they may not even make sense in light of new data; (3) go into this research on neurodevelopment with a complete open mind and no preconceived notions; rather, let the inherent nature of the data surprise you.

The USC Baby Data

We had access to 36 babies, who started the study at about 4 months of age, but 24 of them had been born with complications and some of those were prematurely born. As such, their systems were not as developed as those of the remaining 12 babies that had been born full term, without complications. They came with a label of clinically at risk (CAR), whereas the other 12 full-term babies were labeled as controls (CT). Yet, given the abovementioned stunting data from reports by the FAO, which included both under-weighted and overweighed babies, the physical growth measures provided only part of the story and that part of the story still needed verification in terms of the makings of the WHO—CDC growth charts.

We could in principle have babies that were born prematurely but did not stunt because parents and clinicians provided clever accommodations and compensatory interventions that helped them catch up faster than if nothing was done. We could also have babies that may have been born full term and without complications, but stunted in their develop-ment due to different unknown reasons. One of those reasons could be related to problems with the nervous systems development that a growth chart could not capture in its current form. As such, I preferred to not entirely rely on the clinical pre-labeling and further explore the problem using data-driven methods. I wanted to compare groups according to the CAR versus CT labeling; but also search for self-emerging data-driven groups without any preconceived notions. While Sejal and Caroline searched for further clues on the mak-ing of the growth charts, I initiated a path of trying to understand the development of neu-romotor control in these babies that the USC Smith lab tracked for 5 months.

The infant data could provide an answer to the question concerning the emergence of somatic-motor patterns resembling volition in the newborn. Besides the measures of longi-tudinal growth across three visits, every 2 months, we had data from the APDM sensors affixed to their ankles. These data had been continuously harnessed 9—13 hours each visit.

This was a unique opportunity for us to collaborate toward the discovery of an index of typical neurodevelopment of motor control, but more important yet, given the longitudi-nal nature of the study we had a chance to unveil a pattern revealing risk of stunting. If we found stunted patterns of NSR derived from their acceleration's micro-movements, i.e., like those that we had found in ASD, we may be able to provide a characterization of *risk for ASD* at a much earlier time than presently done. By 3 years of age, the present age for EIP, the nervous systems of a young infant have already undergone major changes leading to changes in the landscape of PDFs characterizing their voluntary speed control (Figs. 1.13 and 1.18).

The SPIBA paired with the micro-movements approach could also inform us about the landscape of PDFs characterizing the amount of involuntary motions in their systems. We had analyzed in the lab open access data from the Autism Brain Imaging Data Exchange project and discovered an evolution in the PDFs characterizing linear speed of *involuntary* head motions harnessed as people laid down in the scanner and were instructed to remain still (Fig. 1.19). Indeed, different typically developing (TD) individuals from groups span-ning ages between 6 and 60 years old have different probability distributions of involun-tary head micro-movements (Fig. 1.20). These micro-movements were extracted from resting state functional magnetic resonance imaging signals harnessed from 1112 people,

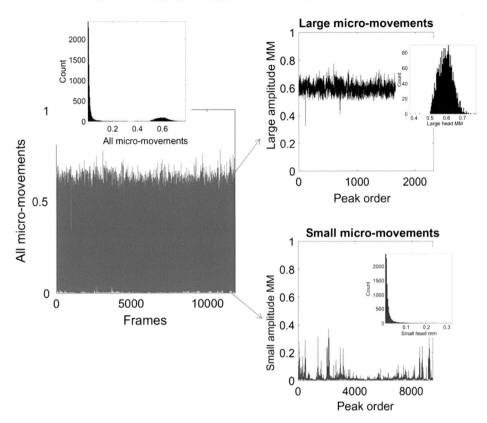

FIGURE 1.19 **Two classes of micro-movements extracted from head motions during resting state.** The head's linear speed of involuntary resting state motions contains fluctuations in amplitude that we examine as spike trains. These micro-movements are properly normalized as described in this chapter and examined in the order in which they occurred in their original frame. Then the micro-movements from amplitudes one standard deviation above the mean (black) are obtained and analyzed separately from those of lower amplitude (red). To that end, they serve as input to a Gamma process where the Gamma shape and scale (NSR) parameters are estimated for each micro-movement type.

including those with a diagnosis of autism-related conditions. The latter not only (involuntarily) moved more; they also had much higher rates of random noise in these speed-dependent head micro-movements.[12]

Too much involuntary motions when the person tries to remain still reflect failure of the brain to control the body at will. It also reflects the type of feedback the PNS is echoing back to the central controllers of the brain. *Would a baby develop volitional control of the brain over the body if the feedback consisted of persistent spontaneous random noise?* It seemed to me that it was fundamental to characterize the transitions from spontaneous random noise to well-structured signals in the self-generated motions the nascent nervous systems of the infants were producing. The wearable accelerometers capture motion continuously throughout the day. All I needed to do was parse the data out as we did with the coma

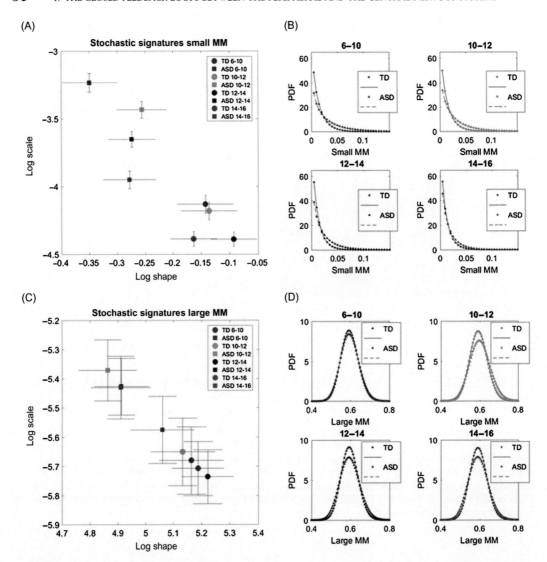

FIGURE 1.20 **Developmental evolution of the PDFs of two classes of involuntary head micro-movements in TD and autistic (ASD) participants.** (A) Small amplitude micro-movements have distinct signatures within a cross section of the TD population included in each given age bracket. They differ from those derived from the ASD matching in age in these cross-sectional data. The differences can be appreciated on the Gamma plane plots where the ASD show higher levels of NSR and skewness than the age-matched TD. Each point represents the empirically estimated shape and scale parameters using the micro-movements in Fig. 1.6 and the Gamma process in the SPIBA framework. (B) Estimated PDFs using the estimated Gamma parameters for each age group show the differences across groups and between TD and ASD. (C) The head micro-movements of larger amplitude.

patient and identify the signatures of the micro-movements transitioning from spontaneous random noise toward well-structured and systematic signals preceding the acquisition of volitional control.

The Pre-Labeled Data Showed Differences in the Rates of Anatomical Increment Across the Babies

Researchers from Beth's lab took anatomical data comprising values of body weight (kg), head circumference (cm), and body length (cm) each visit, approximately every 2 months for the span of 5–6 months. These are similar to the measurements clinicians take during the baby's visit to the pediatrician's office. Their absolute values are used to measure growth from visit to visit. When I pooled these absolute measurements across all babies of the CAR group and all babies of the CT group, I found no statistical differences between babies that had been born with complications (CAR) and the control group (CT) born full term without complications. How could this be if they were already so different in developmental trajectory?

I corrected the absolute values by days since birth at the time of each visit to convert the metrics to a relative quantity individually tracked. Since the preterm babies had started the study later than the full-term babies to allow for their system to stabilize and catch up, we needed to rather use an individualized relative measure. Then the distributions for each of the groups and each of the parameters turned skewed and the degree of dispersion and skewness in the data clearly distinguished the two pre-labeled groups. The two types of data plots can be appreciated in Fig. 1.21, where the insets show the nearly symmetric distributions of the absolute values with no significant statistical differences between the groups. In contrast, the anatomic increment version shows the skewed nature of the data revealing as well the developmental differences already detectable as early as 4 months of age.

Data-Driven Approach

The pre-labeled data was informative from a summative format standpoint. However, pooling the data in that way left out individual differences in each baby developmental trajectory. Given the irregular, nonlinear rates of developmental change and the non-Gaussian nature of the anatomical increment data, I switched to a data-driven approach that examined each baby's trajectory individually and gave me a better sense of self-emerging rather than a priori hand-picked groups.

To build our new data sets based on the variational features of these growth parameters, I iteratively median-ranked the physical growth increment data, until I had groups that were close in size. At one end, I grouped all babies for whom all three parameters were simultaneously changing along the most rapid rate, above the median of each parameter. At the other end, were the babies changing at the slowest rate along all three parameters. The middle group then was the most variable, i.e., with more variable rates along these anatomical increment dimensions. I also used the Alberta Infant Motor Scores (AIMS)[34] that clinicians recorded each visit. This observational test provides a sense of

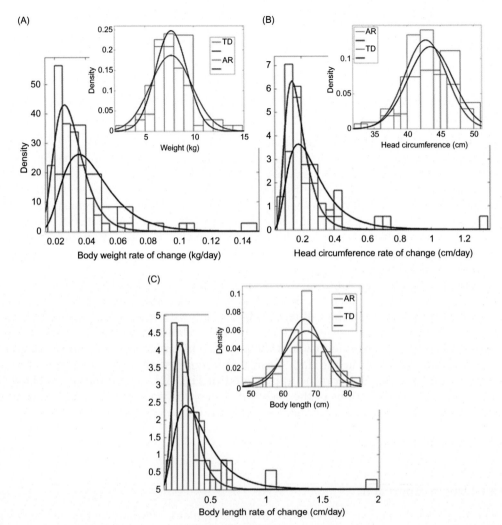

FIGURE 1.21   **Differences between absolute anatomical values and anatomical increment data revealing differences in the rates of growth of the two pre-labeled groups of babies.**

readiness to walk. In turn, such readiness depends on the rates of physical growth. For example, some babies may gain too much weight too fast and fall behind in their ability to control the body to sit, balance the head—trunk posture, and eventually steadily stand upright. The AIMS gives a sense for these other aspects of development that are closely related to the development of neuromotor control. Fig. 1.22 shows the median ranking of the groups. The four ranked groups were then ordered into three clusters and their motor control patterns examined according to the signals extracted from the wearable accelerometers.

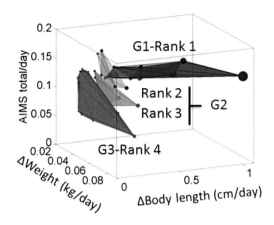

FIGURE 1.22 **Iterative median ranking of the babies revealed four ranked clusters that were further used to form three groups: (group 1 TD, group 2 PAR, and group 3 at HR).** The size of the circle denotes the rate of change in head circumference (cm/day). The z-axis denotes the rate of change in total AIMS scores per visit.

To perform our analyses, we combined the micro-movements we extracted from the acceleration and temperature information with the SPIBA methods. We derived the spike trains from the fluctuations in amplitude of the acceleration signals and input them to the Gamma process to track the frequency and amplitude of the shifts in estimated parameter values and their location on the Gamma plane. Those transitions from the Gamma plane left upper quadrant (Gamma LUQ) to the right lower quadrant (Gamma RLQ) (and vice versa) were systematically quantified. We obtained the frequency and the amplitude of the transitions (as shown in Fig. 1.9) to determine how the amplitudes in acceleration were reaching high signal to noise (lowering the value along the scale axis) and converging to a symmetric distribution (i.e., to the right of the shape axis).

The stochastic signatures of the micro-movements quantifying the fluctuations in the amplitude of the signal served to profile the evolution over time of the transitions from spontaneous random noise to well-structured signals. As the PDFs of the babies changed dispersion and shape from day to day, we were able to individually track the progression in neuromotor control development as a function of the progression in physical growth. Both rates of change were quantified as I searched for a scale-invariant relation between these nonlinear and non-Gaussian processes with very irregular outcomes for each individual baby changing so much from visit to visit.

Fig. 1.23 shows sample traces of the acceleration and the temperature the sensors recorded from the infant's leg. The sensors were located at each ankle. The figure shows the raw acceleration traces where we mark motion occurring at different levels of temperature.

When the temperature is high we know that the battery is releasing more energy because the baby is actively moving more. Those self-generated motions drain the battery faster than passive motions (as when the mom takes the baby in a stroller and the sensors pick up the inertial motion of the stroller.) In such cases, the battery consumption is lower on average. By color-coding the motions according to their temperature regimes, we can infer the levels of noise from the baby's self-generated activity versus those from more passive motions generated by other agents. Fig. 1.24 shows the methods to derive the new data types integrating motion and temperature.

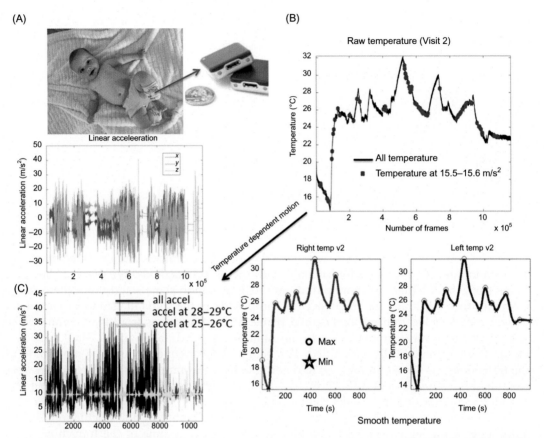

FIGURE 1.23 **Raw signals from wearable sensors harnessing biorhythms from the nervous systems.** (A) Participant baby wearing the sensors in his ankles with picture of the wearables shown to scale relative to a US dime. (B) Raw temperature traces with red dots indicating frames when the acceleration motion reached 15.5–15.6 m/s² and smoothed temperatures with peaks counting the fluctuations in the peaks and valleys. (C) Traces from the triaxial accelerometer and corresponding acceleration with highlighted frames for different temperature intervals.

The SPIBA methods were explained in Fig. 1.9 while Fig. 1.25 produces an example of the stochastic trajectory longitudinally obtained from visit to visit. This is for a TD baby (classified by the iterative median-ranking method described in Fig. 1.22). The changing PDFs from visit to visit provide a sense of the evolution in these signatures. This is a summative graph of the Gamma plane using the temperature-dependent motion data, i.e., estimating for each temperature degree the motion corresponding to that temperature regime. However we can also track the Gamma plane trajectories using the motion data as continuously recorded and color-coding the values for Gamma estimation parameters according to the average temperature they are associated with along this continuum. Fig. 1.26 provides that type of visualization tool on the Gamma plane and on the summary

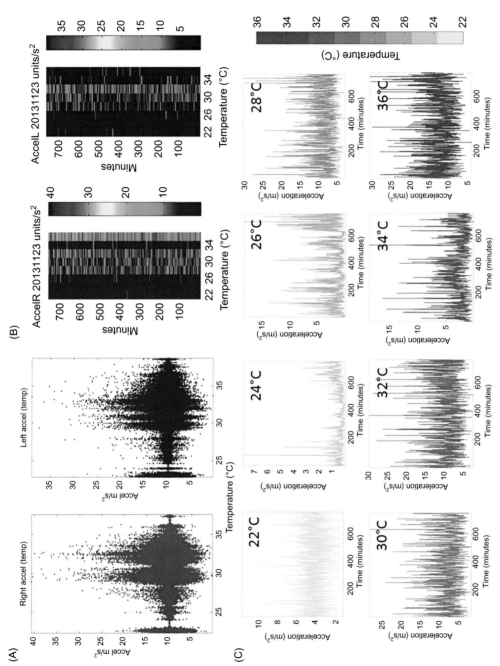

FIGURE 1.24 **New data type integrates motion and temperature.** (A) Matrices for the two sensors gathering all the acceleration data registered for each temperature degree. (B) Color map data matrices obtained from the maximal deviation from the mean acceleration in (A) in the time they were registered at each minute and temperature degree interval. Each interval in this case spans 2°C and the color bar indicates the acceleration range for the maximal deviations from the mean acceleration (m/s²) for each sensor. (C) Each level of maximal deviation from the mean acceleration per temperature interval is displayed color-coded according to the color bar for the number of minutes registered (750 min, 12.5 h).

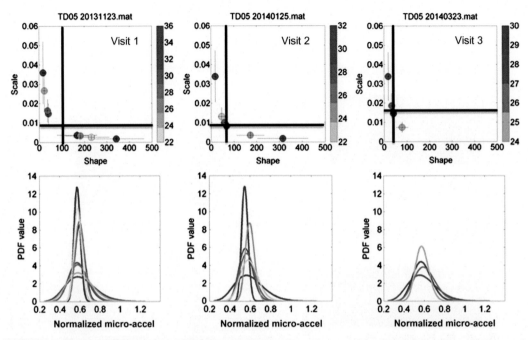

FIGURE 1.25 **Longitudinally tracking the temperature-dependent motion signal from the accelerometer's micro-movements using the SPIBA framework described in** Fig. 1.12. Each color represents the estimated PDF under the temperature interval shown in the color bar. The median shape and median scale across all estimates are used to draw the line dividing the LUQ and RLQ of the Gamma plane. The transitions from noise to signal are tracked on the Gamma plane indicating the changes in shape and dispersion of the PDFs over the day and across visits.

statistics space. There we estimate the Gamma moments and plot them in five dimensions. These are the mean, the variance, and the skewness along the first three dimensions. Then the size of the marker denotes the kurtosis and the color denotes the temperature along a gradient. Further, I divide the color gradient according to the Gamma LUQ and RLQ because they have different statistical inferential meanings. Those colored in shades along a reddish gradient are on the Gamma LUQ denoting distributions that are more skewed, tending toward the memoryless exponential.

The NSR is also higher for the points in this quadrant. In contrast, the points representing the PDFs in the Gamma RLQ are colored in shades along a bluish gradient. They denote PDFs tending toward the symmetric shapes like the Gaussian distribution and with higher signal content (lower NSR) in the micro-movements. This baby starts out with very little signal in the RLQ, mostly light blue denoting passive motions. As time goes by, in the third visit, his signal content has increased as denoted by the higher number of points in the RLQ and the darker shades of blue indicating an abundance of self-generated motions with such predictive signatures. Neuromotor control is emerging in this baby. The idea then was to summarize these transitions and provide an index of neuromotor control change as a function of the changes in anatomical increments.

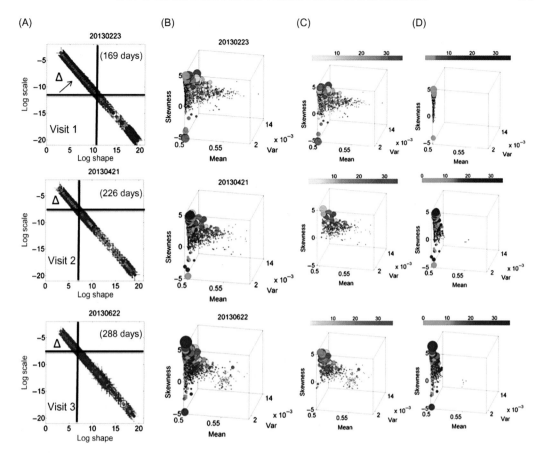

FIGURE 1.26 **Tracking the shifts in Gamma moments as a baby's movement's transition from spontaneous random noise towards signal resembling maturation.** (A) Sample stochastic maps to quantify the rates of change in noise evolution on the Gamma parameter plane for one (CT) baby across three visits spanning 4 months across infancy. Each plot is obtained from the motions registered in one visit with each point representing an estimation of 1200 measurements (built with 1-min blocks, 20 Hz × 60 s values registered over the span of 8 h.) Each colored circle represents the empirically estimated Gamma parameters as in Fig. 1.12. The color represents the average temperature across the *range* in that minute block (see color bars on the three-dimensional plots). The noise range is tracked using the median of the values as a cutoff between higher and lower noise levels. Points above the median scale values (higher noise-to-signal values on the upper-left quadrant) are plotted in shades of red, whereas those in the lower levels of noise (lower-right quadrant) are plotted in shades of blue. (B) Five-dimensional scatter plots involving the estimated mean (*x*-axis), estimated variance (*y*-axis), the skewness (*z*-axis), and the kurtosis (the size of the marker) color-coded with blue or red gradients according the median cutoff levels of noise. (C and D) Separable evolution of the high and low noise levels across visits plotted along with their temperature ranges. Darker colors represent higher values of temperature on average for the corresponding minute-block fluctuations in amplitude that went into the Gamma process estimation. Darker colors of higher temperature on average are from actively generated motions registered by the inertial sensors. These larger motions consume more battery energy and therefore, on average, generate more heat.

The transitions between the LUQ and RLQ were quantified in frequency and amplitude (shown in Fig. 1.27 for two sample babies from the originally pre-labeled groups CAR and CT). We uncovered stationary and nonstationary transitions that revealed marked differences even in the first visit (i.e., when the babies were 4 months old).

I coined transitions that remained changing within the same Gamma plane quadrant *stationary transitions*. The stochastic values would not jump to the other quadrant. A possible extreme scenario would be transitions within high noise and exponential distribution shapes with no jumps to the quadrant with the high signal content and distributions with symmetric shapes.

Persistent noise and randomness across visits would imply very little change in stochastic signatures, mostly remaining in the LUQ. Given the empirical result that growth and motion patterns change their stochastic signatures in an accelerated rate at this age, I could safely infer that no development of neuromotor control was taking place under such hypothetical scenario, or that it was stunted in such cases where no transitions to the RLQ of the Gamma plane were registered. In marked contrast, babies evolving and self-adapting neuromotor patterns from visit to visit would manifest high frequency in the quadrant-to-quadrant transitions. Their patterns would manifest convergence toward the RLQ where PDFs are Gaussian-like, symmetrically shaped, and have high signal content (a tendency to lower NSR values along the scale axis). We have learned from our cross sectional data analyses that absence of points in the RLQ was the feature of the autistic system (Figure 1.13) and that important transitions manifested in the PDFs from 3 to 5 years of age denoting a reduction in the noise and randomness of the kinematic signals (Figure 1.18). Recall here that the Gamma scale, coined by physicists the Fano factor,[35] is also the NSR. Fig. 1.28 shows the summary of these quantifications according to the TD, partially at risk (PAR), and high-risk (HR) groups that the iterative median-ranking method revealed.

Scale-Free Relation Between Two Nonlinear and Non-Gaussian Processes of a Complex System

The examination of changes in the statistical features of the anatomical increments and those in the evolution of the families of probability distributions characterizing the biophysical rhythms of the peripheral limbs gave a sense of maturation or stunting in the development of these bodily parameters. Each set of parameters separately showed adaptive changes from visit to visit. Yet, some babies had faster rates than others and some were nearly not changing at all within a visit or across visits. These irregularities motivated me to examine the rates of change in physical growth as they changed in tandem with the rates of change in neuromotor control. I searched for a rule that enabled me to predict (despite different scales) what the most likely rate of change in motor control would be in the following visit, given the rate of growth in the previous visit (see appendix for the derivation of a first-order updating stochastic rule similar to one I encountered while studying intentionally staged movements in sports routines[36]).

When we examined the rates of change in quantities from growth and motor development, we found that those babies in the group 1, the ones simultaneously growing at the fastest rate along all three dimensions, were also developing neuromotor control

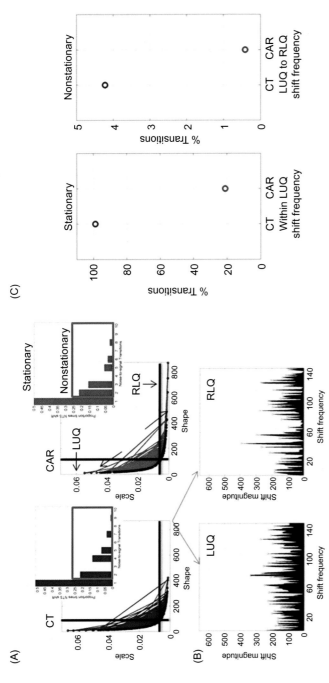

FIGURE 1.27 **Tracking the rate of adaptive change through the noise-to-signal transitions on the Gamma plane.** (A) The stochastic trajectory obtained from two representative infants, one CT and one CAR across 8 h using 1 min blocks (with ½ min sliding window) to continuously estimate the Gamma parameters. The median lines obtained from the median shape and median scale values divide the Gamma plane into quadrants whereby the amplitude and frequency of the transition of points between the LUQ and the RLQ are quantified. Arrows mark segments of the trajectories (change in probability distribution) as they transition from the LUQ of high noise and skewed distributions to the RLQ of low noise and near-symmetric distributions. Insets show the histograms with 10 bins quantifying the proportions of stationary (inner-quadrant) transitions (one transition in first bin) and those quantifying nonstationary (intra-quadrant) transitions (two or more in subsequent bins) highlighted with a rectangle. (B) Area plots of the shifts in the trajectory of (A) corresponding to each quadrant of interest for the CT baby. The graphs are obtained by computing the magnitude of the velocity vector connecting the points positioned on the Gamma parameter plane representing the estimated PDF. The peaks are marked in each plot. (C) Percent frequency of the shifts for each representative CAR and CT baby quantified within a given quadrant (the LUQ in this case) and between quadrants (from the LUQ to the RLQ in this case) denoting stationary and nonstationary shifts respectively. These quantities are the averaged percent values across the three visits.

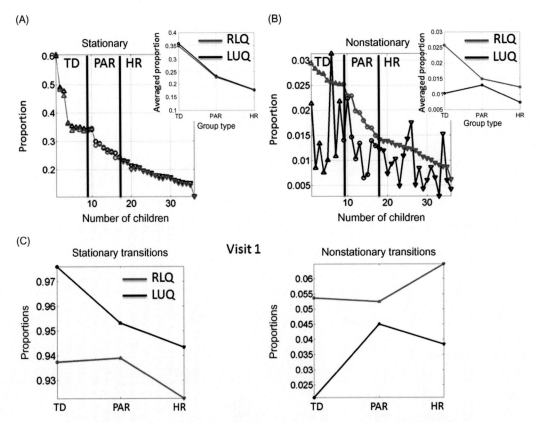

FIGURE 1.28 **Frequency of noise-to-signal transitions distinguishes babies at HR from TD babies.** (A) Stationary (inner-quadrant) transitions sorted according to the proportion of times fluctuating within each quadrant before crossing to the other quadrant. Each dot represents a baby (up-triangles are TD, circles are PAR, and down-triangles are HR). Inset is the median across each group. (B) Nonstationary (inter-quadrants) transitions sorted according to the proportion of times crossing across quadrants from the RLQ to the LUQ. The same index used to plot the opposite direction of transitions shows higher variability when crossing from the LUQ the RLQ. Inset shows the median values per group. (C) Median values of noise-to-signal transitions per group during the first visit already distinguish the groups in both the stationary and the nonstationary cases.

at the fastest rate. Their slope, derived from a linear fit to the scatter, provided the best fit of the three groups suggesting a scale-invariant relation that was violated for those babies in the stunted group 3. Not only was their rate of growth the slowest as a group, their transitions from spontaneous random noise to predictive motor signal were not occurring. Very likely their neuromotor development was stunted. As such, we discovered a putative index of risk for stunting in the development of neuromotor control at the precognitive state. This index may serve to flag failure to scaffold self--generated action execution to develop the ingredients for socio-motor behaviors (Fig. 1.29).

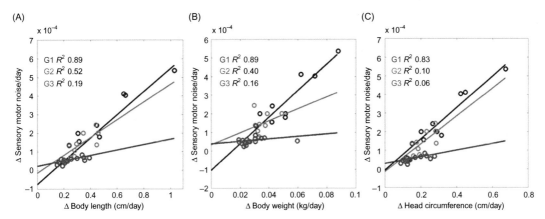

FIGURE 1.29 Index of risk uncovered to automatically flag neurodevelopmental derail or stagnation very early on. (A–C) Linear relation is found between the rates of change in physical growth and the rates of change in the neural control of movements measuring the maximal amplitude of the transition from the LUQ to the RLQ (see main text reporting the goodness of fit parameters for each scatter). The relation degrades and the slope flattens as the baby's body grows slower and does not undergo changes in the signatures of sensory-motor noise derived from fluctuations in motor performance. HR, red scatter (Rank4 group G3); PAR, gray scatter (Rank2, Rank3 G2), and TD, blue scatter (Rank1 G1).

CONCLUSIONS

This chapter provides a sense for the evolution of the stochastic signatures of adaptive processes in neurodevelopment and beyond. These included those processes partaking in neuromotor control and physical growth, but also the processes of regaining volition after brain injury. Both required longitudinal tracking of change in the nervous systems signatures—a major feat that we achieved using noninvasive means at the clinic and the household.

The detective works we did in my lab concerning the making of the growth charts taught us that developmental data from complex systems such as those of newborn humans are highly nonlinear, irregular, and variable over time. As such, we should not pre-impose assumptions of any kind on data self-generated by such complex systems. Furthermore, we learned that despite these irregularities, disparities in scales and individual timelines, the nervous systems and the growth of the babies that develop along a typical path change in tandem according to linear rules. Violation of these rules connecting the rates of change in physical growth and the rates of change in neuromotor control provide an index of stunting in the overall development, thus endowing us with a new tool to dynamically assess development in complex multilayered systems. Whether uncovering it in a nascent nervous system or recovering it in a mature nervous system that underwent injury, this chapter offered a new window into the longitudinal tracking of volitional control.

APPENDIX—FIRST-ORDER STOCHASTIC RULE

Given the time series of the rates of change in NSR (ΔN) and that of the rates of change in physical body growth (ΔB) derived from random fluctuations in the amplitudes of the underlying physiological signals since birth we derive the following relation below:

A first-order stochastic update rule anticipates the ΔN of a future trial (visit) based on the combination of the current shift in ΔN and the current ($\Delta N + \Delta B$) with multiplicative error:

$$\ln\left(\Delta_N^{t+1} + \lambda\left(\Delta_N^t + \Delta_B^t\right)\right) = \ln\left(\Delta\left(\Delta_N^t\right)\right)$$

where λ is a constant of proportionality derived from the entire data set, ΔN is the velocity of the NSR, and $\Delta(\Delta N)$ is the acceleration of the NSR as they evolve in time:

$$A_N^t + \upsilon V_B^t = V_N^{t+1} - V_N^t + \upsilon V_B^t$$
$$= V_N^{t+1} + \lambda\left(V_N^t + V_B^t\right)$$

$$V_N^{t+1} + \lambda\left(V_N^t + V_B^t\right) = m\left(A_N^t\right) + b + \varepsilon$$

$$\ln\left(V_N^{t+1} + \lambda\left(V_N^t + V_B^t\right)\right) = m\,\ln\left(\left(A_N^t\right) + b + \varepsilon\right) \tag{1.1}$$

The timescale in this case is the visit (approximately 2-month intervals). However in actual use it should be more frequently sampled, e.g., sample motor signatures and physical body growth each day of the first month of life. Exponentiation of Eq. (1.1) gives:

$$e^{\ln\left(V_N^{t+1} + \lambda\left(V_N^t + V_B^t\right)\right)} = e^{m\ln\left(\left(A_N^t\right) + b + \varepsilon\right)}$$
$$= e^{m\ln\left(A_N^t\right)}e^{b+\varepsilon} = e^{\ln\left(A_N^t\right)^m}e^{b+\varepsilon} \tag{1.2}$$

Given fitting regression parameters m and b, from the slope values set $m = 1 - \delta$ we can approximate Eq. (1.2) to leading order.

$$V_N^{t+1} = A_N^t\left[1 - \delta\ln A_N^t + O(\delta^2)\right]e^{b+\varepsilon} - \lambda\left(V_N^t + V_B^t\right)$$

As m approaches 1 (line of unity) the delta term becomes vanishingly small and the equation approximates a first-order stochastic rule anticipating the noise-to-signal change in the next visit based on the combination of the shift in the rate of change in noise (noise acceleration) in the current visit and the sum of the rate of change in noise with rate of change in body growth, with multiplicative error, arriving at the predictive rule:

$$V_N^{t+1} + \lambda\left(V_N^t + V_B^t\right) = A_N^t[.]e^{b+\varepsilon}$$

In Δ-notation:

$$\Delta_N^{t+1} + \lambda\left(\Delta_N^t + \Delta_B^t\right) = \Delta\left(\Delta_N^t\right)[.]e^{b+\varepsilon}$$

$$\ln\left(\Delta_N^{t+1} + \lambda\left(\Delta_N^t + \Delta_B^t\right)\right) \simeq \ln\left(\Delta\left(\Delta_N^t\right)\right) + \ln\left(e^{b+\varepsilon}\right) = \ln\left(\Delta\left(\Delta_N^t\right)\right) + b + \varepsilon$$

With m, b, and ε estimated empirically from large longitudinal data sets registered from birth onward.

In addition to incremental charts of physical growth we can now build incremental charts of neural motor control using this rule and empirically estimating the day-to-day rates of change in noise-to-signal transitions from the fluctuations in the baby's motor performance—as we did in the 36 babies of the present cohort.

References

1. Torres EB, et al. Neonatal diagnostics: toward dynamic growth charts of neuromotor control. *Front Pediatr.* 2016;4(121):1−15.
2. Kioussis D, Pachnis V. Immune and nervous systems: more than just a superficial similarity? *Immunity.* 2009;31(5):705−710.
3. Hawgood S, et al. Precision medicine: beyond the inflection point. *Sci Transl Med.* 2015;7(300). 300ps17.
4. Torres EB, Jose JV. Novel Diagnostic Tool to Quantify Signatures of Movement in Subjects With Neurological Disorders, *Autism and Autism Spectrum Disorders*. New Brunswick, NJ: The State University of New Jersey; 2012.
5. Smith BA, et al. Daily quantity of infant leg movement: wearable sensor algorithm and relationship to walking onset. *Sensors (Basel).* 2015;15(8):19006−19020.
6. Torres EB, Lande B. Objective and personalized longitudinal assessment of a pregnant patient with post severe brain trauma. *Front Hum Neurosci.* 2015;9:128.
7. Torres E. *Theoretical framework for the study of sensory-motor integration. Cognitive Science.* San Diego, CA: University of California; 2001:109.
8. Shadmehr R, Wise SP. *The computational neurobiology of reaching and pointing: a foundation for motor learning. Computational Neuroscience.* Cambridge, MA: MIT Press; 2005:575. xvii.
9. Torres EB. Two classes of movements in motor control. *Exp Brain Res..* 2011;215(3−4):269−283.
10. Torres EB, et al. Autism: the micro-movement perspective. *Front Integr Neurosci.* 2013;7:32.
11. Torres EB, et al. Toward precision psychiatry: statistical platform for the personalized characterization of natural behaviors. *Front Neurol.* 2016;7:8.
12. Torres EB, Denisova K. Motor noise is rich signal in autism research and pharmacological treatments. *Sci Rep.* 2016;6:37422.
13. Torres EB, et al. Neural correlates of learning and trajectory planning in the posterior parietal cortex. *Front Integr Neurosci.* 2013;7:39.
14. Lord C, et al. The autism diagnostic observation schedule-generic: a standard measure of social and communication deficits associated with the spectrum of autism. *J Autism Dev Disord.* 2000;30(3):205−223.
15. Lord C, et al. Autism diagnostic observation schedule: a standardized observation of communicative and social behavior. *J Autism Dev Disord.* 1989;19(2):185−212.
16. Kalampratsidou V, Torres EB. *Outcome measures of deliberate and spontaneous motions. MOCO'16 Third International Symposium on Movement and Computing.* Thessaloniki, GA, Greece: ACM; 2016. MOCO'16.
17. Von Holst E, Mittelstaedt H. The principle of reafference: interactions between the central nervous system and the peripheral organs. In: Dodwell PC, ed. *Perceptual Processing: Stimulus Equivalence and Pattern Recognition.* New York: Appleton-Century-Crofts; 1950:41−72.
18. Von Holst E. Relations between the central nervous system and the peripheral organs. *Br J Anim Behav.* 1954;2(3):89−94.
19. Grusser OJ. On the history of the ideas of efference copy and reafference. *Clio Med.* 1995;33:35−55.
20. Torres EB, Zipser D. Reaching to grasp with a multi-jointed arm. I. Computational model. *J Neurophysiol.* 2002;88(5):2355−2367.
21. Kuczmarski RJ, et al. CDC growth charts: United States. *Adv Data.* 2000;314:1−27.
22. Use of World Health Organization and CDC Growth Charts for children aged 0−59 months in the United States. In: Roper WL et al., eds. *Morbidity and Mortality Weekly Report.* Atlanta, GA: Center for Disease Control and Prevention; 2010.
23. Zablotsky B, et al. Estimated prevalence of autism and other developmental disabilities following questionnaire changes in the 2014 national health interview survey. *Natl Health Stat Report.* 2015;(87)1−21.

24. Thelen E, Fisher DM. From spontaneous to instrumental behavior: kinematic analysis of movement changes during very early learning. *Child Dev*. 1983;54(1):129−140.

25. Thelen E, Spencer JP. Postural control during reaching in young infants: a dynamic systems approach. *Neurosci Biobehav Rev*. 1998;22(4):507−514.

26. Thelen E. Grounded in the world: developmental origins of the embodied mind. *Infancy*. 2000;1(1):3−28.

27. Kuczmarski RJ, et al. 2000 CDC Growth Charts for the United States: methods and development. *Vital Health Stat*. 2002;11(246):1−190.

28. Flegal KM, Cole TJ. Construction of LMS parameters for the Centers for Disease Control and Prevention 2000 growth charts. *Natl Health Stat Report*. 2013;(63)1−3.

29. Meredith H. On the distribution of anatomic increment data in early childhood. *Am J Phys Antropol*. 1962;20:516.

30. Garn SM, Rohmann CG, Robinow M. Increments in handwrist ossification. *Am J Phys Anthropol*. 1961;19:45−53.

31. Garn SM, Rohmann CG. On the prevalence of skewness in incremental data. *Am J Phys Anthropol*. 1963;21:235−236.

32. Garn SM, Silverman FN, Rohmann CG. A rational approach to the assessment of skeletal maturation. *Ann Radiol (Paris)*. 1964;7:297−307.

33. Garn SM, Blumenthal T, Rohmann CG. On skewness in the ossification centers of the elbow. *Am J Phys Anthropol..* 1965;23(3):303−304.

34. Piper MC, Darrah J. *Motor Assessment of the Developing Infant*. Philadelphia, PA: Saunders; 1994. xii, 210 pp.

35. Fano U. Ionization yield of radiations. II. The fluctuations of the number of ions. *Phys Rev*. 1947;72(1):26.

36. Torres EB. Signatures of movement variability anticipate hand speed according to levels of intent. *Behav Brain Funct*. 2013;9:10.

37. de Onis M, Onyango AW. The Centers for Disease Control and Prevention 2000 growth charts and the growth of breastfed infants. *Acta Paediatr*. 2003;92(4):413−419.

2

Critical Ingredients for Proper Social Interactions: Rethinking the Mirror Neuron System Theory

There is something shared between our first- and third-person experience of these [social interactions] phenomena: the observer and the observed are both individuals endowed with a **similar brain—body system**. *Vittorio Gallese, Christian Keysers and Giacomo Rizzolatti*[1]

PART I

In 1998, I attended the annual meeting of the Society for Neural Motor Control (NCM) and listened to Giacomo Rizzolatti give a talk about a new type of cortical cells his group had coined "mirror neurons". The lecture talked about the possible roles of this special kind of neurons in creating a representation of the actions of others, while using as reference one's own actions. I thought the concept of a mirror neuron was fascinating. The lecture focused on electrophysiological recordings from cortical areas of the rhesus monkey brain; but the idea of extending this mirroring concept to humans was already implicit in the proposal of the group from Parma (as it came to be known years later). The initial discovery they made was that neurons throughout various areas of the ventral pre-motor cortex of macaque monkeys (area F5)[2] (Fig. 2.1A) discharged both when individuals performed a given motor act, and when individuals observed another person performing a motor act with a similar end goal (e.g., Fig. 2.1B—C).[7] They called this property *mirroring*.

Their original results prompted many new studies to replicate the finding and to possibly extend it to other areas of the brain. The discovery gave rise to a wave of new and thought-provocative questions, and since then, several studies have further delineated

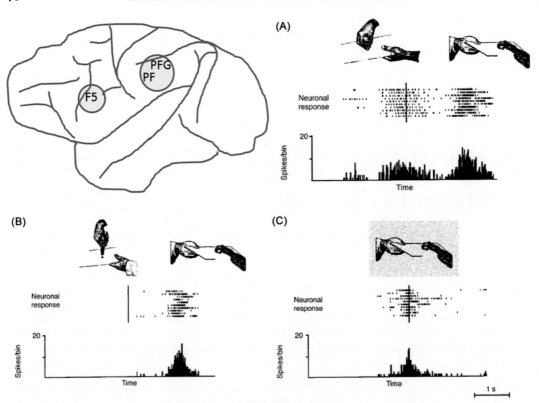

FIGURE 2.1   **Mirror neurons.** Cortical areas in the rhesus monkey brain containing neurons with mirror activity, including area F5 originally discovered by the Parma group (highlighted in[43]). (A) Sample mirror neuron (from[74,75]) firing during grasping of food by the experimenter and then by the animal, when the experimenter brings the food on the tray to the animal. (B) The experimenter grasps the food with a tool and follows similar sequence as in (A). (C) The monkey grasps the food in darkness. Vertical line in (A) and (B) represent the moment when the experimenter grasps the food. Spikes are aligned to this landmark, whereas in (C) spikes are aligned to the estimated time of grasp. Spike binning at 20 ms.

these regions and suggested possible roles in action observation related to the agent performing the action; the object serving as the goal for the action; or the activities related to eye movements—Inevitably required to carry on goal-directed actions[8] (Fig. 2.2A). Fig. 2.2B, (taken from[8]) shows several suggested pathways interconnecting these areas in the macaque brain.

These studies examined the *mirroring* problem from the stand point of action observation and proposed a new framework called the Mirror-Neuron Systems Theory (MNST) to research various aspects of language and social exchange.[1,2,13–15] Within the context of the MNST, it was not clear how the cells' mirroring activities would be modulated by

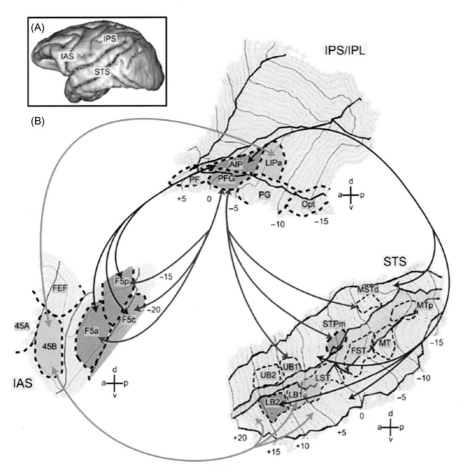

FIGURE 2.2 **Mirror neuron system** (according to[76–78]). (A) Lateral view of the monkey brain showing main sulci and areas involved in action observation: intraparietal, arcuate, and cingulated sulci. Nomenclature: AI, inferior arcuate sulcus; AS, superior arcuate sulcus; C, central sulcus; Ca, calcarine fissure; Cg, cingulated sulcus; DLPF, dorsolateral prefrontal cortex; IO, inferior occipital sulcus; L, lateral fissure; Lu, lunate sulcus; P, principal sulcus; PO, parieto-occipital sulcus; ST, superior temporal sulcus; VLPF, ventrolateral prefrontal cortex. (B) STS–IPL–F5 grasping observation networks in the monkey. Lateral view of a macaque brain showing locations of three regions involved in action observation: inferior arcuate sulcus (IA), intraparietal sulcus and inferior parietal lobule (IP), and superior temporal sulcus (ST). Flattened representation of inferior arcuate, intraparietal, and superior temporal sulci. Visual information on observed actions can be sent from STS through parietal cortex to premotor area F5 along two functional routes: a STPm–PFG–F5c, agent-related action observation pathway (red lines) and a LB2–AIP–F5a /p object-related action observation pathway (blue lines). Visual information from STS can also reach parietal and prefrontal areas involved in oculomotion, through the LB1/LST-LIP-45B oculomotion-related action observation pathway (green lines). The arrows specify the functional routes (taken from[79]).

kinesthetic reafference. This was a question I had all along. This aspect of the problem had been largely neglected, but it seemed relevant when interchangeably using different vantage points to track the dynamics embedded in the kinds of action generation and action observation that take place during social exchange. Questions addressing the use of different frames of references (even today) are not an integral part of the MNST narrative—They are the focus of research concerning other cortical areas for (intentional) action planning.[16]

The MNST in its current form does not address the possible roles that kinesthetic feedback may play on automatic mimicry, entrainment, or synchronous micromotions of the face and body that are known to take place during the types of social dyadic exchange that evoke mirroring activity. In such social contexts, it is known that people unconsciously tend to mirror each other's gestures and bodily rhythms at levels imperceptible to the naked eye.[17] The issue of kinesthetic biofeedback is relevant to the question of autism and the difficulties, the affected individuals may have with social interactions. This is so, because such biofeedback is continuously available to the brain as endoafference (through internally generated motions) and as exoafference (through externally generated motions).[18] It is unknown how the maps linking these forms of sensory feedback mature during neurodevelopment; yet they may be important to scaffold the evolution of social interactions later in life.

One aspect of social interactions is imitation. Imitation can be deliberate (overt), under the person's self-awareness, or it can be spontaneous (covert), largely beneath the person's awareness. According to ongoing research, the mirror neurons do not seem to engage the system in overt imitation while observing a person, nor do they lead to overt mimicry of the observed actions while the actions unfold. It would be a disaster if this actually happened. Imagine for a moment, what the outcome of a social exchange would be if during conversation you overtly imitated the other person's gestures. Surely, the social encounter would not go well. Yet, covertly, beneath awareness, gestural micromovements of the face and body do manifest some automatic engagement between the observer and the observed agent during a social encounter. Indeed, when social behavior is examined with high-grade instruments that pick up such neurophysiological minute fluctuations along the continuous flow of peripheral motions from the face and the body, it is now well-established that people do entrain their biorhythms at different levels of the nervous systems.[17,19] It is unknown though whether the MNS that the Parma group discovered in the neocortex also engages in the modulation of such microlevel activity, or if these spontaneous mimicry is rather modulated by more primitive, subcortical regions of the brain and/or by the peripheral networks of nerves throughout the body.

Along the lines of efferent outputs to move facial muscles, some F5 neurons described in the mirror neuron system are very responsive to orofacial motions related to eating and communication[20]; while others are responsive to sounds associated to particular goal-directed actions such as breaking the peanuts to eat.[21-23] These types of subtle face-related activities stand in stark contrast to the responses of other F5 neurons that become more active during the execution and observation of overt goal-directed bodily actions.[24,25] In monkeys, mirror neurons can fire as well when a grasping action is performed just out of sight[26] and the neurons activity can differentiate whether the action is carried out in the peripersonal or in the extrapersonal space of the animal.[27] Identifying the systems responsible for the sensory-motor feedback modulating these self-generated activities will open new avenues of enquiry in disciplines related to the MNST. For example, the largely underexplored kinesthetic reafferent component of this problem would help in creating new paradigms to address mirroring issues within the realm of social interactions during neurodevelopment.

The implications of the MNST and the possible future questions that we could ask upon incorporating peripheral kinesthetic reafference into the mirroring concept could be quite significant. In particular, we could address new issues of relevance to further understand social constructs such as affect and empathy—Seemingly affected in some neurodevelopmental disorders like the autism spectrum. Along these lines, the neuroanatomy of peripheral ganglia and nerves from the neck up and from the neck down (Fig. 2.3) could already set the stage to probe differential roles of such systems in the modulation and gating of output activities to generate different types of motions with social content. Among these are motions with emotional substance (e.g., expressing affect) vs motions with

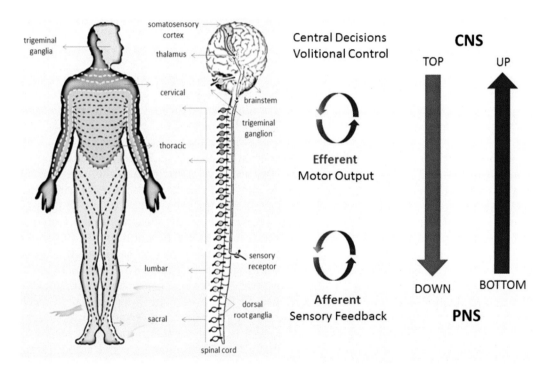

FIGURE 2.3 **New paradigm for MNST:** Proposed approach to study the mirror neuron systems through bottom-up processes that include the peripheral nerves of the face and body. This peripheral network provides reafferent kinesthetic input to the brain from continuously self-generated activity the brain itself causes. The outputs transmitted through efferent nerves result in motions that are controlled by nervous systems from the top down and can be continuously harnessed noninvasively at the periphery. This closed-loop system paired with wearables and new analytics can be used in co-adaptive interactive interfaces that provide sensory augmentation and/or sensory substitution therapies to measure, track, and treat the nervous systems of individuals with an autism diagnosis. This chapter shows examples of the use of such interfaces in children with autism who have difficulties communicating through spoken language. Top-down prompting and instructions are set aside to prioritize bottom-up driven self-discovery of goals, the sensory-motor consequences of their execution, and ultimately the acquisition of agency and volitional control.

intentional content (e.g., controlling purposeful, goal-directed movements). Both types of motions coexist during social dyadic exchange, but we know very little about their interdependencies.

During social interactions, action execution, and action observation can somehow co-occur without interference between dynamically shifting vantage points of the observer and observed agents, and the emotional states of the two social actors, i.e., as they engage in conversation through facial expressions, spoken language, and bodily gestures where overt movements and subtle gestural micromotions may share common pathways. How are the mirror neurons contributing to these different types of motions and to the kinds of sensory-motor transformations they engage at the output (efferent) level? And how do they suppress unintended actions associated to the type of mirroring activities evoked by observation? Understanding these questions may be important to assess the exchange of information between top-down and bottom-up processes of the central and peripheral nervous systems (Fig. 2.3), as well as their contributions to the exquisitely timed events of the social dynamics.

Some of these aspects of mirror neurons have gained attention in recent years.[28–30] Along those lines, new questions on mirror neurons have been partly motivated by the possibility that extracellular neural recordings in F5 (describing the mirroring activities) may be necessarily biased towards large pyramidal tract neurons (PTNs) (easier to record from than the smaller interneurons). Some of the PTNs turned out to have classical mirroring type of activity, while others showed suppressing activity. The latter have been found to be differentially recruited during action observation and as such, were coined "suppression mirror-neurons". They have been proposed to likely participate in the type of inhibitory code that seems to prevent overt mimicry of observed actions during social exchange.[31]

In a series of elegant experiments, Alexander Kraskov and colleagues at the University College London uncovered mirror activity and sensorimotor function in the spinal cord. These group of researchers have proposed that such activity may be necessary to suppress overt bodily action generation when PTNs with mirror properties become activated in response to action observation.[29–31] Their discovery also suggests an important interplay between facilitation and inhibition of activities by the same neuron as a possible mechanism to allow smooth transitions between states of execution and observation of actions[29] (see Fig. 2.4 with permission, showing marked differences in the population's responses). To these conclusions, I would add as well that a type of spontaneous and smooth transition between deliberate portions of continuous complex motions exists in everything we do (e.g., movements that take place during walking, sports, and the performing arts, among others).[32] One additional role for the mirror neurons could be the modulation of the ebb and flow of these coexisting staged and automatic segments of complex bodily movements. Staged segments deliberately performed by the agent are consciously controlled, but segments that spontaneously co-occur remain largely beneath the agent's awareness—Unless an external interlocutor explicitly points them out to the agent.[32] I clearly recall having a conversation about this very topic with Vittorio Gallese in my office at Rutgers Psychology. Gallese came to Rutgers to give a seminar hosted by Philosophy Professor Alvin Goldman. As part of the visit, he visited my lab at the Rutgers University

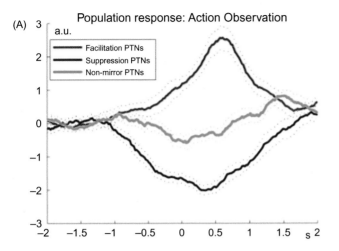

Population response: Action Observation

(A)

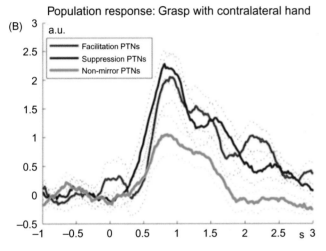

Population response: Grasp with contralateral hand

(B)

FIGURE 2.4 **PTN with mirror neuron activity** (taken from[31]). (A) Population average of PTN firing rates and standard error of the mean (dotted lines). Action observation contrasts with (B) precision grip grasp of a food reward performed by the monkey with contralateral hand. Data plotted separately for mirror PTNs, which showed facilitation (n = 11, red) and suppression effects (n = 14, blue) during action observation, or no effect (nonmirror PTNs, n = 23, green). Alignment at time zero in (A) is to the touch of magnetic sensor by experimenter and in (B) to the cue for the monkey to begin its reach-to-grasp movement. Arbitrary units (normalized firing rates) in vertical axis and time (seconds) along the horizontal axis. Red, facilitation PTNs; blue, suppression PTNs; green, nonmirror PTNs.

A, Population response: Action observation; B, Population response: Grasp with contralateral head.

Center for Cognitive Science (Ruccs). We had an opportunity to meet in person and chat about our ongoing research on embodied cognition and movement classes.[32] I remember he brought up the example of a tennis player trying to upset the serve of his opponent by pointing out "how perfect his wrist motions were to get the serve just right". Clearly, this automated motion of the wrist, so perfectly going on beneath awareness, became totally out of whack, the instant the tennis player brought it up to his opponent's awareness. Now, inevitably attending to the wrist motions transiently upset the tennis serve. The trick worked.

More generally, the smooth transitions between deliberate and spontaneous movements, present in complex sports routines and routines of the performing artists, necessarily occur too in dyadic social exchange. They give rise to patterns of physical entrainment and synergistic activities within the actor's bodily rhythms. In a dyadic exchange, these synergies covertly manifest—A point that we further develop and provide analytics for in Chapter 6. Perhaps, within a social dyad, this form of synergistic control of many degrees of freedom in the body of the actor scaffold the ability to synergistically engage the degrees of freedom in the body of the external agent. Such an ability would allow the spontaneous emergence of synergies between the two participants of the interactive social dyad.[17]

During the type of "social dance" that takes place within the coupled biorhythms of the social dyad[17], both states of action observation and action execution can coexist for any given gesture. One aspect of this problem that we do not know much about is how their macro- and microlevel entrainments share a given action, and how/if their dynamic exchange may or may not relate to mirror neurons. The discovery of facilitation–suppression activities in PTNs of the motor systems broadens our understanding of mirror neurons across cortical and corticospinal structures necessary to better understand the central–peripheral nervous systems exchange. In turn, such broader understanding helps us conceptualize embodied cognitive actions and their synchronous oscillations across diverse social contexts.

THE HUMAN MIRROR NEURON SYSTEMS

In humans, weaving the story of mirror neurons has been a bit more complicated than in monkeys because technical and ethical considerations limit the research to rely on sources of data that are not as accurate as cortical spike activity. These include noninvasive techniques such as fMRI, transcranial magnetic stimulation, and EEG/MEG.[15,33−36] Using such means pre-motor and parietal cortices have been identified as integral parts of the human mirror neuron systems[7](Fig. 2.5A) along with the anterior insula[37] and the anterior cingulate cortex.[38] It is very possible however, that other areas participate in action observation/imagery and action execution, but new techniques will be necessary to address this proposition in humans.

One major setback in human research is that motor artifacts are common across all these techniques that record activities from the brain. As such, large covert movements are difficult (if not impossible) to sample during the experiments. Most of the paradigms are restricted to action observation and motor imagery, whereby responses consist of mouse clicks or self-reports. Naturalistic movements in purposeful actions are not possible to assess without contaminating the very signal under study. As such, to this day, we do not truly understand the kinesthetic efferent–reafferent components of human mirror neuron systems. Such data would be necessary to derive a bottom-up mechanistic explanation on the emergence of "the self" and its dynamic (social) updating in relation to others. Deriving explanations for such adaptive social code would be useful in closed-loop feedback controlled settings of use in therapies and rehabilitation. In such settings, the continuous feedback of biophysical signals' variability across the brain and body can be used to guide the nervous systems towards effective motion-driven communication during social exchange.

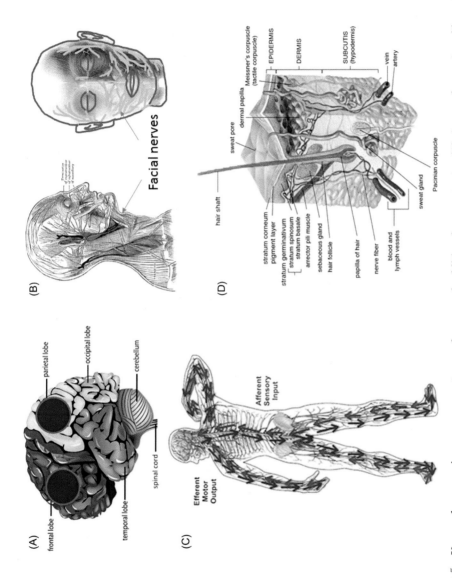

FIGURE 2.5 **Known human mirror neuron system and proposed additional components..** (A) Areas where mirror-like activity has been detected using noninvasive techniques (e.g., fMRI, MEG, and EEG) Areas of the face (B) and body (C) where micromovements mirroring activity should be assessed for efferent and afferent nerves and skin. (D) Skin receptors and layers contributing to sensations of touch, pressure, temperature, pain, proprioception, and kinesthetic feedback.

As noted, Fig. 2.5A shows some mirror systems areas identified in the human brain. But in addition to brain areas, here, we also suggest the idea of a peripheral mirror system distributed across the face (Fig. 2.5B) and body (Fig. 2.5C), whereby the activities of the peripheral receptors and conducting nerve fibers could be broadcasted through high-grade wearables affixed to the skin. In naturalistic behaviors, such activities could be continuously monitored and quantified using a myriad of kinematics and electrophysiological parameters in the time and frequency domains.

The skin is rich in receptors with specificity for different forms of environmental inputs, i.e., sampling across a broad range of frequencies (Fig. 2.5D) that we can use to understand the sense of touch and its role in mirroring and empathizing with others. These include mechanoreceptors to perceive sensations of pressure, vibrations, and texture; thermoreceptors to perceive temperature (e.g., of objects we touch and hold); pain receptors (nociceptors) to perceive pain from injurious or hurtful stimuli, and proprioceptors to perceive one's own body positions in relation to each other and the external surrounding environment. Perhaps, some of these receptors possess a type of code congruent with the code of visual, auditory, and olfactory sensory receptors in the neocortex to form overlapping sensory maps and help the body communicate with the brain.

In neurodevelopmental disorders that affect social exchange, these various receptors of the skin could provide clues to profile their proper functioning and maturation during early developmental stages. Because skin-related activities are accessible with noninvasive means, output related to sweat, temperature, and muscle tone (readily accessible through wearable biosensors) could help us better map the frequency and temporal ranges of bodily sensations that the brain may be having (or not having) access to through afferent channels.

Mirror Neuron Systems Without Reafferent Codes

When I attended the lecture at the 1998 NCM meeting, I was hoping to see some data connecting the bodily actions of the animals, or the movements of the experimenters to the neural activities of the cortical neurons they were recording from, across various brain regions. I was also hoping to see these phenomena explained from two vantage points: The internal vantage point of the nervous systems of the animal being studied and the external vantage point of the observer (the experimenter studying the animal). I was expecting to see at least some behavioral data registered from the dyad interacting in tandem. I saw instead a one-sided account of the phenomena based on the experimenter's interpretation of the spiking activity of these neurons in relation to a description of the animal's or the experimenter averaged hand's motions.

The group from Parma was making the case of a special property of these neurons to recognize one's own actions and the actions of others; but there were no behavioral recordings, e.g., recordings of the underlying physiology of the many layers of bodily biorhythms encompassing the actions of the dyad (i.e., the animal and the experimenter in tandem). The motions of their bodies occur simultaneously at many levels of the nervous systems, including the autonomic nervous systems, the enteric nervous systems, the peripheral nervous systems innervating the muscles of the limbs and face, and the central nervous systems engaging the brain and the spinal cord. The group of peripheral afferent nerves receiving information from skin receptors was not included in their discussion either. When examining the literature, I could see motor physiology in the form of electromyographic (EMG) activities (e.g., from major muscles of the arm) being included in the figures in the form of a grand average. The single trace of averaged activities over many trials obtained from the major muscles was being used to align the spiking data from trial to trial with the motions' initiation and ending. Yet, averaging across repetitions under assumed normal distribution smoothed out fluctuations in these muscles activities that very probably contained signals relevant to the interactions taking place between the animal and the experimenter. At a macrolevel of enquiry, this was not being considered, but zooming into the phenomena would surely uncover other novel findings escaping the naked eye of the observer.

In retrospective, I think it was perhaps too premature to think of the contributions of dyadic-coupled bodily rhythms added by peripheral efferent and afferent nerves to the mirroring code of these neurons. However, it has been nearly 20 years now, since the MNS discovery. The extent of body physiology that we see in the MNS literature is mainly aimed to assess the presence or absence of movements beyond some threshold of large, overt motions. Gestural micromovements remain largely neglected. Additionally, some averaged trace of kinematics (e.g., distance, speed, or some positional parameter, etc.) may also be collected and superimposed on the spikes (in the case of monkeys). Yet, the evolution of continuous minute fluctuations in the bodily biorhythmic motions and their signatures across the nervous systems has not been systematically assessed. Consequently, we do not have any knowledge regarding the interdependencies and codependencies bound to take place among the signals from the central and peripheral nervous systems of the socially interactive dyad during action observation and action execution.

It seemed to me that the MNST was making a case for "disembodied actions". The theory lacked proper multilayered dynamic and stochastic characterization of various movement classes embedded in observed and executed actions. The theory was also fundamentally missing the reafferent component of the continuous kinesthetic flow of motions the biorhythms themselves produce.

In general, neurophysiology for systems neuroscience tends to be confined to studies of the neocortex, particularly in theories proposing, e.g., how, a given brain within a social context may be figuring out what other brain(s) in that social context may be thinking, intending to do, believing, etc. In some contexts of cognitivist theories, the guessing of the intentions and beliefs of others fall under a general umbrella referred to as Theory of Mind (ToM).[39,40] For example, in the context of neurodevelopmental disorders (e.g., autism) that alter the abilities to interact socially, the broken MNS, somewhat related to the ToM, and the ToM itself have been proposed as *deficit* models of the condition.

It is important to point out, however, that such models have not helped building accommodations for the affected individuals. The cognitivist theories leave out the body in motion—i.e., the very thing producing the *behaviors*, the theorists observe to make subjective inferences and interpret data about social mechanism. These opinions lead to strong conclusions regarding the person's overall intellectual abilities and directly impact the person's life. Movements embedded in activities of daily living make up behaviors that are part of the social repertoire, but since such theories cannot be extended beyond the confines of the subjective mind, they lack the proper tools to guide the type of research that would stand a chance to help *physical* social exchange in disorders of the nervous systems that affect social interactions.

I call these "dangerous theories" because they are easy to sell to the public and do make a dent in legislations that directly impact the lives of affected families. They also affect the perception that society at large develops about the affected individuals. Many in our society think the children are "misbehaving" deliberately. People are generally ill-informed by these theoretical ideas that cross-over the mainstream media.

The cognitivist theories do not address many physiological maladies of such conditions. They render them as "comorbidities" impervious to the mental representations the theories use. Consequently, the science the popular media uses to explain autism to the public does not broadcast the types of physical problems that plague the day-to-day existence of the affected children and that produce so much anguish and sadness to their existence.

We have witnessed firsthand in my lab the type of emotional stress such views cause in the affected people (see interviews with children in the spectrum of autism in[41] Chapter 26). While such theories claim that the affected children lack empathy, have "mind blindness" or may even be sociopaths[39,42–44], many of the children are in pain, cannot properly feel their bodies and/or control/feel their motions like their peers do. I can assert this with some degree of certainty because by now, over the course of 9 years, members of my lab and I have personally objectively profiled the nervous systems activities of many children and adolescents with neurodevelopmental disorders. We have also met or corresponded with hundreds of families affected by autism and/or attention deficit hyperactivity disorder (ADHD). And we have interacted one-on-one with their caregivers and learned their day-to-day struggles across the continental United States.

Along those lines, in a serendipitous twist of events, we came across the National Science Foundation Innovative Corps (the NSF I-Corps) and received funding to discover the possible market fit for some analytics we developed to handle data from wearable biosensors. We had developed the technology while doing basic scientific research in autism, funded by the NSF. Under their funding, we were encouraged to find the segment of the population with a critical need for our scientific innovation. Part of the idea was to identify through one-on-one interviews, a sector of the US population that would desperately need the type of nervous systems' outcome measures we created. These biometrics can continuously monitor the somatic-motor signatures of naturalistic behaviors taking place when people interact with the surrounding environment, or when they interact with others during social exchange. It took us some time to discover the market fit for these outcome measures; but after completing 120 interviews with open-ended questions such as "what keeps you awake at night?" and "what do you think is the biggest problem you may face day to day?", etc., we discovered that in the world of autism, this notion of a mental illness or a disembodied cognitive/social disorder has largely contributed to the absence of insurance coverage for therapies that specifically address *sensory-motor issues*. It was a very subtle point—Not so easy to discover. Occupational therapies (OTs), speech therapies, and behavioral therapies are all part of early intervention programs publicly available and covered by insurance. That is until they bring up the issues of sensory-motor disturbances in autism spectrum disorders (ASD) or in any other developmental disorders (like sensory processing disorders). The OTs that focus on pediatrics sensory issues or sensory-motor issues do not have insurance coverage because of the recommendations of the psychological–psychiatric constructs that drive the field (e.g., cognitivists theories) and by the recommendations of the American Pediatrics Academy. Of course, the insurance issue is unique to the United States. European approaches are very different, according to a tradition of universal medical coverage.

In the United States, the prevalence of subjective inventories, the lack of objective outcome measures of behavior, and the claims of "evidence-based" support from surveys, complicate matters for the families touched by a neurodevelopmental disorder. We discovered that the only therapy that receives some form of insurance coverage is the one aimed at reshaping the child's behaviors to make them look socially appropriate. Therapies such as applied behavioral analysis (ABA) use operant conditioning techniques (Quoted from Wikipedia). "Operant conditioning" (also called "instrumental conditioning") is a type of learning in which the strength of a behavior is modified by the behavior's consequences, such as reward or punishment". In ABA, the behavior features are determined by observation by the Board Certified Behavioral Analyst (BCBA). The BCBA has credentials (i.e., it is certified) to reshape the child's behavior. This reshaping takes place, however, without scientific measures of the consequences of the reshaping action on the child's developing nervous systems. They do not measure the functioning of the autistic (coping) nervous systems. Their explanations and practical approach follow the disembodied theories of the mind. The evidence they provide to justify their use to the public is solely based on subjective opinion. Such opinions are derived from observation of behavior without examining any of the physical underpinnings. As such, these methods do not track the sensory-motor *consequences* of reshaping behavior on the nervous systems' being intervened on.

Since the sensory-motor approaches do not receive any support from American Academy of Pediatrics[45], none of the OTs that use these approaches get insurance coverage in the United States. For example, the types of biometrics we describe in this book could indeed serve as outcome measures to occupational and physical therapies (see movies in **Notes[1]**). As such, they could help affected families in general, but in particular they would help low-income families that cannot afford the current rates of alternative therapies such as hippotherapy (involving horse riding guided by OTs), aqua therapy involving interactions in the swimming pool, and other forms of OTs with a focus on the neurodevelopment of sensory-motor issues.

I will talk about these societal problems in more detail in Chapter 7, but it suffices to say here that families in a high-income bracket in the United States (i.e., over 250K combined household income) pay for such therapies with out-of-pocket funds at rates that generally range between $150/h and 350/h. As such, despite their effectiveness, those therapies are inaccessible to most families affected by autism. When systematically practiced, they do improve body awareness, sensorimotor integration, and overall dyadic social exchange. Such therapies target the kinds of sensorimotor feedback loops that MNS would require for establishing the flux of observation and execution of actions in social dyads.

As a young, naïve graduate student, I did not know any of that in 1998, when I attended the NCM conference and learned about MNST and later found out about its

[1]Link to the I-Corps Movies here

direct connection to ToM in the context of autism. To me, as to anyone out there in the public, these theories were just very elegant, scholarly written papers proposing very interesting ideas. At the time, I also wondered about the math, the logic, the computational modeling, and more generally, the hard core, rigorous science missing from those papers, which were mainly supported by surveys and opinions. Back then, (I thought) "may be these fields are such that those other important aspects of the *scientific method* would come later". Those were my thoughts back then. Today, nearly 20 years later, I came face-to-face one-to-many times with the consequences of such irresponsible pseudoscience. Such theoretical frameworks are used to help diagnose and treat disorders of the developing nervous systems, but they do not physically measure anything about the nervous systems. It is all based on opinion.

The platforms in use to diagnose disorders of the nervous systems and to recommend treatments are completely detached from the actual lives of the affected people, and yet so easy to sell. It is mindboggling how any of such theories without a shred of true scientific knowledge about the neurotypical brain may be translated to explain autistic phenomena. It is even more disturbing to see how they may be further used to support behavioral reshaping treatments that do not comfort or support the developing physical body of the child in any way. As the theories behind them, the evaluations of treatment effectiveness are a matter of opinion and biased interpretation. The words *physical measurement*, *objective quantification*, *mathematical modeling*, and *computational simulation* do not exist in the lexicon of such ideas attempting to explain the possible workings of a disembodied brain.

The amount of distress I have witnessed in children that come to my lab upon endless months of such interventions compares to Post-Traumatic Stressed Syndrome observed in adults. I have yet to find a BCBA that allows me to measure the nervous systems biorhythms during the behavioral reshaping procedures. We have tried numerous times and interviewed BCBAs in many such centers, including the very headquarter, without any success. I can say without a doubt that based on our experience, behavioral practice is the most impenetrable frontier in autism to allow us to do any computational-neuroscience driven research.

The year I attended this talk at NCM, I had barely finished my Master's Degree in Cognitive Science with a focus on mathematical models and computational simulations of complex goal-directed movements generated by bodies with many redundant degrees of freedom. I was using artificial neural networks to model possible scenarios bound to emerge in real cortical neurons participating in sensory-motor transformations. Because of that, I had a strong interest in the kinds of philosophical works advanced by ideas like those in the MNST. I have always had such interest. In graduate school, I asked philosopher Pat Churchland of UCSD to be in my thesis committee and was very happy when she accepted. In recent years, I collaborate with philosopher Maria Brincker[46,47] of University of Massachusetts Boston and have delightful and very intellectually stimulating interactions with her.

Philosophical ideas are very important, as they provide a necessary step to brainstorm about the science we do, to question it, and to renovate it. They can be an invigorating force working behind the scenes when they are properly linked to scientific evidence. But when you take ideas (such as those advanced by the ToM paradigm, or those of rote behaviorism) to the next physical level of implementation (e.g., interventions), while skipping the part about physically measuring phenomena, or going as far as substantially biasing the empirical evidence using opinion-based criteria[48], the outcome can be rather tragic.

I could not imagine using theoretical work, e.g., such as that I did for my PhD thesis, to intervene in nascent nervous system without first validating it with empirical data. By the late 1990s, I had trained artificial neural networks using an equation I derived, which I will describe in Chapter 4. Through computer simulations, I had found all sorts of new predictions about possible patterns of neural activity during sensory-motor (co-ordinate) transformations. During goal-directed actions, these co-ordinate transformations are common place, e.g., transforming sensory information globally, from an allocentric to and egocentric vantage point; or locally, as when the brain must code various body parts relative to an egocentric frame anchored somewhere on the body. These were theoretical predictions of an artificial neural network, i.e., worthless notions without actual empirical data. I would not have dared use any of those theoretical ideas to recommend treatment to alleviate some disorder of the nascent nervous system of a child.

It took me years of training in electrophysiology and then much learning of computational methods, signal processing, and the designs of new analytics to empirically characterize physical phenomena, so I would dare propose an idea to the field of basic science.[49,50] It took the empirical examination of data from thousands of people to dare say "perhaps we can translate this to clinical settings?". And yet all along, our work states the caveats of the methods, so others can blindly reproduce our results. I guess my lab and I are still in shock when faced with the approaches we see in the field of autism and the very fact that so much pseudoscience has dictated people's lives for so long.

The ideas of the MNST covered a lot of ground in basic science. They have not yet crossed-over the clinical domain, but seem to suggest a broad range of causal threads pertaining to the evolution of social interactions, ranging from figuring out the origins of empathy to how language may have emerged in humans. I could not, however, see the big leap justifying any of these causal links without examining the body neurophysiological changes underlying the unfolding cells' activities. The neural activities of MNS cells evolved while actions took place either as self-generated by the nervous system where these cells inhabited, or by the sensing of external actions generated by other nervous systems the animal was interacting with or merely observing. As any other work in neuroscience, the MNST was entirely disembodied—Despite theoretical or philosophical claims of embodiment.[3,51] It was fascinating though, to hear about mirror neurons, and that alone justified the talk. I was in awe with Rizzolatti's lecture. It certainly made me think a lot as a young graduate student. It was an inspirational talk, and regardless of the shortcomings one could identify today[52], it did for my research program more than any other theory in motor control ever did.

RETHINKING SIMULATION THEORY

*A crucial element of social cognition is the brain's capacity to directly link the first- and third person experiences of these phenomena (i.e. link 'I do and I feel' with 'he does and he feels'). We will define this mechanism 'simulation'... We will posit that, in our brain, there are neural mechanisms (mirror mechanisms) that allow us to directly understand the meaning of the actions and emotions of others by internally replicating ('**simulating**') them without any explicit reflective mediation.*[1]

The notion is so very powerful. Let's dissect it a little bit and then see how we can measure some of the underlying parts we can identify. But let's do so while keeping in mind that we are inevitably leaving out stuff that we do not have the foggiest idea exists; neither do we know the validity of any of the assumptions we make about these phenomena (e.g., the role of glial cells in all of it; the role of the immune system in all of it, the genetics underpinnings, among other aspects of the "dark matter" of brain science that remains open to questions). To all of that, we must add the caveats of our own methods as well.

Dissecting Simulation Theory

To realize this idea, we first need to understand what is being assumed that the brain can do. It is assumed that the brain can figure out the notion of a frame of reference (that's a tough one!); then the same brain can place at least two frames of references in separate places, one on its own self, and one external to it, to build the first person (self)-view, and the third person (other)-view (e.g., Fig. 2.6). That requires, in the first place, a nervous system having the ability to self-monitor its own existence and somehow understand that it is not someone else's activity that is taking place, but rather one's own activity. In other words, the brain needs to build a notion of self to be able to self-monitor its own activity. But this is highly nontrivial if the brain lacks somehow the ability to distinguish the flow of activity that is self-generated from the flow of extraneous activity (over which the brain has no immediate control). Then, the brain also needs to have the ability to understand the nature of the extraneous activity: Is the ongoing activity a direct consequence of my own self-generated activity in the recent past? Or is it totally independent of it? In other words, it is not just the activity (from the self or the other) that the brain needs to know. It is also the sensory consequences of such self-generated activities that seems so relevant to the idea of self-monitoring. Is the flow causal, such that the past events predict the future events with some degree of certainty? Or is there no well-structured affinity between past and future events? Ultimately, what to keep as a memory and how to retrieve it on demand becomes rather important in all of it.

Then, the system monitoring the self-generated activity must have the ability to cast the information interchangeably from the ego or from the allocentric frame of reference, and it would better be rapidly acquainted with the notions of relative codes, metrics, and similarity measures, as they become very important when we use different frames of reference to navigate through our mental and physical spaces.

It seems that critical to the idea of simulation theory is also the ability to *internally replicate* (they equate this with *simulate*) the actions and emotions of others in what I interpret as *automatically* to mean "without any explicit reflective mediation"—I take it as doing so spontaneously, without explicit instructions, autonomously in some sense. But then, what causes such autonomy? And how can it be externally modulated, to allow for internally driven adaptive exchange with the extraneous flows of information. How is the internal dynamics of the nervous systems relating to the dynamics of the external stimuli? Answering this question may be important to help the system focus on externally driven guiding signals to build error correction codes, e.g., as when an external sensory goal guides movement to attain a specific purpose.

FIGURE 2.6 **Expressing physical phenomena from different vantage points.** (A) In one case, it is possible to define phenomena from the vantage point of the nervous systems of the monkey on the right-hand side of the picture by locating a co-ordinate system somewhere on the body (e.g., the shoulder) and expressing the hand position of the animal in an egocentric reference frame. The hand motions can then be measured relative to that egocentric frame, but also relative to an allocentric frame of reference located, e.g., on the other monkey's hand. Another frame of reference can be situated on a location external to both animals and coded allocentrically in both cases. (B) The monkeys' frames of references can be exchanged and we can then speak of an egocentric frame anchored at the monkey on the left-hand side of the picture coding the shoulder location as an external goal with allocentric properties like those of the object. Navigating interchangeably through these different systems is possible through co-ordinate transformations that code relative positions and reference-frame invariant quantities such as rates of change.

I can safely assume that I am leaving out quite a bit in this attempt to dissect the problem, but a critical explicit assumption I am not leaving out is this: "the observer and the observed are both individuals endowed with a **similar brain−body system**". This is a reasonable assumption but (1) it requires that we measure the dynamically coupled brain−-body systems and build a metric of similarity providing a standardized scale of how entrained the systems are within the self and with others; (2) it requires then the use of the standardized scale and formal distance metrics to measure departures from the norm, to classify pathologies, and their severity along a gradient; (3) once disparities are quantified across the human spectrum, what do we do with the classification? How do we use that information to build corrective codes that do not interfere with the systems' autonomy in the first place? More important of all, how do we dynamically update such code during early neurodevelopment, when motor signatures shift so rapidly from day to day? (Fig. 2.7).[53]

The continuous monitoring of neurodevelopmental trajectories ranging from physical growth to the maturation of neuromotor control are extremely important to help us detect risk of derail from the typical path[47] and rapidly intervene. When no appreciable change in neuromotor variability is quantified, one should worry and seriously consider the baby may be at risk of later developing a nervous system disorder. One population that requires careful monitoring is that of babies born prematurely (Fig. 2.7A) because within a similar number of months (e.g., 3 visits in 5 months), their signatures of motor variability do not change at the same rate as those of a full-term baby (Fig. 2.7B). In such cases, sensory-motor-based therapies should be recommended. Such early interventions do help the body provide feedback to and ultimately connect with the developing brain.

At all levels of the nervous systems, there are dynamic changes that shift stochastic signatures while the person ages and the maturation of different nervous systems processes ensues. In this sense, the current static notion of classification or diagnoses goes against the actual dynamics of human development. We need to altogether disrupt the present *static* notion of diagnosis if we want to make real scientific progress.

The probability distribution functions describing the variability of the nervous systems' biorhythms during development and beyond changes with age and what is more important yet, the rates of change of such empirically estimated stochastic parameters change nonuniformly within different age groups (Fig. 2.8). For example, they are accelerated in neonates.[47] But by 5 years of age, they transit into more steady-state signatures that once again changes dramatically during puberty.[47,55,56] By old age, these signatures plateau in normal aging but rapidly deteriorate with neurodegeneration. Examining the statistical variations of the fluctuations in our motions under a *one size fits all* approach is inadequate to build dynamic outcome measures of nervous systems performance. Indeed, the human probabilistic landscape of biorhythms' fluctuations is "alive and kicking". It changes as we live our lives and age; but it does so dramatically different in those who go on to receive a diagnosis of autism (Fig. 2.8).

When it comes to disorders of the nervous systems—The main theme of this book—We must be extremely careful on how we construct the scales and definitions that will inevitably be used to try and shift the pathology towards normalcy. Our schemes may not always be appropriate, particularly if they interfere at some unforeseen level with the well-being of the person and end-up bringing on additional sources of stress to the person's nervous systems, thus diminishing overall quality of life.

In the case of autism, the neurodevelopmental disorder that has driven my lab's research since 2009, the MNS theoretic assumption that everyone in the social scene has the same brain–body physiology falls apart in ways we have by now measured quite exhaustively in thousands of individuals of all ages (see Fig. 2.8). Specifically, the brain--body loops are functioning differently in nervous systems that underwent some neurodevelopmental glitch. These include individuals in the broad spectrum of autism, a nervous system that manifests elevated levels of random and noisy involuntary motions. So, to be able to use the MNST idea in autism, we need to investigate some of the assumptions the

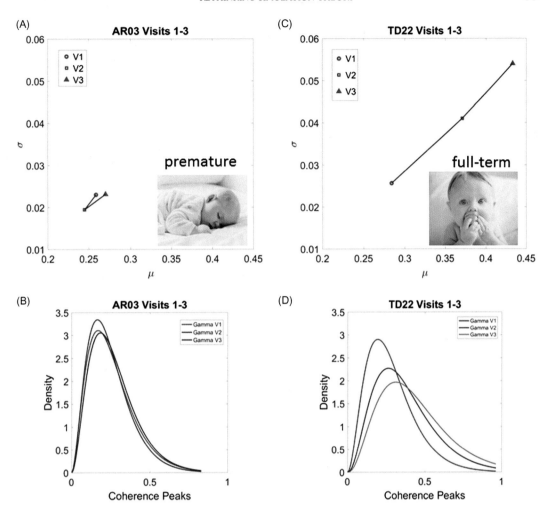

FIGURE 2.7 **Detecting risk for neurodevelopmental stunting of motor control.** Babies born prematurely and full-term babies were recorded with inertial measurement units attached to both arms over the span of 5 months across three visits. In each visit, a minimum of 2 hours of motion were continuously recorded (APDM 128Hz, Portland, OR). Synchronous acceleration patterns of their arms revealed very little change in this premature representative baby whereby the probability distribution function of the coherency peaks did not significantly change from visit to visit in marked contrast to the full-term baby. Likewise, the empirically estimated stochastic trajectories measured by the Gamma moments did not change much in the premature baby, but from visit to visit the full-term baby displayed marked changes in the coherence of the arms. The latter suggesting nascent synergistic patterns of motor coordination absent in the premature baby across similar time span.

theory makes about brain–body system similarities and (at a minimum) understand the assumed ability that the system can identify the self in relation to others and others in relation to the self. How symmetric is this map between one's own body and others under utterly dissimilar efferent motor output and persistently random and noisy reafferent kinesthetic sensing?

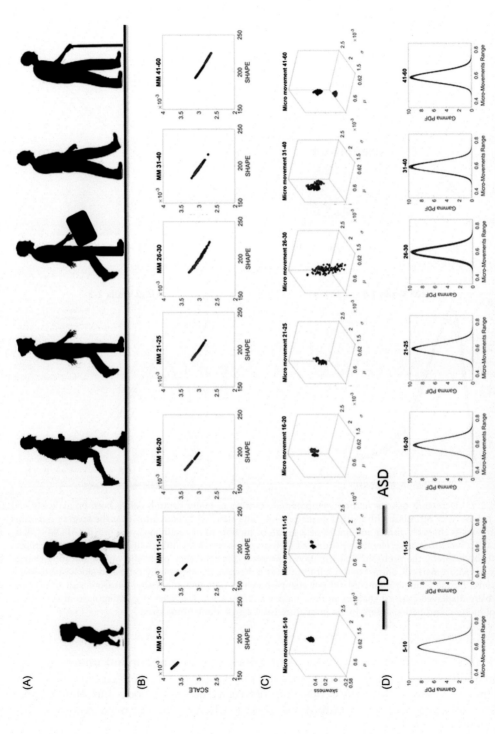

FIGURE 2.8 Age-depending evolution of the probability distribution functions' landscape of human development between 5 years and 60 years of age. (A) Schematic of human aging. (B) Empirically estimated Gamma shape and scale parameters plotted on the Gamma parameter plane for TD controls (TD in blue) and age-matched individuals with autism spectrum disorders (ASD in red). Each group has 100 points randomly selected from

the open access Autism Brain Imaging Data Exchange repository. The distributions are estimated from the fluctuations in the amplitude of the linear speed of involuntary head motions while participants are in resting state fMRI. Head motions are extracted using conventional methods.[54] The first rate of linear displacements is measured as a time series of speed peaks. The moment-by-moment fluctuations in the amplitude of the maxima are converted to a point process (the micromovements data type.) This point process is then examined under the general rubric of a Poisson random process under identically independent distributed (IID) assumption. A Gamma process is assumed to estimate the shape and scale parameters using MLE and 95% confidence intervals. The points cluster and shift at different rates for different age groups, thus indicating the need to not assume a "one size fits all model" for statistical inference and data interpretation. (B) The estimated Gamma moments (mean, variance, skewness and kurtosis) are plotted along the X-, Y-, and Z-axis, and the size of the parameter is the kurtosis. Here, the clusters clearly separate at puberty (11–15) and during the critical ages, when mental disorders like schizophrenia, bipolar, and depression (among others) emerge and become evident (16–20 and 21–25). During the mid-age period (31–40), there is also tight clustering of the ASD group, but more spread of the TD cluster; and from 41 to 60 years of age, there is a clear separation between the two groups according to these involuntary head motion patterns. The stochastic signatures of involuntary head motion variability are one example of biorhythmic parameters (extracted from the nervous systems) with shifting characteristics that evolve with age and do so at irregular rates within different age groups. (D) The evolution in the dispersion and shape of the estimated probability distribution functions (PDFs) using the continuous Gamma family is also evident for a randomly chosen person of a given age group in the general population. These marked differences in stochastic signatures and rates of change in parameter values would be missed under traditional methods.

HOW DO I KNOW IT'S ME MOVING MY OWN BODY WITH AGENCY?

When I joined Rutgers University, the Psychology and the CS Departments that hired me had been awarded an Integrative Graduate Education and Research Traineeship (IGERT) grant with a focus on Perceptual Science. As part of the team, I had to teach a class that brought together students from both departments and engaged them in the type of interdisciplinary collaboration that the award was designed to promote. This was a challenging task because psychology students have very little mathematical and computational training, so the class could not be too math-demanding or require programming. Then at the other extreme, CS students with that sort of mathematical and computational training have very little knowledge about the brain and body neurophysiology and neuroanatomy. The CS students generally have not grasped neuroscientific and neuropsychological ideas when they come to such a class. In other words, half the class had a lot of qualitative, descriptive knowledge but no clue on how to do or build anything; while the other half came with a bag of computational tricks but did not know what to do with that in relation to brain science. If I taught a class on math and programming, I would put the psych students to sleep and bore to death the CS students. If I taught a class on neuroscience and neuropsychology, I would bore to death the psych students and likely, not get the CS fellows entirely engaged.

I decided to design a hands-on class (**Notes**[2]), whereby I taught people mathematics and the computational skills required to analyze motion data through their own doing. That is, they were to collect motion data of themselves and learn how to analyze it and interpret it, while considering some of what we know about neuromotor control in relation to Psychological and Cognitive Neuroscientific principles. In this way, they could see what abstract mathematical objects (like quaternions) meant in the context of measuring brain control over the body, or how gradients, cost functions, and policies (e.g., for reward in the context of reinforcement learning) translated into physical bodies in motion, among other things. By forming interdisciplinary teams and encouraging the various team-members to help each other by sharing their skills and teaching each other, we came up with a great project.

To make the class interactive, I decided that a project had to be proposed by each team and a vote completed to democratically select that best project. By best, here we meant not only advancing a question we had never asked (i.e., it had to be a new question in both fields, psychology and CS), but also, the project had to have the potential of requiring knowledge from both disciplines. Since we had six teams in the class, they competed (through a lot of good-user-friendly arguing) and voted for the best idea for a project. Then, they implemented it from beginning to end. The interesting thing here is that the project we all best liked was precisely addressing some of the above-mentioned questions and challenging several of the assumptions the MNST made. To make a long story short, the team went to Washington, DC and won the Community Choice as the most popular video and poster of the 2012 IGERT competition, which the reader can see here: http:// igert2012.videohall.com/presentations/220

---

[2]Link to the IGERT class link here

We entitled the project "What do we see in each other? How the perception of movement drives social interactions". The making of this project was an incredibly rewarding experience for all of us who participated in it. Polina Yanovich, a very gifted CS student who rotated through the lab for some time, proposed the initial idea, and played the most important role in carrying on what was needed to implement it. She built a very cool avatar that we then used to test whether a person could recognize her/his own motion patterns and distinguish them from others, even when they were distorted in various ways. Fig. 2.9 walks the reader through the pipeline of the making of the experimental stimuli, which was quite labor intensive.

The students collected their own movements during the execution of boxing routines, the tennis serve, and walking. Fig. 2.10 shows examples of the digitized upper body in motion performing the tennis serve (using the Motion Monitor software, InnSport, Chicago, IL); while Fig. 2.11A—B shows examples of motion capture outputs from positional trajectories of both hands performing those motions concurrently. The right hand

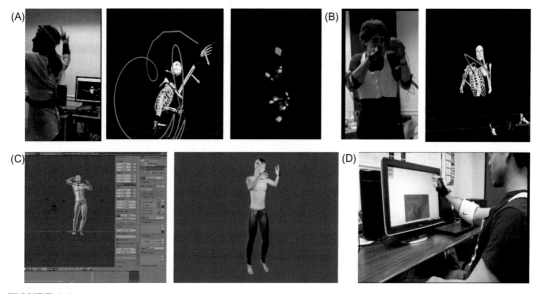

FIGURE 2.9 **Pipeline to build stimuli for the IGERT experiment.** (A) The athlete (and novices) performed the tennis serve and (B) boxing routines while we recorded the body under various feedback conditions. Here, the upper body is shown with positional trajectories of selected sensors on the hands. Sensors recorded positional data at 240 Hz and real-time feedback was used to calibrate the motion caption. In one of the conditions, glowing sticks were attached to the body and the person moved in the dark receiving feedback from the sticks. In another condition, the person relied on proprioception while the eyes were closed. In a different condition yet, the person received visual feedback from the performance via a mirror or an avatar in real time. (C) The students used open access Blender software (left panel) to build an avatar (right panel) then endowed this avatar with the veridical motions of the participant and with noisy variants of it. (D) Participants had to watch video clips of the avatar in motion and decide "Me" vs "Not Me" to indicate if they thought the avatar was the person himself/herself or another person. While they were deciding continuous motion recordings from the upper body were tracked and used in another experimental session to identify during the decisions who was deciding (self vs others) as well as to quantify the stochastic patterns of (embodied) decision-making.

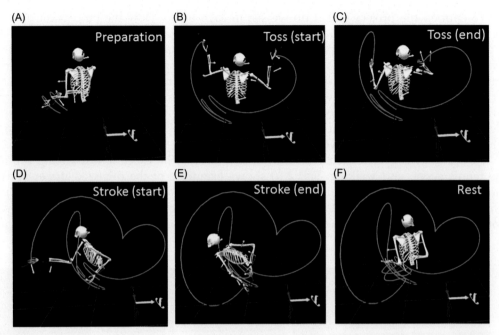

FIGURE 2.10 **Stages of the toss and strike segments of the tennis serve.** (A) The athlete prepares for the serve tossing the ball with the right hand and striking with the left hand. Sensors are shown in selected areas of the upper body including the hand, forearm, trunk, and head, while green traces are the positional trajectories described by the hands. (B) The toss starts by simultaneously moving both hands, the right hand tosses the ball while (C) the left hand helps gain momentum, then the right hand spontaneously goes down from the ball tossing (end of toss), (D) while the left-hand initiates the strike. (E) The strike ends and both hands go back to rest. The full trajectories of the hands are shown.

tossed the ball while the left hand made the strike (in tandem). We could automatically segment the ebb and flow of staged and supplementary segments of the two hands using their concurrent speed profiles. This is shown in Fig. 2.11C–D. The methods to automatically recover deliberate and spontaneous segments of these complex motions are discussed at length for boxing routines in.[32,57]

We used the Polhemus Liberty system recording at 240 Hz to register the motions of 15 body parts while the person walked, performed the Jab-Cross-Hook-Upper Cut boxing routines and played the tennis serve. Sample positional trajectories for each movement type are shown in Fig. 2.12 for selected body segments. To generate movement data and estimate the stochastic signatures corresponding to the continuous family of Gamma probability distributions for each parameter and each individual participant, we used different contexts for each type of routine (walking, boxing, and tennis serving). These routines were performed in full light with and without real-time feedback from the motions (e.g., as the participant in Fig. 2.9A and B did with and without viewing their own digitized version using a skeleton leaving traces of the hands displaying the real-time motion trajectories.) They were also performed in complete darkness with eyes closed (not shown since

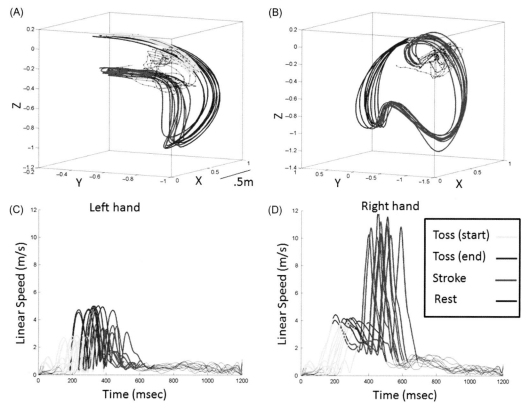

FIGURE 2.11 **Automatic decomposition of the toss and strike in the tennis serve of Fig. 2.10.** (A−B) Left- and right hand trajectories decomposed (see legend) according to the respective speed profiles (C−D). Note that the strike can reach high speed (near 12 m/s) while the toss reaches 5 m/s. The full routine lasts about 600 ms, on average.

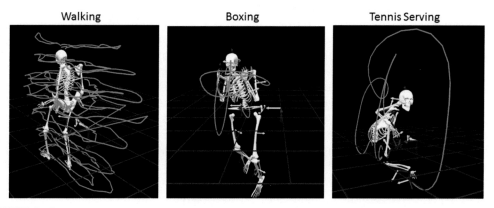

FIGURE 2.12 **Participants performed different movement types.** We used different movements to endow the avatar with and to probe the patterns of the nervous systems under different forms of sensory guidance. The empirically estimated signatures of variability (noise-to-signal ratio) in the spatial and temporal domains were used to distort the veridical movements of the avatar in subtle ways. Temporal parameter was the interpeak-interval-times of the joints' angular speed. Spatial parameter was the amplitude of the angular speed (peak values).

the room was completely dark, as we covered all possible sources of light, including the computers and switches in the room). Then, we attached glowing sticks across the body and recorded the subjects as they watched themselves in front of a mirror, or without the mirror. The third panel in Fig. 2.9A shows an example of one frame of the tennis serve but the reader can see movies in **Notes**[3].

These variations across conditions allowed us to later select the type of noise we wanted to add to the avatar by examining which context and type of sensory guidance was most effective in changing people's perception of the self and others. Each of these manipulations altered the kinematics output of the person and with it shifted the stochastic signatures of the continuous flow of motions. As such, when we plotted the results on the Gamma parameter plane, we could see which condition made the motions noisier and more random (i.e., shifted the estimated signatures to the left upper quadrant of the Gamma parameter plane). We could also contrast that outcome to the context that did the opposite, by moving the signatures to the lower right quadrant where the shapes of the distributions are more symmetric and the scale (noise-to-signal ratio) values are lower. The context also helped us understand the effects of the type of sensory guidance we used (i.e., visual or proprioceptive). For example, moving in the dark without any visual feedback was a context that enhanced proprioceptive guidance and mental imagery, whereas moving in the dark while viewing the body contour in motion through glowing sticks provided visual guidance (aligned with the body sensation). Likewise, moving with vision of oneself in real time, using the digitized version of the upper body, was very different from doing so without visual feedback. These manipulations were great to study and had not ever been sampled in the same person across different sports and naturalistic routines. The motor control we knew from basic scientific research had been centered on goal-directed segments. Their analyses primarily focused on the patterns of variability of the endpoint error, measured as the discrepancy between the target and the location where the hand landed while aiming for the target.[58] Common paradigms used in the field of neuromotor control included the reach-point-grasp families, i.e., a very small subset of things we do in the social scene; and their analyses were primarily constrained to one arm—hand effector. By marked contrast, our new experimental paradigm included complex full-body motions and motions involving an opponent, thus providing a test bed to develop metrics of social dyadic exchange for MNST, whereby the body in motion and its perception by the self and others could be studied in detail.

The system we used registered the positions and orientation trajectories of 15 body parts and (as mentioned), we digitized them using the Motion Monitor software of InnSports (Chicago, IL). Once the positional and rotational trajectories and their first- and second-order derivatives were obtained from these recordings, the students learned about techniques for signal processing and statistical analyses that we had designed in my lab to handle primate motions from bodies with many degrees of freedom. Some of these techniques and the mathematics behind them will be covered in Chapter 4 while Chapters 3 and 6 respectively, delve into the biometrics for dyadic exchange.

The class was very engaging and versatile. In parallel to the class sessions teaching the math and coding techniques directly applicable to the kinematics data they were

---

[3]Link to IGERT movie competition here

collecting, the students built different versions of an avatar to create videos of the avatar in motion (Fig. 2.13). Fig. 2.13B−F shows multiple versions of the video clips of the avatars they saw of the person moving. In one version of the decision-making experiments, we used the avatar in Fig. 2.13B. Later, we expanded this paradigm to explore how the person visualized the self during decision-making and used as well other avatar figures. These ranged from various humanoid versions to stick-figures to point-light displays presented in video clips from different vantage points (without the traces of the trajectories). Sejal made these movies and the idea was to make it really challenging for someone to perceive the identity of the avatar.

During those sessions of the IGERT class, once the stimuli were finished and we ran experiments as in Fig. 2.9D, the students would then have to decide whether the video they were watching was from their own motions or from someone else's motions.

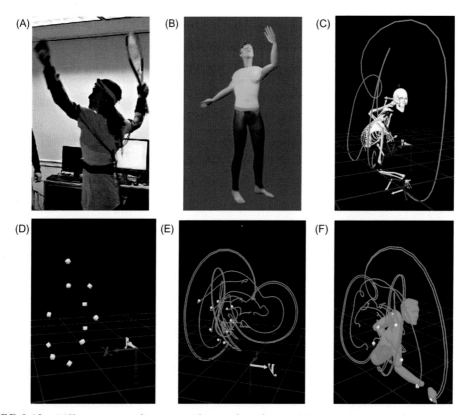

FIGURE 2.13　**Different types of avatar used to explore the participant's perception of self vs others.** (A) The athlete performing the tennis serve. (B) The avatar we used to make the video clips so the participants decided "Me" vs "Not Me". (C) A digitized skeleton of the participant performing the tennis serve and tracing the trajectories of the hand. (D) The sensor locations moving as a point light display in a minimalistic representation of the participant's upper body. (E) The sensor locations plus the trajectories they trace. (F) A humanoid avatar and the sensor locations with the trajectories they trace.

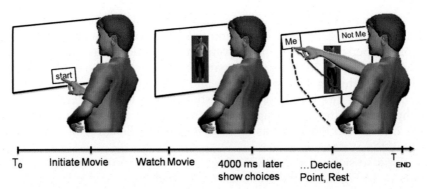

FIGURE 2.14 **Experimental epochs for the decision-making paradigm collecting upper-body kinematics.**
The participant sat comfortably in front of a touch screen (as in Fig. 2.9D). In the first trial, a label with the word
"start" appeared on the center–bottom location of the screen and the participant was instructed to touch the
word on the screen to initiate the flow of trials at his/her own pace. A movie clip appeared on the screen for 4000
ms giving the participant ample time to decide between "Me" vs "Not Me", when the labels appeared on the
upper corners of the touch screen. Trials were randomized to present different orders of avatar type (veridical vs
noisy versions) and to vary as well as the locations of the labels to balance contralateral and ipsilateral hand's
response movements. The participant decided and pointed his decision by touching the label on the screen with
the dominant hand, then returning the hand to rest (without being instructed to do so). The sensors continuously
recorded the upper body, arm, trunk, and the hand. The red trajectory exemplifies the forward pointing motion
to the selected target while the blue trajectory exemplifies the retraction of the hand to rest. The times of the
touches were recorded along with the pixel locations. The physical locations of the screen corners and base were
also recorded (using the positional sensors). The participants rested between trials. The experiment flow was
determined by the personal pace of each participant.

The epochs of the experiment are shown in Fig. 2.14. The participant started out by
touching the screen to activate the trial. To that end, we placed a box with the word
"start" in Trial 1, and informed the participant that this was the location to touch to activate the trial. The computer program running the session would then register the touch
time and the touchscreen pixels' locations. Further, the participant was wired up with
wearable sensors (Polhemus Liberty 240 Hz, Colchester, VT) and we could monitor all the
motions of the upper body while the decision-making was taking place. Once they
touched the screen, the movie presenting the avatar would appear and they would watch
the video clip and then mark their decision by pointing. In 4000 ms after the avatar movie
was presented, the computer program also placed two words on the upper left- (Me) and
right-hand (Not Me) corners of the screen and recorded the times of the presentation of
the movie and of the presentation of the words. Then, the participant could either finish
the full video clip and indicate the decision by pointing to the word Me or Not Me. If the
participant completed the decision before the movie was finished, s/he could simply point
the decision to the targeted choice. Either way, the motions were continuously recorded.
The clips lasted 4 s, which gave plenty of time to decide. All participants decided before
the 4 s expired.

We registered all elements: The touches, the movie presentation times as well as the
movie ending times, if the participant halted it with the decision. We also allowed plenty

of time to rest because the participant controlled the flow of the experiment by initiating the trial when s/he felt was appropriate. The trajectories were complex because they reflected the embodied decision-making of the person. That is, we went beyond mouse clicks so commonly reported in the literature of motion perception. We used embodied decision-making. In addition to the endpoint touches, we recorded somatic-motor physiology by listening to the peripheral nerves' signals as they become amplified by the muscles and output the mental intent of the person. We also recorded the spontaneous retracting hand motions and analyzed their stochastic signatures in relation to those of the mental processes. Because we registered many relevant behavioral landmarks during the stimulus presentation and the decisions, we could study more than the reaction time (RT) of the participant. We registered the motor decision times as well. Fig. 2.15 shows the hand trajectories from the resting position to the choices one female participant made. They are plotted as velocity vector flows in three dimensions. The dots on the trajectories mark the peaks of the velocity along those flows and the arrows indicate that these trajectories are from the hand resting area to the word on the screen indicating the choice (Me vs Not Me.) The inset in the upper right-hand corner shows the speed profiles across all the forward trials to the decision target. The peaks mark the velocity peak corresponding to those on the hand's velocity flow. Several kinematics landmarks allowed automatic identification of the forward decision-making trajectories and separation from the retracting trajectories bringing the hand to rest after the decision was completed. This is important in two ways: (1) To study the remaining of the decision-making process upon decision; (2) to separate these trajectories from those that the hand performed to initiate the trial by touching the word "start" on the screen (Fig. 2.14 presents the epochs of the experiment in schematic form).

The automatic separation between forward and backwards segments was possible because we registered (1) the touches; (2) the distances from the hand to the screen, as the hand moved towards and away from the screen. Since we recorded relevant physical positions of the screen along with the experimental locations of relevance, we could monitor the unfolding behavior of the upper body including the arm and the hand. These landmarks included the physical locations of the "Start" label, the "Me" label, the "Not Me" label, and the four corners of the screen, registered relative to a global frame of reference. We positioned a sensor on the base of the screen to track the physical location of the screen continuously, relative to this global frame. In addition to all physical parameters, we also registered the pixel location of the labels we placed on the touch monitor. In this way, we had the actual touch registered in the local screen co-ordinates and could measure the distance from the touch to the screen targets in screen co-ordinates. The zero-velocity at rest and at the target, combined with the distance information (decreasing as the hand approached the targeted choice and increasing as the hand moved away from it) allowed us to very precisely and *automatically* isolate the trials of deliberate decisions and separate them from the trials of supplementary motions afterwards. The latter were very informative as well because we could measure the noise-to-signal ratio upon decision completion, along with other metrics to anticipate the next trial. Since the flow was continuous, we could learn about the statistical influences of past trials on future decisions, e.g., as we had done in the study of complex boxing routines.[57]

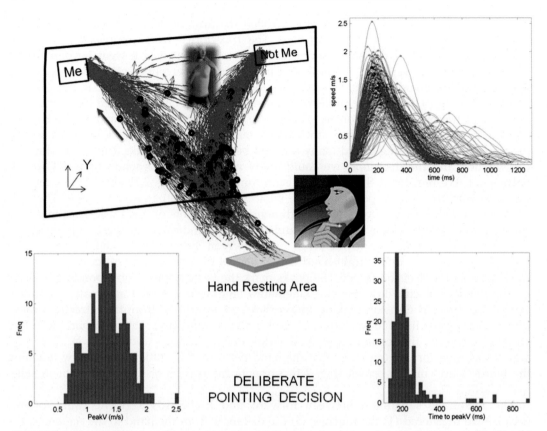

FIGURE 2.15　**Embodied decision-making.** Hand trajectories forward to the chosen target are displayed in the form of velocity fields with landmarks indicating the spatial location of the local peak velocity (i.e., the largest vector of the local flow before each pause in the acceleration−deceleration phases of uncertain decision-making). Speed profiles measuring the length of each velocity vector per time are plotted on the upper right hand corner with the corresponding landmarks. The frequency histograms of the peak speed values and the time to reach those peaks are plotted on the bottom left- and right hand panels, respectively.

Examples of parameters we studied are the peak velocities and their distributions along with the times to reach the first significant peak. We also studied other temporal and spatial aspects of the trajectories of the hand and the rest of the upper body. Fig. 2.16 shows the velocity flows and parameters of interest for the retracting hand trajectories towards rest, once the decision was completed. Here, we note the multimodal nature of the distribution of the times to reach peak velocity and the spread of the speed maxima. Sometimes, the global peak occurred early in the motion but other times, in other trials, the global peak occurred late. Indeed, these variable dynamics of the peaks indicated that even after the decision had been made and communicated through pointing, this participant was doubtful that she had decided correctly about the identity of her own motions or of the motions of others.

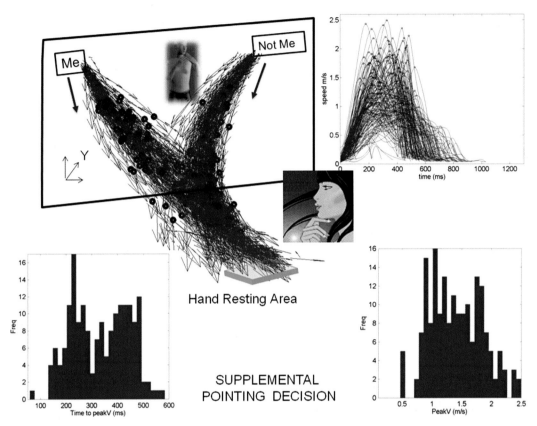

FIGURE 2.16 **Spontaneous hand retractions upon the pointing decision.** Velocity flows described by the hand returning to rest with the landmarks signaling the peaks along the velocity field. Speed profiles from the velocity flows show the multiple acceleration−deceleration phases of the motion. The frequency histogram of the time to peak velocity and of the values of the speed peaks are also plotted on the bottom panels (left- and right hand side, respectively.)

The variability in the response kinematics was expected. The movements we endowed the avatar with were from the veridical motion-capture sessions, whereby the patterns of each participant were used to make the movies. Then, we performed a stochastic analysis and obtained the noise signatures from each person's motions across all those contexts we described above. We added to our database, the stochastic signatures of movements derived from patients with autism and used all different types of signatures in the database to distort the original person's signatures. All these noise profiles were then used to distort the avatar. Thus, in some trials the participants were seeing themselves move (veridical cases); but in other trials, they were seeing their movements plus noise. The video clips they were deciding on also included the veridical motions of others and the motions of others contaminated with noise. It was really confusing to see all those movie clips and decide. In fact, every single participant said they had no certainty, whatsoever

TABLE 2.1   Percent correct outcome on deciding "ME versus Not ME", where the noise to perturb the avatar comes from the temporal dynamics of the bodily motions, i.e., the joint angular speed interpeak-interval time periods.

Subjects	Veridical	Self-Noise	Others' Noise	ASD Noise	Overall Accuracy
T1	100%	75%	73%	75%	81%
T2	100%	100%	100%	100%	100%
T3	50%	55%	27%	25%	40%
B1	100%	85%	75%	75%	84%
B2	100%	60%	63%	43%	67%
B3	80%	75%	71%	71%	74%
AVRG	88.3%	75%	68.2%	64.8%	74.3%

on whether they were right or wrong. During the experiment, they felt their decision was random. But it wasn't! (See Table 2.1).

As it turned out, most participants performed above chance (i.e., above 50% in a two-choice decision). The worst performer oddly did better with self-noise (although not very well) but some did 100% correct in the veridical self-case. Interestingly, the best performer was a top athlete and this prompted us to want to further explore the performance of athletes in relation to that of nonathletes, while using this basic paradigm. It is a question that we want to address in the future, when we assess more participants and include those with a nervous systems' disorder.

A critical aspect of this paradigm is whether the temporal dynamics of the decision-making process mapped in any way to the signatures of their kinesthetic temporal dynamics. Fig. 2.17 shows examples of temporal dynamics for boxing and tennis serve extracted from the motions of one participant. We show on the avatar, the 10 locations we used to place sensors and collect the temporal data consisting of the time series of the interpeak time intervals in the angular speed of the rotating joints. Notice that the histograms are not symmetric, a feature that prompted us to use methods of maximum likelihood estimation (MLE) to approximate the family of probability distributions best characterizing our data. As in other studies involving human kinematic parameters, this turned out to be the continuous Gamma family spanning distributions from the exponential, skewed, and symmetric range that fitted well with 95% confidence the mental decision times and the motor decisions times.

When trying different noise signatures, the one related to the amplitude of the joint angle excursions was not adequate to endow the avatar with because it distorted the movements so much, that it was difficult to preserve the form of the original routine. Since the perturbation of the movements driven by the variability of physical excursions of the joints did not work (i.e., these overtly distorted the geometric path lengths of joint rotations and translations), we tried instead perturbing the movements with variations drawn from the temporal dynamics.

The temporal dynamics of the motions embedded in the peak-to-peak intervals from the joints' angular speed maxima did the trick, as subjects could discern "Me" vs "Not

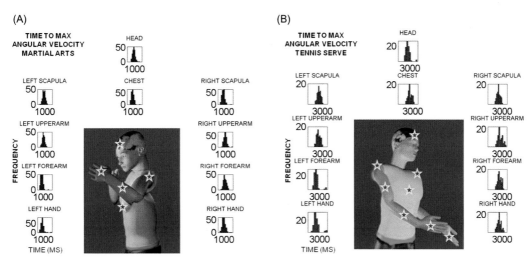

FIGURE 2.17 **Locations of the sensors on the participants' upper body** used to build the avatar and make the video clips for the boxing (A) and tennis serve (B) sports routines. 10 locations on the head (1), shoulders (2), upper arms (2), forearms (2), hands (2), and trunk (1). Frequency histograms of the interpeak time interval times for the angular speed are shown for all locations.

Me" above chance. Using this type of distortion of the motions' timing was subtle enough that it did not entirely distort the form of the routines; but did change the rhythms in delicate ways. At a conscious level, one could not really tell who was who with any degree of certainty. It was all very confusing. Yet, beneath the participant's awareness, somehow the choice was correct with such high accuracy that it surprised us all. Fig. 2.18 shows the congruency between the temporal dynamics of mental and physical processes during the decision-making task, where the noise was based on the fluctuations in the interpeak time intervals of the angular speed.

This graph depicts the performance underlying the participants' responses by sports type, as well as the summary signatures of the movement decision times. It was truly remarkable that the temporal dynamics of the decision-making process aligned well to the temporal dynamics of the bodily motions we endowed the avatar with. The temporal dynamics of the movements of the male and female participants had slight differences. When we plotted the estimated Gamma parameters on the log—log Gamma parameter plane, we found a power fit with different exponents (i.e., different slopes and intercepts of the power relation) for each of the male and female groups. This was not the case for the summary points of temporal dynamics of the decision-making. In that case, all summary points across both sports routines were well-fit by the same line. Somehow, the decisions about self vs other was independent of the differences in temporal body dynamics inherent to each sports routine the participants were deciding on. However, the stochastic signatures empirically derived from the fluctuations of the temporal information did have *separable* slopes and intercepts on the log—log Gamma parameter plane (Fig. 2.18).

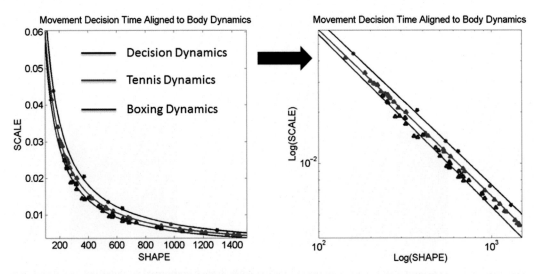

FIGURE 2.18 **Mental and physical decisions.** Congruent relations between the stochastic signatures of the temporal dynamics of mental decisions and physical bodily motions the participants had to decide on. Black, Decision dynamics; Red, Tennis dynamics; Blue, Boxing dynamics.

In future versions of this work, we aim at further exploring questions related to potential differences and similarities between males and females. We are also interested in differences across sports routines when athletes vs novices perform different segments: Would an athlete or a performing artist know the difference between a novice person and a person well-trained in sports and the performing arts? One interesting aspect of the problem was to learn a bit more about the differences in performance when endowing the avatar with noise derived from the motions' temporal dynamics, or with noise derived from the stochastic signatures of fluctuations in the *amplitude* of the angular speed. Once again, we addressed this question within the context of a decision-making experiment; but we hereby designed a new way to assess decision-making tasks.

A Different Paradigm to Study Decision-Making

Studies of the Psychological and Cognitive Sciences often rely on decision-making. Many studies address questions related to action–perception loops using paradigms, whereby the subjects must decide between two or more stimuli. The decision is often communicated verbally, through key presses or through mouse clicks. On occasion, a joystick may be used to communicate the decision. In most cases, the parameters of interest are decision accuracy and RT. Sometimes, the endpoint positional errors of the hand at the target are also used as a measure of performance. Yet, continuous motions from the arm–-hand communicating the decision or from the bodily rhythms registered in tandem with the abstract cognitive decisions are rarely assessed.

We decided to increase a notch, the complexity of the data type generated in a decision-making experiment and recorded all motions of the upper body of the participants while they were deciding "Me" vs "Not Me" in the previous experiment. In this way, we could go beyond endpoint errors, RT, and accuracy of the clicks and tap into other questions about bodily involvement during cognitive decisions as they unfold, as they occur, and even after they are made.

Sejal Mistry, the student from biomathematics that we mentioned in Chapter 1, led the effort and explored different aspects of this problem. She used different avatar types such as those three types shown in Fig. 2.19A for upper-body motions and Fig. 2.19B for full-body walking motions. The video clips of these avatars in Fig. 2.19A making decisions with the right hand were presented to the participant from different viewpoints. We used three different designs in increasing levels of closeness to a humanoid avatar, whereby the head and arm motions (but not the trunk motions) were shown during the decision. This economical version of the upper body provided enough detail to address our questions on whether the fluctuations of the persons' upper-body motions during decision-making could distinguish the type of noise (temporal-based vs amplitude-based) and the type of avatar figure.

The task was difficult. The video clip in one trial contained only one avatar type recorded from one vantage point and endowed with one motion signature. As in the previous experiment involving sports routines, this motor signature could be veridical or noisy and within those noisy versions, the noise could be derived from the participant deciding, or from other participants in the group. As before, we also used a complex movement such as walking intermixed with the upper body decision-making movements (Fig. 2.19B). The individual did not see the actual experiments or the actual motions of others or self.

Sejal made these movies, so the person trying to discern "Me" vs "Not Me" was naïve to all the manipulations we tried. In this version of the avatar-driven experiment, the level of accuracy in identifying "Me" vs "Not Me" was secondary to the goals of our enquiry regarding the possibility that temporal dynamics-based noise may be easier to identify than amplitude-based noise; or that the avatar figure type could be blindly distinguished in the variability patterns of the decision-making motions.

By varying the avatar type and exploring the noise in relation to the joint angular excursions, we hoped to make the decision far more challenging and hypothesized that the participants would degrade their performance in relation to the performance that used the noise derived from the temporal dynamics of the inter-peak-intervals times of the joint's angular speed.

Sejal Mistry presented her results at the Society for Neuroscience in a Nanosymposium talk. She did an amazing job and the abstract of that talk prompted Elsevier to invite me to write this book. The idea of rethinking the MNST from the bottom-up and measuring it noninvasively in the peripheral bodily biorhythms is indeed useful to open new lines of enquiry in this problem domain. For the first time, we had an experimental paradigm paired with a unifying statistical platform that enabled us to study the stochastic patterns of biorhythms from the periphery while centrally driven decisions were being executed.

As I explained in the introductory chapter of the book, the new analytics convert the continuous analogue signals from the bodily biorhythms (e.g., kinematics, heart rate,

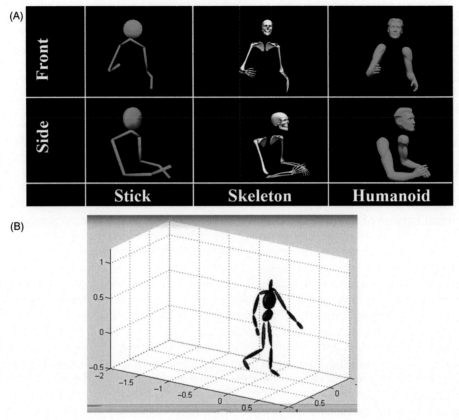

FIGURE 2.19   **Deciding on patterns of self vs others in embodied decision-making.** (A) Different avatar types were used to explore their effects on the decisions of "Me" vs "Not Me" when using movie clips involving these avatars endowed with veridical motions and noisy variants of them. The noise consisted of fluctuations in the amplitudes of the angular excursions involving angular speed peaks of the 10 joints in Fig. 2.17. (B) Avatar to interleave video clips of embodied decision-making with video clips of walking patterns of the person and others. Veridical and noisy variants were also used to endow the avatar with.

temperature, EMG, etc.), registered with high sampling resolution, into a point process and treats the waveform as spike trains. As such, this platform unifies the analyses of the type of data derived from spike trains that neuroscientists routinely record extracellularly with the types of bodily fluctuations that we can harness across multiple layers of the peripheral nervous systems. These biorhythms can be recorded with noninvasive means today. In this sense, we have created a new way to study embodied cognitive processes— Such as decision-making—During naturalistic and continuously unfolding behaviors. Jill had published several papers on related issues involving visual illusions[59–61], so we had begun the path of building a series of experimental paradigms where these methods could be used in standard ways.

The ability to record and analyze in tandem the central and peripheral biorhythms, thus opens many new avenues to explore the MNST in more systematic ways in humans. This is particularly useful in humans with a pathology of the nervous system that results in problems with social interactions. While the analyses we describe here are centered on the participant, we show in Chapter 6, a different type of analysis amenable to examine the real-time co-adaptive dyadic interactions between the two members of the social dyad, or between the human participant and the avatar while they interact in real time.

Sejal tested eight participants in the decision-making task using the avatars in Fig. 2.19. We defined three layers of decision-making parameters and explored their stochastic signatures using the fluctuations in the amplitude of the angular excursions and two modes of execution: deliberate vs spontaneous. The deliberate mode focused on the epochs of the decision involving the forward pointing motion to the targeted choice. These included parameters of the upper body, including the hand's trajectory, parameters of the walking motion the subjects performed and observed to decide "Me" vs "Not Me" in the avatar-driven gait experiment and parameters of the abstract decisions. The latter included the RT and the movement decision times. These temporal parameters comprised the time from the onset of the movie presentation to the onset of the hand motion lifting from the resting position in route to the targeted choice; then, we also examined the period from the onset of the hand motion (speed above zero) to the first speed peak. Fig. 2.20A shows the stochastic signatures of the deliberate segments from these three layers of decision-making parameters, while Fig. 2.20B shows the results corresponding to the spontaneous segments. As before, here we used the log−log of the Gamma parameter plane to plot the various layers of parameters for the noisy (n) and veridical (v) versions denoted as such for each subject in the legend of the figure. In the panels, "ABSTRACT" refers to the distributions of the decision-making times; "GAIT" refers to the distributions of the fluctuations in amplitude of the angular excursions during walking; and "HAND" refers to the distributions of the fluctuations in amplitude of the upper body's and hand's angular speed while making decisions (because we tracked multiple points on the upper body, including the hand). Across the two panels, the "HAND" patterns differ because they correspond to the deliberate segments of the forward pointing motion (Fig. 2.20A) and to the spontaneous (uninstructed) retractions of the hand back to rest, upon executing the pointing decision (Fig. 2.20B). The other two "ABSTRACT" and "GAIT" are the same in both pictures and merely used as reference to visualize the differences in "HAND" patterns between deliberate and spontaneous modes of the decision.

Importantly, the noisy versions of this avatar were made by deriving the noise from the amplitude of the angular excursions instead of using the temporal dynamics of the interpeak intervals timing. In this version of the experiment, this type of noise did not distort the avatar as it did in the sports routines (where the form of the routine was too unrecognizable to use as stimuli). The accuracy of the performance was highest for the noisy versions of the decision-making avatar when the noise was derived from the person's motion (self-noise in Table 2.2). This contrasts with the first experiment's outcome using the noise derived from the motions' temporal dynamics. In that case, we found that the highest performance was based on the avatar endowed with the veridical motions of the person (Table 2.1).

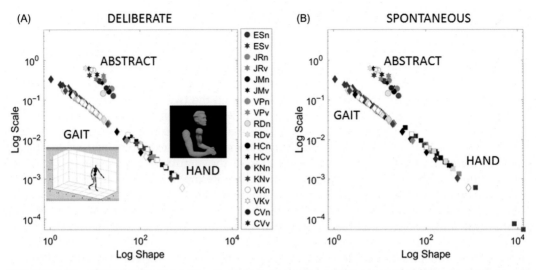

FIGURE 2.20 **Differentiation of stochastic signatures across conditions according to the movements' variability for deliberate and spontaneous aspects of the decision-making process.** (A) Gamma parameter estimates of fluctuations in "ABSTRACT" decision timing for noise (n) and veridical (v) cases, "GAIT" from walking patterns, and "HAND" from pointing to signal the decisions while engaging the upper body. Signatures of the deliberate segments of the pointing decision (forward) stand in contrast to the spontaneous decision segments (retractions of the hand to rest). Legend denotes the colors used for each subject across conditions. Markers denote the conditions (circles and stars for "ABSTRACT"; diamonds for "GAIT" and squares for "HAND").

TABLE 2.2   Percent correct outcome on deciding "ME versus Not ME", where the noise to perturb the avatar comes from the amplitude of the angular rotation peaks.

Subjects	Veridical	Self-Noise	Others-Noise
S1	97%	94%	97%
S2	61%	98%	34%
S3	62%	51%	52%
S4	55%	98%	22%
S5	63%	97%	46%
S6	62%	97%	45%
S7	60%	96%	44%
S8	40%	95%	18%
AVRG	63%	91%	45%

One additional layer of interest is the hand signatures because during the retraction of the hand towards rest, the person would still hesitate (e.g., Fig. 2.16) even though, the execution of the decision had been already completed when the hand arrived at the target of choice. Although participants had already decided, their spontaneous retractions broadcasted the uncertainties of their prior decisions.

The question of identifying one's own motion amidst the noise was inspired by the very notion of mirror neurons, but in this case, we extended this notion to the peripheral network of efferent and afferent nerves and to the skin receptors providing kinesthetic feedback sensed from moment-to-moment fluctuations in the movements. How good would the person be at relating self-executed decisions to the observation of subtle variations in those very motions? In our version of this problem, we went a step further to ask if the person could distinguish subtle kinesthetic differences between observation of self-generated motions and other's motions; and to further probe whether this was possible, despite different types of noise, including noise from the systems with a diagnosis of autism—Which we had discovered had elevated levels of noise and randomness across voluntary, involuntary, and autonomic systems.[17,55,56,62]

Fig. 2.20 focuses on self-noise, since that type of perturbation was the one with the highest accuracy during the performance (Table 2.2). This figure examines the parameters for the deliberate and spontaneous segments of each of the hand movement segments (forward to the targeted choice and backwards retracting to rest). The map of points on the Gamma parameter plane reflect higher variability in the abstract decision times for the veridical version of the avatar than for the noisy variant (for self-noise). Further the upper-body variability, including the variability from the fluctuations in the amplitude of angular excursions of the hand in motion communicating the decision through the pointing motion, have higher signal-to-noise ratio according to the lower values of the Gamma scale parameter, and more symmetric distributions (according to the higher values of the estimated Gamma shape parameter). The variability of the "GAIT" condition was higher than that detected for the unfolding kinematics of the upper body and the distributions of the fluctuations in amplitude of the angular excursions were more skewed. As such, we asked if the variability alone could help us blindly separate these three layers of parameters relevant to the decision-making process.

BLINDLY PREDICTING THE TASK FROM THE PERFORMANCE VARIABILITY

A simple linear classifier helped us address this question. We used three conditions, "ABSTRACT" representing the timings of the decision-making process; "GAIT" representing the fluctuations of the angular speed amplitude during the walking patterns that subjects performed (i.e., the patterns they saw in the video clips to make the decision); and the "HAND" condition representing the fluctuations in the angular speed amplitude of the upper-body joints (including the hand) as the movements pointing the decisions unfolded. For each condition above, we had the veridical and the noisy variants, where the noise referred to the self-noise—Using the signatures derived from the person's own motions. We used 60 trials per each of the 8 participants and 3 conditions for a total of 1440 trials per each of the 2 modes, veridical, and noise. The trials came from the performance in response to the humanoid avatar, since this was the most accurate of all three in Fig. 2.19A.

To test the linear classifier, we first built the proper data type. Since we originally had different units (ms for the decision timings, cm/s for the hand linear speed amplitude and deg/s for the bodily angular excursions amplitude), we used the unitless micromovements normalized between 0 and 1. Trials were represented in an m-dimensional space, each coordinate corresponding to the parameter of choice input to the "leave-one-out" decoding algorithm for each of the conditions and modes. One at a time, data from each trial picked at random were used to predict the parameter variability from the condition type and mode type, based on the parameter distributions derived from all the remaining trials (leave-one-out cross-validation), and were assigned to the class of its nearest neighbor in the m-dimensional space using Euclidean distance.[63] For assessing statistical significance of the decoding results, a value of 1 was assigned to correctly predicted trials and a value of 0 to the incorrectly predicted ones. The mean of the sequences of correctly and incorrectly classified trials was compared statistically using a nonparametric Wilcoxon rank test and represented graphically as confusion matrices where each column of the matrix represents the instances in the predicted class while each row of the matrix represents the instances in the actual class. For example, in Fig. 2.21A for the "HAND" condition, the matrix entry corresponding to the first row and first column gives a probability of 0.75 for classifying a noise trial when it was actually a noise trial, while entry in the second row and second column of the matrix gives a probability of 0.51 for classifying a randomly selected trial as veridical when it was actually veridical. The average across the diagonal is 0.63 reflecting the probability of correctly classifying trials. The off-diagonal entries reflect the confusing cases, i.e., entry in first row and second column gives the probability of classifying as a noise trial, a trial that was actually veridical, while entry in the first row and first column gives the probability of classifying the trial as veridical when it was actually a noise trial. Here chance is 0.5, since we have two cases: veridical and noise.

When participants were deciding whether the movie clip they were watching was "Me" vs "Not Me", the variability in the timing of their decisions could better predict trials that were veridical than trials that were noisy. However, the variability in the fluctuations of the angular speed amplitude (for the walking patterns and for the patterns of hand pointing motions that they were deciding on) predicted a given trial best for the condition involving the avatar with self-noise.

When examining instead all the three conditions in tandem (chance 1/3) for the noise and veridical cases separately, the classification was always correct for the "ABSTRACT" condition. The walking patterns comprising the "GAIT" condition and the hand pointing patterns comprising the "HAND" condition showed higher confusion, with better classification for the veridical case. This result is shown in Fig. 2.21B. It suggests that the trial-to-trial variability for each of these conditions predicted the condition best when the avatar was endowed with veridical motions of the participants deciding "Me" vs "Not Me". When the avatar was endowed with movements perturbed with noise derived from the fluctuations in the amplitude of their own self-generated bodily angular speed, the participants did degrade their performance in comparison to the version of the experiment where the avatar was endowed with noise derived from the temporal dynamics of the peak to peak timings. Somehow, not only the accuracy of deciding but also the inherent variability of the mental and physical performance of these participants were superior at

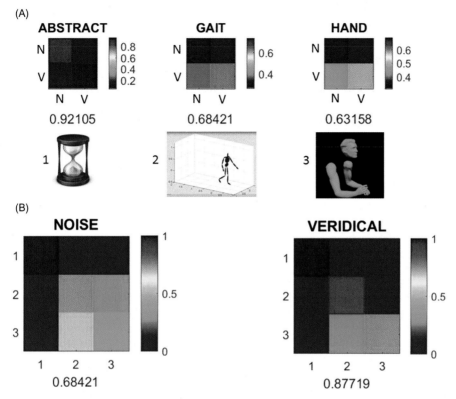

FIGURE 2.21 **Blind trial classification of mental and physical variability.** (A) Linear classifier can distinguish within each condition the noise vs veridical trials above chance. (B) Linear classifier can distinguish above chance a randomly chosen trial across conditions for each of the veridical and noisy cases.

correctly predicting trials when the noise used in the avatars was derived from the temporal dynamics of the bodily actions they themselves generated.

In this version of the experiment, the avatar movements were walking or deciding (by pointing to a target). Six of the participants took part in both experiments (i.e., the experiment that used the movements from the sports routines and the experiments that used the decision-making movements). In both cases, the participants reported high uncertainty while deciding. They unanimously thought their decisions were random. We were shocked to see that this was not the case. In fact, they performed with accuracy well above chance. Furthermore, the stochastic variations in their motions during decision-making could blindly separate the condition ("ABSTRACT", "GAIT", "HAND") for the veridical and the self-noise cases.

This experimental paradigm affords many variations and should be repeated with far more subjects than we used. The methods do provide new ways to explore the MNST from the bottom-up to learn the contributions of kinesthetic reafference to the execution and the perception of actions. One of the lessons learned was the need to understand the person's nervous systems intrinsic dynamics. By this, I do not just mean the time intervals of the events taking place (e.g., the peaks of the speed or acceleration signals), but also the order of those events and their interdependent statistics. The next 5 years in my lab will be devoted to the development of models that relax the identically independent distributed assumption we adopted in these earlier works. One of the reasons for this stem from the influences of past events on present and future events that we could observe and quantify in the stochastic signatures of these various parameters. In Chapter 5, we will retake this issue in the context of sports performance in autism, along with various departures from the temporal dynamics and statistical orders of neurotypical peers.

SOME ADDITIONAL THOUGHTS

The field of Psychology has had a long history of investigating how people perceive motions in the form of point light displays (like those in Fig. 2.13D). That body of work has revealed that a person can tell when point light displays represent biological motion vs when they display random motions.[64,65] Even typically developing (TD) infants can do this, unlike infants who already received a diagnosis of autism.[66,67] And within that popular point-light-displays paradigm, we can guess the emotional state of the person performing the motions displayed on the screen, i.e., if the person is angry, sad, or happy; identify the sex of the person, etc. Indeed, there have been many variations on this theme within the field of psychology.[66,67] The basic point-lights display paradigm exploits the top-down visual recognition of the motions that we externally perceive in others. That body of work always asked something about the brain's visual perception of motion. Yet, when in 2012, our IGERT class was designing the project described in this chapter, there had not been a single piece of work asking about the kinesthetic perception of our own self-generated motions.

We felt it was important to investigate self-perception as a bottom-up process emerging from our physical bodies in motion, i.e., through the powerful concept of kinesthetic reafference (the sensory feedback we get from the consequences of continuously self-generated biological motions). We had extensively studied this construct across disorders of the nervous systems and had found selective signatures of noise-to-signal ratio for different nervous systems' biorhythms variability.[18,68] This finding poses a more general question on whether persistently sensing back corrupted self-generated motor noise may have an impact on how the person perceives the bodily motions of others, particularly when those motions occur within a social context. If this was the case, neurodevelopmental disorders such as autism, defined as social deficits with an emphasis on ToM constructs, would have to include deficits as well in bodily kinesthetic sensations. Furthermore, kinesthetic sensations from self-generated actions would have to be included as an integral part of action observation and mental inferences derived from action observation. At present, this is not the case. General conceptual frameworks such as ToM or broken MNS (attempting to explain disorders such as autism) fail to recognize the contributions of bodily sensations and sensory-motor control to the scaffolding of the child's ability to perceive the self and its social surroundings, and to develop a notion of its intrinsic dynamics.

In our version of MNST, to build this experimental paradigm we adopt a bottom-up approach to motion perception. This approach is grounded on our proposed notion that our brain builds maps of sensory consequences from self-generated movements (as suggested in Fig. 2.3) and uses those maps when building kinesthetic percepts that involve audiovisual observation of the motions of others. This concept is very different from that assumed by the MNST because it investigates how we may arrive at a somatic-motor percept of actions' consequences from the bottom up, i.e., by sensing our body and forming *causal* maps of the effects these very sensations may have on our actions. We do not fully see ourselves in three dimensions (as we would the bodies in motion of others in a social scene), so to recognize whether or not others move like us, we need to form a multidimensional percept of our bodies in motion. This multidimensional percept would then have to be projected to the three-dimensional visual percept we form of the motions of others in the social scene.

Our proposition assumes that it is possible to build this kinesthetic map using an internally defined (egocentric) vantage point to represent not only the kinesthetic sensations of our own bodies in motion, but also to build the map of the consequences of such actions. In the absence of kinesthetic maps, the maps of the self-generated actions' consequences would also be absent. An intact visual system would perhaps recognize motions, even distinguish biological from nonbiological motions; but to more generally recognize the motions' impending consequences, and make inferences about antecedent causes, we would need a different type of recognition system. Such a system would have to have visited at some point, the actions preceding the consequences and accumulated enough information to make statistical inferences about the probable presence of one given the other. During social exchange, when temporal dynamics are so important to keep the flow of conversation, to physically entrain, and to detect (largely beneath awareness) subtle nuances of gestural micromotions; such a code would be relevant for timely responses, i.e., to compensate for inherent transduction and transmission sensory-motor processing delays throughout the central and the peripheral nervous systems.

Visually recognizing the actions of others would be necessary, yet insufficient to infer their intentions. To infer the intentions of others from the unfolding actions, one would need to have access to the corresponding underlying causal map of probable consequences of one's own actions. That map would require in the first place to have representations of one's own actions. But representation of one's own actions would require having sensed those actions in the first place: A sensation that arises from kinesthetic reafference derived from self-generated motions, rather than from vision alone. Indeed, we can simultaneously feel and see our hands and feet in motion within the reachable peripersonal space; but we cannot do so in the space behind us. We can only feel (but not see) other parts of our body in motion as they fall out of sight. To fill in that visual void and fully perceive our body, we need to compensate with other sensations. In this sense, *the returning stream of afference that the self-generated actions themselves cause* seems critical to build the notion of cause and effect instantiated in the association of the actions with their sensory consequences.

Unlike traditional theories of Cognitive Psychology (e.g., the two streams theory of the visual ventral and dorsal pathways[69]), our approach to the problem of motion perception does not separate "vision for action" and "vision from perception" as two independent streams of information impervious to each other. Instead, we assume a hybrid interconnected construct where kinesthetic reafference serves as a multidimensional bridge to help perceiving ourselves "moving" in the motions of others. From an analytical stand point, this construct brings to the MNST more than just action—movement. More broadly speaking, it adds other motion dimensions to the notion of an action; whereby biorhythms of the multilayered nervous systems are also bound to play a role in the characterization of the actions the person performs and observes. In this sense, the recognition of the self-in-action in the unfolding actions of others would require recognition of more than the overt movements underlying the action. It would require recognition of variability signatures as well, in biorhythms that are hidden to the naked eye, e.g., harnessed from voluntary, involuntary, and autonomic motions.

The advent of wearable biosensing instrumentation registering signals harnessed from the autonomic systems and from the peripheral afferent—efferent bodily and facial nerves can provide the corpus of data necessary to build new models of statistical inferences of actions and their consequences in more general, broader terms than has ever been posed in mirroring theories. Such corpus of data would also include instances when the person performs the types of subtle spontaneous motions that transpire largely beneath awareness and lend fluidity to purposeful actions.

Given the marked shifts in probability distributions derived from such biorhythms that we had found in the autonomic and involuntary motor outputs of individuals with autism during early development (e.g., Fig. 2.8)[55,56,70], and given that these corrupted motions then give rise to corrupted motor reafference; we could safely assume that in a person with autism, the kinesthetic-to-visual map may be somewhat altered. We know from the literature that the visual perception of biological motions is atypical in some children with autism, from an early age.[71,72] We also know that at young ages, there are atypical manifestations of how some children with autism decode the intentions of others[39] or represent "pretend play". But none of the work, thus, far has assessed the kinesthetic component of actions physically sensed by the person as the person experiences the self-generated actions and their consequences.

I felt it was necessary to assess the congruency between a presumed internal map of our bodily kinesthetic sensations, their associated consequences, and an external projection of that map onto others (e.g., through observation using vision or other senses). To do so, we quantified the uncertainties of recognizing self vs others in an avatar endowed with the continuous motor flow of the person. But unlike most experiments in psychology, where the decisions are quantified through mouse clicks or self-reports; we recorded the continuous flow of motions of the person as the decision unfolded, such that the person also saw the deciding avatar and tried to discern "Me" vs "Not Me" within the decision-making context.

Methodologically speaking, this was a completely different way to study decision-making, shifting the focus from a disembodied decision-making framework to a truly embodied one. From the standpoint of MNST, the new paradigm offers a new way to study the map of the consequences of self-generated bodily activity in response to the brain's perception of self vs others—A fundamental missing piece in the MNST story.

PART II: A SMALL PEDAGOGICAL PARENTHESIS ABOUT TAKING MOTOR CONTROL TO THE CLASSROOM IN A DO-IT-YOURSELF SETTING

The MNST was very well received by the IGERT students and provoked ample room for discussion and brainstorming. All students levitated towards the project that involved the MNST—A sign that this theory is very thought-provoking. The notion of the "kinesthetic self"—Missing from this theory in its current form, drove the students to learn and willingly expand their knowledge base. Those in CS found the psychology and neuroscience-related ideas very attractive when they anchored those ideas to the social scene and the notion of sensing their own bodies in motion. Those in psychology took a tremendous interest in the mathematics behind kinematics analyses. Indeed, they were willing to learn about the special rotation group SO3, so commonly used to build analytics. They were willing to learn the necessary tools to plot the trajectories and derive the higher-order parameters from the raw data and they were willing to learn elements of signal processing to clean up the signals. Lastly, they were certainly willing to learn statistical techniques they never saw before and very much welcomed the notion of empirical estimation of distributions from data. This notion, so important to other disciplines in physics, applied mathematics and engineering is unknown to the psychology student. They are simply taught that a theoretical distribution like the Gaussian is sufficient to do all analyses. Their surprise (and joy) when they learned to estimate the inherent stochastic signatures of the data from their own motions was noted. This was a two-way street exercise, as I too learned what to teach these students and how to do it effectively to keep them curious.

We were lucky because several undergraduate students taking the class in the Psychology Department were already participating in our sports-related studies assessing boxing routines and the tennis serve. Besides our undergraduate, karate, and kick-boxing experts, one of our very own graduate students taking the class was a semiprofessional tennis player. We could enlist the expert athletes to guide the naïve participants in the study. The recording sessions were fun and we always ordered food at the end, so there was some bonding "cooking" in the background throughout the whole semester. By the end of the project, we felt like a big family and those of us who could not attend the IGERT competition in Washington, DC, were glued to the livestreaming session of it. When they announced that we had won and were voted up the most popular videoposter (termed community choice prize), we were ecstatic! (See **Notes**[3])

We learned a great deal from this project because we did not know what the outcome was going to be. Imagine this for a moment; you do not see yourself in three dimensions! You have no access to the level of micromovement detail we were using to endow the avatar with and yet, the participants performed above chance for the most part. When presented with the movie of the avatar moving with or without noise, they felt their decision-making was totally random and the outcome would likely be wrong most of the time. Their choice was clearly beyond visual recognition of biological motion. It tapped into the kinesthetic sensations of the self-generated motions as they distinguished those from the motions self-generated by others.

The overall results provided evidence in support of a new notion required to support part of the underlying assumptions of the MNST, namely the implicit assumption that we have a bodily notion of the self and that we have systems with similar (yet separable) brain–body patterns of action–generation and action–recognition to those of others we observe or interact with. Up to this point, the MNST had been a mentally brain driven theory. The work presented in this chapter adds the physically body driven component to this theory. In the process of designing and implementing the class, we created a general paradigm to evoke and objectively quantify biorhythms of decision-making in both the mental and the physical domains. For the first time, the measurements and analyses included the body's unfolding continuous motion, thus providing a unifying mind–body paradigm-shifting framework for embodied cognition.

We can see that the continuous flow of information from our own self-generated biorhythms is perceived both visually and kinesthetically, and that typically there is congruence between the two. Indeed, it was possible for most participants we tested to distinguish (with high certainty) whether what they were seeing were entirely allocentrically based (performed by others) or egocentrically based (self-driven), even when we mixed the two by adding noise from others to the veridical version of the motions of the person, or by adding noise derived from the person's movements to the avatar endowed with the movements of others. Interestingly, despite the subtle changes in the avatar motions, people could immediately know the presence of allocentric motor noise when the signatures were derived from the autistic systems. So, in a way, by visually experiencing autistic micromovements' motor-noise, people could tell something was different from neurotypical micromovements' motor noise—Even at that level of unconscious perception.

PHYSICAL MOVEMENT KINEMATICS AS A WINDOW INTO MENTAL DECISIONS

When participants were performing the experiment to decide between "Me" vs "Not Me", we recorded the upper-body motions and could examine those signatures as well, so we could tell the amount of hesitation present in the decision-making process by examining the hand motion trajectories and their rich kinematics signatures, e.g., the acceleration–deceleration profiles. While the computer program could time-stamp precise segments of the stimulus presentation flow, the touches of the participant on the touchscreen to activate movies in each trial and to mark their decisions; the unfolding kinematics landmarks could automatically inform us, in a data-driven way, of other aspects of mental processes. Such mental processes (e.g., decisions, hesitations, change of mind, among others) were funneled out through the movement features. For example, the near-zero velocity could inform us of different states of the hand at rest, and when combined with distance-to-target information, it could tell us which resting state the hand was in: (1) Resting at the target within minimal distance to it because it had either reached the choice it intended to on the top corners of the touchscreen, or it had reached the location indicating to initiate the start of a trial in center–bottom portion of the touch screen; (2) resting away from the computer screen upon returning the hand back after traveling in increasing distance away from the targets' area; (3) accelerating and decelerating continuously during hesitations that slowed-down or sped-up the hand motions during the decision-making process.

The combination of computerized landmarks and dynamically unfolding data-driven epochs identifiable from the hand kinematics provided the decision movement time, which we separated from the traditionally studied RT in mouse-clicks. The RT spanned from the time of the video onset (triggered by the hand touch of the screen), the onset of hand motion signaled by the departure of near-zero speed of the hand to the initiation of the acceleration phase of the motion. This RT was unique to each subject and on the order of 1200 ms from video presentation to hand motion onset (velocity increasing away from 0 m/s) across the group. It gave us a sense of mental chronology of this self vs others identification process.

WHERE TO GO WITH THIS?

Learning that people can have a good sense of the bodily self vs others' body inspired us to design the type of interactive co-adaptive interface we have today in the lab to evoke social awareness. In neurodevelopmental disorders that give rise to issues with social interactions, we will (1) assess the extent to which a person with a diagnosis of ASD can identify his/her own kinesthetic noise; (2) determine the biorhythms through which this identification process is easiest (e.g., the heart rate variability, the bodily kinematics variability, or the EEG-based variability); (3) ascertain the extent to which we can change the motor noise signatures using sound and/or visual input using these kinesthetic information; and (4) gain insights into the extent to which we can influence the cross-talk among the autonomic, peripheral, and central nervous systems as assessed by the interactions among the deliberate, spontaneous, and inevitable nervous systems' processes that we have identified and measured noninvasively. We will expand on these concepts on Chapters 3 and 6, respectively.

SELF-SUPERVISION: DISTINGUISHING MY SELF-GENERATED KINESTHETIC NOISE FROM EXTRANEOUS NOISE

As the biorhythms' parameters dynamically unfold in time, they broadcast some probabilistic information on what the brain may be thinking while it is deciding. The stochastic signatures inherently present in the variability of orderly events and their temporal interdependencies were derived and mapped onto those of the actual motions of the avatar the person was watching—The ones distorted and the veridical ones derived from self-motions and from the motions of others (**Notes**[4]). In this sense, we could address the relationship between the first person (egocentric) vantage point and the third person (allocentric) vantage point as they interchangeably switched in the participant's mind. These (seemingly) compatible vantage points were being used to imagine what it was like to walk or play sports routines in three-dimensional space as if looking at oneself from

[4]Link to movies of veridical vs. distorted avatar

outside the body. It was mindblowing to quantify the level of accuracy most participants had on this distinction of self vs others. It was also surprising to find such congruency between the temporal dynamics of the mental thoughts (deciding about movement patterns of self-generated motor actions vs motor actions of others) and the temporal dynamics of the physical execution of those motor actions.

The participants were not only imagining themselves from an allocentric vantage point; they were imagining others as well. And more importantly, they had to have a very robust notion of relative values because these decisions implicitly required mental rotations and relative computations when comparing the differences and similarities across the various versions of the avatar (veridical Me, veridical Not Me, noisy Me with my own noise; noisy Me with the noise of others, noisy Not Me, and noisy Not Me with noise from Not Me). Here, Not Me represents others. All these possible combinations seemed to blend together after a few video clips, but they were in fact very clear at some unconscious level, when people were making such decisions. They were implicitly certain, at a level transpiring largely beneath awareness. Yet, at the explicit level of conscious decisions, participants had no idea if they were right or wrong.

These insights alerted us of the possibility of using this closed-loop interactive paradigm and extending it to a real-time, co-adaptive interface to assess the interactions of a person with an avatar (Fig. 2.22). In this case, we would use more than one sensory modality to evoke the interactive co-adaptation between the end-user and the avatar. But before deploying this idea in the world of autism, we needed to investigate if the motor-noise signatures would shift with interactive, co-adaptive interfaces engaging the child. More precisely, we needed to understand how to best engage the child in the spectrum of autism without having to instruct him/her what to do. In the absence of verbal communication, we needed to create new means of communication that would open a window into what they liked, what they were interested in, and what their inherent sensory-motor predispositions were.

SELF-DISCOVERING CAUSE AND EFFECT: AUTONOMY FROM THE BOTTOM-UP

Goal specification is something we take for granted in motor control research because we instruct the person what to do: "here is the target, reach for it". But we cannot explicitly tell a neonate what the goal is. The nascent brain must self-discover what the goal ought to be in any given context, so that the body can successfully and autonomously act on it and do something purposeful under self-control. For example, feeding in the first days of life requires controlling the airflow and pressure to not choke while suctioning and swallowing the milk from the mom's breast. These well-controlled biorhythms are critical for survival. They may be precursors of social exchange with the mom in the early days of the newborn baby. This type of nascent volition scaffolds other forms of deliberate volition—e.g., bound to appear later, when the same apparatus that developed for safe-feeding purposes has successfully adapted to co-articulate muscles and produce sounds.

For humans, there must be nothing more rewarding than having self-control at will and the capacity of exerting that volitional controlon demand from birth onwards. In fact, I

FIGURE 2.22 **Interactive co-adaptive human—avatar interface to evoke self-discovery of a goal and modulation of the person's volitional control over the consequences of impending goal-directed actions.** (A) The participant wearing a suit with light emitting diodes (LED) moves around the room as cameras on the ceiling track the motions at 480 Hz (Phase Space, CA). The motions are projected on an avatar on the screen which moves in tandem with the person. The avatar can move can move faster or slower than the real-time motion signal, so we can measure awareness of such delays and their influences on the person's perception of the self in motion. The task is to discover a musical spot in a volume of the space that we have programmed so that when the hips enter the volume (vRoI), the music plays. The participant does not know where the region is. This vRoI is globally defined relative to an allocentric frame of reference somewhere in the room. (B) The participant self-discovers the vRoI and music plays; then the program switches to a relative code, whereby the position of the hands are now modulating the speed of the music—playback based on distance from the hips. The hips are now the (egocentric) frame of reference of interest and the participant ought to self-discover the proper distance to play back the music at the proper speed. Here, she just discovered what that proper distance was and rejoices in her ability to control the music playback. This sonification of the movements lead to better awareness of body parts once the person gains control of the actions and their sensory (sound) consequences. The avatar is then used to further engage the participant in dyadic exchange resulting in the modulation of other music qualities (e.g., pitch and timber) through closed audiovisual stochastically controlled feedback loops.

would go as far as claiming that our awareness and deliberate use of this power to control actions autonomously (including speech) is what makes humans so unique. Indeed, having volitional control over something and transforming it to subserve our purposes, to effectively achieve our objectives on a timely manner give us tremendous joy.

At the level of a goal-directed action over an object, a sequence of instructed and attained goals may merely resemble a motor control experiment. Yet, at the social level, there are additional conditions that we need to consider because the goals are not always obvious. A conversation that spontaneously occurs during a social encounter does not exactly come with instructions. The process of self-discovering the main point of the conversation to establish a successful rapport with the speaker at the other end is far from trivial. It requires mastery and co-ordination of very subtle cues that a person learns to identify over time, starting from a very early age. Such a self-realization is a bottom-up process, whereby the self-discovery of goals is spontaneous rather than explicitly instructed from the top down.

We will provide an example in Chapter 6, of such a process, and will briefly describe the philosophy of our approach here. We built a set-up to evoke motivation for self-initiated exploration to eventually evoke the self-discovery of agency in children that had been labeled as low-functioning ASD. These children did not yet possess spoken language as a form of communication. As such, we could not instruct them what to do or how to accomplish a given set of goals. In fact, attempting to do so would visibly cause anxiety and stress in the children. We decided to follow a different route, whereby we would spontaneously evoke interactions between the children and our team using as a proxy computer media, without instructing them what to do, but rather steering their curiosity in implicit ways. To that end, we created an interactive interface between the child's body and a computer, whereby the computer would trigger media playing states using the real-time motions of the child's body. All we needed to do was to program the computer to have certain states of the child's motions match certain states of the peripersonal space of the child and when that matching occurred, the media would play.

For example, the child would spontaneously move the hand in the peripersonal space, and we would create a volume nearby the child's hand. Once the child's hand passed within a distance to the center of the volume's location in three-dimensional space, the media would start playing (Fig. 2.23). This playing media evoked curiosity in the child. It triggered a search self-initiated by the child- for "the magic spot" that caused the media to play. The volume, which we coined virtual region of interest (vRoI) could shrink into a more precise target-region, once the child engaged in the process of self-discovery that the first (unprompted) media-triggering evoked (Fig. 2.24). Then, we could shift the volume location and have the child find it again. We could also change the media type and use audio, video, movies of the child's preference and even a video camera facing the child, which would turn ON when the hand entered the vRoI and project the child's face and upper body on a screen in front of the child. In the meantime, our algorithms processed the variability of the movement parameters and tracked their evolution (as in Fig. 1.9 of Chapter 1) to determine the media inducing the fastest rate of change in the stochastic signatures of the child's motions. That media was then rendered as the media type that the nervous systems of the child "preferred".

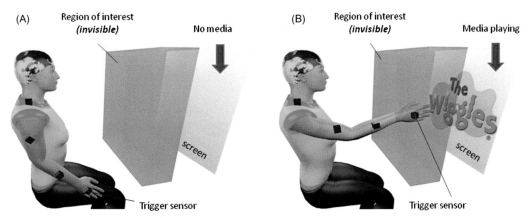

FIGURE 2.23 **Schematics of a simple set-up to habilitate volition from the bottom up through the sponta-neous self-discovery of implicitly defined goals.** (A) The person seats comfortably in front of a computer screen and wears sensors on the upper body (shoulder, upper arm, forearm, and hand). The sensors sample continuously the positions of these body parts. We define a volume in peripersonal space (a vRoI). If the hand accidentally crosses the volume, the media plays on the screen. The goal the person is to self-discover that the hand must be continuously kept inside a vRoI for the media to play continuously (we can shrink the volume and move it around to generalize this type of engagement across the space.)

The engagement of the child with the media was rather spontaneous, much as when a curious baby flails his arms until one his hands touches a toy. The baby may slowly come to discover that it is his hand touching the toy that makes the toy move. He may also discover that his arm moving the hand which causes the toy to move belongs to his body, so that if he moves his body and with it his hand, he can reach the toy and or move it too. That sense of agency "it is me doing that", which we so commonly see emerge in babies, was our source of inspiration to design our experimental task. We hoped to quantitatively track the evolution of the process of self-discovery of goals to evoke the learning of *cause and effect*: "I move my hand and cause the media to play". In doing so, we were trying to capture and objectively quantify the stochastic evolution of the self-discovery process—As the process unfolded in real time.

What surprised us all about the interactive interface we built for ASD children, who could not verbally communicate with us was the consistency all participants showed in the overall evolution of this process of self-realization. In the end, all children consistently self-discovered the vRoI and controlled the media play at will.

We try to capture the sequence of events that consistently took place at the school we first tested this set-up in Fig. 2.25. In addition to the children with ASD, we tested TD children to try and understand how a TD nervous system would solve the problem. Each child sat comfortably in front of a computer screen—As they regularly did in class. They wore sensors on the hand, arm, and trunk, and whenever possible on the head. Some children are too sensitive in the head area and did not tolerate wearing the sensors or having us touch them near the face. In those cases, we did not use the head sensor. We

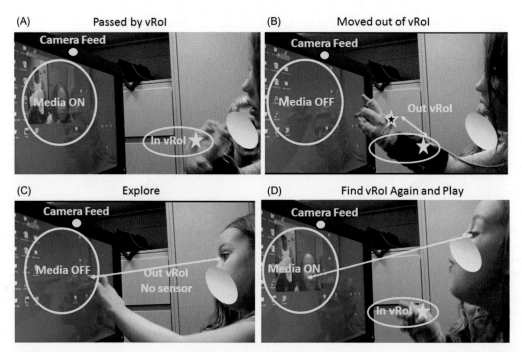

FIGURE 2.24 **Engaging a child with autism spectrum disorders (from the bottom-up) without prompting or instructions.** (A) The child passes the hand accidentally by the vRoI and triggers the camera feed facing her. (B) She then moves the hand outside the vRoI and the media turns-off. (C) She is curious and explores the space with the other hand if the motion in that region has the same consequences as the other hand (i.e., she can see herself in the screen). She touches the screen and nothing happens because there is no sensor in that hand and the other hand is not in the vRoI. (D) Upon exploration, she finds the vRoI with the appropriate hand and triggers the real-time image of herself, thus engaging in playful interactions with it (facial gestures, vocalizations, and sustained eye contact.) (Ellipse blocks the facial features so the child cannot be identified but leaves out the eyes and mouth to show the dramatic effects of this interface on the child's self-motivation and agency over the situation.)

interchangeably used two modes of engagement by tracking position from a global frame of reference anchored somewhere in the room and from a local frame of reference anchored somewhere on the child's body. The global frame of reference defined the vRoI with the origin set at a table in the room. The local frame of reference anchored the origin to the trunk or to the shoulder of the dominant hand wearing the sensor. We did not place a sensor on the other hand on purpose. We wanted the child to explore the possibility of using that other hand to search for the vRoI and self-discover that there would be no outcome from that exploration. This was a way to evoke the self-realization that it was the other hand causing the media to play.

We could engage the child and media-playing states, using simple distance-based criteria defined by the real-time position of the hand (registered by the sensor) and the center of the vRoI defined both globally and locally. The volume size was also programmable, so we could make the targeted goal at its center more precise by shrinking the volume, or easier to attain by allowing the volume to cover a broader region in

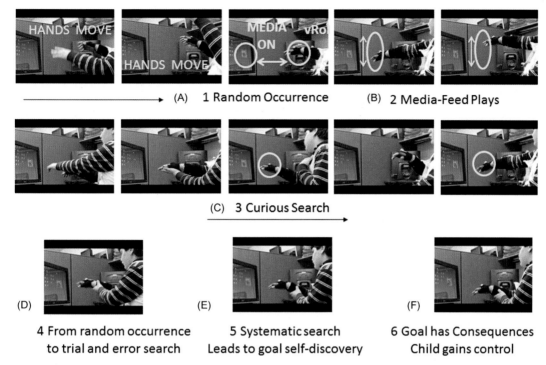

(A) 1 Random Occurrence   (B) 2 Media-Feed Plays

(C) 3 Curious Search

(D)   (E)   (F)

4 From random occurrence   5 Systematic search   6 Goal has Consequences
to trial and error search   Leads to goal self-discovery   Child gains control

FIGURE 2.25 **General sequence of this paradigm found in all TD and ASD children we tested.** The child in the picture is labeled by his school "low functioning" ASD because he has no spoken language abilities. In our eyes, and according to the outcome of our biometrics, this child is incredibly capable and his nervous system is very plastic. The noise patterns and the probability distribution functions that we estimated in real time from his motions became rather anticipatory of the consequences of his searching actions. (A) The hands motions trigger media (video and audio in this case). (B) The media feed plays for sustained times as the child explores the volume containing the vRoI. (C) He explores the peripersonal space using the other hand and other body parts— Very common across all children in the spectrum that tried this set-up. Sometimes, the media plays (if the hand hits the vRoI) and sometimes it does not, but the children do not give up; they keep searching. (D) Random search turns into systematic trial and error. (E) This systematic search leads to the self-discovery of the goal and its consequences. (F) The child discovers cause and effect, thus gaining self-control of the situation. The child bridges mental intention with physical action and its consequences. He acquires volitional control over the situation and that is extremely rewarding, but the reward is internally generated from the bottom-up processes involved. This set-up contrasts with the type of top-down approach that instructs the child and externally rewards him without measuring the sensory-motor consequences of that process.

the child's peripersonal space. All along, we hid behind the scene, watched the child's performance remotely on video and using our computer program, we steered the child's behavior from the distance, without directly interfering with his/her curiosity and motivation.

Across all children, TD and ASD alike, the sequence described in Fig. 2.25 was quite common. When on occasion the child had some trouble initiating movement—Very common in ASD—We moved the global frame of reference [and with it the volume (vRoI)] near to or on the child's hand. We did this so that the slightest motion the child made would lead the hand to enter the vRoI and trigger the media. Then, the media would play and surprise the child. In such cases, their natural curiosity about the seemingly magically triggering of the media, led these children to actively explore their peripersonal space until they understood that their hand motions within a specific location was causing the media to play. Once they figured out this causal connection on their own, they would proceed to hold the hand wearing the sensor with the other hand to watch the full movie. In some cases, they would see themselves on the screen, because we would use a camera-feed on the monitor facing the child and would capture the real-time image of the child and project it on the screen when the child made contact with the vRoI. Some children truly loved this and engaged more with their own image than with the media.

Because they were wearing sensors outputting the motor signals and we were landmarking all behavioral events of relevance, we could implicitly steer the self-discovery process, to lead the child to make the connection between cause (goal-directed hand motion) and effect (media play). We could do so implicitly, without any prompting or instruction, without interfering with the goal—Self-discovery process. All throughout the unfolding of this co-adaptive process between external media playing and self-generated motions, we could quantify systematic trends in the stochastic changes that took place across all the children we tested. Such changes and their rates occurred independently of verbal proficiency, diagnosis, and age.[73] They demonstrated to us that the sensory-motor systems of the developing child are extremely plastic and can be successfully steered in a direction that leads to volitional control of the child's brain over the child's body.

We tested neurotypical children and they too followed this process. The only marked difference between them and their peers with ASD was that they verbalized what was happening and once they located the vRoI they were done with it, i.e., they would stop altogether. Because of this, in the TD cases, we had to actively shift the volume around more often than in the ASD cases, or the TD children would be bored with the game. The joy that all children experienced from the process of self-discovery was far more evident in the ASD children we tested. They seemed to be in awe with their own (newly found) volition. Some of them even vocalized full phrases for the first time, while others expressed their joy through body gestures. The ASD children took longer to explore the vRoI and convince themselves "Ok, it's me causing that music (or video) to play".

More formally, the process unfolded as follows:

1. *Spontaneous (unprompted) media triggering*: The child randomly moved the hand and accidentally passed by the vRoiI that we defined in the volume located near the child's trunk (within the child's peripersonal space). Once the hand moved into the vRoI and this near-zero distance between the hand and the center of the volume triggered the media, the child became curious (Fig. 2.25A);
2. Curious search initiated to find the vRoI (Fig. 2.25B—C);
3. Trial-and-error took place and moved the hand around to systematically trigger the media (Fig. 2.25C—D);

4. Media playing occurred more often and with it, the self-discovery of the goal: "if I place my hand in this region I get the media to play" (Fig. 2.25E);
5. Systematic goal achievement helped the child define the correct space position with high certainty (correctly predicting the consequences of his actions) and deliberately holding the hand in the center of the volume to watch the media playing continuously (Fig. 2.25F).

Some children did not particularly engage with the video projecting their own face in real time. They were more curious and better engaged with general audiovisual media. Other children were really attuned to their own faces and played with the image by making different faces and engaging in playful mimicry. In all cases, we tracked the activities of the sensors and could automatically infer the child's interest from the stochastic patterns of motion. Some parameters of interest were the amount of time they spent exploring inside the vRoI vs the amount of time they spent exploring outside the vRoI. Parameter identification for inside- and outside-vRoI was possible because we continuously tracked all motions and could compute the distance from the hand's current position to the volumetric vRoI that we defined.

See, e.g., a vRoI and the hand trajectories along with the hand's angular speed profiles in Fig. 2.26A. See a metric we defined in Fig. 2.26B to quantify and validate the preferences of a given media during the session. Fig. 2.27B shows ways to selectively find the media type which shifted the stochastic signatures of the hand's speed micromovements towards more predictive states with higher signal-to-noise content. More specifically, by tracking the shifts in the empirically estimated Gamma parameters towards the lower-hand corner of the plane, we can identify media that changes at the highest rate towards low scale values (i.e., low noise-to-signal ratio) and high shape values (i.e., symmetric distributions). This is the stochastic regime we want to visit because we know that typical children start random and noisy but move the stochastic signatures rather fast to predictable and well-structured signal. We could steer the stochastic signatures of the hand searching for the vRoI from spontaneous random noise to well-structured and highly predictable signal. Fig. 2.28 gives examples of the evolution in the stochastic changes that we quantified per session in neurotypicals (Fig. 2.28AB) and ASD (Fig. 2.28CD). The scales of motion are normalized between 0 and 1 to account for anatomical differences across the children and to provide a statistical standard metric that captures *change* robustly in these nonstationary signals.

A remarkable outcome from these sessions was that some children vocalized during the interactions with their own face as they became engaged with it. Others vocalized during their interactions with media as they explored the space, and specially as they found the region, and held the hand in place to play the media continuously. The sheer accomplishment of self-control and the acquisition of agency produced these unusual vocalizations and denoted their sudden awareness of cause and effect leading to the volitional control of their own self-generated motions and their sensory consequences.

We did not reward the children with food or tokens. We did not actually need to do that. The control of their actions and their own discovery of cause and effect relations between their self-generated motions and the consequences of those actions were powerful

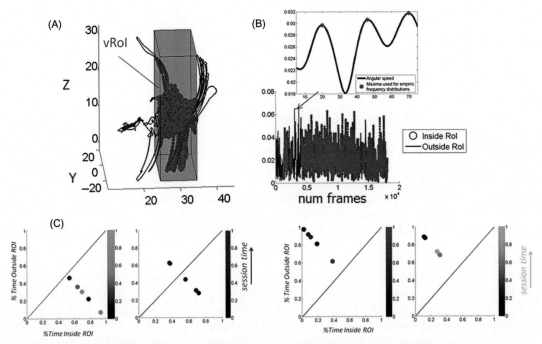

FIGURE 2.26 **Real-time continuous data collection.** (A) Three-dimensional trajectories from the hand are plotted along with the volume (vRoI) in red. Red portions inside the vRoI are automatically marked based on the distance, while blue portions fall out in the search space. (B) The angular speed profiles of the hand automatically colored-coded according to the distance from the vRoI. Peaks are marked by magenta asterisks. (C) Temporal metric to automatically quantify the amount of time the child spent searching inside or outside the vRoI with color code on the quality of the motion according to normalized values of noise (e.g., higher values indicate high noise vs low noise in darker values). Different examples obtained from the same child across different media thus revealing media preferences. Each point is an estimate across 1-minute windows (240 Hz x 60 frames) yielding tight 95% confidence regions. The entire process takes less than 15 min.

enough to bring together several systems and regulate them so that children that were otherwise nonverbal actually articulated full meaningful phrases "what's happening?...happening now?"

THE BRIDGE BETWEEN MENTAL INTENT AND PHYSICAL VOLITION

We have the video tapes of these sessions but to honor the children's privacy, we will not release them in full. We will do so by blurring the faces and voices in the cases where we obtained University Institutional Review Board and parental consent to do so (see **Notes**[5]). This, however, masks somehow the amazing moments of curiosity and self-

[5]Link to movie about the sensory substitution in ASD

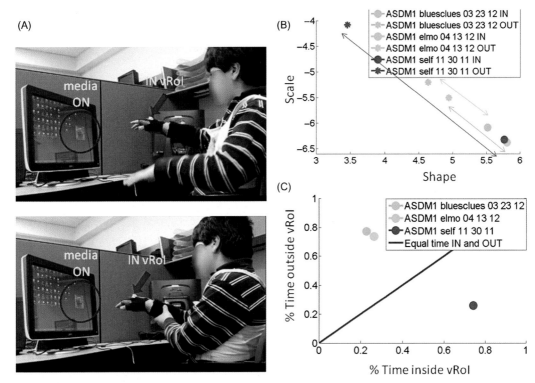

FIGURE 2.27 **Media selection based on outcome measure of sensory-motor physiological changes in stochastic signatures.** (A) Child exploring and playing with the media. (B) Estimated Gamma shape and scale parameters plotted on the Gamma parameter plane as they evolve (arrow shows the change towards lower noise and more symmetric shapes of the evolving distributions). Brown marker indicating (self-viewing from the camera feed) results in the most dramatic shift of the motor variability of the hand measured in terms of the moment by moment fluctuations in the amplitude of the angular speed. The self-video is followed in preference by the elmo video and lastly by the bluesclues video. (C) The self-video is also rendered as the favorite by the amount of time the child spent inside the vRoI playing it continuously vs the amount of time exploring for elmo and blues clues cases.

discovery the children had during those experimental sessions. Their faces lit up and we could hear the whispering of vocalizations of entire phrases. Indeed, these were very powerful moments that anyone could see unfolding in real time during those days of initial testing. We could not believe our eyes. We thought that perhaps we tapped into something fundamentally human that was driving these phenomena: The self-discovery and acquisition of volitional control over intended actions and their consequences, thus evoking the sense of agency.

Many in the spectrum had communicated to us that their body "has a mind of its own". They would say "my brain intends to do something but it takes the scenic route to get there".[74] To an external observer describing the person's behavior in a social context, this disconnect between the desire to act with a purpose and the outcome of performing

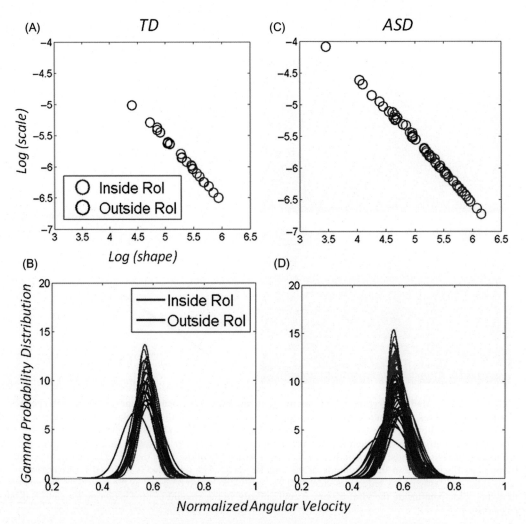

FIGURE 2.28   **Plasticity of the autistic sensory-motor signatures.** (A) Evolution in the stochastic signatures of the hand's fluctuations in the amplitude of the angular speed while moving from the search outside the vRoI to the systematic goal-directed motions inside the vRoI (to play the media continuously.) Distributions lower the dispersion and become more symmetric as the child self-discovers the goals and its consequences. TD children verbalized their discovery and explained to us what was taking place: "Hey!, if I move my hand right here I can see myself on the screen" and then they proceeded to show us how. (B) Children in the spectrum of autism also followed the path of self-discovery but could not verbalize it with spoken language. They used body language and gestures to show us their discovery. Some children vocalized phrases. All children showed their sensory-motor systems have plasticity as they transitioned their signatures from spontaneous random noise to well-structured signals. Red, Inside RoI; Blue, Inside RoI.

the intended action may look as *lack of intent*. This may be particularly the case if that observer describing the person with autism has no training in motor control and cannot understand that physical volition and mental intent must go hand-in-hand for the person to be able to interact socially. If there is a problem with neural control of movement, it will be difficult to establish the bridge between mental intention and physical volition.

Such problems are difficult to identify because they can be due to poor sensory and somatic motor feedback, which are invisible to the naked eye. They can be due to limitations in the communication across the neural networks of the brain and the periphery (e.g., when pre- and post-synaptic terminals are not establishing communication throughout the nervous systems)—Also difficult to quantify noninvasively. They can be due to sensory-motor integration issues preventing the simultaneous perception of one's own body and the body of others in the social environment—A proposition that is hard to test because it requires longitudinal monitoring of a process that takes years to evolve. There may be many other potential reasons underlying the difficulties a nervous system may have in building a proper bridge between mental intent and physical volition over the intended action. However, following a research program that ignores the body in motion and relies exclusively on representational (guessed) constructs will not shed light on the issues.

A case in point are cognitive theories of autism. Such theories have, yet, to consider the role of the physical body in action and the feedback that continuously self-generated motions provide to the brain. They have yet, to understand the importance of such reafferent feedback to estimate and anticipate sensory consequences important to engage and drive one's own body during social exchange.

The disembodied treatment of the brain by Cognitive Neuroscience, Psychological Sciences, and Psychiatry in general, is a paramount obstacle to scientific progress. In the realm of autism this problem, without a doubt has stalled the discovery of new philosophical concepts to drive the design of new therapies and the development of biometrics that assess the effectiveness of such therapies.

In the United States (unlike the United Kingdom), e.g., there is a prevalence of ABA—A method that reshapes behaviors in human children without measuring the sensory-motor consequences of the reshaping process on the child's coping nervous systems (See **Notes** [6]). This type of therapy so prevalent across the United States, intervenes in a nascent nervous system without measuring the physiological impact such stressful demands (e.g., performing thousands of discrete trials such as pointing to an object in exchange for an external reward) may have on the development of stress-coping mechanisms of the child subject to such effortful task.

Without objectively measuring the nervous systems output, it is not possible to quantify on a proper scale the effectiveness or the risk of such behavioral reshaping therapies. This is particularly so, because behavior in this case is not defined from the egocentric vantage point of the nervous systems of the person being treated. It is rather described from the vantage point of an external observer with social expectations that do not necessarily accommodate the individual needs of the affected person—A person with a uniquely evolving (coping) nervous system.

[6]On ABA only being applied in the US

In our example here, we followed an intuitive notion from neonatal neurodevelopment and we measured the outcome of the nervous systems objectively by examining its unfolding biorhythms moment by moment, in a multilayered manner; i.e., by profiling their signatures and the shifts that performance itself produced on them. Offering this new way to evoke the self-discovery of environmental goals—As a neonate would, we implicitly (without instructions) guided the child toward a path of spontaneous self-realization of *cause and effect*.[75] The reward in this case internally emerged from the attainment of self-agency and volitional control over the child's self-generated actions and their (now well-mapped) sensory consequences.

HABILITATION AND ENHANCEMENT OF VOLITION TO EVOKE VOCALIZATION

Given the success of this experiment, we decided to extend the paradigm and explore new avenues to engage the child in social exchange with the overarching goal of evoking vocalization and verbalization of spoken language utterances.[76]We were fortunate to receive funding from the Nancy Lurie Marks Family Foundation to carry out this project.

We combined the results from the IGERT project and the results from our paradigm to spontaneously habilitate and rehabilitate volition in ASD and created a new interactive set-up. This set-up then led us to develop a co-adaptive human—avatar interface to try and evoke various types of interactions between the child and audiovisual media, while also encouraging interactive co-adaptation between the child and the avatar. As in the IGERT experiment, we endowed the avatar with veridical motions from the end-user biorhythms. We then explored which sensory channel and context were the most adequate to engage the person. We identified which biorhythm shifted the stochastic signatures maximally towards regions on the Gamma parameter plane with the lowest noise and the highest symmetric shape in the distribution (as in Figs. 2.27–2.28). To that end, we manipulated the feedback the end-user received from sounds, but also from the visualization of the avatar itself, which we endowed with various noisy versions of the person's real-time motions, as we explained above (See movies in **Notes**[7]).

We could stream live the positional data from the LEDs and derive various kinematics signatures in the temporal and in the spatial domain—As we did with the IGERT project above. In the new set-up, we could identify as well real-time changes in the bodily motions leading to better body awareness. We adapted network analytic tools[77] traditionally used to study brain connectivity[78] to the study of connectivity in the peripheral network of the body as well as the interconnectivity between bodies in motion. Using graph theory and network connectivity analyses, we could derive new ways to express synergistic patterns within the end-user's body. We extended the use of these new tools to the study of self-emerging patters across the end-user and the avatar's bodies. Since it is possible to play the avatar feedback with a lag (slower) or lead (faster) in relation to the veridical real-time execution of the end-user motions; and since we could also endow the avatar with veridical or noisy variants of the end-user movements; we were able to explore coupled synergies that would have

[7]Link to movie of person—avatar interactions.

been otherwise invisible to the naked eye. Fig. 2.29 shows different contexts we exposed the participant to, with the purpose of investigating the sensory modality and context where the noise-to-signal ratio of the bodily fluctuations of various motion biorhythms was at its minimum and the shape of the estimated probability distribution function was most symmetric. Upon identification of the motor regime suggesting the source of sensory guidance that would best enhance kinesthetic reafferent feedback, we proceeded to study the interactions with the avatar using those sources of guidance and context conducive of low noise and highly predictable kinesthetic signal.

Some labs interested in embodied cognition had investigated people's reaction to self-generated motions incongruent with expected internal kinesthetic flow vs external motions incongruent with expected visually perceived motions.[79] Their findings pointed at people's awareness of such incongruences. This result paired with our IGERT findings of people's self-awareness of their own motions and their abilities to distinguish them from those of others—Even in the presence of noise, gave us proof of concept that we could perhaps enhance body awareness in children with neurodevelopmental disorders such as ASD.

We decided to use our stochastic methods to help close the loop and develop a noise-dampening technique. As in Fig. 2.28, we quantified the reduction in the somatic-motor noise by dynamically identifying the sensory channel and the biorhythms that best led the stochastic signatures of the real-time motions towards systematically well-structured regimes easy to predict and high in signal-to-noise ratio using the various contexts of Fig. 2.29. Further, since we could also identify the sensory channel with the most random

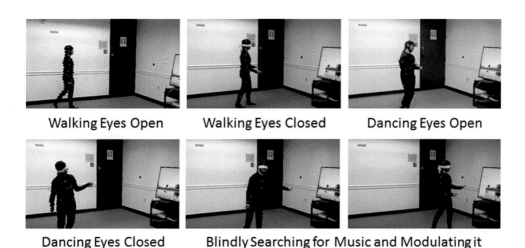

Walking Eyes Open   Walking Eyes Closed   Dancing Eyes Open

Dancing Eyes Closed   Blindly Searching for Music and Modulating it

FIGURE 2.29 **Exploration of the evolution of stochastic signatures in response to various stimuli and contexts.** Stimuli included music and visual feedback. Context included evoking reliance on proprioceptive feedback by blindfolding the participant, and evoking different families of movements through activities like walking and dancing. From the estimated stochastic signatures, we could selectively evaluate the condition, context, and type of sensory guidance that minimized the noise-to-signal ratio, maximized the symmetry of the probability distributions describing the fluctuations in micromovements' amplitude and that overall made the self-generated nervous systems' activity anticipatory of its sensory consequences.

and the noisiest signatures, we could develop a sensory-substitution technique that selectively used a favorable sensory channel to guide the co-adaptive interaction as an entry point to later enable and improve communication with the least favorable channel.

Given that we developed the means to have the participants spontaneously evoke media-playing and control the play, we saw an opportunity to explore their own preferred channel for dyadic engagement with the avatar. This could be auditory, e.g., through the sonification of motion (including motions of the heart, body, and EEG biorhythms); visual input through the triggering of nice visuals (including cartoons, movies, and the child's own image next to the avatar image). We are at present, even exploring the idea of including touch as another channel complementing kinesthetic input from ongoing motions through the use of vests that vibrate[80], as soon as they are available in the market.

Fig. 2.30 shows examples of a participant interacting with the avatar endowed with their own veridical motions, while exploring a region in space that produces music as they pass by it (see links to movies in **Notes**[7]). In this case, we also establish a local-distance based map (i.e., egocentrically anchored at the hips) that associates different

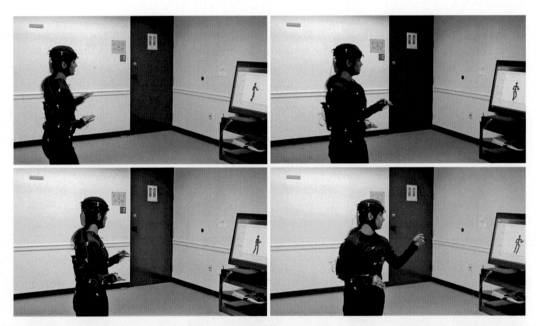

FIGURE 2.30   **Brain body avatar interface (BBAI) tool.** Upon determination of the form of sensory guidance that led to the highest signal-to-noise ratios and to the most predictable stochastic regimes, the user–avatar interactions are initiated to further probe bodily awareness. In tandem with the real-time feedback from bodily kinematics, the participant wears an electroencephalographic head cap containing an accelerometer to measure head motions and EEG-activities in tandem with bodily biorhythms. Yellow wireless device (Neuroelectrics 500 Hz, Barcelona, Spain) transmits EEG-activity to a computer in the room for near-real- time monitoring and analyses to feed the interface as another channel of adaptable nervous system output. One of the EEG leads records heart rate so that we simultaneously sample multiple layers of the nervous systems, central, and peripheral activities including autonomic responses. The set-up is currently used to provide a comprehensive profiling of the CNS–PNS interactions and to monitor the plasticity of the nervous systems in children with autism, cerebral palsy, and attention-deficit hyperactive disorder.

speeds in the music playback with different distances between the hand and the vRoI moving with person. Clearly, we also define the motion as a function of allocentric distance measured relative to a global frame of reference located somewhere in the room and can monitor the information interchangeably using these two frames of references. The important point is that we have the control of these aspects of the paradigm. As such, we can steer the person's behavior beneath the person's awareness, thus minimizing possible interference with the performance, likely emerging from the end-user's reactions, or awareness of our manipulations. We act as a hidden force guiding the person towards regimes of anticipation that can be quite effective because we are controlling the consequences of the person's impending actions, a step ahead of the game.

Once again, we emphasize the importance of utilizing two vantage points to express our problems: The vantage point of an external observer and the vantage point of the internal nervous systems of the person revealing its intrinsic dynamics. Our work was in part motivated by a missing part of the MNST: The vantage points and the metrics a nervous system would internally define based on kinesthetic reafference and the continuous cross-talk among the central, peripheral, and autonomic nervous systems. Unlike the MNST, we did not assume that in a social dyadic exchange "the observer and the observed are both individuals endowed with a **similar brain–body system**".

Whether using an actual social dyadic exchange between two humans, or a simulated one between a human and an artificial agent, we first profiled the stochastic signatures of the individual bodies in motion along with those of their coupled dynamics, and then co-adapted them based on our empirical findings. In fact, our paradigm assumes that the two actors do not necessarily have a *similar brain–body system*. Instead, we build our solution from the bottom-up, discovering the sensory capabilities and predispositions of the person's nervous systems.

We first find key ways to communicate with the nervous system that is neurologically different, and then evoke the spontaneous emergence of similarity between that system and the other actor's system. In the computerized enacted social setting that we built, the avatar's system was endowed with various biorhythmic dynamics. In this way, we *simulated* possible interactive scenarios involving coupled dynamics bound to emerge during dyadic social exchange. As always, our take-home message was to not assume anything obvious about the observed phenomena, but rather measure it, objectively quantify it and guide any guesses by empirical estimation. This prescription has worked wonders with the ASD children!

References

1. Gallese V, Keysers C, Rizzolatti G. A unifying view of the basis of social cognition. *Trends Cogn Sci.* 2004;8 (9):396–403.
2. Rizzolatti G, Craighero L. The mirror–neuron system. *Annu Rev Neurosci.* 2004;27:169–192.
3. Iacoboni M. Imitation, empathy, and mirror neurons. *Annu Rev Psychol.* 2009;60:653–670.
4. Rizzolatti G, et al. Neurons related to reaching-grasping arm movements in the rostral part of area 6 (area 6a beta). *Exp Brain Res.* 1990;82(2):337–350.
5. Rizzolatti G, et al. Functional organization of inferior area 6 in the macaque monkey. II. Area F5 and the control of distal movements. *Exp Brain Res.* 1988;71(3):491–507.

6. Umilta MA, et al. When pliers become fingers in the monkey motor system. *Proc Natl Acad Sci USA*. 2008;105 (6):2209–2213.

7. Rizzolatti G, Fogassi L. The mirror mechanism: recent findings and perspectives. *Philos Trans R Soc Lond B Biol Sci*. 2014;369(1644):20130420.

8. Rozzi S, Coude G. Grasping actions and social interaction: neural bases and anatomical circuitry in the monkey. *Front Psychol*. 2015;6:973.

9. Luppino G, et al. Multiple representations of body movements in mesial area 6 and the adjacent cingulate cortex: an intracortical microstimulation study in the macaque monkey. *J Comp Neurol*. 1991;311(4):463–482.

10. Matelli M, Luppino G, Rizzolatti G. Architecture of superior and mesial area 6 and the adjacent cingulate cortex in the macaque monkey. *J Comp Neurol*. 1991;311(4):445–462.

11. Caretta A, et al. Characterization and regional distribution of a class of synapses with highly concentrated cAMP binding sites in the rat brain. *Eur J Neurosci*. 1991;3(7):669–687.

12. Nelissen K, et al. Action observation circuits in the macaque monkey cortex. *J Neurosci*. 2011;31 (10):3743–3756.

13. Arbib MA. From monkey-like action recognition to human language: an evolutionary framework for neurolinguistics. *Behav Brain Sci*. 2005;28(2):105–124. discussion 125-67.

14. Fogassi L, et al. Parietal lobe: from action organization to intention understanding. *Science*. 2005;308 (5722):662–667.

15. Iacoboni M, et al. Grasping the intentions of others with one's own mirror neuron system. *PLoS Biol*. 2005;3 (3):e79.

16. Andersen RA, Buneo CA. Sensorimotor integration in posterior parietal cortex. *Adv Neurol*. 2003;93:159–177.

17. Whyatt C, Torres EB. *The social-dance: decomposing naturalistic dyadic interaction dynamics to the 'micro-level'*. Fourth International Symposium on Movement and Computing, MOCO'17. London, UK: ACM; 2017.

18. Von Holst E, Mittelstaedt H. The principle of reafference: interactions between the central nervous system and the peripheral organs. In: Dodwell PC, ed. *Perceptual Processing: Stimulus equivalence and pattern recognition*. New York: Appleton-Century-Crofts; 1950:41–72.

19. Kalampratsidou V, Torres EB. *Outcome measures of deliberate and spontaneous motions. Third International Symposium on Movement and Computing*, MOCO'16. Thessaloniki, GA, Greece: ACM; 2016.

20. Ferrari PF, et al. Mirror neurons responding to the observation of ingestive and communicative mouth actions in the monkey ventral premotor cortex. *Eur J Neurosci*. 2003;17(8):1703–1714.

21. Lahav A, Saltzman E, Schlaug G. Action representation of sound: audiomotor recognition network while listening to newly acquired actions. *J Neurosci*. 2007;27(2):308–314.

22. Keysers C, et al. Audiovisual mirror neurons and action recognition. *Exp Brain Res*. 2003;153(4):628–636.

23. Kohler E, et al. Hearing sounds, understanding actions: action representation in mirror neurons. *Science*. 2002;297(5582):846–848.

24. Gallese V, et al. Action recognition in the premotor cortex. *Brain*. 1996;119(Pt 2):593–609.

25. Rizzolatti G, et al. Premotor cortex and the recognition of motor actions. *Brain Res Cogn Brain Res*. 1996;3 (2):131–141.

26. Umilta MA, et al. I know what you are doing. a neurophysiological study. *Neuron*. 2001;31(1):155–165.

27. Caggiano V, et al. Mirror neurons differentially encode the peripersonal and extrapersonal space of monkeys. *Science*. 2009;324(5925):403–406.

28. Waldert S, et al. Modulation of the intracortical LFP during action execution and observation. *J Neurosci*. 2015;35(22):8451–8461.

29. Kraskov A, et al. Corticospinal mirror neurons. *Philos Trans R Soc Lond B Biol Sci*. 2014;369(1644). 20130174.

30. Kilner JM, Kraskov A, Lemon RN. Do monkey F5 mirror neurons show changes in firing rate during repeated observation of natural actions? *J Neurophysiol*. 2014;111(6):1214–1226.

31. Kraskov A, et al. Corticospinal neurons in macaque ventral premotor cortex with mirror properties: a potential mechanism for action suppression? *Neuron*. 2009;64(6):922–930.

32. Torres EB. Two classes of movements in motor control. *Exp Brain Res*. 2011;215(3–4):269–283.

33. Fadiga L, et al. Motor facilitation during action observation: a magnetic stimulation study. *J Neurophysiol*. 1995;73(6):2608–2611.

34. Grafton ST, et al. Localization of grasp representations in humans by positron emission tomography. 2. Observation compared with imagination. *Exp Brain Res*. 1996;112(1):103–111.

35. Rizzolatti G, et al. Localization of grasp representations in humans by PET: 1. Observation versus execution. *Exp Brain Res.* 1996;111(2):246—252.
36. Iacoboni M. Neural mechanisms of imitation. *Curr Opin Neurobiol.* 2005;15(6):632—637.
37. Wicker B, et al. Both of us disgusted in My insula: the common neural basis of seeing and feeling disgust. *Neuron.* 2003;40(3):655—664.
38. Caruana F, et al. Mirth and laughter elicited by electrical stimulation of the human anterior cingulate cortex. *Cortex.* 2015;71:323—331.
39. Baron-Cohen S, Leslie AM, Frith U. Does the autistic child have a "theory of mind"? *Cognition.* 1985;21(1):37—46.
40. Leekam SR, Perner J. Does the autistic child have a metarepresentational deficit? *Cognition.* 1991;40(3):203—218.
41. Mistry S, Whyatt CP. Autism: a bullying perspective. In: Torres EB, Whyatt CP, eds. *Autism: The Movement Sensing Perspective.* Taylor and Francis: CRC Press; 2017.
42. Baron-Cohen S, Cosmides L, Toobv J. *Mindblindness: an essay on autism and theory of mind.* MIT Press; 1995.
43. Baron-Cohen S, Wheelwright S. The empathy quotient: an investigation of adults with Asperger syndrome or high functioning autism, and normal sex differences. *J Autism Dev Disord.* 2004;34(2):163—175.
44. Nentjes L, et al. Examining the influence of psychopathy, hostility biases, and automatic processing on criminal offenders' Theory of Mind. *Int J Law Psychiatry.* 2015;38:92—99.
45. Section On C, et al. Sensory integration therapies for children with developmental and behavioral disorders. *Pediatrics.* 2012;129(6):1186—1189.
46. Brincker M, Torres EB. Why study movement variability in autism? In: Torres EB, Whyatt CP, eds. *Autism: The Movement Sensing Approach.* Taylor and Francis: US: CRC Press; 2017:1—37.
47. Brincker M, Torres EB. Noise from the periphery in autism. *Front Integr Neurosci.* 2013;7:34.
48. Yong E. Replication studies: bad copy. *Nature.* 2012;485(7398):298—300.
49. Torres EB, et al. Neural correlates of learning and trajectory planning in the posterior parietal cortex. *Front Integr Neurosci.* 2013;7:39.
50. Torres E, Andersen R. Space-time separation during obstacle-avoidance learning in monkeys. *J Neurophysiol.* 2006;96(5):2613—2632.
51. Gallese V, Caruana F. Embodied simulation: beyond the expression/experience dualism of emotions. *Trends Cogn Sci.* 2016;20(6):397—398.
52. Hickok G. *The myth of mirror neurons : the real neuroscience of communication and cognition.* First edition New York: W. W. Norton & Company; 2014:292.
53. Torres EB, et al. Neonatal diagnostics: toward dynamic growth charts of neuromotor control. *Front Pediat.* 2016;4(121):1—15.
54. Friston KJ, et al. Analysis of fMRI time-series revisited. *Neuroimage.* 1995;2(1):45—53.
55. Torres EB, et al. Toward precision psychiatry: statistical platform for the personalized characterization of natural behaviors. *Front Neurol.* 2016;7:8.
56. Torres EB, et al. Autism: the micro-movement perspective. *Front Integr Neurosci.* 2013;7:32.
57. Torres EB. Signatures of movement variability anticipate hand speed according to levels of intent. *Behavioral and Brain Functions.* 2013;9:10.
58. van Beers RJ, Haggard P, Wolpert DM. The role of execution noise in movement variability. *J Neurophysiol.* 2004;91(2):1050—1063.
59. Nguyen J, et al. Automatically characterizing sensory-motor patterns underlying reach-to-grasp movements on a physical depth inversion illusion. *Front Hum Neurosci.* 2015;9:694.
60. Nguyen, J., et al.*Characterization of visuomotor behavior in patients with schizophrenia under a 3D-depth inversion illusion.*in *The Annual Meeting of the Society for Neuroscience.* 2014. Washington DC.
61. Nguyen J, et al. Methods to explore the influence of top-down visual processes on motor behavior. *J Vis Exp.* 2014;(86).
62. Torres EB, et al. Stochastic signatures of involuntary head micro-movements can be used to classify females of ABIDE into different subtypes of 3 neurodevelopmental disorders. *Front Integr Neurosci.* 2017;11(10):1—17.
63. Duda RO, Hart PE, Stork DG. *Pattern classification.* 2nd ed. New York: Wiley; 2001:654. xx.
64. Pinto J, Shiffrar M. Subconfigurations of the human form in the perception of biological motion displays. *Acta Psychol (Amst).* 1999;102(2—3):293—318.

65. Shiffrar M, Lichtey L, Heptulla Chatterjee S. The perception of biological motion across apertures. *Percept Psychophys*. 1997;59(1):51−59.
66. Galazka MA, et al. Human infants detect other people's interactions based on complex patterns of kinematic information. *PLoS One*. 2014;9(11). e112432.
67. Nackaerts E, et al. Recognizing biological motion and emotions from point-light displays in autism spectrum disorders. *PLoS One*. 2012;7(9). e44473.
68. Torres EB, Cole J, Poizner H. Motor output variability, deafferentation, and putative deficits in kinesthetic reafference in Parkinson's disease. *Front Hum Neurosci*. 2014;8:823.
69. Goodale MA, Milner AD. Separate visual pathways for perception and action. *Trends Neurosci*. 1992;15 (1):20−25.
70. Torres EB, et al. Characterization of the statistical signatures of micro-movements underlying natural gait patterns in children with Phelan McDermid syndrome: towards precision-phenotyping of behavior in ASD. *Front Integr Neurosci*. 2016;10:22.
71. Klin A, Shultz S, Jones W. Social visual engagement in infants and toddlers with autism: early developmental transitions and a model of pathogenesis. *Neurosci Biobehav Rev*. 2015;50:189−203.
72. Klin A, et al. Two-year-olds with autism orient to non-social contingencies rather than biological motion. *Nature*. 2009;459(7244):257−261.
73. Torres EB, Yanovich P, Metaxas DN. Give spontaneity and self-discovery a chance in ASD: spontaneous peripheral limb variability as a proxy to evoke centrally driven intentional acts. *Front Integr Neurosci*. 2013;7:46.
74. Amos P. *Rhythm and timing in autism: learning to dance*. Frontiers in Integrative. *Neuroscience*. 2013;7(27):1−15.
75. Torres EB. Rethinking the study of volition for clinical use. In: Lazcko J, Latash M, eds. *Progress in Motor Control: Theories and Translations*. New York: Springer; 2016.
76. Kalampratsidou, V. and E.B.Torres. *Body−brain−avatar interface: a tool to study sensory-motor integration and neuroplasticity*. in*Fourth International Symposium on Movement and Computing, MOCO'17*. 2017. London, UK.
77. Newman MEJ. *Networks : an introduction*. xi. Oxford ; New York: Oxford University Press; 2010:772.
78. Sporns O. *Networks of the brain.*. xi. Cambridge, Mass: MIT Press; 2011. 412 p., 8 p. of plates.
79. Padrao G, et al. Violating body movement semantics: neural signatures of self-generated and external-generated errors. *Neuroimage*. 2016;124(Pt A):147−156.
80. Novich S, Eagleman DM. Using space and time to encode vibrotactile information: toward an estimate of the skin's achievable throughput. *Experimental Brain Research*. 2015;233(10):2777−2788.

3

The Case of Autism Spectrum Disorders: When One Cannot Properly Feel the Body and Its Motions From the Start of Life

There are in fact two things, science and opinion; the former begets knowledge, the latter ignorance **Hippocrates.**

PART I: HOW DO YOU MEASURE AUTISM?

The year 2008 was an interesting one for me. Early in February, after 50 years in power, Fidel Castro officially retired as President of Cuba, the place where I was born and raised for a little over 20 years completely oblivious to racism. I never consciously experienced racial differences in my *umwelt* while growing up back there; but since my arrival in the United States in 1990, I had become fully aware of races in a way that is sad to explain. Thus, when later in August of that same year, Barack Obama became the first African—American man to be nominated by a major political party for President of the United States, I felt a sense of progress and the hope that potential for change was coming. I felt that once you consciously make note of something you were unaware of and *that something* enters your perceptual bubble, there is no going back to your "ignorance is a bliss" state of mind. You feel the need to act and change the course of events for the better. That happened to me with autism right around the same time when the United States and Cuba were about to enter a new historic era of governance.

In the late summer days of 2008, I had just arrived at Rutgers University to take a tenure-track post and begin the arduous path of building a research program, maintaining a lab and helping advance a problem in a given field. I was happily carrying a tall pile of books and boxes that obstructed my view ahead while trying to find my new office, when

Objective Biometric Methods for the Diagnosis and Treatment of Nervous System Disorders
DOI: https://doi.org/10.1016/B978-0-12-804082-9.00003-4

I accidentally bumped and dropped the whole pile. This was on the hallway of the second floor of the Psychology Department at the Busch campus of Rutgers. I had just, at that very instant, found my office! As I started to collect my books and papers from the floor, right in front of my new office, Professor Sandy Harris passed by and helped me out. She introduced herself and very warmly welcomed me to the Dept. She came in to help me settle the boxes on my new desk and asked me, *what do you work on?* This can be a major mistake sometimes, as I tend to go on autopilot mode forever and shock the person on the other end of the long monologue with a "core dump" of research stuff (very socially inappropriate, I realize now after many years of obliviously ruining social interactions). Luckily, this time around, I had a poster that I hung right there on the wall and used as an anchor for my explanation. The poster was among the things I was carrying in the boxes and the byproduct of a workshop I had recently attended at the Gulbenkian Institute for Neuroscience in Portugal. I had all these sensory-motor measures and research-related content illustrated in it. I had put that poster together for a class I taught for a little over a week while visiting their beautiful neuroscience center in Oeiras. So, this time I was succinct and to the point. I was able to reply with the same question: So, what do you work on Sandy? *Autism*, she replied.

Much as racism had remained out of my environmental bubble for over half of my life at that point, autism too was this esoteric thing that every now and then popped up in the news. So, I asked Sandy, *how do you measure autism?*

Our conversation did not go very far, but I got an invitation to visit her school, the Douglass Developmental Disability Center (the DDDC) associated with Rutgers, where she was the Executive Director. The school ran a program for autism spectrum disorders (ASD) helping participants across their lifespan get better integrated into society, hold a job, and socialize in general. I accepted her invitation and visited her school within the next week or so. What I saw at that school shook me to my core. It sent me into a path of research that I have not been able to leave ever since. To begin with, I felt that I connected with those children at a level I cannot explain, but that reassured me I was one of them and had been one of them my entire life. I did not have to talk to them to know this. They knew of my presence as much as I knew of their presence, at a level that is not apparent to the naked eye; i.e., somewhere beneath the type of conscious awareness that social interactions (as we know it) requires.

For the following 3 months, I did nothing but read about autism research. I watched every talk about it, scrutinized every paper out there, and discovered that the type of modeling and quantitative work I did in sensory-motor control was nonexistent in autism. Sandy could not exactly tell me how they measured autism because there is no way to do so. How do you measure a nervous system changing by the day and coping with all sorts of physiological and psychological issues? I realized there was room to invent a new field altogether. There was progress to be made and change to be implemented in the near horizon.

By October, I heard about a call for proposals at the NSF for their Cyber-Enabled Discovery Type I Idea. The call concerned bold theoretical ideas that had no empirical evidence yet, but with the potential for transformative change. The submission deadline was due on December, so I had not much time left to prepare it considering I was new to autism, the topic I chose to address. I used all of my prior knowledge and ideas for motor control research (see also Chapter 4, in this volume) and combined them with what I had learned about autism those previous 3 months. I essentially proposed a new theoretical framework for the study of social interactions with direct applications to autism research and clinical practices. They awarded me the grant by May 2009. It was part of the American Recovery and Reinvestment Act funds from an economic stimulus package enacted by Congress and signed into law by President Obama on February 17, 2009. The proposal got awarded $670,000.00 and with those funds, I was ready to initiate the path of a new research program in my lab. I had found my call (in life) somehow that 2008. I knew what to devote my energy to and work hard towards for the rest of my existence. Considering that one can go throughout life without having a purpose, I was grateful beyond belief for knowing my purpose in life.

MEASURING ASD OBJECTIVELY, THAT IS

As I began to set-up the clinical study to start characterizing cerebrocortical and somatic-motor patterns of ASD, I decided to run a different set of studies in parallel. I reasoned that I could use complex movements in sports and the performing arts as a springboard to begin the path of developing biometrics of natural behaviors for real-time use, particularly dyadic behaviors so common in the social realm. The idea was to develop methods to be used at home, at the clinic, or the school, rather than constraining the science to a rather artificial lab setting. In the process of assessing if proficiency in one sport would facilitate or hinder motor learning in another sport, I came across my very first participant with ASD.

I had a professional tennis player, a member of the Rutgers lacrosse team and a karate competitor enlisted in the study. I also had others who were completely naïve to sports but occasionally exercised so they were well-fit and healthy. They had to learn a full routine of Jab—Cross—Hook—Uppercut and other more complex routines mixing karate with kick-boxing moves. One day, while I was testing one of the participants during the last session of practice, she asked her brother (I will call him MJ) if he would like to try out the boxing routines and get recorded. MJ had accompanied her that day. He was 17 years old and had a diagnosis of ASD. He said "yes" and we proceeded to suit him up with the 15 sensors (electromagnetic system Polhemus Liberty 240 Hz) (Fig. 3.1) that we used in tandem with the Motion Monitor software package (InnSports, Chicago, IL) to digitize the

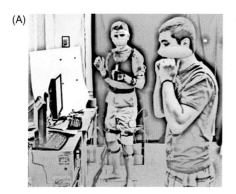

FIGURE 3.1 **Tracking complex boxing motions.** (A) Martial arts instructor with a young adolescent ASD participant wearing 15 sensors across the body to record continuous boxing behaviors. (B) Surprising performance without practice by the ASD participant. He could mirror ahead the movements of the instructor, performing the full long routine, while it took all other participants several sessions to glue together all the subroutines into a fluid performance.

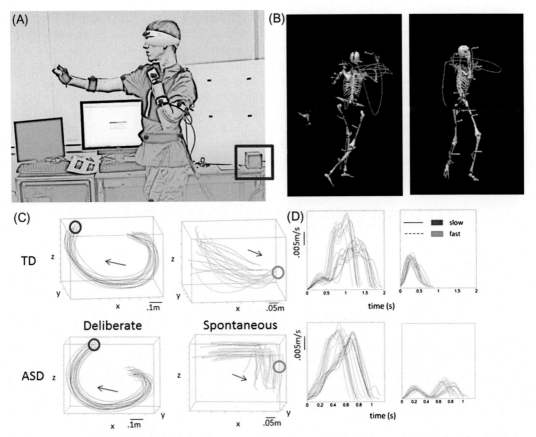

FIGURE 3.2   **The set-up for these experiments.** (A) The tethered system Polhemus Liberty 240 Hz with 15 sensors marked in red recorded the positions and orientations of body parts relative to the global frame marked by a red square. The positions were used to build a kinematics model and digitize the full body so as to continuously track the motions. (B) Sample avatar digitized from the participant's motor output and used to calibrate and track the experiments. (C) Sample trajectories from the upper-cut routine registered from a representative naïve control and the ASD participant performed at different speeds (continuous line represents slow, discontinuous line fast). The speeds were instructed at random by a computer interface. Black trajectories are deliberate and colored trajectories are uninstructed retractions the participants spontaneously performed reportedly without awareness they were doing them. Note the differences in the retraction segments between the control and ASD. Whereas the shape of the forward upper-cut aimed at an imaginary opponent is similar between the control and ASD participant, the retractions are not. Notably the fast retractions (green) are highly curved, while the slow retractions (red) are straighter. Red circle marks the approximate location of the target (imaginary) opponent and blue circle marks the approximate location near the face where the hand retracted. (D) Corresponding speed profiles also indicate fundamental differences between the control and ASD participant.

participant's body and track their motions. Fig. 3.2 shows an example of the avatar we can build using this system to study the kinematics of unconstrained *full-body* complex movements in three dimensions and some of the sample hand movement trajectories and speed profiles from MJ and his sister.

The interesting thing about MJ was that throughout the session, as his sister practiced and performed the routines, he did not look at her once. He was busy playing video games. It took each participant more than one visit to master the full routine fluidly and "glue" together all the pieces in a smooth, continuous motion. Yet, when we asked MJ to get ready and were about to give him instructions, he surprised us all by performing the entire routine in one fluid, continuous motion from beginning to end. Somehow, MJ must have peripherally registered the entire session, rehearsed it in his mind and output it at once without having to practice it at all. He must have had photographic memory of the room, where to stand, how to face the martial artist, and how to do everything we thought we had to explain to him. He was the one who explained it to us through the routines gestures, without a single word. We were in awe.

As someone who had been studying motor control for over a decade then, I noticed however, that something was very different with MJ movements in relation to all other participants, athletes, and naïve subjects included. His repeated motions seemed identical from trial-to-trial (with the precision and high consistency of the movements you would program a robot to do). They were very identical in form, with a kind of *odd perfection*. I could not perceive from trial-to-trial, the type of variability I typically perceived in others, as they were learning the routine. But then I thought, whereas MJ did it all at once in one day, all others required practice, so the performance was bound to be different. I just wondered how different and in what sense (beyond the obvious observations I was making from his overt movements).

Movement variability can be very informative of the nervous systems performance, but because the body has so many degrees of freedom (DOF) and there are many layers of control systems, we need to be careful how we assess variability, so we do not incur in gross data loss. Along these lines, I always try to identify from the barrage of information we gather with sensors, the layers with high likelihood of providing information regarding physiological issues. These often tend to hide beneath awareness. I decided to examine these data very closely to try and understand what was so different that I could not exactly put my finger on but could definitely perceive.

To my dismay, when I plotted MJ's distribution of trial-to-trial peak velocities from his hand trajectories, they were not well-fit by the log-normal distribution (Fig. 3.3A) as the others were (Fig. 3.3B). The shape of the frequency histogram was much skewed with a long tail and was best-fit by the exponential distribution (Fig. 3.3A). I was very intrigued by this because I had not come across such distribution in typical human data. I had seen high skewness in data from patients with neurodegenerative disorders like Parkinson's disease (PD) in advanced stages, where there is an abundance of small speed peaks from the tremor. Somehow, the small peaks in the PD patients I had studied accumulated differently than the small peaks in this case with ASD.

As I came across this very intriguing result, I was invited to give a talk at Indiana University, where I took the occasion to meet with Professor Jorge V José, a theoretical physicist that I had met at the COSYNE meeting back in 2006. As it turned out, during my time at CALTECH, I had been auditing various classes in control dynamical systems. One of the textbooks recommended was "Classical dynamics: a contemporary approach" by José and Saletan[1] I had studied several chapters in depth for that class and applied various concepts to my own research. Nonetheless, during our conversation at the COSYNE

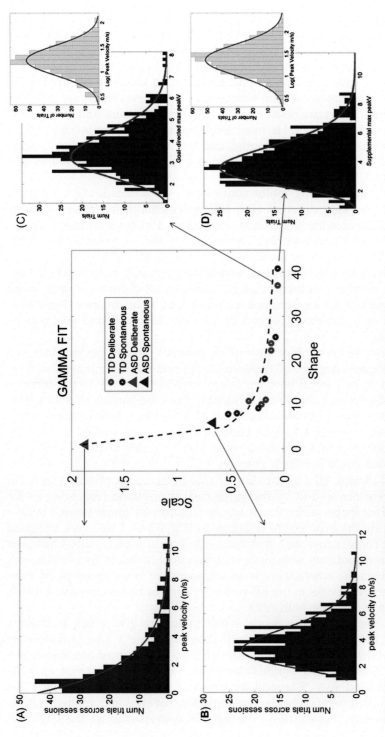

FIGURE 3.3 **Beyond the log-normal distribution.** (A) Exponential fit to the frequency histogram speed maxima in the deliberate segment of the JAB registered in the forward movements aimed at an imaginary opponent by the participant with ASD. (B) Gamma fit to the frequency histogram of hand speed maxima corresponding to the retracting JAB segments of the ASD participant. (C–D) Gamma fits to the frequency histograms of hand speed maxima corresponding to the forward and retracting JAB segments, respectively, of the control participant. Insets are the log-transform of the hand speed data generally well fit by a log-normal distribution according to MLE. Central panel is the Gamma parameter plane localizing each individual person according to the deliberate and spontaneous segments of their JAB routine. Note the striking difference between the ASD signatures for deliberate segments and the deliberate signatures of the controls. Contrast this to the signatures of the spontaneous segments of ASD closer to those of controls. TD deliberate; TD spontaneous; ASD deliberate; ASD spontaneous.

meeting the focus was on spike data analyses. We were comparing notes on spike data from area V4 and attentional models Jorge had worked on vs spike data from the Posterior Parietal Cortex and models I had developed. Because of that, I did not make the connection between the book and the person I was talking with. It was a good while before I realized he was the author of that popular textbook. By then, we had initiated a path of collaborative work that still continues today and has led to a number of very interesting discoveries.

While at Bloomington, I brought along some results on spike data analyses to discuss with Jorge, but also brought the recent result of the exponential distribution, so we could perhaps find some time to discuss it. When I mentioned the latter, Jorge immediately suggested that I use the continuous family of Gamma probability distributions because the exponential was a special case of the Gamma family. Up to that point, I was using the log-normal distribution to characterize speed maxima in primate motions, because the frequency histograms of the peak speed values from the hand trajectories across primates were skewed. When performing the 95%-range check including the mean + /- two standard deviations, the range of the assumed errors reached negative values; but the actual data range did not (Fig. 3.4). These values derived from the kinematics are positive with a minimum at 0—value (e.g., when the hand is at rest and the speed is 0, or when the hand reaches the target and the distance to the target is 0, etc.). They were positive values with 0 at the lowest limit. As such, the negative range reached by the negative two-times standard deviation from the assumed theoretical mean made the normal distribution ill-suited for this type of data.

I had realized this feature of the kinematic data at some point while analyzing the unconstrained, three-dimensional movements harnessed from nonhuman primates. Since then, I had been careful not to blindly use parametric models and assumptions of normality when analyzing such data. Yet, the log-normal distribution that I was using to help me interpret the data from the typical fellows[2] did not include the exponential case. Further, the log-normal distribution models a class of multiplicative random process that contrasts to other distributions used to model additive random processes (Fig. 3.4)[3] (though there could be some exceptions[4]). *How could I start empirically examining these differences in a more systematic way?*

I mentioned to Jorge in our conversation the case of a patient with a stroke in the left posterior parietal cortex that Professor Howard Poizner had brought to me when I was still at CALTECH. The distributions of speed maxima in this patient were not unimodal.[5] Further, they could not be well-fit by a mixture of Gaussians, so I had tried a mixture of Gamma probability distribution functions (PDFs) to fit the multiple bumps. Even in this stroke patient, I did not find evidence indicative of an exponential fit for his distributions across multiple movement kinematic parameters.

Something about these data from the ASD participant was very different. I tried the continuous Gamma family on the data from ASD and controls to have them all under the same statistical umbrella (Fig. 3.3) but decided to hunt for more papers and information on the differences between these random processes as they related to human data. During our exchange, Jorge shared a few papers with me on the topic of distributions that are present in natural processes, including one that his brother had shared with him on the log-normal distribution and its ubiquitous presence in nature.[6] Reading through the

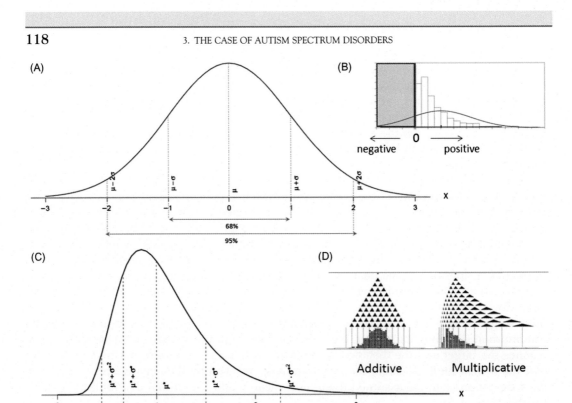

FIGURE 3.4   **The 95% range-check to identify proper distributions to represent the data.** (A) Gaussian distribution, with areas under the curve marked by the *arithmetic mean* $\mu \pm \sigma$ (68%) and $\mu \pm 2\sigma$ (95%). (B) When $\mu \pm 2\sigma$ (95%) range-check is performed on primate kinematic data such as the distance or speed with positive values, the negative range is also covered by the Gaussian distribution. In such case a skewed distribution (e.g., C) is more appropriate. The *geometric mean* μ^* is also shown for the 68% and the 95% areas more appropriate for, e.g., the positive kinematic data with values ranging from 0 onwards, limited by the physical body constraints. (D) Schematic configurations of the Galton board representing examples of additive (e.g., symmetric Gaussian) and multiplicative (e.g., skewed log-normal) processes to portray different types of scientific data.

papers and thinking about next steps made my trip back to New Jersey very short. Next thing I knew, I was landing in Newark and my head was exploding with new ideas and questions to pursue.

ONE PLAIN RANDOM PROCESS IN NATURE VS MANY RUNNING IN TANDEM WITHIN INTERCONNECTED SYSTEMS POISED TO DEVELOP INTELLIGENCE

Most notably, what stood out from my notes of the trip back home was a question I asked myself; the statistical patterns of processes in nature that these papers talked about (e.g., predicting the number of molecules in a concentration gradient, predicting the weather, or forecasting the amount of rain in one region of the planet, etc.) could very well be fundamentally different from those of a *living nervous system*.

From the conception of life, a nervous system is endowed with self-supervision from, e.g., neuroimmune processes that keep the fetus from being rejected by the mother's immune system. And when the various parts of the nascent nervous systems of the fetus survive and go on to develop somatic-motor autonomy, the one thing they need to have so they can *adapt* and continue to learn and develop autonomous motor control, is self-supervision and self-corrective abilities. As such, it did matter what type of information the sensory-motor systems had access to when confronted with their own self-generated variability vs when accessing extraneous variability from external sources of sensory input. If the random process characterizing the variability of the motions the system self-generated was exponential, that implied that the flow of sensory information from kinesthetic reafference was such that prior events did not contribute to the prediction of future events any more than present events did. In other words, the sensory input from such a source of variability did as good as the "here and now"—Because of the memoryless property of the exponential distribution. The type of historicity random processes need in order to make inferences about possible consequences of the future actions and decisions was not going to be found in such re-entrant sensory flow resembling a memoryless random process.

As I pondered these ideas, I came across an extremely interesting finding. The exponential distribution I had just found in the participant with ASD was also present in the frequency histograms of speed maxima from the goal-directed movements of Ian Waterman (Fig. 3.5). Ian is the special case that lost his proprioception from the neck down at the age of 19 years old.[7] Yet, Ian had managed to self-discover how to compensate for this "here and now" statistics in his motion patterns and consequently close the sensory-motor feedback loops using vision and motor imagery. He had mastered new representational maps of his motions and their consequences so he could move in a controlled way again. The fitted exponential for IW's hand speed maxima is shown in Fig. 3.5A for the forward pointing to targets and in Fig. 3.5B for retraction motions away from the target. I fitted other distributions as well to compare the adequacy of the exponential fit vs the log-normal and Gamma fits. As in the ASD participant, the log-normal was inadequate, so an exponential family worked better.

Unlike typical participants with physical kinesthetic reafference who reportedly performed the retraction beneath awareness; for IW, this retracting movement is also very deliberate. I wondered what Ian's brain looked like when engaging in the type of motor imagery replacing kinesthetic reafference, but Ian lives in the United Kingdom. The kinematics data Poizner gave me to analyze had been collected in the late 1990s during a tour Ian gave throughout the United States, part of which had been featured in a BBC documentary (**Notes**[1]).

A strike of luck brought Ian Waterman to my lab. I had remained in touch with Dr. Jonathan Cole, Ian's discoverer and author of the wonderful book "Pride in a daily

[1]BBC documentary about Ian Waterman's case.http://www.dailymotion.com/video/x12647t_the-man-who-lost-his-body-bbc-documentary_tech

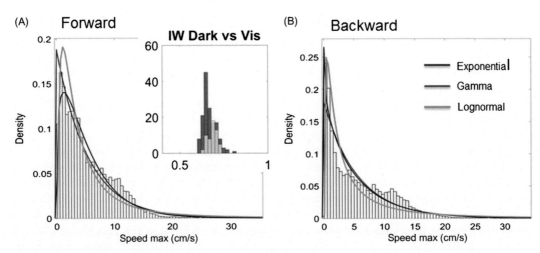

FIGURE 3.5 **Kinematic data in the form of speed maxima from physically deafferented subject Ian Waterman is well-fit by the memoryless exponential distribution.** (A) Frequency histogram representing the distribution of speed maxima for the forward goal-directed segments aimed at a visual target. Inset shows the normalized parameters ranging between 0 and 1 gathered into a frequency histogram for movements in the dark and with visual feedback of the target. Different distribution fits using MLE and fitting the exponential, Gamma and log-normal distributions. (B) Same as in (A) for the backwards movements when the hand spontaneously retracts to rest. Red, exponential; Blue, Gamma; Green, Log-normal.

marathon" that I devoured when Poizner gave me Ian's data and I analyzed it. Since then, my fascination with kinesthetic reafference and sensory substitution and augmentation techniques has not stopped. As it turned out, it was Dr. Oliver Sacks 80th anniversary. To celebrate his birthday, his colleagues were having a festival and bringing all the special cases that over the years Sacks had written about in many books. Ian Waterman, "The man without proprioception", was one of those famous cases. When Cole told me they would be in New York City I was thrilled. Imagine meeting a person whose data you have modeled, analyzed, and used to theorize about motor control for years!

I organized a talk by Cole and took the opportunity to have Ian visit the lab to perform an experiment. He loves science and helping advance knowledge in every way possible, so he was very happy to help us understand what goes on in his brain, when he does motor imagery.

We ran a brain—computer interface (BCI) task where the participant had to imagine moving the cursor to the left or to the right, as prompted by an interface in Fig. 3.6A—B. Upon closed loop adaptation, and tracking and adapting the responses of the person's brain and those of the computer, the participants were capable to move the cursor (the blue bar in Fig. 3.6A) on their own, by mere thought. We published this work in typical controls[8], but expanded it to test Ian Waterman on it.

In the case of Ian, who engages in motor imagery and deliberately uses visual reafference in all his movements[7], this experiment engaged his entire surface cortical map when

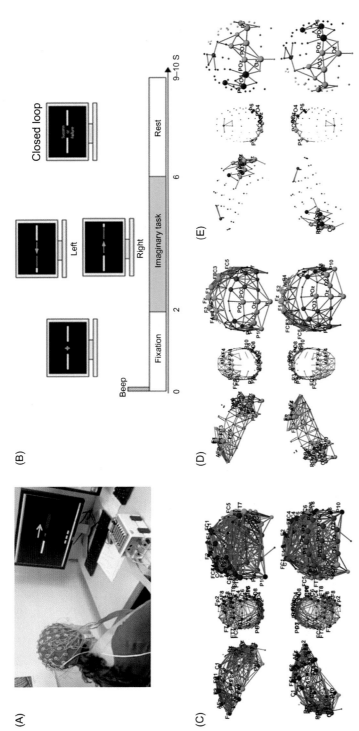

FIGURE 3.6 Surface cortical activity during motor imagery and target visual feedback differs between deafferented subject Ian Waterman and controls. (A) Experimental task within the context of a closed-loop BCI. The interface displays a target direction (green arrow) for the subject to follow. The task is to move the blue colored bar to the white colored bar according to the instructed target direction by mere thought. Participants train first in open loop mode, where the performance of the participant at trial t is not considered at trial t + 1. Then, they train in closed loop mode, where an algorithm considers the performance from prior trials to adjust the learning for subsequent trials. Eventually the participant's surface cortical activity is such that by mere thought the desired direction of motion is attained and the blue bar "moves" to cover the white bar. (B) Epochs of each trial last 10 s. There is a beep, followed by 2 s of fixation, and then the desired target direction is presented. The participants are instructed to use motor imagery as they practice and attempt to move the arrow in the instructed direction (4 s). The brain activity and prior outcome are used to update the current output balancing two modes: Machine (dominated by an algorithm) and human (dominated by the brain activity). In the last few minutes, the participant receives feedback on the performance and rests. (C) Ian Waterman recruits the entire cortex during the imaginary task in closed loop. This is in contrast to these 2 (out of 20) representative controls (D) and (E). All controls, as did these representative participants displayed sparser activation and fewer network modules.

he thought of the movement direction and used motor imagery to learn the BCI task (Fig. 3.6C). In marked contrast, neurotypical controls activated only a few regions of the parietal−occipital pre-frontal networks (Fig. 3.6D−E).[8,9] According to phase locking value metrics evaluating the pairwise synchronicity of the fluctuations in brain-waveform amplitude across 64 EEG leads, some controls showed coupled engagement between the front and back of their brain (Fig. 3.6D); while others showed engagement of these areas in a disconnected way (Fig. 3.6E). Yet, none of the 20 folks we tested showed the dense connectivity patterns that Ian's brain had.

The neocortical regions of IW's brain must have evolved to utilize resources we typically do not use at that level. Most of our actions transpire quite automatically and beneath awareness. They are likely controlled by subcortical structures inaccessible with the EEG leads. As such, surface electrodes may not pick up those fluctuations and the code is sparser in controls than in Ian's. Somehow, the nervous system of IW learned to use exoafference from vision or auditory inputs to replace the missing *physical endoafference* from movement, touch, and pressure. However, he did develop *mental endoafference* from motor imagery. Ian reportedly uses motor imagery at all times. Somehow, his brain operates *physically disembodied* from the continuous physical afferent input the bodily movements provide, but it does so *virtually embodied* by motor imagery and intentional thoughts externally sustained. It was remarkable to see that the patterns of activity in his cortex are not just constrained to parietal−frontal networks but rather engaging the entire cortical surface that we measured.

Along these lines, the principle of reafference provided me with a way to think about distinguishing endo- from exoafference in the flow of variability from sensory-motor activities. On the other hand, my own thesis work to address Bernstein's DOF problem (**see** Chapter 4) provided me with the conceptualization of the high dimensionality of the body as a means to inevitably have a system in motion understand and route inherently different types of variability that self-emerged from the motion dynamics. In physically afferented people, the contributions from endogeneous and exogeneous sensory inputs to dynamically and prospectively reshape the efferent motor variability must differ from those with deafferentation. I started to develop a notion of *physical deafferentation* (as in the case of Ian Waterman) vs one of *virtual deafferentation*, as in patients with advanced PD and the participant with ASD here. I called virtual deafferentation the persistently corrupted motor output that would re-enter the system as noisy and random sensory-motor input, thus providing faulty updating sensory feedback and likely obfuscating the development of a predictive code. In such cases, my bet was that exogenous input (like visual and auditory inputs) would help them restore prospective motor function towards typical levels.

In Ian, exogenous sources seemed to play a heavy role in prospectively updating the efferent motor output variability. How the external sensory guidance from vision was provided made a difference in the stochastic signatures of Ian's speed maxima (Fig. 3.7A−D). It changed the frequency histograms of the moment by moment fluctuations in speed peak amplitude. And that difference was selectively shifted between forward and backwards motions, thus implying subtle differential influences on the type of control strategies that Ian's brain may be using. More specifically, when the target was continuously ON and he pointed to it in the dark, his distribution for the forward reach was skewed and had many

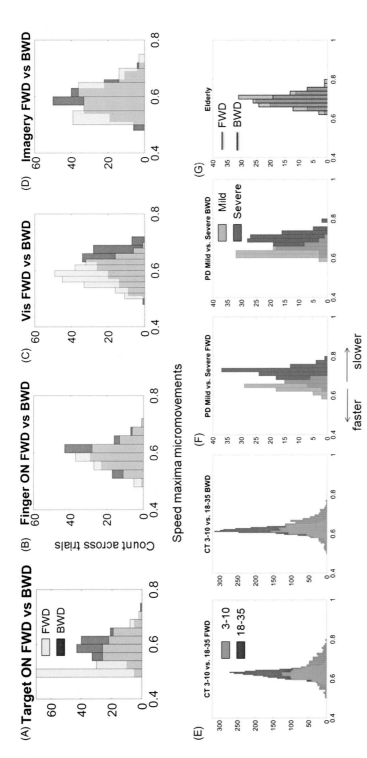

FIGURE 3.7 **Different sources of sensory guidance affect the stochastic signatures of the moment-by-moment fluctuations in speed maxima during continuous pointing behavior.** (A–D) Performance of deafferented subject Ian Waterman during (A) continuous visual feedback of the target while pointing in the dark without visual feedback from his moving limb. Forward reaches, goal-directed segments (yellow) result in a skewed frequency histogram of speed maxima micromovements (using Eq. (3.3) in Table 3.1, whereby lower values indicate faster speeds on average). Backward segments from spontaneous (uninstructed) retractions have comparable signatures to those of controls. (B) Visual feedback of his pointing finger moving in the dark without target vision shows beneficial effects on both the forward and backwards segments. (C) Complete visual feedback of the target and moving finger in full light provides the best performance (closest to controls) and it is the usual way in which IW operates. (D) Reliance on motor imagery in the absence of visual guidance produces skewed frequency histograms for the forward reaches towards the target, whereas the backward segments have characteristics closer to the controls. (E) The performance of control participants spanning ages from 3 to 35 years old for the forward and backwards segments. (F) The performance of PD patients with different degrees of severity also for the forward and backward segments. (G) The performance of elderly participants in forward and backward segments.

small peaks from small corrective motions (Fig. 3.7A). When he was provided with visual feedback of his moving hand and could rely on a memory of the target, his motions in the dark along both the forward and backwards directions were close in shape and dispersion to the patterns registered in typical age-matched controls, only slower on average than those in the neurotypical controls (Fig. 3.7B). If he was to move in full light with simultaneous feedback from the target location and the moving hand, he also had patterns that resembled those of neurotypical controls in the forward segment but markedly different patterns in the retraction (Fig. 3.7C). Lastly, in a different condition where he could only rely on motor imagery and the memory of the target, his forward movements were very different from those of neurotypical controls, but the backwards ones were closer in signature to normative data (Fig. 3.7D). The panels in Fig. 3.7E show the patterns of controls across different ages as they point in full light with simultaneous visual feedback of the target and the moving hand. They also show the differential patterns of PD patients of different degree of severity.

Given the subtle but important differences that the exact same biomechanical movement manifested on the hand trajectories to the target, and the effects that systematic manipulations of the external source of sensory guidance had in IW; I then wondered how visual guidance could help the motor control of those patients with *virtual deafferentation*. In other words, IW had no physical afferent feedback from the continuous flow of motion. The viral attack did spare some pain and temperature afferent networks, so he could feel some of that. Yet that source of sensory input was not enough to help him centrally control the body in motion. In the absence of the type of continuous feedback from kinesthetic reafference, how was he closing the motor feedback loops with vision? Patients with the type of physical afferent feedback that was persistently corrupted by noise and randomness may find external sensory guidance useful to help overcompensate for the loss of bodily sensation and boost their motor control. Along these lines, I thought that external inputs such as visual, auditory, and tactile feedback could be individually parameterized according to the person's unique motor output signatures. In turn, such channels could serve as noise cancellation networks when the appropriate frequencies to shift the motor output signatures were uncovered. In due time, this turned out to be a feasible hypothesis to test, since we eventually were able to measure the motor output systematically and in near real time using noninvasive means.

I had initiated the study of such questions in the PD population and learned that as in the case of Ian Waterman, they too heavily relied on continuous visual feedback from their moving limb[10] (See details in Chapter 4). Somehow aligning the continuous vision of the target with vision of the physically moving hand helped the PD patients more than merely providing the exogenous visual input of the target (as in Fig. 3.7A for IW), or the visual feedback of the physically moving hand alone (as in Fig. 3.7B for IW). They were not physically deafferented, as IW, yet they too had high noise-to-signal ratios and higher randomness in their patterns of fluctuations in speed maxima. This was also the case for MJ, the young ASD individual. As with IW, MJ reliance on visual feedback resulted in more accurate motions than moving in the dark. In all cases. I had found that their stochastic patterns of variability shifted within the session as a function of changes in extraneous sensory inputs.[5,10,11] *How could I extend these results to probe the ASD patterns in a more systematic way?*

Using Ian Waterman's data and the data from MJ, I had identified a critical need to better characterize and track the statistical shifts in this self-generated and self-corrected variability. I needed a new statistical framework that allowed me to do this tracking continuously and visualize it in near real time. In order to learn if the nervous system as a whole was capable of developing inferential mechanisms about the consequences of its self-generated motions, I could not adopt the methods in the extant literature. Those methods were static and assumed stationarity, linearity, and normality in the data. The *one size fits all* approach in the literature could at best only tell me if someone in a very homogeneous handpicked group was different above chance from someone else in another very homogeneous group that was purposely handpicked to be different. Being the human population so heterogeneous, I needed a personalized assessment that also allowed me to track *change*.

I was not aware of self-corrective processes in nature using the output as re-entrant feedback input to predict future consequences and correct future outputs. Perhaps taking a different vantage point and zooming way out there in the universe could reveal such closed feedback loop systems with the potential to make intelligent decisions, yet locally, here on Earth, it was hard to think of the statistics required to foretell rainfall along the same lines as those of the statistics required to quantify and infer body self-awareness and mental introspection.

My new goal was then to identify different classes of random processes appropriate to capture closed loop interactions within the central nervous systems and the peripheral nervous systems, such that I could design effective interventions to steer the nervous systems' performance with minimal interference *in real time*. This required my characterization of such processes across different levels of autonomy and awareness because such different levels co-exist across the nervous systems. I needed to characterize such statistics for voluntary (deliberate) and for voluntary (automatic) processes, but also for involuntary activity of the kind that spontaneously happen (as when we sleep) and for the kind that spontaneously happens in spite of being instructed not to move. The latter could be a good marker to measure levels of volitional control at a microlevel of movements. But most important, I was highly intrigued by the spontaneous segments of complex movements, those which happen largely beneath awareness. These were the ones I could use in therapies without interfering too much with the active and reactive nature of the nervous systems in motion. Yet, these processes occur at different time scales and frequencies. *How could I build a unifying framework that could handle variability at all such disparate levels of the nervous systems?*

I reasoned that if I could tap into those spontaneous aspects of naturally self-generated motions, I could use them as a *back door* to access and change the systems' autonomy beneath the systems' full awareness (i.e., without natural resistance emerging from the system's reaction to the intervention). If I achieved that, I could reduce the amount of stress I witnessed those children with ASD undergo during the behavioral reshaping therapies that prevailed in the clinical field as a form of treatment.

One aspect of the problem that worried me was the sort of inevitable autonomic processes, namely those that are set from birth and impervious to volition. These included the heartbeat, respiration, and digestive patterns. I imagined that for reasons of survival, they may have a very narrow room for change. They are what they are and better remain steady within narrow bandwidth of variability range if the organism is to survive. How the changes in deliberate and spontaneous processes may influence such inevitable processes? Or the other way around, *how constraining the signatures of inevitable processes would be, when trying to reshape deliberate and spontaneous processes through noise-cancellation techniques?*

AN AHA! MOMENT

As soon as I got back from Bloomington and settled in the lab, I brought out the kinematics data spanning several years of empirical analyses and reanalyzed it through the new lens of an exponential family like the Gamma. Besides the data from neurotypical subjects, my sets included patients with PD (e.g., Fig. 3.7E) and the above-mentioned deafferented subject, Ian Waterman. Because of the similarities of some of the motor signatures between IW's and all ASD participants that I later profiled, Ian's case ended up playing a fundamental role in our research program, particularly the program devoted to the design and testing of sensory-motor substitution and augmentation and noise-cancellation techniques. I will refer in Chapter 6, to the enormous contribution his data made to our understanding of autism (in my lab). His self-discovered solution to deafferentation and centralized motor control helped us develop possible ways to habilitate volition in a subset of the affected population.

I normalized all the data, since there were different anatomical sizes across different age groups. To that end, I tried several log transforms on the data. I also tried several normalization formulae [e.g., Eqs. (3.1)−(3.3) below] and plotted the data in histogram form (Figs. 3.7−3.8) to visualize it in various ways. I settled for a scale from 0 to 1 and eventually adopted the normalization using Eq. (3.3), which Jorge's student Di Wu suggested from morphology-related research dealing with the issues of disparity in human anatomy and their potential allometric effects.[12] With regards to the speed, the differences in limb size were important to consider when making a population statement as they influence the amount of distance covered per unit time within the midline region of peripersonal space where these experiments took place.

In the case of the stroke patient that I had mentioned, I obtained bimodal distributions[5] and fitted a mixture of Gamma distributions to his data. That worked out well, so based on the empirical evidence, I decided to try and fit the single Gamma PDF to the unimodal data of each of these subjects, a subset of which I plot in Fig. 3.8A−D.

When I spread out the data plots on a table in front of me so as to zoom out from the phenomena, I was able to acquire a bird's view of all of it at once. This was very informative. I immediately noticed that each subject had a slightly different PDF with a slightly different shape and dispersion, so I wondered how I could build a map to visualize or capture this notion a bit better and in a more compressed form. I contacted Jorge again, who by then had acquired an interest in the problem. I shared my results with him and requested his advice on the statistical patterns I was seeing. I revisited the issue of the exponential distribution showing up again in Ian's data (Fig. 3.5), but put that on hold for

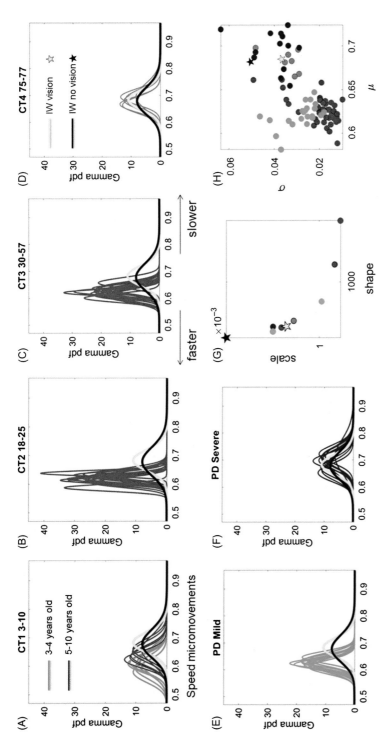

FIGURE 3.8 **Empirically estimated age-dependent evolution of the probability density function best characterizing the fluctuations in amplitude of the hand speed maxima during pointing movements to visual targets.** A Gamma process under the assumption of independent identically distributed (IID) random variables is used to model spike trains of fluctuations in amplitude of the normalized speed maxima waveform (using normalization Eq. (3.3) in Table 3.1). (A) Young children spanning ages from 3 to 10 years old produce skewed distributions with wide dispersion and a significant shift when transitioning from 3–4 years old (green) to older ages above 5 years old (red). Empirically estimated signatures of deafferented subject Ian Waterman are shown for the cases of visual feedback (yellow PDF) and during no visual feedback pointing in the dark under reliance on motor imagery (black PDF). (B) Empirically estimated PDFs as in (A) for 18–25-year-old participants. (C) Empirically estimated PDFs as in (A) for 30–57-year-old participants. (D) Empirically estimated PDFs as in (A) for elderly 75–77-year-old participants. (E) Empirically estimated PDFs as in (A) for patients with PD at their mild stage still with independent mobility. (F) Different PD patients at a severe stage with no independent mobility. (G) Gamma plane map of empirically estimated shape and scale (dispersion) parameters from A-F obtained by pooling across each group. Each dot represents the overall signature of each group, plotted in relation to the empirically estimated signatures of the deafferented participant IW. (H) Empirically estimated Gamma moments (mean and variance) for each of the participants in A-F. Each dot corresponds to a subject's signatures. Yellow line, IW vision; black line, IW no vision.

the time-being (since the story I had in mind about rainfall, self-emerging intelligence, and the deliberate control of autonomy was complicated to articulate over a short phone call). So, I asked if he knew of a more compressed way to visualize these data. Since I was seeing these subtle differences across the typical subjects and markedly different shapes in the pathological cases, I was hoping to build a map between variability and pathology, much as the taxonomy I had designed to connect control levels and variability types across movement classes.[2]

Jorge suggested using *the Gamma-parameter plane* to plot the estimated Gamma parameters, the shape and the scale, with their respective confidence intervals (CIs). In this way, I could visualize both the subtle differences across controls but also emphasize the more pronounced differences across patients. He mentioned that in particle physics, they use this and other parameter spaces to represent data involving phase transitions (e.g., from liquid to gas), uncover the possible presence of power laws and more generally help interpret phenomena through rigorous statistical inference.

It was an incredibly useful piece of advice that I implemented right away. Fig. 3.8G shows a summary graph of a subset of the data for which the individual PDFs are shown in Figs. 3.8A–F. In Fig. 3.8H, the points represent the estimated Gamma moments of these participants. This type of visualization tool provided me with a way to map human motor performance using a common standardized way so as to build the types of maps I was searching for.

Using all of these data accumulated over a decade, I was able to gain a number of new insights about the variability inherently present in the trial-by-trial fluctuations in speed amplitude and timing. This opened a number of new possibilities to analyze kinematic data in general (Fig. 3.9). It was an "AHA!" moment that came entirely from empirical data, rather than from a theoretical formulation driven by an equation or a model.

Empirically Driven Ideas Inspired on Biorhythms' Statistics

First, I could immediately see subtle differences across age groups during neurodevelopment that I could not possibly see under the "one size fits all" parametric model that prevailed across the field (Fig. 3.9A).[13] The current statistical paradigm generally assumes an ideal theoretical distribution (e.g., the normal) across the population and imposes averaging that smooths out as noise important fluctuations in signals from the nervous systems. These methods incur in gross behavioral data loss by discarding spontaneous aspects of behavior as a nuisance, or simply by not acknowledging their contribution to voluntary control.

The stochastic profiling under the new individualized umbrella immediately showed me that each participant was indeed different. In particular, the cross-sectional data from neurotypical participants of different ages offered a different notion for the evolving course of neuromotor development that stood apart from that of participants with disorders of the nervous systems. These differences were manifested across different shapes

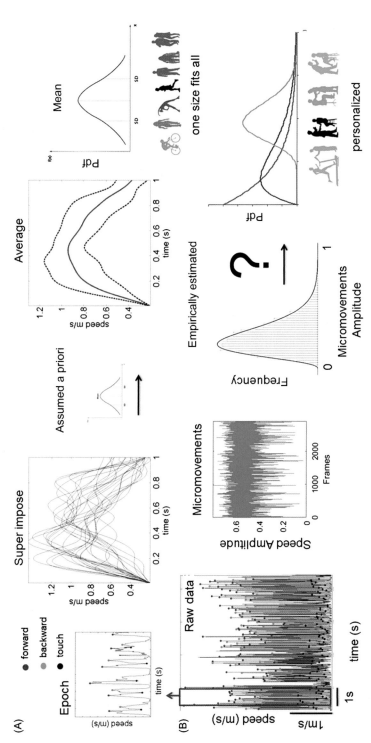

FIGURE 3.9 **Comparison of the current statistical model following a "one size fits all" approach and the new proposed statistical platform for the individualized behavioral analysis (SPIBA).** (A) Traditional approaches build epochs of kinematic data (e.g., here speed profiles from pointing behavior) obtained for discrete goal-directed segments without considering the continuity of the behavior into spontaneous retractions of the hand. These discrete data are then averaged under the a priori assumed theoretical Gaussian distribution, thus incurring in data loss of the minute fluctuations in amplitude and timing the original raw data has. This model produces a one size fits all treatment across the population, thus disregarding the heterogeneity and the individuality of the patients. This is the state-of-the-art analysis as of 2017. (B) SPIBA (proposed in 2011) considers the continuous flow of motion across different biorhythms and processes taking place across the various nervous systems' layers. Here, the same data as in (A) is not taken discretely with an exclusive focus on the deliberate processes of goal-directed pointing. Instead, the data is continuously acquired and tracked to automatically separate the deliberate and spontaneous segments of the behavior. A new waveform capturing (in this case) the moment by moment fluctuations in amplitude is then built for each process type (for clarity shown here only for deliberate segments) and a frequency histogram used to fit the best family of probability distributions describing the random point process under consideration. MLE and Kolmogorov–Smirnov test for empirically estimated CDF's yield, the continuous Gamma family as a good candidate to characterize these micromovements in all human population across the human lifespan. As such, they are then estimated individually for each person within a new personalized framework suitable for Precision Medicine.

TABLE 3.1   Different types of normalization

$V_{i,[0,1]} = \frac{V_i - V_{min}}{V_{max} - V_{min}}$	1
$V_{i,[-1,1]} = \frac{V_i - \left(\frac{V_{max} + V_{min}}{2}\right)}{\left(\frac{V_{max} - V_{min}}{2}\right)}$	2
$V_{i,[0,1]} = \frac{V_{max}^i}{V_{max}^i + V_{avrg}^{[i-1,i+1]}}$	3

and dispersions of the empirically estimated PDFs shown in Fig. 3.8A−F. They were also manifested across the different locations corresponding to each group on the Gamma parameter plane (Fig. 3.8G) and different types of variability (Fig. 3.8H). Along those lines, the PDFs from young children in Fig. 3.8A had the highest dispersion and were closest in shape to the normalized profiles of Ian Waterman's signatures under full continuous visual feedback (yellow trace). The children's fluctuations changed much faster than IW's, a feature appreciated in Fig. 3.8H, where the estimated mean values of the children's fluctuations in the speed amplitude are lower. (Recall here that lower in this case indicates faster local speed on average, as the term in the denominator of the ratio in Table 3.1 Case 3, has higher magnitude than the local peak, thus making the magnitude of the ratio smaller). Further, children had higher variance than college-age controls. In a sense, this was *good variability* of a system poised to take off and change in a transformative route to develop and mature. In contrast, the college age folks had very low dispersion and Gaussian like shapes. The dispersion increased after 35 years of age. Further, the signatures from old age participants, those over 75 years of age, regressed to the high dispersion manifested in the young children, albeit much slower rates. The differences concerning the slower speeds in the elderly were consistent with those of bradykinetic PD patients. The cross-sectional trajectory of typical aging can be summarized in Fig. 3.8G, where we plot the shapes and scale (dispersion) parameters form Figs. 3.8A−3.8F on the Gamma parameter plane. Note here and in Fig. 3.8A, the shifts in the distributions from the 3−4-year-old group to the 8−10-year-old group. For reference, we also plot the signatures of Ian Waterman. This plot shows that natural aging does change the PDFs. Specifically, the plots characterize the maturation process of voluntary control during early neurodevelopment and contrast to the decline of signal-to-noise ratios with old age.

I also examined patients with PD across a range of severities. In the case of severe PD participants, they too tended to have signatures located near IW. This was puzzling as elderly and severe PD patients stood apart from controls and mild PD patients. In advanced PD, we know of bradykinesia (seen here in the slowness of their mean speed micromovements), yet the fact that they concentrated near the signature of IW when he had visual feedback made me wonder (1) if these patients suffered from a type of *representational deafferentation*. In other words, they could not update maps of sensory consequences for impending action plans. In my conceptualization of such maps, these deficits would emerge from a brain persistently receiving corrupted kinesthetic reafference due to continuous high levels of noise and randomness in the peripheral signals; (2) I also wondered if (as in IW's case of deafferentation, visual feedback could help them. The answer to the latter is explained in detail in Chapter 4. Indeed, as with IW in Fig. 3.7C, we found that under continuous vision of the moving finger; these patients considerably restored their kinematic performance.[2,5]

I then pondered whether the use of external sensory guidance could make a difference in the motor control of the children with noisy motor output variability. The question of whether the estimated Gamma signatures would shift with the different types of visual guidance was answered in the data from IW, Fig. 3.10.[14] That data provided the proof of concept to build a co-adaptive visuo–motor interface for ASD to be discussed in Chapter 8.

Second, I could see there were appreciable differences in the rates of change of these cross-sectional data encompassing different neural stages across the lifespan. Whereas children sharply transitioned the shape and dispersion of the PDFs from 3 to 8 years of age and then dramatically changed the signatures by college age; the adults seemed to have plateaued after 60. With old age, they were changing at a slower pace and tending towards the patterns of neurodegeneration. This observation was important as it prompted other lines of reasoning that I later used in neonatal research, where the rates of change in physical growth and motor control are rather nonlinear and accelerated. Indeed, they are appreciable even on a time scale of days (see **Part II in** Chapter 1) that slows down by 5 years of age and then seems to peak between puberty and the early 1920s. The new plots showed me that relative to the 3–6-year-old children, the elderly participants were much slower to change. Furthermore, they tended to regress to high dispersion (in marked contrast to the college age) (Fig. 3.8A–D).

I thought these were important empirical revelations in the data that I would have never guessed using a theoretical model. In this sense, I learned that normative data was not a static concept—As it is assumed in the extant literature. It changed over time at *different rates* within *different epochs of the human lifespan.*

Back to the Drawing Board?

This empirical finding on the evolution of stochastic signatures with natural aging and pathology were very interesting in light of the fact that major case studies in neuroscience had shaped the course of entire lines of enquiry with theoretical propositions. Several such cases had been followed over the span of 15–20 years under a common assumption of an ideal Gaussian distribution without considering (1) the *skewness* and dispersion of the data; (2) the *shifts* in PDF shape and dispersion with aging; (3) *the variable rates of change in the shifts* in shape and dispersion with aging.

A case in point was patient DF with apperceptive agnosia. Her data launched one of the major brain-science hypotheses of perception–action. This is a hypothesis that divides the so-called vision for action and vision for perception pathways into the ventral and dorsal streams, respectively. The research along those lines has been very contentious and at times I dare say ridiculously irrational. Yet, the authors arguing one way or another have never put their statistical results through a lens that considers the shifts in PDFs and their variable rates of change as a function of aging. In fact, the famous patient, DF was tested

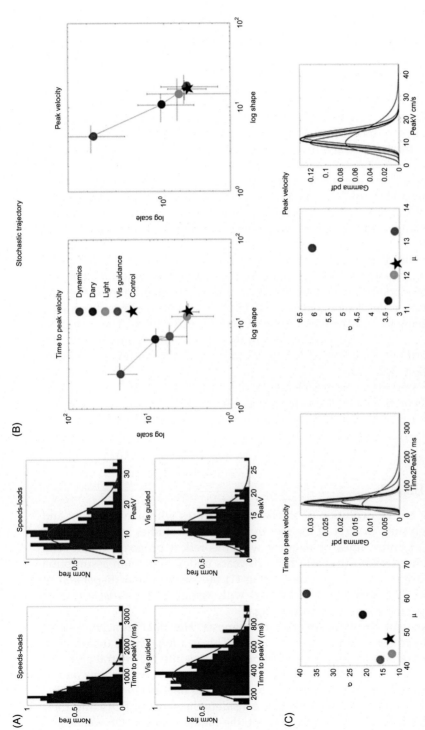

FIGURE 3.10 Shifts in stochastic signatures of speed amplitude and timing as a function of sensory-feedback manipulations quantified for deafferented subject Ian Waterman in relation to age-matched control. Manipulations are differently labeled in the legend: "dynamics" consists of changes in instructed speed and carrying loads on the moving arm; "dark" consists on pointing in the dark towards a remembered target while relying on motor imagery; "light" consists of a regular condition with lights ON in the room and relying on full feedback of the visual target and the moving hand; "visual guidance" consists of continuous visual feedback from the target position while pointing in the dark with no feedback from the moving limb; "control" is the activity for the age-match control participant pointing in the regular light condition. (A) Frequency histograms of the time to peak velocity and peak velocity obtained from speed profiles of hand pointing movements towards a visual target in peripersonal space. Top panels are under dynamic manipulations (a very challenging experiment for deafferented subjects in general, lacking the proprioception, and kinesthetic feedback). Bottom panel refers to pointing under visual guidance. (B) Stochastic trajectory on the Gamma parameter plane given by the shifts in the empirically estimated Gamma shape and scale parameters for the time to peak velocity and the peak velocity values under different sensory-feedback conditions (see legend). (C) Corresponding empirically estimated Gamma moments (mean and variance) and probability density functions (same color code as legend) for each condition. Note the shifts are within the time scale of the experimental session on the order of minutes. Red, dynamics; Blue, dark; Green, light; Pink, vis-guidance; Black star, control.

over the span of 15 years to provide evidence for this two-stream theory of visual perception.[15–19] Not once, the empirical PDFs from the kinematic parameters of the motions underlying her reaching and grasping behaviors were shown in any of the papers that are blindly cited as empirical evidence that these two pathways are impervious to each other's information.[15,20–23]

Upon my empirical discoveries that day, I doubted that entire line of research altogether. When I saw the empirically estimated PDFs from over a decade's worth of data in front of me, I really wished to one day be able to get my hands over entire data sets that across the years have been used to tell us *how the brain works*. The amount of arrogance I had witnessed and even personally experienced now somehow felt really unjustified. I recalled the time when my then student Jill Nguyen (now an analyst somewhere in NYC working on issues related to the stock market) presented her work at the Annual Meeting of the Visual Science Society (VSS) and suggested there may be cross-talk between the ventral and dorsal visual streams. Oh boy! She was threatened (and very loudly too) that she would never be able to publish her work, ever! Coming from biomedical engineering with high quantitative skills, Jill was shocked and could not understand what they were saying; but as it is common in my lab, we laughed it off and though "these poor fellows really need to get a clue and learn some manners while they're at it". She did publish her work[24–27] and even ended up giving a talk at VSS the following year on the very topic[28]—A highly nontrivial thing do be able to accomplish if you are entirely new to that crowd.

Somehow, after seeing the evolution of these empirically derived histograms across the human lifespan, I got the funny feeling that we are going to have to go back to the drawing board *a lot*.

Anchors of Known Etiology and Longitudinal Dynamic Diagnoses

Another vantage point I took was to examine normative data relative to disorders of the nervous systems with known etiology. These included for instance, the deafferented case of Ian Waterman whose signatures fell in markedly different areas of the Gamma parameter plane when he was using continuous visual feedback than when he was pointing in the dark (Fig. 3.7A–D) and Fig. 3.10. Besides learning about the distances from these patients to the normative data points on the Gamma plane (Fig. 3.8G), the locations of the patient points could provide an anchor to reference other patients with idiopathic nervous systems disorders. In my mind, this idea opened a new avenue to help characterize disorders that up to then had been described exclusively using clinical pencil-and-paper inventories. We could then recast these disorders in a very different way. It was a new avenue to potentially build taxonomy of neurological disorders, rather than exclusively refer to them as psychiatric disorders of some categorical type using, e.g., Diagnostics Statistical Manual criteria, which was mostly based on opinion and not driven by physiology.[29,30] At last, we could characterize central, somatic, and autonomic physiology across a multitude of nervous systems disorders and begin the path of objective classification based on physical data acquired noninvasively with biosensors.

Third, in plotting the longitudinal data that I had for a given patient, performing the same task across more than one session, I realized the Gamma signatures shifted on the Gamma parameter plane. This allowed me to think of these motor output variability phenomena as a dynamically changing rather than a static problem (Fig. 3.10).

It is worthwhile expanding a bit on this third point, as it changed my way to do behavioral analyses in general. This observation on the dynamic nature of the stochastic signatures extracted from the motor output data came partly from cross-sectional group data and partly from longitudinal data registered from a same person across multiple visits to the lab to perform the same task. These data features helped me more generally reconceptualize the notion of a static diagnosis into one of a *dynamic diagnosis*. In this sense, the new dynamic concept was more suitable to the inherent nature of early neurodevelopment, a process that—As we describe in **Part II of** Chapter 1—Is occurring at rather nonlinear, accelerated rates. They were also more amenable to characterize the coping nervous systems, rapidly evolving owing to the high plasticity of nascent systems.

I reasoned that the motor output I had been harnessing noninvasively from the kinematic metrics is read-out from the efferent motor flow. Yet that signal is also (at least partly) an assessment of the type of afferent kinesthetic sensory input that re-enters the system combined with other sources of sensory input. We could now begin to *visualize* external environmental influences on the reafferent motor code through the shifts quantified in stochastic signatures of ensuing self-generated movements. Recall here the principle of reafference[31] *"Voluntary movements show themselves to be dependent on the returning stream of afference which they themselves cause"* helping the nervous system distinguish sensory flow generated by internal sources from sensory flow generated by external sources. This is to say that the system causes motions under different inputs and I was harnessing (in a closed loop form) the subtle yet, quantifiable, effects of the movement itself on the subsequent information due to the various inputs. As such, different variations in the sources of sensory input could induce different (and *separable*) variations in the self-generated actions that I could systematically measure (in closed loop) once again at the output. I could then isolate which specific sources of sensory guidance maximally changed the signatures of the motor pattern with the highest certainty.

Given a patient within an age group, those sources of sensory input for which the rates of change in the motor stochastic patterns systematically resulted in PDFs within typically registered ranges could then give me a sense of how to use sensory guidance to help me steer the stochastic signatures of a patient of a certain age group towards the typical regimes. In this sense, the visualization tool I had in front of me could help me track development and disorder progression *dynamically*. This dynamic tracking could be done in relation to expected typical rates for a given age group or disorder stage (e.g., time length, since disorder onset). Fig. 3.10 illustrates this idea empirically using actual data from IW in relation to age-matched control.

The dynamic evolution of the stochastic signatures derived from the fluctuations in the amplitude and timing of speed maxima was a major revelation for me, given that such type of data is generally treated as stationary. However, given the age-dependent evolving signatures and the possibilities they offered as outcome metrics of nervous systems' performance, I thought that PDF shifts could potentially open a new era for intervention therapies in general. Not only had we designed a dynamic outcome measure, but more importantly, we had designed a measure that allowed near real-time parameterization of the output. This motor output could then be fed back as reparameterized input. We will see an example of this in Chapter 6. The unique feature of this new metric of closed loop interaction was that it was based on the physiological signatures of the nervous systems as they fluctuated and changed in real time. Those fluctuations that up to now had been discarded as noise by averaging (Fig. 3.9A) were now the signal of interest (Fig. 3.9B) to guide us in identifying adequate sources of sensory guidance for each person.

Part of my prior work with adult patients (see details in Chapter 4) had helped me identify the optimal source of sensory guidance to help restore control of their upper limb during voluntary movement. Yet, that was a type of a posteriori analyses. Once I collected the data, I could see which source of sensory guidance (external vs internal) was the best for a patient type. Here, I could do more than that. Given a context, I could see in near real time which source of sensory guidance *most likely* would shift the motor variability towards neurotypical regimes and systematically parameterize the motor output, and then reparameterize the sensory–kinesthetic motor input, i.e., the motor input integrated with other sensory modalities, to close the somatic-motor loops in a well-informed manner. Further, since part of the sensory input could be not just light or sounds or touch, etc. but rather more elaborate cognitive loads; we could also probe the *central, somatic, and autonomic* systems (Fig. 3.11A) in response to varying degrees of cognitive loads. This would give us a way to more formally examine the nervous systems maps between sensory consequences and inherent variability across levels of control.[2]

MIND THE GAP: BEHAVIOR IS MUCH MORE THAN MOUSE CLICKS

I immediately thought about the possibilities that this would open for the nascent field of embodied cognition and for the field of Psychology in general. Every single experiment in the fields of Cognitive Science and Psychology that explores the functioning of the mind requires *behavior*. In their paradigms, motor behaviors are reduced to mouse clicks, targeted joystick motions, or targeted motions excluding spontaneously co-occurring motions that transpire largely beneath awareness (e.g., involuntary micromovements, motion segments incidental to goal-directed actions, among others.) Behavior, thus conceived, has been very discrete, leaving out gaps of critical information directly generated by the nervous systems.

More generally, behavior really means *continuous motion*; moving with intent, moving automatically even with low to no awareness of it, moving spontaneously totally oblivious to what you are doing—As when there is no meaning or purpose to be conveyed, or the meaning is ambiguously embedded in the co-occurring goal directed segments of your "behavior". Besides overt movements, there are also the autonomic biorhythmical motions of the heartbeat, digestion, and respiration patterns embedded in all of those movements

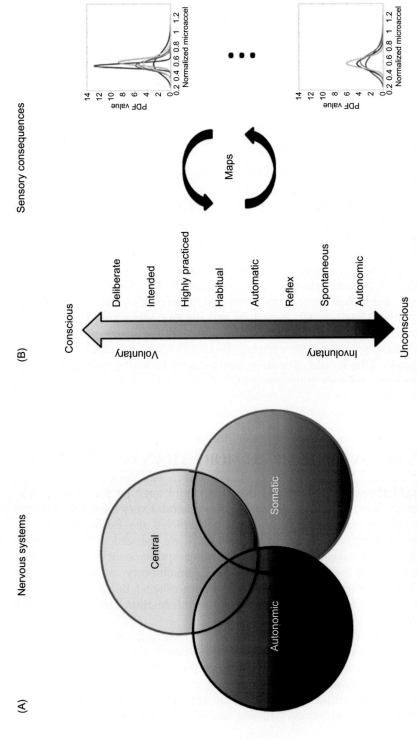

FIGURE 3.11 **Proposed taxonomy to address nervous systems' typical and atypical development as well as normal aging and aging with pathological conditions (caused by degeneration or injury).** (A) Three interacting nervous systems with different but interrelated functions give rise to different physical biorhythms. (B) Different control levels map to different levels of adaptive variability. Interactions among these levels give rise to different somatic, autonomic and central maps, the latter including as well predictive maps of sensory consequences for optimal execution of impending actions derived probabilistically from kinesthetic reafferent inputs across the multilayered systems.

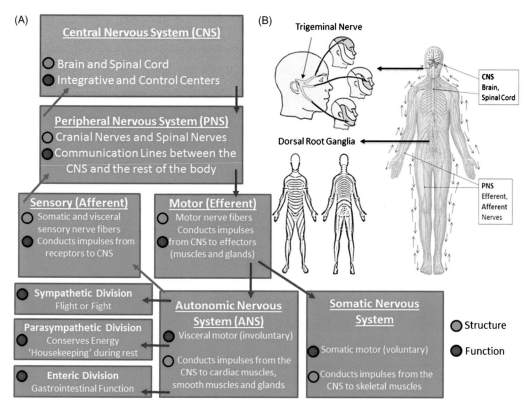

FIGURE 3.12   **Nervous systems anatomy and physiology.** (A) Structure and function of the autonomic, somatic, and central nervous systems. (B) Anatomical substrate with physical maps of facial (cranial) and bodily (spinal) nerves involved in efferent and afferent flow of information between the central and peripheral nervous systems are proposed to give rise to representational probabilistic maps of sensory consequences of impending actions. Green, structure; Red, Function.

(Fig. 3.12A) across the face and body (Fig. 3.12B). Indeed, the cranial and spinal nerves conducting efferent and afferent information across the nervous systems (Fig. 3.12B) play a fundamental role in scaffolding individual functions of the nervous systems (e.g., causally mapping its own intrinsic dynamics onto consequential sensory dynamics). Ultimately, the contribute to build social functions within the organism and the social collective this organism is part of.

Together, structure and function of the nervous systems give rise to a family of processes that I have coined "inevitable" to underscore that there is an autonomous portion of the motions they produce that lies entirely beyond our volition (Fig. 3.13). Indeed, I make a distinction between the terms *deliberate autonomy* and *inevitable autonomy* when I refer to our ability to control certain classes of rhythms generated by our nervous systems in a central vs peripheral fashion, respectively (Fig. 3.13). In this sense, the autonomic and enteric

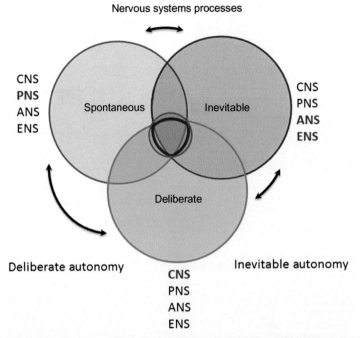

FIGURE 3.13 Cross-talk among three fundamental processes of the nervous systems. The deliberate, spontaneous, and inevitable processes with unique rates of change in stochastic signatures and range of values. Deliberate processes manifest through overt movements executed during goal-directed actions (e.g., instructed pointing to a visual target). They are systematic in nature and well-characterized by low variability and robustness to changes in dynamics. Spontaneous processes manifest through highly automatic, covert movements that have not been instructed and pursue no goal (e.g. retracting, the hand from a target upon completion of the goal-directed phase of a pointing task). They co-exist with and are incidental to goal-directed segments of complex motions. They are highly variable, highly susceptible to environmental cues and changes in motion dynamics. They provide fluidity to behavior at large and occur largely beneath the person's awareness. However, they can be brought up to conscious control when instructed. The inevitable processes are generated by autonomic activity such as the heartbeat. They have much narrower bandwidth of range of change and unlike the deliberate and spontaneous processes, cannot be volitionally controlled or highly perturbed by environmental cues. They are robust and provide a unique signature of the person's nervous systems.

nervous systems (the ANS and ENS) as part of the PNS, are bound to play fundamental roles in the development and well-balance of these separable and yet interrelated classes of processes and the motions they give rise to. In Chapter 6, we further explore these ideas within a new framework of connectivity network analyses applied to peripheral *intelligent* nodes serving as hubs of semi-independent networks efficiently gating and routing information back and forth throughout the body. We will show different analytical tools to visualize interactions within the central networks harnessed from EEG activities; the interactions within the peripheral networks gathered throughout the body with a grid of high resolution wearables and cameras; and the heart activity through pairwise coupled networks demonstrating the interdependencies and "handshakes" among these synergistic interactions. As we begin the path of translating basic science to clinical applications, we need proper analytics. We also need good, noninvasive instrumentation to address these types of multilayered interactions within and across parts of the nervous systems. These signals and the ability of the systems that generate them to communicate and integrate information are at the heart of what is needed for proper dynamics—A key ingredient in social exchange.

Neuroscience, however, assesses behaviors using discrete means. Most times the assessment is not even registered with instrumentation. It is often described by words and/or using pencil-and-paper methods. Biased subjective opinions are commonly called "objective analyses" and correlated (in the best case scenario) with other elements of the experiment. In clinical psychology, there are entire accreditation programs and even a Board of Certification program for so called Behavioral Analysts that examine behavior based on opinions. The saddest truth in all of it is that such nonscientific practices have been *officially* set as the gold standard to treat autism. It is a major tragedy!

The extent of the data in such practices may be discrete events registering conscious responses. These may be self-generated by the participant (e.g., through button presses, joystick motion to select a response, mouse clicks, etc.) or they may be acquired by the experimenter (or clinician) using, e.g., timestamps attained with clickers, etc. These data represent unambiguous events according to some arbitrary criteria set by the experiment or criteria set by the user's conscious selections. When *continuously* recording physiological data such as heartbeat, respiration, temperature, and including as well spontaneously co-occurring covert movements that we are not even aware of, the gaps between these discretely registered presses can be filled. These more complete data reveal a wealth of invaluable information that we have yet to systematically quantify, characterize, and more fully describe because they constitute—As it turns out, most of what we do. As such, the existing cognitive theories of how the brain works are substantiated by skewed data from conscious processes, i.e., the stuff subjects and experimenters explicitly think about. Such theories do not cover any of the spontaneous and autonomous phenomena occurring largely beneath awareness. Unknowingly, we have built a theory of brain functioning that is severely incomplete. This reflects as well in the types of diagnoses and treatment's recommendations we have in the medical fields. The importance of filling in those gaps is noted in the disorders of neurodevelopment, where the spontaneous activities we are now neglecting conveys most of the information we need to track. There is where the adaptiveness of those coping nervous systems will be reflected.

For example, the extent of visuo–motor behavior addressed in scientific peer-reviewed papers is restricted to parameters like reaction times (RTs), or time-series of those discrete events the participant or the experimenter consciously selects. These data are analyzed with parametric or nonparametric models, depending on how skewed the distributions may be. In the case of RT, nonnormality in the data is clearly delineated by the long tail of infrequently long responses; but in most other cases the Gaussian assumption is imposed, or a power transformation performed to squash the tail of the frequency histogram and enforce normality. Regardless of the methods used to analyze the discrete data, there is no data capturing the gaps between the presses. As such, those subconscious aspects of behavior remain unknown to Psychology, Cognitive, and Behavioral neuroscience at large. All the micromotions of the behaving body remain unexplored, yet as it happened with the microsaccadic motions of the eye during fixation tasks[32], the micromovements of *continuously* acquired physiological data (even during resting state) someday will be very revealing of the inner workings of the nervous systems —Particularly of its dynamics.

I dare say that given the different ganglia in our nervous systems (Fig. 3.12B), behavioral analyses should be further refined to separately consider the elements of kinesthetic reafference above and below the neck. Across facial and bodily structures, respectively, innervated by cranial and spinal nerves, research on elements of behavior that fall largely beneath awareness may be beneficial. For example, this would include spontaneous, automatic, and autonomic segments of our motions, including those that serve as input to the vestibular system for standing upright and balancing up against gravity; those externally experienced from the optic flow and those internally experienced though actively self-generated motions.

The use of the new proposed statistical lens to capture fluctuations in the amplitude and timing of multilayered biophysical signals *continuously* harnessed from the nervous systems helped us launch a research program in the lab aimed at creating a unifying platform to analyze the motion these processes in Fig. 3.13 give rise to in a very integrative manner. Further, the various layers of voluntary, automatic, and autonomic control could be mapped onto different layers of variability, so that behavioral analyses of the individual were: (1) Comprehensive in nature; (2) truly objective; (3) standardized to a common scale; and (4) based on physical biorhythms harnessed from multiple layers of the nervous systems. More exciting yet was the prospective that we could do all of it noninvasively with high resolution wearable sensing technology that listened to our biophysical rhythms with high-enough sampling resolution to allow for the tracking of *physiologically relevant changes* in real time. Critical to this new option was then the ability to separate signal from noise as we tracked their ebb and flow within natural settings. The new vision of this research program was congruent with the outlook of a Precision Medicine[33] and mobile-Health agenda aimed at the development of a new model for Precision Psychiatry.[13] This model responded to the new research domain criteria launched by the National Institute of Mental Health in 2010.[34,35]

SCIENCE MEETS TECHNOLOGY TRANSFER AND COMMERCIALIZATION

I managed to publish my first couple of papers on the movement classes, their mapping onto different signatures of motor variability[2] and the case of MJ with ASD.[11] But before that, I had a serendipitous encounter that later led to the commercialization path of the new technology we created in the lab. My department encouraged me to attend the inauguration of the Brain-Health Science Institute at the Life Science Building of the Busch Campus at Rutgers. I went to present a poster on our recent work ("it's good for tenure Liz, show people what you work on, they said") so I went.

I, of course, got lost. While the Life Science building is less than a block away from my Dept, I drove aimlessly looking for it around campus—A typical and rather annoying property of my brain that I somehow cannot get rid of. I arrived late, took a poster board and ended up in the back of the room where no one would ever come anyways. The room was full of work on genetics and molecular-based science. Behavioral analyses and motor control are not even considered science in these circles. They think of it as "motion caption stuff", so I felt it was a big waste of my time—Incidentally, a person on tenure track needs days that last 48 h, at least.

As I finished putting my poster up, a gentleman came by and very nicely engaged in conversation. He asked me several very interesting questions, including many of the issues I had thought about (e.g., those mentioned above) but not yet discussed with anyone. I had confirmed several of these hypothetical scenarios of what we could do with these methods in real time, but there was much more work to do ahead, so I was waiting for more solid proof of concept before publishing any of it. The fact that someone became interested in my work was surreal. In Chapter 4, you may perhaps come to appreciate why this seemed so odd to me. I had been trying to publish my work forever and had had a very bad experience with the scientific community. I had entire hard drives full of data, analyses, and biometric results that to this date are still waiting to be published because people in my field did not have a use for any of it. They correlated spikes with very crude measures of behavior and had no use for mathematical models of continuous motion generation in high dimensions, or for anything different from what they did in a two dimensional, rather discrete world. So, to have someone actually interested on what I had done was remarkable. And more interesting yet, how did he know what to ask? These were things I had not talked to people about yet, how could someone else have a clear vision of what could be done with these new methods? In the middle of the conversation, he stopped abruptly and asked me "have you presented this yet?" to what I responded "no, I got here late". "OK, if you were so kind as to roll that up and follow me to the Office of Technology Transfer and Commercialization (the OTC)?" He was Stuart Palmer, the Director of the OTC (today no longer at Rutgers). He helped me with the process of disclosing, protecting and in due time filing a patent for the methods. I remember amidst my total shock that on my way out of his office he said "This can be a game changer and you don't even know it yet".

I am forever grateful to Palmer because more than one time, people I thought were colleagues have tried to run away with it. Some went as far as breaching our Confidentiality Disclosure Agreement, trying to appropriate our methods and publishing the methods and analyses as theirs. It is crazy what goes on out in the underbelly of the prestigious academic world, even that of top universities! (I give one concrete example in Chapter 7, but the same thing has happened by now so often that the lab (sadly) now sees these behaviors as common place.)

The day I met Palmer, I immediately called Jorge after my exchange with the Rutgers OTC and proposed that we shared the rights to the invention 50–50, since he introduced me to the idea of the Gamma parameter plane. This in turn made my head spin for several hours on that flight back home from Indiana and consequently, paved the way to other ideas I was able to discover empirically. We did file a patent as coin-ventors[36] and have continued collaborating since then; although with other additional interests since the numbers of questions we can now ask using these new methods are nearly infinite.

Essentially, as living creatures, we start moving at birth and if we survive birth and go on to live, we only stop moving when we die. To the nervous systems, everything is motion. *And we had discovered how to quantify it continuously, parameterize it and reparameterize it in real time!* Palmer was absolutely right. It took me some time to see it all in front of me, but then I did. And once you do, you do not want to ever go back to the way behavioral analyses were being done up to then. That means: *all* of the behavioral analyses that scaffolds the "science" of the brain and health sciences as we know them today.

PARAMETERIZING CHANGES IN THE NERVOUS SYSTEMS IN NEAR REAL TIME

This new personalized statistical platform paired with a modified version of von Holst principle of reafference to cover what I coined spontaneous and inevitable autonomous processes opened new possibilities to intervene in the developing neuromotor system and inflict change well-informed by the systems' output. We could now go in with a better idea of the sorts of sensory guidance, we ought to be using to assist rehabilitation in a personalized manner.

This possibility meant that we could positively impact neurodevelopment and steer it in the right direction. In essence, we now had a dynamic outcome measure of behavior that allowed us to habilitate and rehabilitate motor control without having to instruct the person to do anything. The person just had to behave naturally. The next step was to come up with an idea to help us in fact habilitate and rehabilitate *volition*, agency and body ownership in general. We saw some of this in Chapter 2, but will revisit these possibilities in Chapter 6.

In addition to the Gamma parameter plane, I found other parameter spaces with classification power. Points in two dimensions (e.g., confined to the Gamma parameter plane) could be confounded as they could be thought of as a projection from other higher dimensions. Yet, lifting the parameter plane to higher dimensions by augmenting it with other parameters could help separate the data (often in unexpected ways.) This was a very different approach to examine kinematics data in general. The new approach was applicable to any of the biorhythms of the nervous systems that we could harness (noninvasively) with a plethora of instruments, spanning from wearables, EEG, electromyography (EMG), ECG, and various MOCAP systems, among other means. The idea was then to integrate these data and create new data types, probe different degrees of control across a gradient of voluntary, automatic, involuntary, and autonomic levels using instrumentation at our disposal to figure out how to record various biorhythms in tandem, among other many technical questions that we now had.

Most of my old data had been collected from upper body limbs in the class of reaching and reach-to-grasp movements. Perhaps other exponential families would better characterize full-body movements of higher complexity involving as well the co-ordination of the head and the trunk with the limbs. To address this question, I used a combination of maximum likelihood estimation (MLE) and the Kolmogorov–Smirnov test for empirical cumulative distribution functions (CDFs) to evaluate other distributions derived from the time series biophysical data. These included time series of speed-maxima and acceleration maxima data from complex movements in sports performance. Experiments included boxing routines, ballet routines, and the tennis serve. I found that the Gamma family was indeed the best-fit across all subjects and data sets. I could separately fit each person's signature with high confidence, thus creating a scatter. This automatically could give me a group-pattern, whereby I could also plot the estimated Gamma parameters with their respective 95% CIs. I started out by using this and other visualization plots to summarize the ensemble data using individual points derived from each patient and adult group for two classes of movements in pointing behavior: *Deliberately* moving the hand towards the visual target and *spontaneously* retracting it away from the target to rest before the next trial.[2,5,10] This was a first step towards adapting our basic pointing task to a task more appropriate to address problems relevant to the field of Cognitive Science.

I reasoned that the preliminary results on the basic pointing task came from a motor-control experiment, but ASD is defined as a cognitive problem. Thus, I decided to extend and adapt the basic pointing paradigm to address cognitive aspects of decision-making in ASD participants. Here, the idea was to be able to visualize how cognitive loads affected the signatures of variability in motor performance and how in turn, the adaptive shifts in motor learning affected the variability of parameters quantifying cognitive decision-making. Unlike those experimental paradigms that focused on discrete responses (i.e., conscious decisions), here we would add a continuous layer of somatic-motor physiology to fill in the gaps between clicks (or screen touches) denoting the execution of the decisions. Those gaps would now contain motor output activity in the form of a continuous point process capturing the random fluctuations in the amplitude and timing of critical events derived from motor output parameters (e.g., position, velocity, acceleration). They would also contain micromovements from other physiological signals like heartbeat, respiration, and even electroencephalography (EEG). We could start systematically expressing mental decisions as a direct function of continuously self-generated actions. Only in this case, those actions were not restricted to discrete goal-directed segments. They included everything peripheral to or complementing the goal-directed part.

The neuroscience of autism had been restricted to the study of the brain, but the peripheral nervous systems are very plastic, have autonomy, and contribute with important feedback to brain development. In fact, owing to their autonomy and efficient coding, several components of the peripheral nervous systems are thought to have a brain of their own. These include the enteric nervous system of the gut[37] and the peripheral nociceptive and thermal afferent networks across the body.[38] These are rather primitive structures preserved and modified throughout evolution, but certainly preceding the appearance of the neocortex. As such, to pose this problem from the bottom-up, we may need to go through the phylogenetically orderly route that in the neonate, begins to develop and mature earlier than the neocortex. The PNS structures provide life support and autonomy to the nascent organism in the pre-cognitive stage of neurodevelopment. *How was the peripheral nervous system evolving during that scaffolding stage?*

For the first time, I could start asking these questions using a common framework that closed these CNS-PNS loops within an experimental setting. I could probe the system at both the motor and the cognitive ends to see how one caused changes in the other. This was a step above mere correlation between the brain controlling the body in motion and the body in motion feeding back updated information to the brain. In our new scenario, the statistics revealed by the activities of the PNS contained important information of use to both the CNS and the PNS. By having control of the output at each end, which we could now parameterize, we could also have control of the input. We could reparameterize the combined sensory inputs from internal kinesthetic sources and external sensory sources in ways that were rather well-informed by the *continuous* dynamics of physiological somatic-motor responses. As such, we could voluntarily steer the nervous systems' performance away from noise and randomness, towards predictive regimes of higher certainty.

Along these lines, three fundamental properties of the peripheral output that we could reparameterize as input are (1) the high dimensionality of the space (see also Chapter 4, providing a general solution to Bernstein degrees of freedom problem), (2) the subconscious nature of automatic, involuntary, and spontaneous movements in general, i.e., including movements that we had now characterized in a variety of practical ways (see Chapter 5), and (3) the different levels of autonomy embedded in the PNS (e.g., deliberate autonomy under conscious control vs inevitable autonomy beyond volitional control), which we will illustrate and contrast with actual data in Chapter 6. The full characterization of these three important components of self-generated bodily motions would allow me to *reshape behavior* in a completely different way than ever before, i.e., by enabling the child to control the flow of behavior at will, from the bottom-up, with minimal resistance. This contrasted with traditional methods imposing the changes from the top-down, without understanding the source of sensory input that at any given time would drive the child's nervous systems most effectively. The one size fits all model of traditional methods was also static. Across the field, there was this notion of a magic control parameter or a magic task to do it all. Faced with such complex dynamics and irregular rates of change from day–to-day, how could we expect to have such a "magic bullet"?

While watching behavioral reshaping therapies so commonly used in autism, one of my worries was the obvious stress they caused to the children. The reactions to the therapy would at times escalate to tantrums and show their pernicious effects through very distressed behavior. Clearly, the children could not verbalize their feelings to the clinician; but all signs of stress and anxiety, as we intuitively know them, were there. If the clinician chose to ignore those signs to advance some agenda, that personal goal could not be equated with the measurement of behavior in any scientific way; much less with the success of such treatments, a success that so often heralded by these practitioners. And, yet the claims that led to legislation and tax payers' coverage of such methods continue to be along the lines of "evidence-based" successful outcomes based on "objective measures". I honestly cannot see the evidence or the objectivity in such practices. I do see the stress from animal-conditioning-based approaches and the large profits that these generate.

PART II: PHYSIOLOGICAL SIGNALS UNDERLYING COGNITIVE DECISIONS

Designing Appropriate Cognitive-Motor Tasks

I arrived at the DDDC to meet with the personnel who works with the children every day. They were the only ones who could orient me as to what was feasible. I could see immediately that none of the lab set-ups would work there given the plethora of sensory and motor issues these children visibly had. I decided to take some time to think about effective ways to design an experimental paradigm suitable to their classroom environment. I requested the school to let me quietly sit in the back of the classroom to observe the interactions that took place. From that daily experience spanning 3 weeks in a row, I gathered enough elements to design a study and carry on research *on the go*. I then had the funds to get the mobile equipment I needed to that end and recruit people to help me carry out the research plan.

The NSF-grant allocated funds for a post-doctoral fellow. I hired Dr. Robert W Isenhower, freshly graduated from the Action–Perception Psychology program of the University of Connecticut. I hired Rob (as we dearly called him) to spearhead the autism project in the lab. He played a fundamental role in that study. Later on, as a result of his work in this project, he became a full-time research faculty at the DDDC and also acquired his Board Certification as a Behavioral Analyst. I think that he may have found his passion in life through this project.

Rob and I formed a good team that worked hand-in-hand to design the study, pilot it, collect, and analyze the data under the new statistical lens I had tailored to this research program. I had the skills to model and analyze kinematic data in general. I could program our mobile set-up and since I was developing the new statistical platform as the study evolved, I took over the analyses while Rob, who had superb experimental skills, focused on the logistics, collection, and organization of the Big Data, we were generating day-by-day. Day-by-day, we talked after each session and planned new contingencies for the next day. In the span of 2 years, together, we made a number of fundamental discoveries that helped shape the research program I carry on today.

The new experiment consisted of a redesign of my old basic pointing task enhanced by adding a variety of cognitive loads to the target stimuli and converting it to a decision-making task. The child would communicate the decision by pointing to the stimulus. One fundamental difference here was that we did not just collect the discrete mouse clicks or the touch responses. In addition to those, we continuously collected all the efferent motor output that the sampling resolution of our wearable biosensors afforded us (at 240 Hz). The children wore these sensors on their skin. As such, the sensors listened to the peripheral network of nerves producing and regulating the various synergies to co-ordinate their complex movements. During the years prior to this study,

I had developed the basic pointing task to study the balance between deliberate and spontaneous aspects of goal directed actions. This paradigm allowed me to collect the continuous motor data in Figs. 3.5–3.10. As I explain later in Chapter 4, I could also test my computational model of path generation in high-dimensional spaces (e.g., postures space) using primarily the pointing and the reach-to-grasp movement families as their trajectories projected to three, four, and five dimensions.[2,5,10,39,40] But the stochastic layer of analyses I had added to this research program brought in new interesting questions.

The original basic pointing task consisted of a forward and retracting hand movement loop that the participant had to perform continuously at his/her own pace (Fig. 3.14A). To that end, a green dot would appear on the black background of a touch screen and the subject had to point to the dot, touch it, upon which the dot disappeared. This was a deliberate, goal-directed movement that ended up accomplishing the goal and advancing the task to the next trial (at the subject's own pace). Then, the subject would spontaneously (without instruction) retract the hand to rest. The subtle aspect of this seemingly simple task was that I did not impose time restrictions for the duration of the reach. This paradigm stood in stark contrast to typical motor control experiments, whereby a duration time of the trajectory was a priori modeled and used for empirical assessment[41]. I discuss this aspect of the reaching problem in great detail in Chapter 4.

In my version of the pointing task, the participant could take as long as s/he desired while I continuously captured the motion with sensors. Consequently, without instructing the participant, the hand would spontaneously retract to the original position or to other position away from the target to rest between trials. I was continuously registering the motions; hence I was capturing that retracting segment as well. As such, motor output data was being registered also during those gaps connecting conscious pointing decisions. This gave us a window into the part of the pointing movement that is rather automatically occurring beneath the person's awareness.

The set-up and experiment types are shown in Fig. 3.14A for the basic pointing task. The modification of this task to tap into cognitive issues is shown in Fig. 3.14B. This new version involves decision-making on stimuli with increasing levels of cognitive loads.

In the adults, using the basic pointing task, we could automatically parse out the forward and backward segments from trial-to-trial, because they have very specific kinematic characteristics: The forward segment ends at the target location (which we also measured), so at the end of the forward motion the distance from the hand to the target is zero. Further, the velocity is also zero, as it is at the start, right before the movement onset. Likewise, the backwards segment is trivial to detect because the hand leaves the target by gradually increasing the distance away from it and accelerating to reach a peak before decelerating again towards the resting position. In adults, the trajectories and speed profiles were very consistent. This is shown in Fig. 3.15 for a representative adult. His hand trajectories from the basic pointing task are shown in Fig. 3.15A for the forward and retracting cases. The arrows indicate the flow of the motion towards (or away from) the target location, while Fig. 3.15B shows the corresponding speed profiles (mostly bell-shaped) with the speed maxima marked for each trial. This kinematic landmark of the speed maxima is also plotted on the hand trajectories so the reader can gain a sense for

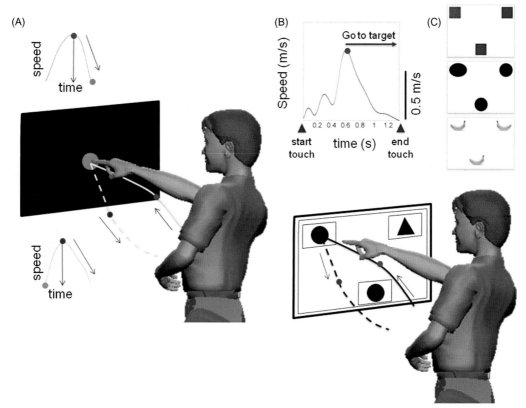

FIGURE 3.14 **Experimental paradigm to study natural movements in the context of basic pointing and decision-making.** (A) Basic pointing task involves a visual target to point to and a continuous loop forward to the target and backwards away from it. The forward segment is instructed and as such performed deliberately to touch the target. The backwards segment may or may not be instructed. When instructed, a postural position may be required for the return back, thus making it a goal-directed action as the forward one. When the motion is not instructed (as in the experiments described in the chapter) the segment is spontaneously performed and reported by the subjects to occur largely beneath their awareness. Arrows indicate the directions of the continuous flow of motion. Insets are the speed profiles with corresponding speed maxima also marked on the hand positional trajectories. (B) Adaptation of the basic pointing task to decision-making under variable cognitive loads. The MTS paradigm can be used, whereby the participant first points straight to a screen location to evoke the sample (e.g., the black circle); then, two alternatives are shown and the participant at its own pace selects the correct match, then without instruction, returns the hand back to rest in preparation for the next self-evoked trial. The participant is in control of the experimental flow. In the case of the children, a picture rewarding the action is shown at the end of the touch regardless of the correctness in the response. If the response is correct, the picture is from a preferred set predetermined with the input from the child, parents, and teachers. If the response is incorrect the computer program selects the picture from a neutral set. This MTS version of the task is performed with identical biomechanical structure as the basic pointing task. However, the fluctuations in speed amplitude and timing change differently with the cognitive stimuli and decision-making needs for each segment type. As such, it is possible to probe the system and characterize changes in somatic-motor performance under the kinesthetic reafference principle extended to consider the cross-talk among the three kinds of processes outlined in Fig. 3.13. The speed profile of one touch to the target under cognitive stimuli and decision-making is shown with landmarks of the deliberate behavior. Note the multimodal speed profile due to hesitation during the decision funneled out through the motor output. (C) Sample stimuli used in the experiments described in the chapter.

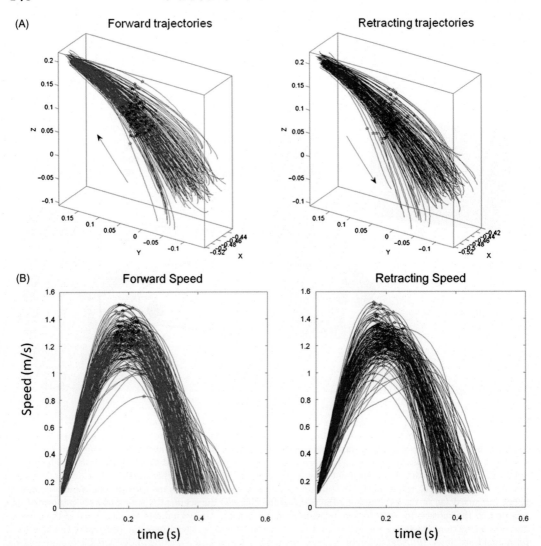

FIGURE 3.15 Basic pointing task output from the continuously repeated hand movement trajectories and their corresponding speed profiles. (A) Forward trajectories to the visual target with the speed maxima location marked along the path. Arrow marks the flow of motion. Uninstructed trajectory backwards, spontaneously retracting the hand to rest. (B) Corresponding speed profiles automatically parsed out from the continuous flow of repeated trials using the appropriate kinematics features of pointing behavior (zero distance to target and zero velocity at rest in the start and ending of the motion). These landmarks were complemented with the screen touches registered as well by our interface.

where in space they occur in relation to the target. The spread of these points is what interests us, as from moment to moment it contains the variability of the peak amplitudes and timings corresponding to these self-generated actions. This variability changes with the context of the task, the stimuli and other factors, even when the basic biomechanical structure of the movement is the same. Speed and other kinematic parameters change their stochastic signatures consistently with the nature of the task demands. We can then continuously measure this change from trial-to-trial, both when the person is intending to reach the target and when the person incidentally retracts the hand away from it.

I would like to revisit why this new way of analyzing kinematics data is different from the "one size fits all" approach. To that end, I use continuous data from this very pointing task in Fig. 3.9. I underscore here that the figure highlights the contrast between using the traditional approach and the new one.

In the traditional way of analyzing time series data from hand kinematics (Fig. 3.9A), one can build epochs within the data according to some criterion (perhaps driven by the goals of the task); then one can superimpose the trials and take the average, which I show with + /- standard deviation lines as it is done in each paper using kinematics to measure movements. I emphasize the a priori assumed *theoretical* Gaussian distribution to obtain the theoretical mean and build the grand average curve. This approach assumes ideal population's mean and casts the person's variability relative to this assumed ideal parameter (but it does not *empirically* tests that theoretical assumption). This method incurs in gross data loss as it smooths out (through Gaussian averaging) the minute fluctuations in the amplitude of the speed maxima. These fluctuations are considered noise and lost in this process. Everyone's performance is thus measured by this assumed *Gaussian ruler*. But, we just saw that even the process of normally aging is accompanied by shifts in the parameters defining the PDFs. That is, we cannot a priori impose the same ruler for a 4-year-old child and a 20-year-old college student. It is not just that their means and variance are different. The random processes we represent these time series with are also different.

The alternative method we are discussing in panel **9B** is fundamentally different from traditional approaches. Here, we continuously estimate the PDF that best characterizes the fluctuations in the amplitude and timing of the speed signal. This is done as the person intentionally does something and as the person automatically or spontaneously interleaves other segments of behavior with the goal-directed segments. In the example used to illustrate the differences between the two methods, we use the basic pointing task, forward to the target, and retracting away from it without being instructed to do so. The sampling resolution of these sensors is 240 Hz, so we can make robust estimations and update the signatures every minute, using a sliding window of, e.g., a second long interval that our sensors afford us. Clearly, the sampling resolution of the sensors will be important to consider to not incur in gross data loss. Data loss is inevitable but we can compensate for that by using tasks that continuously sample from all movement classes and levels of control. We can also exploit a grid of sensors and probe different body locations during different naturalistic tasks. Sensor placement is important to harness activity from body parts that are relevant to a task. Likewise, the level of the nervous systems that we sample from plays a role on the activities we will be able to study. For example, today we combine EEG with bodily kinematics and heart rate because these signals can now come integrated in one biosensor. But at the time of these experiments, such biosensors were not available to research. These are all considerations we need to make when designing a task and the type of biosensors we will be using to try and capture key elements properly quantifying relevant physiological phenomena.

Any biorhythm contains such ebb and flow of deliberate and spontaneous segments. As such, we can apply similar methods to other data sets from different instrumentations. These would include, e.g., EEG, EMG, electrodermal activity (EDA), kinematics from wearables, etc. All such rhythms fluctuate in time and in amplitude. All we need to do is bring them to a common sampling resolution and standardize the wave form of interest so as to place all data from various instruments and disparate sampling resolutions along a similar scale.

We created a standardized statistical scale by defining the micromovements waveform using a normalization that sets the fluctuations in amplitude to a scale between 0 and 1 (as in Fig. 3.9B). This parameterization is, however, not a binary scale. We still retain information between the peaks. Our scale is real numbers. This is so, because in addition to the local maxima (the peaks of the data), we consider all points registered between the local minima surrounding the peaks. We described three different types of normalization in Table 3.1. In the data we described here, we settled for the third one. In this way, across different instrumentations and sampling resolutions we could measure the shifts in noise-to-signal ratios. We began the process of extracting noise that is physiologically relevant and separating it from instrumentation noise. For example, we sampled the sensors in idle mode, or in cases where the motion was not directly generated by a human (e.g., place the sensors in a cart with a remote control and move them with randomly generated regimes). Those cases and others are worthwhile exploring to learn about various types of noise too (e.g., purposefully distort the electromagnetic field of the polhemus sensors, move the EEG cap to explore mechanical artifacts, etc.)

MOTOR RESEARCH WITH CHILDREN IS MORE CHALLENGING SO IT REQUIRES NEW METHODS

Research with young children is different from research with adults. Adults can sustain their attention span longer and move in more organized ways than young children. For example, adapting the basic pointing task to include decision-making made the hand trajectories more variable. This was so because when registering the data continuously, the motor output reflects the mental states. As such, even the hesitation in mental thoughts while considering multiple choices, is reflected in the hand trajectories and their corresponding velocity profiles.

Despite the visible increase in variability (compare Fig. 3.15 from the basic pointing task vs Fig. 3.16 from decision-making in the Match to Sample, MTS task), the trajectories of the adults were easy to parse and automatically separated forward and retracting segments. In Fig. 3.16, I omit for clarity the straight pointing trajectories towards the bottom-center location of the touch screen that activates the sample to be matched. Those pointing motions to initiate the task trial are close to the basic pointing trajectory of Fig. 3.15A−B, since they too involve pointing to a visual target without the decision-making element. The hand trajectories to the two possible matching targets are shown in the form of velocity fields flowing from the resting position of the hand to each target in Fig. 3.16A, top panel and then backwards during the spontaneous hand retraction towards the hand resting area shown in Fig. 3.14B, top panel. The black dots on the hand velocity

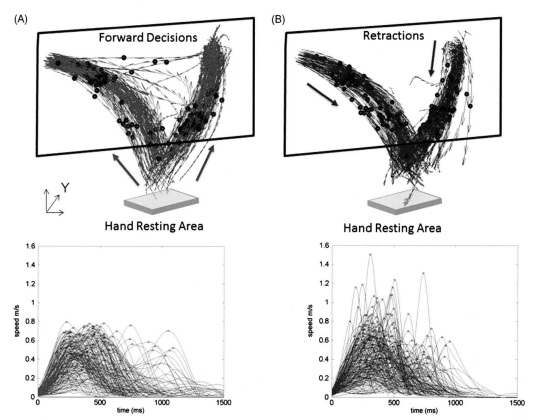

FIGURE 3.16   **Performance of the MTS task, pointing to a selected choice in the context of decision-making by a neurotypical young adult control.** (A) The velocity vector flows plotted for the two possible choices are shown with the landmarks reflecting critical points (maxima) of the speed. Arrows mark the flow of the motions, away from the hand resting area towards the matching sample locations. Corresponding speed profiles are shown on the bottom panel with the speed maxima marked (same as those black dots on the velocity flows). (B) Velocity fields from spontaneous retractions and corresponding speed profiles on the bottom panel. Note the segment trajectories are automatically parsed out using the touchscreens and kinematics features described in Fig. 3.15.

flows represent velocity peaks (like the plots in Chapter 2, Figs. 3.15 and 3.16). These are critical points where the slope of the speed profile changes signs from positive to negative. They are also marked on the speed profiles of the bottom panels of Fig. 3.16A (forward) and Fig. 3.16B (backwards).

When piloting the study at the school, we immediately realized that children are very playful and get easily bored. The sessions ought to be shorter than experiments involving adults. They should also be very flexible to allow for engaging interactions with the adults in the room. Under these conditions, we needed to develop ways to time-stamp the data automatically by creating informative landmarks of when a trial started and ended, while allowing flexibility in natural behaviors. We needed to be able to separate the deliberate

segments to the target from the automatic retractions that the participants would spontaneously perform without instructions. To that end, a real nice touchscreen came in handy
(Dell with 3 ms-delay touch registration). The touchscreen robustly registered the coordinates of the target as well as the touches and the times of the behavioral events. In
later years, the iPad could also do it and we found numerous Apps that allowed all sorts
of communications between the iPad and our computers, so it all became very trivial to
implement. As mentioned, the motions right after the touch, away from the target position
were spontaneous. They were not instructed. I had discovered the spontaneous nature of
retracting segments during my studies of sports routines and had further pursued their
characterization within the context of pointing behavior. The field of motor control
neglects these segments as a nuisance. They turned out to play a very important role in
helping me understand levels of mental intent and levels of volition in the physical realization of intentional acts. They also lend continuity to the behavioral analyses, since along
a continuum past events on those segments influence future events in deliberate segments.
They contribute to adaptive behavior by supporting and reshaping the variability of goal-
directed segments.[2,25,42]

It was a real challenge to extract these segments from the continuous flow of decision-
making behavior in the children. We needed to separate them from the goal-directed
segments. Along these lines, our conversion of the basic pointing task into a decision-
making task considered all these nuances. This task which depends on different levels of
cognitive loads, is illustrated in Fig. 3.14B. In this set-up, we managed to automatically
parse out deliberate and spontaneous movements using the kinematic criteria mentioned
above, and the touch screen information. Fig. 3.17A shows the hand trajectories of a typical child, while the child is engaged in the decision-making task. These are embedded in
the continuous flow of motions the child produces during the experiment. Fig. 3.17B
shows the flow of speed with the marked curves corresponding to the trajectories in
Fig. 3.17A. In Figs. 3.17C-D, we zoom in the speed flow and plot the speed profiles.
These were automatically selected from the continuum of the overall decision-making
behavior. For the first time, we could actually study natural behaviors in children as the
behavior was continuously flowing, without pre-imposing discrete criteria that would
throw away the precious information contained in the spontaneous processes occurring
largely beneath awareness.

It is important to point out that the spontaneous segments had never before been studied or quantified in complex and naturalistic behaviors of human participants; yet the
amount of information they carried when I looked into their statistics was phenomenal. I
will show this more precisely in Chapters 6–7, but here it should suffice to say that in the
case of pointing behavior, the spontaneous movements played a very critical role in helping us understand the data of the autistic system in relation to the data from other disorders of the nervous system across the human lifespan. They also helped us see how such
automatic segments within a given routine could turn rather deliberate on command and
disrupt their automaticity —As I mentioned in Chapter 2, when one may sneakily complement a tennis player on the tennis serve to purposefully make him disrupt it and lose the
serve. It is a simple trick known to experienced players. That trick allowed me to track in

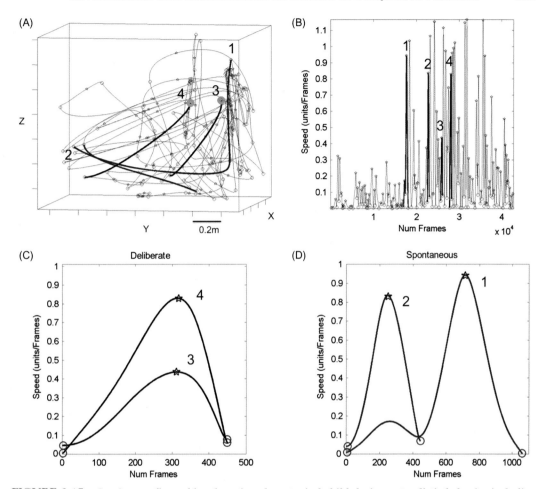

FIGURE 3.17 **Continuous flow of hand motions by a typical child during naturalistic behavior including making a decision and pointing to indicate the outcome of a thought process.** (A) Hand trajectories to the touchscreen with landmarks corresponding to touched spots (green) registered on the screen to mark the selected choices of sample stimuli. Black trajectories are deliberately executed towards the target while blue trajectories are automatic retractions spontaneously performed away from it. (B) Continuous flow of speed profiles corresponding to the hand trajectories in (A). Numbers and colors correspond to those in (A). (C-D) Zooming into the speed flow highlights automatically selected speeds in (A-B) forward to the target and away from it.

skilled athletes, the flexibility of their nervous system while it readily switched between deliberate-conscious and automatic-subconscious levels of performance. In a way, we could *"embody"* that subconsciousness in automatic actions through the use of kinematic analyses and the characterization of their signatures of variability as we perturbed the system in various ways.

I had been quantifying these interplays in the conscious vs subconscious motor signatures for quite some time and had been able to publish those papers as the DDDC study unfolded.[2,5,10,11] These peer-reviewed publications helped me a bit in making my case for the need to study such motions in general. Nonetheless, since they were unknown to people in those health and brain science fields, and had not yet been systematically studied in the field of motor control; it was difficult to even refer to them. I called these movement segments "*spontaneous*" because they were not instructed. Yet, one could argue they are supplemental to goal-directed segments, or are generally incidental to deliberate segments, or are unintended, less intended, etc. The main feature I had quantified in these spontaneous movement segments was their extreme sensitivity to changes in body and environmental dynamics, e.g., to changes in speed, to manipulations involving adaptation to loads, to changes in the light of the room, sound provided through a metronome, etc. Their sensitivity to external and internal environmental dynamic influences also stood in marked contrast to the robustness of deliberate segments to such dynamic manipulations. In fact, we will see in Chapter 5, that even a small change in the context, e.g., turning-off the lights in the room or setting-up a mirror to provide visual feedback to the participant, deeply impacts the noise-to-signal ratios in the kinematic parameters of this class of spontaneous movements. These influences are fundamentally different in the deliberate segments aimed at a goal. Another important feature to consider was their flexibility and malleability on demand. The invisible spontaneous movements seemed to "listen" and make the deliberate movements more visible to the system (i.e., sharpen their signal-to-noise ratio). The evolution of their variability gave us a sense for the plasticity of the nervous system of the person and its flexibility to switch between modes of action *on command*.

The conversion of the basic pointing experiment to the school settings of the DDDC was very important to further advance our research program and make it an integral part of the *health mobile* concept. Most pointing behavior in the motor control field were studied while constraining the arm to move on a plane within very strict temporal demands (see Chapter 4). By adopting more natural pointing actions in three dimensions without temporal demands—Where the subject set his/her own pace, we learned a great deal about typical behavior in children. This approach to developmental research gave us more rigorous ways to study the rapidly changing somatic-sensory-motor systems of children in general. Further, we could do so in fun, playful ways at the school settings, or at the home environment. This was important for the ASD community, since the lab environment can be at times overwhelming for the children. We added the caregiver, the teacher, and the researcher into this picture and started to build a framework that allowed the study of coupled dynamics of dyadic interactions. These new methods gave us unprecedented details of the coupled social dyad, while using noninvasive means and harnessing the activities in real time. Chapter 6 shows examples of the new paradigms we invented to that end.

Considering each child's sensory-motor needs was very important to keep the group engaged within the relatively short amount of time we had at our disposal. To maintain the steady flow of participants while creating minimal interference with their daily school routines, we built a mobile system (using a Polhemus Liberty 240 Hz, Colchester VT and touchscreen DELL laptops) that we took with us every day to the DDDC.

During the class breaks, the personnel from the DDDC would help us rapidly set-up the sensors and the touchscreen, so the experiments could be done in the span of 15 min. This time scale fitted well within the context of a game-like environment between classes and other school activities. Rob's role on this portion of the study was fundamental. He had a special rapport with the children. He knew how to be playful, kind, and respectful of their needs. Rob had a special intuition and *the touch of an angel* to work with the children in the spectrum. Indeed, they loved playing with "Mr. Rob". The DDDC had an ideal setting for the study as well, because the school program mixed typical peers with the children in the spectrum. In this sense, we had age-matched controls for each child in the spectrum. Furthermore, all children from a given class were engaged, so the game-like task during the break felt like another school activity that all the children participated in.

As the children day-by-day gradually felt comfortable with the basic pointing paradigm and we felt confident they enjoyed the task, we decided to gradually introduce them to the cognitive version of this task, where the stimuli had some more cognitive complexity. To that end, we engaged the children in a type of well-structured decision-making routine. Their accuracy and timing were evaluated through the pointing task they used to communicate their decision. This new task gradually increased the levels of cognitive load of the stimuli, while we measured their responses at the motor output. We designed a MTS task, where the samples the child had to match to the given stimulus incrementally gained in complexity. The cognitive stimuli consisted of three levels represented in (Fig. 3.14C):

In *level 1*, we presented similar shapes of different colors so the child had to match the given sample of a shape and color. For example, given a blue square as the sample to be matched, when the child touched it, two alternatives colored squares showed up, one would match the sample (blue) and the other would not (e.g., red) (Fig. 3.14C top). The task of the child was to point to the one matching the sample, while we continuously recorded the motions. Since the children were used to receiving food rewards or tokens after the completion of discrete trials training (DTT, see **Notes**[2]) in their applied behavioral analysis (ABA) school paradigm; we incorporated rewarding pictures upon the correct completion of the task. These consisted of still frames from cartoon movies they preferred. We learned of their preferences from their teachers, school aids, and parents, so we could engage the child more easily. Nonpreferred or neutral pictures were presented in the error trials implicitly providing a form of feedback on the outcome of the action and reinforcing the correct over the incorrect trials. This was in tune with the "Skinner Box-like" ABA principles but with the fundamental difference that in this case, we were objectively quantifying the outcome of the sensory-motor nervous systems.

In *level 2*, the geometric shapes were colored in black, and had a degree of ambiguity that made the task more challenging (Fig. 3.14C, middle). For example, we could present as the sample shape a circle and then as the alternatives to decide on, we would show a circle and an ellipse close in dimensions to the circle. This made the choice more difficult as the child had to think before making the correct decision. The

[2]Discrete Trial Training (DTT) in Applied Behavioral Analysis (ABA)[51,52]

picture reward provided a good incentive to the child and there were plenty of cheerful exclamations to provide instant feedback that the task was going well. But even when mistakes were made, we would cheer the child up "yeah! Next time we'll get it!"

In *level 3*, the shape of the stimulus was rendered in three dimensions and consisted of a fruit oriented differently, thus requiring some mental rotation to decide if it matched the given sample (Fig. 3.14C, bottom). For example, a banana could be presented in such a way that it required a rotation of 180° around the *z-axis* to bring it to an orientation congruent with the sample banana. These stimuli made the task more challenging than those employed in the cases explained in Eq. (3.1) and (3.2). However, the children were able to perform all of them. They initially did so with assistance and made plenty of errors, but eventually, they became quite proficient and independently performed the task on their own (see movies online).

The pointing trajectories of a child are very different from those of an adult; compare (Figs 3.16–3.17). This is evident as the additional element of decision-making is incorporated into the task. An example from the performance of a neurotypical child showing the speed fluctuations extracted from the continuous flow of the trajectories is displayed in Fig. 3.17B. These are the traces of the speed derived from unconstrained hand motion trajectories in a naturalistic setting at the classroom.

We managed to parse out the trajectories relevant to the task using a variety of landmarks aided by the touchscreen. First, the co-ordinates of the touchscreen were marked in three dimensions, as we had a positional sensor on the relevant location. Second, the touches were registered so we had the screen pixels' locations and the timestamps of the touch. Third, we registered in tandem the hand kinematics relative to a common frame of reference, also tracking the target position. The hand sensor gave us the speed (zero-velocity at the target and zero-velocity at rest) and we tracked the distances to the targets for which the positions were also continuously registered in three dimensions. These distances reached near-zero value as well when the hand touched the target. This information allowed automatic detection of the relevant trials and highlighted the distinction between levels of intent. The continuous speed profiles of the MTS pointing task trajectories are shown in **17 A**, with colors corresponding to the trajectories parsed out in **17 A**, towards the green targets (black) and away from them (blue). The ebb and flow of intended and spontaneous fluctuations had different signatures that could be well-characterized and shifted with the cognitive loads.

The histograms derived from the moment-by-moment fluctuations in linear speed maxima and the Gamma parameter plane are shown in Fig. 3.18A. The stochastic trajectories in response to variations in cognitive loads are shown for these neurotypical participants in Fig. 3.18B. These are color coded for each representative age group we studied. There we could immediately see that the youngest group of 3–4-year-old neurotypical children had fluctuations in speed with higher noise and more skewed shapes than the 4–5-years old. Within a year or so, the children of school age had stochastic signatures that shifted down and to the right of the Gamma parameter plane. By college age, these shifts denoted the presence of more symmetric shapes in the PDFs and much lower noise-to-signal ratio. In neurotypicals, a maturation process was evident in the motor control signatures. These signatures shifted with the changes in cognitive stimuli (Fig. 3.18C).

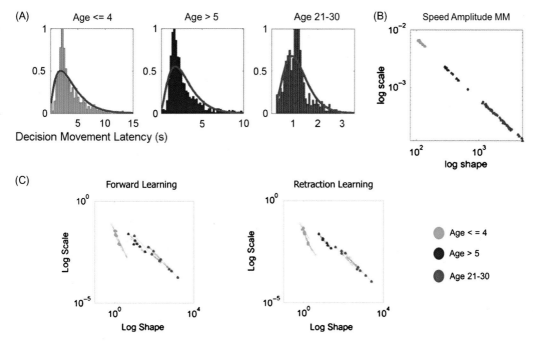

FIGURE 3.18   Differentiation of the stochastic signatures of moment-to-moment fluctuations in decision-movement latency and hand movement speed amplitude across age groups of neurotypically developing individuals. (A) Frequency histograms of the decision movement latency determined between the stimulus onset and the initiation of the reach up to the first significant velocity peak, i.e., before the first deceleration of the motion. (B) Micromovements (MM) of hand speed amplitude as they fluctuate from moment-to-moment (MM normalized between 0 and 1 using Eq. (3.3) on Table 3.1). Age-dependent transitions of stochastic signatures of MM plotted on the log-log of the Gamma parameter plane from different age groups (see legend). This power relation denotes a systematic age-dependent relationship. As the participant typically ages, there is a decrease of the noise to signal ratio (the scale Gamma parameter) accompanied by an increase in the value of the shape parameter as the PDF turns symmetric. (C) Forward and retracting motions give rise to shifts in the stochastic signatures of the raw speed maxima. These shifts manifest different rates of change and variable stochastic trajectories for each age group. Green, Age < = 4; Blue, Age> 5; Red, Age 21−30.

Increase in Cognitive Loads Induces Adaptive Motor Learning

Examination of the kinematics data revealed different PDFs for straight pointing motor behavior and decision-making motor behavior. Both movements contained identical biomechanical structure, but resulted in very different kinematics owing to the hesitation and delays the nervous systems incur while processing sensory information, integrating the sensory information from the external stimuli dynamics with the internally generated motor-sensory dynamics. Further, this process entailed performing sensory-motor transformations while deciding under different levels of cognitive loads. All these factors made the evolutions of the children's performance very rich to study and quantify the amount of noise-to-signal ratio inherently present in the trial-to-trial adaptive fluctuations of their speed profiles.

The cognitive influences of the different loads were continuously channeled out through the motor output responses. Gathering the trajectories for different motions towards (Fig. 3.19A) and away from the decision-target (Fig. 3.19B) shows this adaptive learning. Systematically in all children, practice led to a clear transition from multimodal to unimodal speed profiles. On average, the initial learning demands resulted in slower speed profiles that reflected hesitation. They were accompanied by higher error rates. Yet, over a few trials the children's movements turned smoother and ballistic across both the deliberate and the spontaneous segments of the reach (Fig. 3.19).

I underscore here that our interest is on the stochastic signatures of the moment by moment fluctuations in the amplitude and timing of the speed maxima. The extent to which prior speed-dependent events would or would not contribute to the prediction of future events was of interest to us as well. The stochastic shifts we could extract between the two main tasks were indicative of the amount of adaptive learning the cognitive loaded task required. That type of motor evolution in turn gave us a sense for the plasticity of the peripheral system executing the brain's decision commands. Fig. 3.20 shows the shift from mere "biomechanical" to cognitive pointing for each child (Fig. 3.20A) and for the group (Fig. 3.20B), where we pooled the data in each group to estimate the group's signature. This figure shows larger shifts in the college age participants and more variable responses of the kindergarten children. The latter, as a group, increased the noise and shifted to more skewed PDFs closer to the exponential fit (Gamma shape value of 1) when the cognitive loads were introduced.

The new statistical analyses could capture rather well the subtle statistical shifts that would have been smoothed out as noise through averaging, under the traditional parametric approaches used to analyze kinematics (illustrated in Fig. 3.9A). These new methods could very well serve as outcome measures of ABA-based training. Within the ABA paradigm, the DTT serves to create structured motions but leaves out the spontaneous retracting segments of behavior invisible to the observer. These spontaneous segments would fill in the gap segments between discrete conscious selections of unambiguous segments of behavior. The ABA DTT methods would train children as one would program a robot, i.e., by programming the goal-directed segments from the hand to the target, but leaving out the connecting spontaneous segments that lend fluidity to the motion. The spontaneous segments in fact do more than lending fluidity. We call them the "listening movements" because during motor learning their variability carves out the deliberate movements' variability and helps identify voluntary control in a system with redundant DOF, such as the human body. *The invisible spontaneous movements make visible the deliberate ones.* We will cover these issues in great detail in Chapter 4, and revisit the invaluable role of spontaneous motions for feedback-based control in Chapter 6.

Despite (arguably) being the main intervention in autism and certainly the one covered by health insurance, the ABA methods do not quantify somatic-motor performance using objective means. In this sense, they cannot tell how the continuous physiological motor signal is evolving with the intervention. Unfortunately, the types of objective quantitative analyses that we describe here are not taught in the Clinical Psychology curricula that certifies behavioral analysts across the country. We discovered this in a series of interviews we did

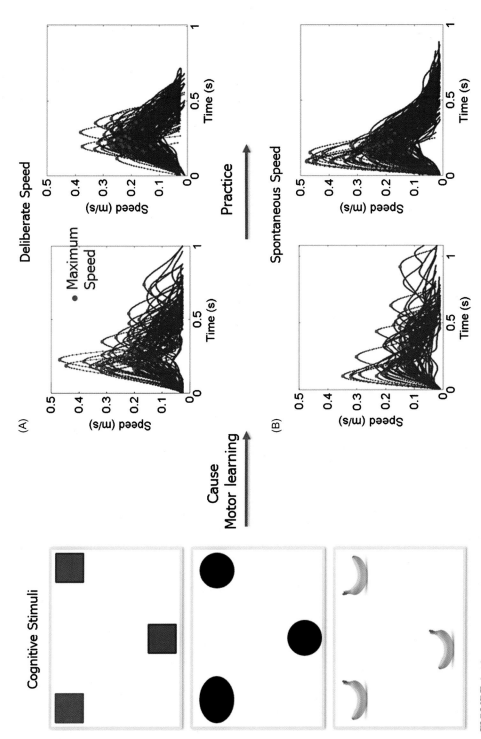

FIGURE 3.19 Changes in cognitive stimuli evoke motor learning. The basic pointing task where motions are ballistic and well-characterized by bell-shaped speed profiles (e.g., see Fig. 3.15) turn to multimodal speed profiles with the decision-making MTS task involving different types of cognitive loads in the stimuli. Sample stimuli are shown along with consequential speed evolution from multimodal to unimodal. Motor learning accompanies cognitive decisions that over practice result in faster and more ballistic motions with lower decision time and higher accuracy.

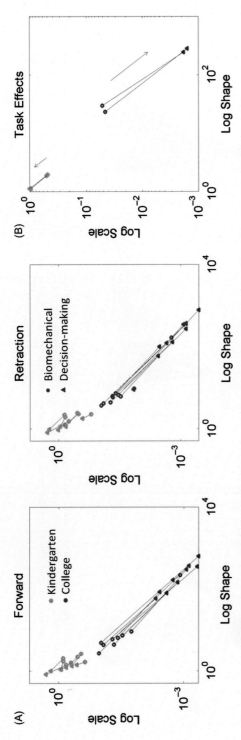

FIGURE 3.20 **Quantifying the stochastic shifts between fluctuations in basic pointing and fluctuations in pointing that requires decision-making.** (A) Forward and retraction segments evoke different signatures and rates of change for young kindergarten children relative to young college-age adults. The young adults manifest a broader range of change of PDFs and fastest rate of change tending towards the symmetric distributions on the Gaussian range of the Gamma family, with lower dispersion. The young children have larger variability, higher noise-to-signal ratio and more skewed distributions tending to the exponential range of the continuous Gamma family (with shape parameter value of 1). Biomechanical; Decision-making.

in 2014—15 under the NSF funding of the Innovation Corps. We also discovered that underlying the certification process to attain accreditation for the ABA method, there is a superb infrastructure to deliver knowledge. As such, my hope is to be able to team-up with people in the Clinical Psychology field and initiate the path of objectivity. We could teach the new generations of psychologists in the clinical field how to quantify the stress levels in any interventions that young children undergo. If post-traumatic stress syndrome in adults affects their nervous systems in negative ways—A nervous system that is already mature—I wonder what stress does to a nascent nervous system, particularly to one that is trying to simultaneously develop and cope with a developmental glitch.

In this sense, I would like to underscore the importance of attaining such physiological measures across the three types of processes depicted in Fig. 3.13: Deliberate, spontaneous, and inevitable. Understanding their cross-talk would be particularly relevant at interventional settings. As we can see here in an experimental setting, such types of DTT tasks do inflict transient changes. These changes are quantifiable within the time scale of minutes in the rapidly developing nervous systems of a child (with or without ASD). I am concerned about possible permanent damage to the developing nervous systems if the levels of stress in the child are being systematically raised beyond certain threshold. We need to determine limiting physiological threshold values to adequately bracket the amount of time allowed each day for such therapies in order to avoid (unintentionally) damaging a child's nervous system while trying to do our best to help that child.

If within a 15 min time-interval once per week for 4 weeks, we were able to cause changes in response patterns of the nervous systems of these children, imagine then what may be happening to their nervous systems when they undergo 40 h of ABA—DTT per week for years. In fact, we have seen the outcome of this regime of training in many children that come to our lab. We have also been contacted by many parents who feel other forms of therapy are needed in ASD.

The issue of therapeutic interventions in ASD is a very political one. We will return to it in Chapter 7, but it suffices to say that the Occupational and Physical Therapists (OT/PT) who include aquatic interventions, horse therapies, and other forms of sensory-motor therapies in their treatments do not have access to insurance coverage for the families they treat. They lack the objective outcome measures that insurance companies require to track progress, to assess benefits, and risks of the intervention, to justify coverage. With a prevalence of coverage for therapies that *make the child socially acceptable*, without treating the underlying sensory and motor issues, there is no way to reach the path of diversification of therapies to treat ASD. This is particularly the case for the low-income sector of the population. At ongoing rates ranging anywhere between \$150/h and \$350/h it is simply impossible to afford such a luxury for the parents of the large majority of affected children. Here we are, discussing outcome measures that would precisely address this issue and help provide the coverage they need. Unfortunately, transforming scientific results into technological advances takes time. At least through systematically interviewing over 120 people in the ecosystem of autism, we were able to identify this critical need for sensory-motor based OT/PT coverage in pediatrics. The good news is that more than one solution already exists. It is only a matter of time, before we can team-up with companies willing to implement the solution into an affordable product.

Adaptive Motor Learning Helps Decision Accuracy

The decrease in the children's decision times, their increase in accuracy of responses, and the overall gain in the organization of the speed profiles from their motions along with their decrease in movement times were remarkable. These changes gave us a sense for the learning they were experiencing day-by-day. Without a doubt, we were also able to distinguish the basic pointing task from the pointing task that required the mastering of higher cognitive loads embedded in the stimuli the child had to decide on. As such, this task proved to be a useful paradigm to build towards a different type of intervention in ASD, i.e., one that allowed the child to self-discover what the goal of the task was. But it was also a good paradigm to investigate the intersection between motor control and cognitive performance. The task had all the required ingredients to do so in a very rigorous manner. It informed us not only of the adaptive motor control abilities of the child but also on the abilities to make decisions based on cognitively loaded stimuli. In this sense, the cognitive stimuli could be easily parameterized for use in basic science.

The methods discussed here are also amenable to clinical settings where cognitively driven interventions were being used to influence or "reshape" behavior. Unknowingly, such interventions did change the motor patterns underlying behavior. As we examined such evolutions in the stochastic signatures of the children's movements (Fig. 3.20A−B), we could see the differences across ages and identify through the evolution of the movement signatures when a child had become proficient at performing the cognitive task (Fig. 3.21A−E). We could do this for both the intentional movements and for the movements the child spontaneously performed. Each movement type revealed the evolution of a different class of nervous systems' process (Fig. 3.13).

SENSING THROUGH MOVEMENT: WHEN NOT ALL NOISE IS CREATED EQUAL

As we later expressed in our paper entitled "Noise from the periphery"[43], there are different types of variability associated with different levels of control.[2] As such, they inform us of the different types of kinesthetic feedback the brain is likely receiving when we continuously move in the physical environment. Given that systematically increasing the levels of uncertainty in sensory stimuli consistently led to higher levels of noise and randomness in the peripheral signals of these very young typically developing (TD) children (Fig. 3.20); I then wondered about possible interdependencies between these *deliberate* and *spontaneous* processes and the *inevitable* processes emerging from the ANS responses (Fig. 3.13). More specifically, if a mere increase in cognitive load during our experiment had clear and quantifiable somatic-motor manifestations in the hand kinematics of the youngest children; I wondered the extent to which such increase in uncertainty would affect the autonomic performance. To that end, we initiated the path of examining the rhythms of the heart. *Was there cross-talk between the cognitive loads and the autonomic nervous system?*

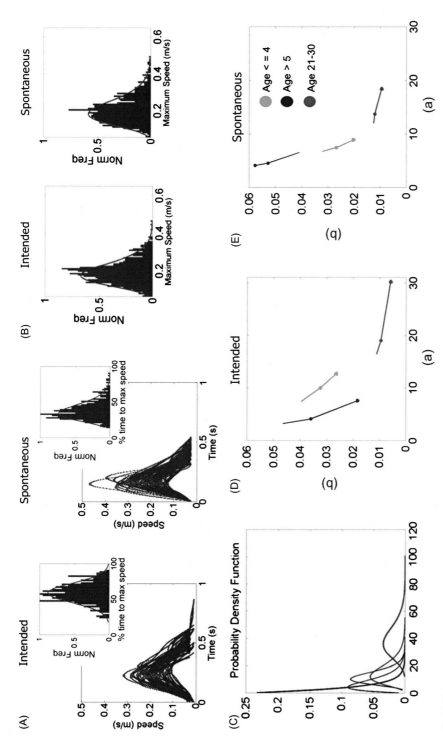

FIGURE 3.21 Consequent motor performance upon cognitive variations in the stimuli. (A) Intended and spontaneous motions generate speed profiles that with practice tend to be ballistic (bell-shaped) and nearly symmetric indicating proportional acceleration and deceleration phase around a unique speed maximum. In each case, insets show the distributions of the time to the speed maximum trial-by-trial. (B) Trial-by-trial variations in the value of speed maxima for the forward hand motions intended to the target and for the spontaneous retractions away from it. (C) Exponential family of continuous Gamma PDFs ranging across different shapes and dispersions to characterize the 0 to positive ranges of speed and distance values. (D) Stochastic shifts of the speed-dependent signatures as a function of cognitive stimuli for the forward reaches intended to the target. (E) Stochastic shifts for the uninstructed backward motions spontaneously retracting the hand away from it. Each color represents a different age group (see legend). Green, Age < = 4; Blue, Age> 5; Red, Age 21–30.

From the Brain to the Heart

Jihye Ryu, a doctorate student in my lab decided to undertake the task of investigating the possible cross-talk between the heart and the brain. She designed an experiment where she could also increase the cognitive load of a basic pointing task in various ways. One was by pointing while simultaneously counting (forward or backwards) with different levels of complexity. The other one was by estimating the duration of a tone and pointing on a scale, displayed on the screen, the guessed amount of perceived tone time duration. As in the cases above, she examined these tasks in relation to the continuous motor performance during the basic pointing task. She used the continuous forward and backwards segments of the pointing motions to assess the balance between deliberate and spontaneous processes. As in the previous experiments, she did not instruct the retraction. The participants spontaneously brought the hand back to rest and as before, she used a touch screen to establish well-defined behavioral landmarks. As in the other experiments thus far explained, the registration of the data was continuous, using the Polhemus Liberty system recording at 240 Hz to record the positional trajectories of a grid of biosensors across the body.

Her data incorporated another type of motion from the ANS. These were the motions generated by the biorhythms continuously harnessed from the heartbeat. The heart signals were obtained via ECG from a wireless Nexus-10 device (Mind Media BV, The Netherlands) and Nexus 10 software Biotrace (Version 2015B) at a sampling rate of 256 Hz. Three electrodes were placed on the chest according to the standardized lead II method, and were attached with adhesive tape. A typical ECG data includes a set of QRS complexes, and detecting R-peaks (within the QRS complex) is essential, as the heart rate metrics needed for this study focuses on the oscillation of intervals between consecutive heartbeats. In order to remove any baseline wandering and to accurately detect the R-peaks, Ji preprocessed the ECG data using the Butterworth IIR band pass filter for 5–30 Hz at second order. She selected the range of the band pass filter based on the finding that a QRS complex is present in the frequency range of 5–30 Hz.[44] To retrieve the time between R-peaks (i.e., interbeat intervals, IBI) from the preprocessed ECG data, we used peak detection methods similar to those we used thus far in the kinematics speed data.

The temporal fluctuations in IBI were very revealing of the interactions between these inevitable processes and the deliberate and spontaneous processes we previously defined using the fluctuations in the amplitude of the kinematics signal. She found that when faced with higher cognitive loads, the heartbeat speeds up and becomes more variable. And such subtle variations were systematically quantifiable in each of the eight subjects she tested. This can be seen in Fig. 3.22A, where the summary statistics from the empirically estimated IBI Gamma moments are displayed for low- and high-cognitive loads, with the arrow denoting the order of the task (first low, then high). Fig. 3.22B shows the corresponding empirically estimated IBI-PDF's and Fig. 3.22C shows the estimated Gamma parameters with 95% CIs.

She also found that more generally, examining voluntary movements in these neurotypical participants of college age, the stochastic signatures of the angular speed micromovements were very informative. These signatures quantified the fluctuations in amplitude of

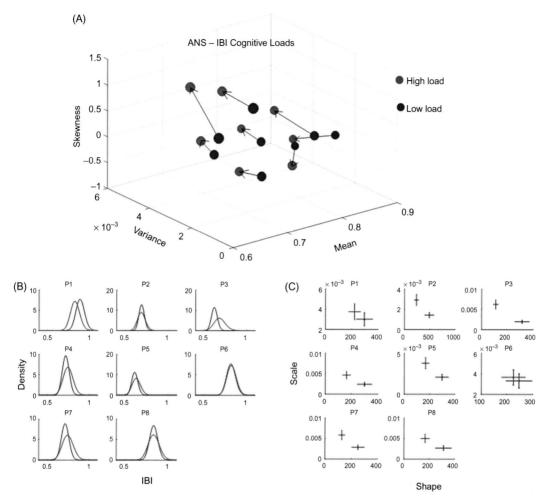

FIGURE 3.22 **Cross-talk between deliberate, spontaneous, and inevitable processes under different cognitive loads reflected through shifts in the interheart-beat-interval timings of the autonomous nervous systems.** (A) Empirically estimated Gamma moments (mean, variance, skewness, and kurtosis) derived from IBI biorhythms during a basic pointing task augmented with different cognitive loads (i.e., counting forward and backwards). Each dot pair represents a participant's signature shift. Arrows denote the order of the task and mark the shift in the stochastic signatures. (B) Empirically estimated continuous Gamma probability density functions for each of the cognitive load conditions. (C) The empirically estimated Gamma shape and scale parameters with 95% CIs from the MLE process. Data reported for 8 young neurotypical participants (young adults ranging 19–13 years old). Red, high load; Blue, low load.

the rate of joints' rotations from moment to moment. They reflected the effects of cognitive loads on the goal-directed segments (the forward motions to the target). The shifts in signatures can be appreciated in Fig. 3.23 showing changes in the PDFs when transitioning from basic pointing to pointing while doing time estimation.

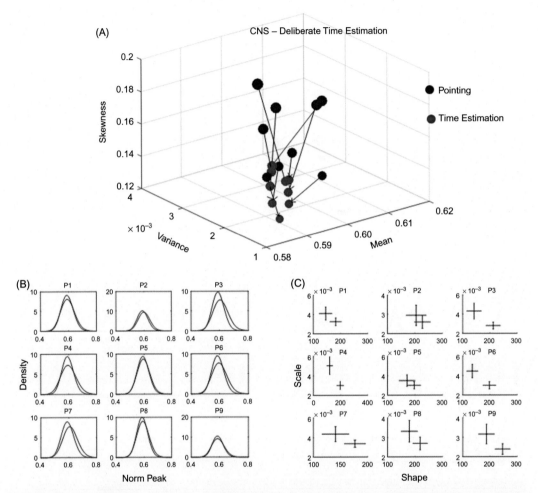

FIGURE 3.23   **Deliberate processes quantified through moment by moment fluctuations in the angular-acceleration amplitude micromovements during basic pointing vs pointing during the estimation of the time of a tone.** (A) Empirically estimated Gamma moments shifting from the basic pointing task to the time estimation pointing task. Each dot pair represents a participant (task order marked by the arrows). (B) Empirically estimated PDFs with colors corresponding to legend in (A). (C) The empirically estimated Gamma parameters on the parameter plane with 95% CIs. Red, time estimation; Blue, pointing.

In contrast to the marked effects, the goal-directed tasks had on the speed amplitude, the spontaneous retractions channeled out these effects best through temporal fluctuations registered in the variability of the interspeed-peak times. This can be seen in Fig. 3.24 using the same format as in Figs. 3.22 and 3.23. The reader can see that such subtle effects could not possibly be captured by clickers or mouse clicks. One really needs to register the *continuous flow* of physiological parameters as they change position at some rate in some space. In other words, when we employ the word *motion*, we really mean change of

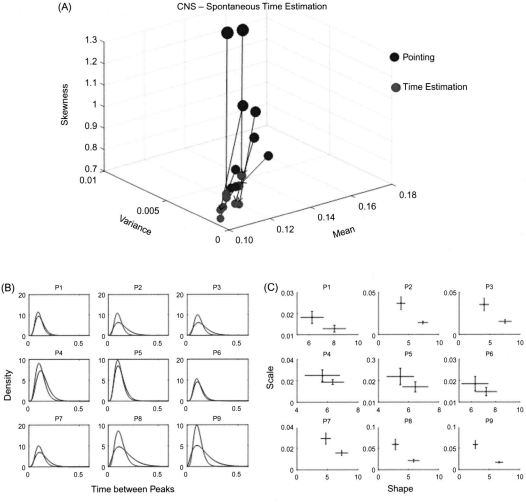

FIGURE 3.24 **Spontaneous processes quantified through the peak-to-peak interval timings of the angular acceleration and their shifts in stochastic signatures from basic pointing to pointing during time estimation.** (A) Empirically estimated Gamma moments based on timing information from the hand pointing motions in basic pointing vs pointing during time estimation (see legend). Each dot pair corresponds to a participant. Arrow marks the task order. (B) Empirically estimated PDFs for the two tasks [color coded as in the legend of (A)]. (C) Corresponding estimated Gamma shape and scale parameters plotted on the Gamma parameter plane with 95% CIs. Red, time estimation; Blue, pointing.

position over time of some parameter in some parameter space. This is a far more general definition than motion referring exclusively to overt movements.

Traditional analytical approaches under the "one size fits all" model would have missed these subtle effects owing to the homogenization of the variability in the data and the assumption of normality pre-imposing a priori Gaussian population statistics. The individualized approach to empirically estimate the statistical distribution parameters for each person was very revealing of the shifts of the Gamma moments derived from fluctuations in amplitude and timing of the angular acceleration signals.

Instead of focusing exclusively on the mean and variance, we examined as well higher moments. The shifts in skewness and kurtosis of the estimated Gamma process along with the shifts in mean and variance were estimated for each task relative to the basic pointing task. All nine points (representing nine participants) in Fig. 3.23 shifted their signatures to lower noise and more symmetrically shaped distributions. Further, the amplitudes of their angular accelerations fluctuated with less variability implying that as they were deciding about the duration of a tone and pointing to indicate the decision, their hand motions were highly controlled. We could safely infer from the statistical signatures that they were more focused during the time estimation task than when merely pointing. As they retracted the hand to rest, the variations in the rates of rotational amplitude were not channeling these subtle differences between the tasks with the same level of certainty that did the interpeak-interval timings. The length of time spanned from maxima to maxima along the angular acceleration signaled whether the person was merely pointing to a visual target, or pointing to indicate a decision on the temporal estimation of the tone. All participants shifted towards symmetric shapes of the distribution with lower dispersion, thus manifesting less variable patterns and faster timing (Fig. 3.23).

The hand was steadily rotating when returning back to rest from the time estimation task. These results gave us confidence that the cognitive load phenomena under study using continuous pointing were indeed quantifiable and highly informative of more than one type of process in the nervous systems. We could track deliberate processes through the shifts in the signatures of the voluntary goal-directed (forward) motions. We could track spontaneous processes at the automatic level through the uninstructed retractions that subjects performed reportedly beneath awareness. They did not even realize they were performing these motions. Lastly, we could track inevitable processes in the ANS activities from the heart. There was definite cross-talk among deliberate and spontaneous processes with the inevitable patterns of the heart. Such subtle effects at the autonomic level were funneled out through the IBI fluctuations. The stochastic signature of the timing between heartbeats, as they shifted with the levels of cognitive loads or automatic performance indicated for each person the level of tolerance their variability had. Armed with this type of methodological proof of concept, we then proceeded to study the autistic somatic-motor system across all these levels of control and all these layers of the nervous systems using noninvasive means.

THE AUTISTIC SOMATIC-MOTOR PHENOTYPE

At the DDDC, we were able to find at least 15 children spanning ages from 3—14 years old and then 5 more adults in their early 20s. They were all perfectly capable of performing our task. We were able to test age-matching controls as the literature requested.

However, it was clear to me that the age-matching idea was an odd requirement in that field, when the neural development of a coping nervous system clearly follows rather mysterious longitudinal developmental trajectories. The expected developmental milestones would most likely not be manifesting at the expected ages. Each paper treated age as an absolute measure across the population, rather than in relation to the individual child's development. As we showed in **Part II of** Chapter 1, even during neurotypical development the stochastic evolution of the somatic-motor signatures and the signatures of physical growth follow a very irregular and nonlinear path. Yet, I did comply with the field's expectations and made sure to recruit enough participants matching the age range of the ASD participants.

The children, adolescents, and young adults we tested at the DDDC were able to perform this task. I think their DTT at their ABA program helped them acquire the proficiency needed to perform the basic pointing task. Yet, generalizing to the MTS was challenging. This task required decision-making and faced them with different cognitive stimuli. We spent some time assisting the ASD participants until they became comfortable and could perform the task independently, without support.

Something that we noticed in the process of teaching them the task was that they did much better and learned much faster if we allowed enough time for them to realize what the purpose of the task was. This realization happened when they correctly matched the sample to the appropriate target and got to see a picture of their favorite cartoon. This rewarding experience was not evoked from the top-down, meaning that the action was not instructed and was not *externally reinforced* with food or a token reward paired with the action completion. Indeed, our version of the task was rather evoking the reward from the bottom-up, meaning that the end goal leading to the rewarding picture was spontaneously self-discovered by the child, as the child controlled the flow of the task. They had to make a decision on their own and had 50–50 chance of being right or wrong. Their reward was *internally reinforced*. Over time, it became very clear to them when their performance was correct. Somehow, without our intervention, all children with ASD got *autonomously* proficient at this task. This type of autonomy was deliberate because they were in control to choose the answer; they were truly executing their free will without external pressure or assistance.

This evolution towards self-control and agency gave me a sense of what would work better than the types of top-down approaches that did not leave room for the self-discovery element. It occurred to me that if we could evoke the self-discovery element, the children would learn faster and be in control rather than being controlled by an external agent. This element of promoting volition and self-agency seemed very important to me. We will see in Chapter 6, that such philosophy did wonders for these children and inspired other set-ups that extended these ideas to the design of new child-driven therapeutic interventions for ASD.

As the children adapted to the task, their distributions did shift signatures and their motions became faster. The evolution of their hand speed profiles can be appreciated in Fig. 3.25A for a representative nonverbal child with ASD. The interesting finding about these data from the children, adolescents, and young adults that we tested was that as the case study of the 17-year-old MJ with ASD, they too manifested the tendency towards the

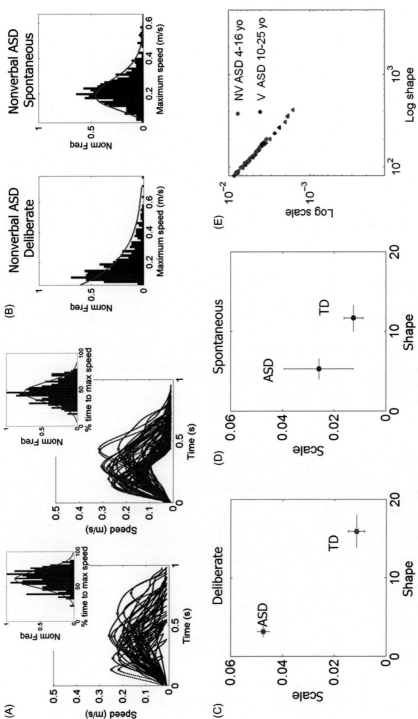

FIGURE 3.25 **Estimating the signatures of individuals with ASD.** (A) Representative trial-by-trial speed activity from a child with ASD. Trials were automatically extracted from the MTS task (as explained in Figs. 3.13–3.15). Insets are frequency histograms of the time from the movement onset to the first velocity peak for the forward and retracting motions. (B) Exponential distribution fit to the frequency histogram of the speed maxima corresponding to the hand forward motions deliberately pointing to the target in the context of decision-making for the MTS task. Skewed Gamma distribution fit to the frequency histogram of the trial-by-trial speed maxima form the uninstructed spontaneous hand retractions extracted from the continuous flow of behavior. (C) Comparison of group signatures obtained by pooling all data from ASD children and age-matched controls in both the deliberate and the spontaneous processes under investigation. (E) The micromovement analyses of the ASD individuals across ages and verbal abilities ranging from 4 to 25 years old. Each point represents a participant (50). They remain on the LUQ of the Gamma parameter plane with high dispersion and skewed distribution tending towards the memoryless exponential. This is in contrast to controls (see Fig. 3.18B) that make the maturational transition towards the right lower quadrant after 5 years of age. This transition indicates lower dispersion (higher signal-to-noise ratio) and more symmetric shapes tending towards the Gaussian distribution. Pink, NV ASD 4–16 years; Black, V ASD 10–25 years.

exponential distribution in the fluctuations of the speed peak amplitude of their goal-directed reaches. This can be appreciated in Fig. 3.25B for this ASD representative child. It was confirmed for the group data pooling across the young children spanning 3−10 years of age in relation to TD age-matched peers. The plots on the Gamma parameter plane of Figs. 3.25C−D show the unambiguous statistical differences in their speed-dependent data. Fig. 3.23E shows the signatures of the fluctuations in speed amplitude micromovements for all ASD participants we examined. Contrast this figure with the evolution in the typical signatures of speed−amplitude micromovements of Fig. 3.18B. Both figures are plotted using similar log-log scale. The trend found in controls whereby the stochastic signatures shift down and to the right as the child ages does not hold in ASD neurodevelopment. In ASD, from 4−25, whether verbal or nonverbal, the somatic-motor signatures do not evolve. They are stagnant and remain confined to the left upper quadrant (LUQ) of the Gamma parameter plane (as shown in Fig.3.25E for participants with ASD spanning ages between 3 and 25 years old).

As in the case of the controls, I also examined in these ASD participants the transient shifts of their stochastic signatures during the decision-making process that the MTS task required. Individually their motor patterns adapted (Fig. 3.26A−B) for both the forward and retracting motions, a feature that we quantified across the group as well (Fig. 3.26C). In this figure, we color code the participants by the reported IQ scores to gain a general sense of possible relations between adaptive motor learning (plasticity in the somatic-motor signal) and IQ. Given the cognitive component of the task, we wanted to know if motor performance was in anyway matching cognitive decision-making. When we examined their decision movement latencies (Fig. 3.26D) for each of the IQ groups, we quantified in ASD a systematic decrease in motor decision latency that matched well the behavior of the TD participants. Independent of age, participants with ASD and lowest IQ matched the performance of the youngest TD group (3−4 years old). As the IQ increased in the second ASD group, their decision-making performance decreased. The latter was quantified by the latencies between the stimulus onset and the initiation of the reach (the time up to the first velocity peak). Lastly, the group with ASD and highest IQ matched the performance of the controls in the corresponding group of college age students. The variability of these patterns also differentiated each group with an overall more variable pattern in ASD. Fig. 3.27A shows the frequency histograms of decision movement latency sorted by IQ and age, while Fig. 3.27B shows the locations of each group on the Gamma parameter plane corresponding to the distributions estimated for the time series of this decision-making parameter.

An interesting finding appeared when we examined these groups split by sex, namely different patterns emerged between males and females in ASD.[45] The result was interesting considering the 5:1 ratio of males to females in the ASD diagnosis.[46,47] Whereas clinical pencil-and-paper inventories prevalent in the field fail to detect the female phenotype, here a simple objective physical metric of transient motor performance during decision-making could easily separate the signatures of variability in the latencies of these decisions as the children pointed to the target. There was a clear separation of these patterns,

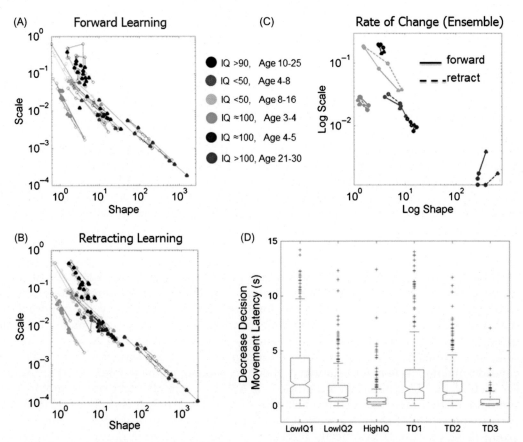

FIGURE 3.26 Stochastic trajectories of all participants with the shifts in Gamma signatures as a function of cognitive stimuli. (A-B) Individual points represent individual participants with lines connecting the shifts for that person corresponding to the shape and dispersion of the distributions as a function of the change in the cognitive stimuli. Colors denote the age and IQ in the legend. (C) Pooled data across different IQ-age groups also reveal the rates of change across three different cognitive stimuli. (D) The evolution on the decrease of decision time of ASD individuals grouped by IQ (independent of age) in relation to the evolution of control groups separated by age. All comparisons are statistically significantly different according to the Kruskall−Wallis nonparametric ANOVA. Note that although the decision-making times improve in ASD, the stochastic signatures for the deliberate and spontaneous processes remain on the LUQ, high in noise and towards skewed distributions tending to the memoryless exponential of the 3-year-old participants. Even the 8−16 participants with ASD (maize) remain with higher noise and more skewed shapes than the 4−5-year-old controls (blue). Further, the 10−25-year-old group with ASD remains at the highest noise level with very skewed distributions. Bold line, forward; dashed line, retract. Black, IQ > 90, Age 10−25; Pink, IQ < 50, Age 4−8; Yellow, IQ < 50, Age 8−16; Green, IQ ≈ 90, Age 3−4; Blue, IQ ≈ 100, Age 4−5; Red, IQ > 100, Age 21−30.

particularly in reference to sex for very early age (Fig. 3.28A vs 3.28B). Yet, these subtle differences escape the naked eye. As such, they remain to this day in 2017 a tremendous challenge for the clinical inventories clinicians perform. Chapter 7, retakes this issue using spontaneous involuntary motions of the head during resting state activity of the fMRI.

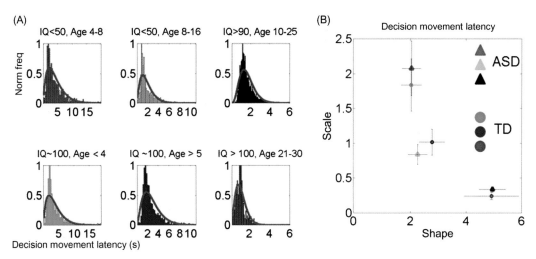

FIGURE 3.27 **Stochastic evolution of the decision movement latency (marked by the time spanned between stimulus onset and movement initiation up to the first significant velocity peak, prior to the first deceleration phase or the hand.)** (A) Frequency histograms of the decision movement latency for each age group and IQ. (B) Gamma parameter plane map of empirically estimated shape and scale parameters denoting the skewness and dispersion of each group's estimated PDF. Legend colors correspond to the colors of the histograms.

There, we were able to examine 300 females and a more clear picture with great statistical power emerges from the results of our analyses.

A summary from these data comparing different IQs and age groups on the Gamma parameter plane is shown in Fig. 3.29. For the first time, we could automatically separate ASD from neurotypical controls using objective means. This separation was clear at the somatic-motor control level but also at the level of cognitive performance. Importantly, the overall message from the results demonstrated that the children with ASD could learn to perform decision-making tasks and become proficient at it. We were excited in the lab since research involving objective assessment of somatic-motor issues was nonexistent. However, it was rather disappointing to see that the road to report these results was rather arduous. It took us three full years of rejection from several autism-related journals, before we found a way to communicate the findings to the scientific community, the parents, and the ASD self-advocates.

FORTY SUBJECTS ARE NOT ENOUGH

One of the major stumbling blocks to publish our work was the unfamiliarity of the scientific community and the clinicians with the notion of empirically estimating the probability distributions rather than assuming a theoretical mean from a theoretical distribution. It took us some time before we realized that outside the fields of Physics, Applied Math, and Engineering, the subject of statistics is taught differently. Every reviewer and editor

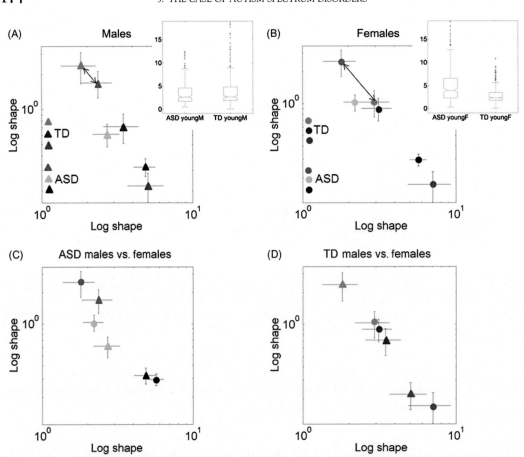

FIGURE 3.28 **Differentiating females and males in ASD and neurotypical controls according to the stochastic signatures of speed parameters extracted from pointing motions during decision-making.** (A) Male participants are compared across the ASD and control groups for each age group. (B) Female participants are compared across the ASD and control groups for each age group. (C) ASD males vs females differentiated within each age group. (D) Control males vs females differentiated maximally within the youngest group.

we came across could not understand that our methods were not assuming the normal distribution across the population but rather individually estimating for each participant the family of probability distributions that most likely fit the inherent variability of the data. The statistical power was not in the group size but rather in the thousands events we could accumulate in an experimental session of the individual. The individual variability changed patterns moment by moment as registered by continuous behavior. This personalized approach was a departure from the traditional "significant hypothesis testing" paradigm that these reviewers from the Psychology and Cognitive Neuroscience fields were demanding from us.

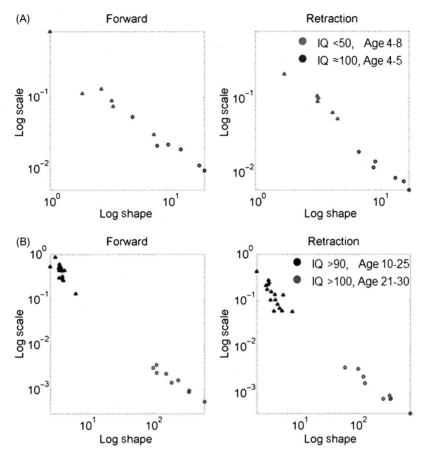

FIGURE 3.29 **Data-driven separation of ASD vs control participants across different ages and IQ scores.**
(A) Young participants with ASD and lower IQ scores segregate from age controls with higher IQ according to
the stochastic signatures derived from hand speed amplitude fluctuations (deliberate) and backwards (spontaneous) movements of the MTS task. (B) Same as in (A) for older participants with ASD and controls with overlapping ages. Each marker represents the shape and scale (dispersion) of an individual participant. Pink, IQ < 50,
Age 4−8; Blue, IQ ≈ 100, Age 4−5; Black, IQ > 90, Age 10−25; Red, IQ > 100, Age 21−30.

The unfamiliarity of the field with the registration of natural behaviors as they continuously occur was also a factor against us. They could not see that we were actually filling
in the gaps between touches or key presses and tapping into processes transpiring largely
beneath awareness. As such, they asked us for more subjects because we did not have two
homogeneous and different groups to run the "ANOVA". They asked us to perform linear
regression and correlate discrete ordinal clinical scores that had no standardized metric
scale (or had never been derived for normative data) with continuous physiological signals
for which we had created a (unitless) standardized normalized scale mapped to physical
data. Ours was a rather continuous real-valued scale ranging between 0 and 1 values

useful across different nervous systems structures, different processes, and independent of the instruments' units and sampling rates.

It was really frustrating to see how much ignorance a field had built upon, while leading a fundamental problem in brain research and directly impacting the lives of many. The whole approach was flawed from beginning to end. It was put together in a very odd way, polluted by high conflicts of interests. There was nothing we could do because the system was rigged. *How do you fight such a system?* None of the reviews were constructive. They were reviews from people claiming there was nothing wrong with the motor systems of ASD individuals. *They could move just fine*, had no sensory-motor issues and the ANOVAs were missing from our paper. And yes, 40 subjects were not enough!

WHEN OPEN ACCESS CAME TO THE RESCUE

It was not the first time in my academic career that I found resistance to new ideas. In fact, this was the norm for me. As such, I decided to go ahead and open the paper for discussion in a forum of people that had an interest in motor and sensory issues in ASD. This was not an easy task, but it was not impossible either.

I had accepted in 2012 an invitation from Professor Sidney Simon of Duke University to become Associate Editor of Frontiers in Integrative Neuroscience. As part of my new role, I had the opportunity to publish an inaugural paper in this peer-reviewed forum. Unlike other journals in science, Frontiers was not only open access, it also had a system whereby the review process was interactive and at the end of the revisions, if the paper was accepted the names of the reviewers were revealed. This method had many advantages as reviewers would curtail disrespectful tones and be openly monitored by the community of editors as well. Furthermore, Frontiers encouraged the launching of Research Topics, including controversial or cutting edge topics that were hard to open for discussion in other more traditional journals. All I needed to launch a Research Topic was to gather 10 potential contributors to agree on the topic and commit to a contribution.

Sensory-motor issues in autism had been well-documented since the 1980s but very rarely measured with objective means. There were many controversies surrounding movements and sensory processing because Psychologists and Psychiatrists wanted to keep the diagnoses free of motor and sensory issues. As such, the diagnoses of autism and related disorders remain primarily confined to cognitive-social phenomena. In my mind, that made no sense whatsoever.

Everything in the autism literature was described by behavioral inventories, yet the notion of behavior in that field was very far from the notion of behavior in the fields that study sensory-motor processes: i.e., transduction, transmission, integration, and transformations of neural signals across sensory-motor domains. Further, to us in the fields of neuromotor control, ecological (Action—Perception) psychology, and movement neuroscience at large, the dichotomy between sensory and motor domains was rather blurred. Motor was also sensory input via the kinesthetic (movement-based) feedback arriving at the brain centers through afferent channels. Sadly, we realized that leading labs in autism research had no sense of the neuroanatomy and neurophysiology of the nervous systems, much less of those aspects of scientific research in the larger context of natural social behaviors. *How do you envision social exchange in a disembodied fashion?*

The oddest piece of the autism science puzzle was the fact that all inventories to measure social deficits were a one-sided description expressing the opinion of what a person (being paid to do so) thought a child lacked, i.e., was impaired at. Nowhere in these inventories was the opposite notion of what a nascent nervous system coping with a neurodevelopmental glitch had already accomplished. In other words, *what were the predispositions and capabilities of that child*? What were those positive potentials that enabled him or her to overcome the horrifyingly uncertain universe of noise and randomness that we readily saw in their bodily micromotions, systemically, across all levels of the PNS and CNS?

After an extensive literature review on the matter, I decided to create dynamic measures of dyadic interactions for true assessment of social exchange. I will speak of those in Chapter 6. I also decided to organize the research Topic in such a way that I could include parents and self-advocates. To that end, I contacted Professor Anne Donnellan of the University of San Diego. I did not know Anne personally. I had never heard her name until I did a google search on "autism sensory-motor systems". Her name popped-up in many feeds. One of them had her phone number at the University of San Diego. I just called her up. She answered the phone and we spoke as if we had known each other our entire lives. I flew to San Diego that week to meet her in person.

Anne is someone who has written extensively in favor of the affected individuals and their families.[48-50] She had developed a pioneering vision about sensory-motor issues and their use to develop accommodations to help people live fruitful lives and thrive with social support. Hers was a fundamental missing piece in this science puzzle. It was a piece about the civil and social rights of individuals with disabilities in general. But in the case of autism, such issues were particularly poignant because of the early age in which these sensory-motor problems began manifesting and becoming obvious to the naked eye. Without civil and human rights to receive an education, the children's future were already doomed by a diagnosis that did not address their health and physiology in the first place. Despite many medical issues, autism was treated as a social-interaction deficit. It also remained under the implicit social perception that the children could be reshaped into more socially-acceptable beings -without considering the implications of such reshaping for their nascent and coping nervous systems.

Science and clinical practices have been rather out of touch with the reality of the lives of people in the spectrum, a fact that I confirmed when my student Jill Nguyen and I interviewed over 120 stakeholders in the ecosystem of autism. These included parents, affected individuals, behavioral therapists, politicians steering legislation, litigators, researchers, entrepreneurs, insurance companies, and many others who unveiled for us the critical needs of this community.

After I gathered about 20 contributors, I invited Anne to host the topic with me and she came up with 20 more people interested in participating. In total, we had 99 contributors, far exceeding the original requirement of 10. Frontiers then sponsored several of the papers by waiving the publication fees to parents and self-advocates. It was an incredible gesture to help us promote a topic that was rather new and controversial. I could then appreciate the power of the Frontiers platform in helping science at large; but more importantly in helping humanity. They opened science to the world.

While we organized the topic, and entitled it "Autism: The Movement Perspective", I teamed-up with Jorge to deploy our experiments at Indiana University. He had met Dr. John Nurnberger at the Christian Sarkine Autism Treatment Center and proposed to join forces and recruit families for our study. A well-known geneticist, Dr. Nurnberger had conducted studies with families in the spectrum for years and had kept in touch with many of them. His lab recruited over 25 families and we were able to record not only the children and adolescents but also their parents. The results were remarkable and allowed us to reach a large number of participants. We could then convince the reviewers that the sensory-motor phenomena we were systematically quantifying in autism was real. More important yet, the excess noise was unique to autism, as we were able to see when we placed all participants on a standardized parameter space, where we compared them with other neurological and neuropsychiatric disorders. Autism stood apart with the highest noise-to-signal ratio and the most random signatures of the speed maxima, its amplitude, and its timing. The map was later published in Frontiers in the context of Precision Psychiatry and the original paper explaining the micromovements perspective appeared in the Research Topic. We had managed to more than double the N for the autism study and could now publish the paper with high confidence.

Fig. 3.30 shows the map of the 176 participants we gathered, including the parents of the participants with ASD from the Indiana University group. Shockingly for us, they fell with the severe PD patients and the elderly. But they were young fellows in their late 20s and early to mid-30s. The oldest person in the parents' group was 42 years old. *Where were these bradykinetic movements with such high noise coming from?* We do not have an answer yet, but were finally able to report the empirical results to the neuroscience community and to the world at large.

The manuscript had such a spike within days, that the Nature Publishing Group which had by then partnered with Frontiers sponsored the Press Release of the paper. The impact of the Research Topic since 2013 has been quite significant as shown by Fig. 3.31. Four years later, we are fast approaching 1/2 million views and have well over 59,000 actual downloads from the topic across the world, with hubs in the United States, the UK, and China. I think that we inadvertently created a new field altogether from a latent field that was sparsely hidden across the world. Frontiers brought them together and I suspect, we

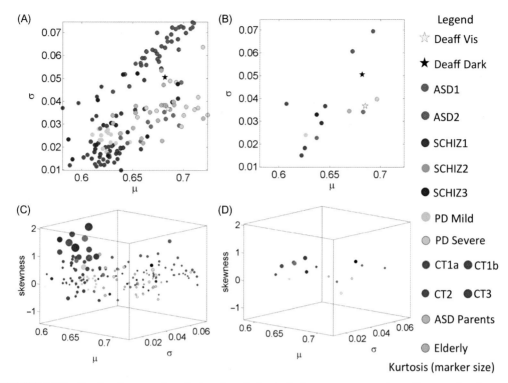

FIGURE 3.30 **Stochastic map of typical and pathological signatures across time.** Cross-sectional population ranging from 3 to 77 years of age and including ASD, PD of different severities, deafferentation, schizophrenia, and controls (See legend). Map also includes young parents of ASD participants. (A) Empirically estimated Gamma moments (mean and variance) from the micromovements in speed amplitude fluctuations. These minute fluctuations are extracted from trial-to-trial as the basic pointing task behavior naturally and continuously unfold according to the individual's own pace. (C) Additional Gamma moments (skewness along the z-axis) and kurtosis represented in the size of the marker. (D) Group data pooled across all subjects with a clinical label. Notice the ASD individuals separate from the controls with highest noise than the severe PD and at the level of noise and variability of the physically deafferented participant moving in the dark. Deaff vis; Deaff dark ; ASD1; ASD2; SCHIZ1; SCHIZ2; SCHIZ3; PD mild; PD severe; CT1a; CT1b; CT2; CT3; ASD parents; Elderly.

are well on our way to make a positive contribution to the lives of affected families. This was just the start of a long road leading to a series of biometrics that will enable the creation of dynamic outcome measures to quantify social exchange in personalized manner but also in groups of people interacting in real time. We pick this up again in Chapter 6, to show the actual applications from theoretical ideas that have become physical reality.

Our secret was to not give up and pursue the objective path of science above and beyond the opinions of many.

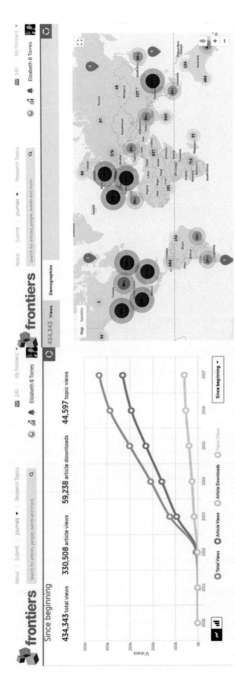

FIGURE 3.31 **Defying the odds: After 3 years of rejection by the traditional peer-reviewed system the Open Access Platform of Frontiers helped us create a new Autism Sensory-Motor Control community.** Our research topic, Autism, the movement perspective brought together hubs of researchers, clinicians, parents, and self-advocates to openly discuss the sensory-motor issues that traditional research and clinical practices had swept under the rug in favor or a narrative that did not help the people affected by ASD. Our paper, Autism: The Micromovement perspective set a turning point to the use of biometrics and initiated a new avenue towards the implementation of Precision Psychiatry and the introduction of mobile health elements to the research and clinical practices. The new dynamic lens is aimed at longitudinally tracking the progression of the disorder, its evolution with treatments and overall the translation of technology to commercialization and affordability for the affected families. As of 2017, over one-fourth of a million people have read our Research Topic and over 30,000 downloaded our papers. This was a communal effort with unprecedented support from an Editorial House.

References

1. José JV, Saletan EJ. *Classical dynamics : A contemporary approach.*. Cambridge, UK; New York: Cambridge University Press; 1998:670. xxv.
2. Torres EB. Two classes of movements in motor control. *Exp Brain Res*. 2011;215(3−4):269−283.
3. Limpert E, Stahel WA. Problems with using the normal distribution--and ways to improve quality and efficiency of data analysis. *PLoS ONE*. 2011;6(7):e21403.
4. Mouri H. Log-normal distribution from a process that is not multiplicative but is additive. *Phys Rev E Stat Nonlin Soft Matter Phys*. 2013;88(4):042124.
5. Torres EB, et al. Sensory-spatial transformations in the left posterior parietal cortex may contribute to reach timing. *J Neurophysiol*. 2010;104(5):2375−2388.
6. Limpert E, Stahel WA, Abbt M. Log-normal distributions across the sciences: keys and clues. *BioScience*. 2001;51(5):341−352.
7. Cole J. *Pride and a daily marathon*. Cambridge, Mass: 1st MIT Press; 1995:194. MIT Press. xx.
8. Choi K, Torres EB. Intentional signal in prefrontal cortex generalizes across different sensory modalities. *J Neurophysiol*. 2014;112(1):61−80.
9. Torres EB, Cole J. A neural correlate of disembodied intention? *J Cognitive Sci*. 2014. **under revision**.
10. Torres EB, Heilman KM, Poizner H. Impaired endogenously evoked automated reaching in Parkinson's disease. *J Neurosci*. 2011;31(49):17848−17863.
11. Torres EB. Atypical signatures of motor variability found in an individual with ASD. *Neurocase: The Neural Basis of Cognition*. 2012;1:1−16.
12. Lleonart J, Salat J, Torres GJ. Removing allometric effects of body size in morphological analysis. *J Theor Biol*. 2000;205(1):85−93.
13. Torres EB, et al. Toward precision psychiatry: statistical platform for the personalized characterization of natural behaviors. *Front Neurol*. 2016;7:8.
14. Torres EB, Cole J, Poizner H. Motor output variability, deafferentation, and putative deficits in kinesthetic reafference in Parkinson's disease. *Front Hum Neurosci*. 2014;8:823.
15. Goodale MA, Milner AD. Separate visual pathways for perception and action. *Trends Neurosci*. 1992;15(1):20−25.
16. Goodale MA, Westwood DA, Milner AD. Two distinct modes of control for object-directed action. *Prog Brain Res*. 2004;144:131−144.
17. Milner AD, Ganel T, Goodale MA. Does grasping in patient D.F. depend on vision? *Trends Cogn Sci*. 2012;16(5):256−257. discussion258-9.
18. Whitwell RL, et al. DF's visual brain in action: the role of tactile cues. *Neuropsychologia*. 2014;55:41−50.
19. Whitwell RL, et al. Patient DF's visual brain in action: Visual feedforward control in visual form agnosia. *Vision Res*. 2015;110(Pt B):265−276.
20. Goodale MA, et al. Kinematic analysis of limb movements in neuropsychological research: subtle deficits and recovery of function. *Can J Psychol*. 1990;44(2):180−195.
21. Milner AD, Goodale MA. Visual pathways to perception and action. *Prog Brain Res*. 1993;95:317−337.
22. Whitwell RL, Milner AD, Goodale MA. The two visual systems hypothesis: new challenges and insights from visual form agnosic patient DF. *Front Neurol*. 2014;5:255.
23. Milner AD. How do the two visual streams interact with each other? *Exp Brain Res*. 2017.
24. Nguyen J, et al. Schizophrenia: The micro-movements perspective. *Neuropsychologia*. 2016;85:310−326.
25. Nguyen J, et al. Automatically characterizing sensory-motor patterns underlying reach-to-grasp movements on a physical depth inversion illusion. *Front Hum Neurosci*. 2015;9:694.
26. Nguyen J, et al. Methods to explore the influence of top-down visual processes on motor behavior. *J Vis Exp*. 2014;(86).
27. Nguyen, J., et al.*Characterization of visuomotor behavior in patients with schizophrenia under a 3D-depth inversion illusion*. in *The Annual Meeting of the Society for Neuroscience*. 2014. Washington DC.
28. Nguyen J, et al. *Quantifying changes in the kinesthetic percept under a 3D perspective visual illusion. Vision Science Society*. Naples: Fla; 2013.
29. American Psychiatric Association and American Psychiatric Association. *Neurodevelopmental disorders : DSM-5 selections.*. Arlington, VA: American Psychiatric Association Publishing; 2016:182. xiv.

30. American Psychiatric Association and American Psychiatric Association. *DSM-5 Task Force. Diagnostic and statistical manual of mental disorders: DSM-5*. 5th ed. Washington, D.C: American Psychiatric Association; 2013:947. xliv.

31. Von Holst E, Mittelstaedt H. The principle of reafference: Interactions between the central nervous system and the peripheral organs. In: Dodwell PC, ed. *Perceptual Processing: Stimulus Equivalence and Pattern Recognition*. New York: Appleton-Century-Crofts; 1950:41–72.

32. Martinez-Conde S, Macknik SL, Hubel DH. The role of fixational eye movements in visual perception. *Nat Rev Neurosci*. 2004;5(3):229–240.

33. Hawgood S, et al. Precision medicine: Beyond the inflection point. *Sci Transl Med*. 2015;7(300). p. 300ps17.

34. Insel T, et al. Research domain criteria (RDoC): toward a new classification framework for research on mental disorders. *Am J Psychiatry*. 2010;167(7):748–751.

35. Insel TR. The NIMH Research Domain Criteria (RDoC) project: precision medicine for psychiatry. *Am J Psychiatry*. 2014;171(4):395–397.

36. Torres EB, Jose JV. *Novel diagnostic tool to quantify signatures of movement in subjects with neurological disorders, autism and autism spectrum disorders*. US: R. University; 2012.

37. Gershon MD. *The second brain: The scientific basis of gut instinct and a groundbreaking new understanding of nervous disorders of the stomach and intestine*. 1st ed. New York, NY: HarperCollinsPublishers; 1998:314. xvi.

38. Du X, et al. Local GABAergic signaling within sensory ganglia controls peripheral nociceptive transmission. *J Clin Invest*. 2017.

39. Torres EB, Zipser D. Reaching to grasp with a multi-jointed arm. I. Computational model. *J Neurophysiol*. 2002;88(5):2355–2367.

40. Torres EB, Zipser D. Simultaneous control of hand displacements and rotations in orientation-matching experiments. *J Appl Physiol (1985)*. 2004;96(5):1978–1987.

41. Flash T, Hogan N. The coordination of arm movements: An experimentally confirmed mathematical model. *J Neurosci*. 1985;5(7):1688–1703.

42. Torres EB. Signatures of movement variability anticipate hand speed according to levels of intent. *Behav Brain Funct*. 2013;9:10.

43. Brincker M, Torres EB. Noise from the periphery in autism. *Front Integr Neurosci*. 2013;7:34.

44. Kathirvel P, et al. An efficient R-peak detection based on new nonlinear transformation and first-order Gaussian differentiator. *Cardiovas Eng Technol*. 2011;2(4):408–425.

45. Torres EB, et al. Strategies to develop putative biomarkers to characterize the female phenotype with autism spectrum disorders. *J Neurophysiol*. 2013;110(7):1646–1662.

46. Volkmar FR, Szatmari P, Sparrow SS. Sex differences in pervasive developmental disorders. *J Autism Dev Disord*. 1993;23(4):579–591.

47. Mandy W, et al. Sex differences in autism spectrum disorder: Evidence from a large sample of children and adolescents. *J Autism Dev Disord*. 2012;42(7):1304–1313.

48. Donnellan AM. *Progress without punishment: Effective approaches for learners with behavior problems. Special Education Series.*. New York: Teachers College Press; 1988:168. xi.

49. Donnellan AM, Leary MR. *Movement differences and diversity in autism/mental retardation : Appreciating and accommodating people with communication and behavior challenges. Moving on Series.*. Madison, WI: DRI Press; 1995:107.

50. Donnellan AM, Hill DA, Leary MR. Rethinking autism: Implications of sensory and movement differences for understanding and support. *Front Integr Neurosci*. 2012;6:124.

51. Leaf JB, et al. Applied behavior analysis is a science and, therefore, progressive. *J Autism Dev Disord*. 2016;46(2):720–731.

52. Choutka CM, Doloughtty PT, Zirkel PA. The "discrete trials" of applied behavior analysis for children with autism: Outcome-related factors in the case law. *J Special Educ*. 2004;38(2):95–103.

4

The Case of Schizophrenia: Is that My Arm Moving on Purpose or Spontaneously Passing by?

This is the principle of least action, a principle so wise and so worthy of the supreme Being, and intrinsic to all natural phenomena... [1744] When a change occurs in Nature, the quantity of action necessary for change is the smallest possible. The quantity of action is the product of the mass of the body times its velocity and the distance it moves [1746]. **Pierre-Louis Moreau de Maupertuis**

PART I   PREDICTIVE REACH: FROM BROWNIES TO A PDE

"The generation of goal-directed movements requires the solution of many difficult computational problems".[1] I know this very well now, but back in 1995, when David Zipser introduced me to this problem, I had no idea how important trying to find a general solution to deal with intentionality in a system with redundant degrees of freedom (DoF) would become throughout my research career. This problem has indeed dominated most of what I have worked on, including how to spot spontaneous aspects of behavior that follow no particular goal or instruction.

At the time that I was introduced to the problem of intentionality in movement generation, I was a freshly arrived graduate student admitted to the Cognitive Science program of University of California, San Diego (UCSD). I had completed a year-long stay at the National Institutes of Health (NIH) after finishing a combined degree in Mathematics and Computer Science. In my internship at the NIH I had acquired a major enthusiasm for developing computational applications geared toward clinical work. As it turns out I owe that to Dr. Harold Varmus. Unknowingly, during a welcoming reception to the interns, I spent about half an hour talking with Dr. Varmus. I learned coincidentally, during a very informal conversation preceding the opening remarks of the meeting, that he had attended medical school at USCF with Dr. Larry Williams, the husband of my English teacher, Judith Williams. I mentioned to him during the reception that I was not sure what to do with my degree, but I wanted to apply my original math knowledge-base to medicine. I had assumed that he was a bystander waiting for the speaker to deliver the welcoming keynote for the interns that summer and we were just chatting away about the Bay Area, where I came from. But in the middle of our conversation he asked me to *"hold that thought"* and next thing I knew, he was announced as the Nobel laureate and Director of the NIH, and delivered a great speech about interdisciplinary collaborative efforts and the need to diversify research scholars.

I ended up working as part of the Baltimore Longitudinal Study of Aging (**Notes** [1]), a large effort of the National Institute on Aging tracking all aspects of human aging since 1958. I assisted with image processing in Alzheimer's related research under the guidance of Dr. Alan Zonderman and Dr. Sue Resnick. There I developed an interest for Brain Science and the thought of tentatively pursuing a PhD in that area. That idea became a reality in the early months of 1995, when I was admitted to the UCSD Cognitive Science Dept., the place that witnessed the birth of the field of Cognitive Science and parallel distributing processing. But unlike my first-year peers at UCSD, I had no clue which lab I wanted to join, or who I wanted to work with, or what any of the Cognitive Science research was about. I was very ignorant and would ask very naïve questions like *"...so, how do these neurons -you talk about- communicate? Do they move at night when we sleep?"* I would get laughed at pretty often and to top that off, I could not parse out all conversational English just yet. The only thing I was comfortable with was the math and the coding. Everything else around me was more or less noise.

While searching for a lab, I recall a nearly 2-hour exchange with Marty Sereno. He was explaining his research and during the conversation I found myself thinking at some point *"oh my God, this guy is so smart!"*; *"how can someone know so much?"* The entire first-year class was in awe with his knowledge and research. I decided to try and do a rotation in his lab. Marty was starting to develop what we know today as Voyager, a tool for brain-image analyses. I reasoned that the skills I already had on surface-curvature analyses and other elements of differential geometry might help me understand some of what he was trying to accomplish. But then, toward the end of that week, on Friday afternoon, I got lost on the Department's second floor and randomly wondered into a lab meeting. They did not throw me out, so I tuned in for a while and then decided to stay.

It was the weekly meeting of the Z-lab, the Zipser group. This lab built computational models of the brain at neural systems level and in a very relaxed manner discussed questions that others had avoided or swept under the rug. Being at that meeting was like reading the methods section of a paper in a completely different way. It was like looking at the details of how things had actually been done to get the results the paper claimed; so we could learn what exactly was left out of the enquiry that could entirely change the claim of the paper if we were to re-do it. Unknowingly, this style of scrutinizing the methods section of each paper I read stuck forever. At that meeting they served freshly baked brownies, a treat I had never had before.

[1]https://www.blsa.nih.gov/)

I'm not sure if it was the brownies they served that day, or the intellectual atmosphere of that group, or the overall sense of how cool doing research suddenly was; but ever since then, the exciting part of my research is actually finding what has not been done yet that could fundamentally change what we think we knew. I owe that to David Zipser (*and the brownies*). Dave (as we dearly called him) taught me to think in a fun way, in a way that would maximally free my mind so that creativity and out-of-the-box questions could come easy and make a dent on my scientific enquiry. Since then, repeating or doing an epsilon-modification on what someone else already did or thought about seems utterly boring. Inevitably, the scientific line of enquiry of my lab today tends to levitate toward phenomena for which not much is known.

The first assignment I faced at the Z-lab meeting back in 1995 was to present to the group some contemporary research. They asked me to explain and implement the cutting edge models of neuromotor control that were driving the empirical research at the time. These were the minimum jerk model[2] and the minimum torque-change model[3] derived in the mid-to-late eighties by independent labs at Massachusetts Institute of Technology (MIT) and Advanced Telecommunications Research Institute International (ATR), Japan, respectively. They were designed to mimic the following problem: given a visual target and some configuration of the arm at time t_0; move the hand along a trajectory to reach the target at some given final time t_f. Figure 4.1 shows the arm reaching problem whereby a certain initial position of the hand can be attained using a multitude of arm postures. Likewise, given a target, many different motion paths can connect the initial position of the hand and the target in the three-dimensional space. Thus, one question was, which path would the brain choose?

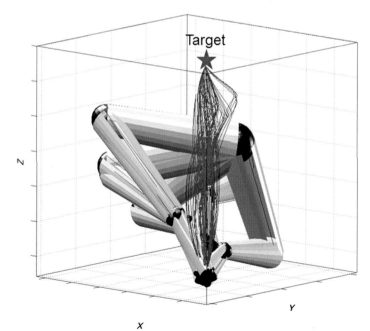

FIGURE 4.1 **The Bernstein DoF problem:** Given a visual target to reach, and an initial position of the hand, how does the brain choose a unique postural path that consistently travels through a hand path and reaches the target? Since there are many more DoF than dimensions in the space where the hand moves, there are many possible solution paths to the same target, how does the brain select the most adequate one for a given situation?

FINDING A PROBLEM WORTHY OF A PHD THESIS IN COGNITIVE SCIENCE

The models of reaching movement at that time were framed under the general rubric of optimal control theory.[4] Optimal control addresses the problem of finding a control law governing a system such that some optimality criterion is met, for example, the hand moves continuously along the path in such a way that it maximizes the smoothness of the motion (or minimizes the jerk, the third derivative of position with respect to time). The control problem involves a continuous time-cost functional. This is a function of the state and control variables that has a value at each location of the space (the cost) for a given target. For example, Figure 4.2A shows a surface representing a simple cost function mapped across the space of joint angles for points $q \in Q \subset R^2$. This Q-space is often called the internal configuration space, since at each location the arm has some postural configuration of the joint angles. In the context of brain functioning and somatic-motor parameters, this space would be used to represent proprioception: postures of the arm (static configurations) sensed back internally through joint receptors. This internally sensed information would also be dynamically updated and sensed back through kinesthetic reafference from self-generated motions (a concept that we visited in the previous chapters of the book).

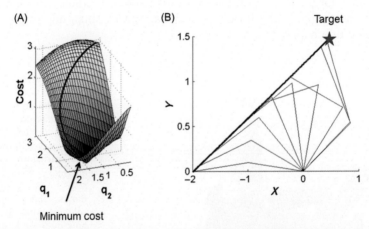

FIGURE 4.2 **Exploring optimality principles to solve the reaching problem.** (A) A simple toy model of two dimensional reaches with two joints of the arm denoting the shoulder (q_1) and elbow (q_2) motions in the Q-space of postural configurations that map to the space X where the hand has to reach the target. The surface represents the value of the cost function at each point in posture space, that is, for each (q_1, q_2) possible configuration of the arm mapping to a set of (x_1, x_2) points representing the hand position in (B). The cost in this case is the remaining distance to the target in X. The blue curve is a connected path from the initial posture $\left(q_1^{initial}, q_2^{initial}\right)$ to the final posture at the target, where the hand is $\left(x_1^{target}, x_2^{target}\right)$ and the cost is at 0 distance. This cost has a unique global minimum at 0, but there are many possible paths in Q that can lead to it while the hand moves in X. Some paths in Q add up to higher total cumulative cost, while one will be the one with minimal cumulative cost. The question is how to attain the unique one? (B) The Q-path maps to its corresponding X-path, a straight line in this case, the shortest among all possible X-paths connecting $\left(x_1^{initial}, x_2^{initial}\right)$ and $\left(x_1^{target}, x_2^{target}\right)$.

The values of this particular cost in Figure 4.2A are based on the distance from the hand (the end effector of the arm) to the target as measured in the external space. That is, the space where the sources of sensory input are external to the body (represented in Figure 4.2B), as in the case of visual input about the target location. At each location, given by two of the joint angles of the arm at the shoulder (q_1) and elbow (q_2), the cost takes a value representing how far the target location in the external visual space is when the arm is at that configuration. In this example, the cost is at its minimum of zero value at (1, 1.5) radians, where in the external space in Figure 4.2B the hand reached the target by traveling along the straight line, that is, the shortest distance path connecting the initial position of the hand to the target position. This is a simple example where there is no dimensional disparity between the space of joint angles and the hand space (as in the more realistic case represented in Figure 4.1). In the context of the nervous systems though, the problem becomes very complex because there are many more joints in the whole body than dimensions in the external environment where the body moves. The specific case of the arm in Figure 4.1 illustrates the question: given a target to reach and a starting posture, how to choose a consistent solution hand path to that target out of the many possible angular excursions that different postural configurations of the arm would allow.

In the optimal control framework, the choice of the cost and control parameters determines how one assumes the brain may prioritize sensory-motor information to plan and execute the movement trajectory to complete a goal directed movement. The goal could involve pointing to, reaching for, or grasping a target object in the visual environment. The mechanical formulation of the problem I encountered in the motor control literature had left out the dimensional disparity issue. This issue was relevant to the problem of possible cognitive strategies the brain may use. The solution these models offered was useful to study the mechanical properties of an artificial system subject to certain known dynamics. However, as we saw in the **Part II of** Chapter 1, the human nervous system develops and grows rapidly with rapidly changing dynamics in its nascent stage, gradually gaining stability as the infant ages. How the optimal control solution would scale for such a complex rapidly-changing dynamics was not clear to me. Furthermore, designing other questions related to possible cognitive issues became problematic. I had to come up with a different formulation altogether, one that was more in tune with the problems relevant to the field of Cognitive Science, rather than to a problem in mechanics.

The solution that these papers offered defined a set of differential equations describing the instant by instant trajectory of the control variables of choice, as they change to minimize the cost function of choice. For example, in the case of maximizing the smoothness of the path, one would suppose that the brain decides to move the hand 15 cm in the span of 800 ms (.8 s). The first thing to note then is that at time t_0 the velocity and the acceleration are all 0 because the hand is at rest. At t_f this is also the case for the velocity and acceleration because the hand stopped at the target and is once again at rest; but if the hand moved to the target, it traveled the desired distance, thus $x(.8) = 15 \ cm$. To find the trajectory that moved the hand along the optimal path (e.g., the path with minimum jerk) one would assign the cost to each possible trajectory and find the one with the minimum cost: a very difficult problem indeed when the arm's actual DoF are considered.

The minimizing trajectory—if one exists—completely and simultaneously must specify both the geometric path and the time course of the motion. In the more realistic case considering the arm's numerous DoF, the problem is very complicated owing to the path under specificity and postural redundancy (Figure 4.1). How to simultaneously attain both the spatial path and the time course seemed very convoluted to me, particularly so when extending this problem to the whole body and considering the interactions across all body parts would make it intractable. The problem of redundancy had been postulated much earlier by Nikolai Bernstein[5], the founding father of the field I was about to get acquainted with when the Z-lab group asked me to present work from the field of neural motor control.

At a first glance, both the minimum jerk and the minimum torque-change models could accomplish this difficult problem provided that no redundant DoF were involved, that is, the hand moved on a plane and so did the arm. Only two of the arm joints were allowed to rotate and change while all others were not considered (as in Figure 4.2B). It was a simplification that allowed exploration of some of the computational aspects of the mechanical problem that were not apparent to the naked eye.

Every paper I read debating over the models' philosophy; however, was concerned with the control parameters of choice and the pre-selection of the cost for the optimization problem that ultimately dictated the solution trajectory. The concern, as I understood it from the papers, was which control parameter the brain was using to solve this problem (if the brain were indeed solving it this way). Some papers advocated for the extrinsic space and the path-smoothness maximization using kinematic criteria; while others advocated for the intrinsic joint angles space of arm configurations and the criteria driven by the dynamics (the forces) of the system. Importantly, the questions these models posed were also debated at the level of neuronal recordings (in the mid-nineties these involved single cell extracellular recordings in awake-behaving monkeys). As such, some experiments would explore cortical regions with dynamics-invariance of directional signal denoting the cell's preferred direction in extrinsic space.[6,7] Others would explore cortical regions that contained neurons sensitive to the changes in loads, mass, etc. related to the forces the arm had to exert to move the hand toward the target.[8] Others compared these regions with various manipulations.[9–11] The problem of which space (extrinsic *vs.* intrinsic) or mode (kinematics *vs.* dynamics/forces) the cells in the neocortex were tuned to in order to plan and execute goal directed movements was highly contentious.[7,12,13] In many occasions the debates took a personal tone[14–16] that puzzled me tremendously when in reality the empirical evidence pointed to both possibilities at once. That is, you had cells that were more or less tuned to kinematics independent of dynamics; you had cells that were very driven by the changes in dynamics and you had cells that were tuned to both! Within any given region you had all kinds of cells active. The argument of which region did what based on which cell was tuned to what dimension of the stimuli seemed strange, particularly because the animals in these experiments were so highly constrained to move and the motions they were allowed to make were so over practiced that very likely the cells' responses and the researchers' interpretations were at odds. I wondered about all the other aspects of the problem that were either assumed to be of a certain kind, hidden to the naked eye and/or altogether purposely "swept under the rug" to simplify the problem.

As a naïve student my concern was entirely different than kinematics *vs.* dynamics. I just thought that a baby having to learn to generate these optimal trajectories would be faced with the problem of having to know their duration beforehand. To me, that seemed entirely backward. The time duration and time course of our movements is not something we are even aware of as adults. Imagine then, a baby thinking "I need to move my hand for exactly 800 Ms to get to my toy 15 cm away"? During neurodevelopment it is likely that myelination and its development have an impact on the transduction and transmission times across the peripheral nervous system (PNS) and central nervous system (CNS). As such, part of what the nascent nervous systems are learning has (necessarily) something to do with internal time delays and movement durations expressed as a function of sensory processing and transduction-transmission delays. I did not think that a time-dependent formulation was adequate to model the developmental processes, or even the processes that take place when a mature system adapts to a new situation, or acquires a motor skill from scratch. In my mind time precisely depended on the adaptive sensory processing and its consequences for the learning of successful action. And yet, all the extant literature involved already-acquired motions, that is, movements that were already highly automatic and over practiced.

As much as we need to simplify a complex problem to be able to build a computational model, I thought the issue of time-dependence and over-training before the experiments took place in these approaches was puzzling (to say the least). The notion of time as an orderly process already resolved was such a major assumption (i.e., taken for granted) that no lab was even considering the question of a predefined movement duration as a basic problem that could not be explained by these models.

The critical details of how the experiments were actually done across the field of neural motor control (at the neural level and at the behavioral level) to empirically validate the computational models remained deeply buried in the methods. The fine print of the claims on "how the brain was doing it" (which were not modest) was hidden in the time-dependent assumptions. In every single case the subjects' arm was constrained to move on a plane, so there was no real chance to learn about Bernstein DoF problem.[5] This planar-motion constrain was perpetuated at all levels of research (from neural to behavioral) by instrumentation that was designed, built and sold to motor control labs. When reading the methods section of top papers from leading labs in the field one could actually see a profound conflict of interest and circular reasoning going hand in hand. One could see that many proponents of such ideas and leaders of that field were also profiting in more than one way: promoting their work and selling equipment to promote it too! Under those circumstances I wondered: how could movement neuroscience make any real progress toward our understanding of unconstrained, natural movements?

Besides the limitations the prevalent experimental apparatus imposed on the body, the instructions to perform a task involving the apparatus also constrained the possible outcomes. During the experiments, the participants would have to practice over many trials and achieve the *magic* length of time to have the hand travel along a predefined path length from the start to the target. Only then, after the highly automatic skilled motion had been acquired, the investigator would carry on the experiment, and *lo and behold!* The planar path of the over-practiced trajectory (as performed with the arm constrained to 2 DoF) was quasi straight, and the temporal speed profile followed a nearly perfect bell-shaped curve.

I could not count how many papers I read before my presentation to the Z-lab, how many talks I attended later on, over the course of those next couple of years, and how many people I met along the way who assured me that this is the way the brain does it. And *"just there is no other possible way"* -I was lectured on these issues by most (including very famous) people in that field.

In the meantime I started to think about the more fun DoF problem Bernstein had posed.[5] Within the context of the arm, how do you manage to attain a consistent *hand path* in space and simultaneously maintain a consistent *joint angle path* in the arm space, when the arm has many more DoF than the space in which the hand moves? What caught my attention was the consistency of the arm path, that is, the consistency in the *length* and *shape* of their excursions. Was that consistency still there if I did not train the arm system to move within a given time duration? In other words, if the temporal profile was highly variable from trial to trial, would the geometry of the path still be consistent? Furthermore, if I were to explicitly instruct variations in speed, would the geometric path remain conserved? I was curious to learn about possible differences between instructed speed (which required top-down intention to modulate the motion timing) and evoking speed differences without instructions (e.g., using obstacles to change the path curvature). The latter would require striking a balance between the top-down intent on avoidance and the management of highly automated bottom-up processes. These included the kinesthetic reafference integrating variations of prior experiences with contextual sensory processing and transmission delays across disparate time scales.

In 1995, I could not find any research addressing these issues because everyone in that field was going by the "fine print" instructions hidden somewhere in the methods section of those papers: train the subject until the hand reaches the temporal goal, and then test to confirm same old boring hypothesis everyone seemed to confirm over and over. But somewhere at MIT, Chistopher Atkeson and James Hollerbach had performed a very interesting experiment published in 1985.[17] This paper empirically demonstrated that the paths of three-dimensional hand trajectories from *unconstrained* arm movements performed at different speeds remains invariant to speed. This was the only paper I found that simultaneously (1) did not constrain the arm to move on a horizontal plane and (2) varied the speed of the motion and with it the temporal profiles and durations of the reaches. So, at least at that level of enquiry we had some empirical evidence to being thinking about a model of the arm with redundant DoF producing speed-invariant paths. The challenge then was how to build such a model in a—as much as possible—biologically plausible sense. More importantly, I wondered how to build a model that allowed me to explore these questions without ascribing to the notion that the brain solved the problem like the model did.

RETHINKING SOME QUESTIONS TO ASK IN NEURAL MOTOR CONTROL

I always like to take the perspective of the nascent nervous systems of a newborn and think how the internal and external world would look like to the various parts of the nervous systems involved in self-discovering a problem to be solved, finding the solution to that problem and possibly storing it for later retrieval and use. What is the baby's nervous system doing at that pre-cognitive time?

Babies do a lot of (seemingly) spontaneous random motions, but as we saw in the previous chapter, these spontaneous rhythms eventually help trigger the process of actively self-generating exploratory movements that eventually acquire a consistent stochastic signature. Whether the nascent nervous systems have *conscious volition* at that stage may get rather entangled in semantics. Yet the very fact that the stochastic signatures evolve and their evolution can be quantified on a time scale of hours within the day provides evidence that they are not innately reflexive in nature. They are rather adaptable. Longitudinal tracking of these evolutions over 5 months then confirm the path toward volition and intentional control in neonates,[18] but also in a coma state,[19] as we saw in Chapter 1. Whether acquiring anew or regaining it, the nervous systems seem to strive for its volitional control.

In the case of reaching, one could think of the baby's arm flailing as a possible strategy of the nascent nervous systems to actively sample both the intrinsic sensory spaces of the joints (muscles) motions and the extrinsic space of the hand motions. Assuming that the baby senses these self-generated movements, balancing spontaneous and active exploration would serve to eventually build a rich **map** between sensory information that originates internally and sensory information that originates externally. More importantly, the spontaneous variations would then continue to serve as a form of background noise supporting self-emergent systematic variations that would carve out a subspace of goal-directedness. They would constitute the "listening motions", hidden in the background (as in a radio) dialing for sensory frequencies of relevance to a given context, in the precise sense of evoking sensory goals for action (i.e., beyond passive perception of those goals).

In building and adapting this map, each hand position affords many possible configurations of the arm's joints (Figure 4.1), such that many postures map to a given hand position. This many-to-one map endows the arm system with ample flexibility to allow the hand to accomplish the same task in many different ways. The problem is, given a task with some constraints and goals, how is the baby eventually going to deploy a consistent solution that allows the arm's excursion to move continuously along an internal path of varying configurations, such that the hand does so in continuous correspondence? This external path becomes visible to the baby at some point (i.e., when the hand enters the baby's fovea) and then, the baby can adjust it relative to the target. The model to mimic this problem should also handle other situations, as the baby sometimes may face obstacles and deviate from the originally intended path to the target; as when the target transiently moves and the baby adjusts the path in flight. How would the baby come up with a solution that flexibly accommodates all these situations, and recruits and releases DoF on demand, rather than pre-planning the full spatiotemporal trajectory a priori?

It seemed as though building the map that associates postures and positions is not enough. To endow the model with the ability to provide a prospective code, I needed to integrate an iterative solution with a recursive one. The solution would be recursive in the sense that it would automatically have to pull the hand to the target and stop moving (on its own) when the target was reached. At the same time, the solution had to allow for flexible updates of the path on demand, as needed in real time. When faced with some transient change (e.g., a sudden change in posture, a sudden obstacle or a change in the target position) the model had to be able to resolve the change in flight, prioritizing a local computation to take care of that sudden change, while still keeping the overall global goal somehow.

The problem seemed to involve different spaces, disparate coordinate systems, different maps and transformations to navigate such maps. In that sense, the framework of differential geometry seemed suitable to represent sensory-motor transformations. But more importantly, the choice of an intermediate geometric stage between goal determination and action execution was amenable to uncover law-like relations in the cognitive realm of representations. The idea was to try and find relations that were not a mere kinematic byproduct of the system's biomechanics, as those found in a robot with similar biomechanical (anthropomorphic) architecture to that of the human arm system. Instead, I wanted to uncover emerging invariances under brain control, for example, they would break with pathologies of the nervous system but could be restored back to normal levels under some form of sensory guidance that the spared sub-systems of the injured brain could still process and integrate into the motion dynamics. As such, these invariants had to be defined by the movement's geometry, independent of the movement dynamics, that is, they had to maintain their signature even if the dynamics were changing. For example, as when the speed varied or a load was added to the arm. In all those cases the time duration of the movement was not a parameter dictating the solution trajectory. It was rather a byproduct of a complex system's integration of information from multiple channels to produced organized behavior subject to fluctuations in the signals internally transduced and transmitted throughout the nervous systems.

In my conceptualization of this problem time was not centrally controlled as an independent predefined parameter (e.g., a centralized ticking clock dictating the orderly evolution of position, velocity, and acceleration along a trajectory.) Instead, I thought of the notion of time as another sensory dimension, possibly linked to peripheral sources in the pre-cognitive stage (e.g., perhaps guided by the rhythms evoked by the pace maker of the autonomic heartbeat). At the central level, my notion of time necessarily had to be adaptable during neurodevelopment, when neural nerve growth and myelination are occurring. Time treated as another "sensory-space dimension" would be related to the integration of disparate and variable sensory transduction and transmission delays arising from multiple sensory modalities across the PNS and CNS. Owing to the various sensory transducers sampling sensory stimuli within diverse frequency ranges, the notion of time had to emerge from probabilistic computations where bottom up processes scaffold the ability to predict the likelihood of sensory-driven events during neurodevelopment. As such, time would be the "sensed" dimension experienced through active motor exploration that made it possible to simultaneously experience all sensorial inputs into a concrete physical realm, including as well the sensations caused by self-generated movements.

The maps between external and internal sensory-motor worlds had to include this dimension of time experienced though self-generated motion as a function of sensory consequences. In light of von Holst's principle of reafference,[20] sensory consequences are linked to reafference from self-generated actions that the system self-monitors. As such, they are also adaptable. They provide the notion of *cause and effect* so the system can register their occurrence above and beyond mere fluctuations in background noise.

In this regard, the type of temporal information I thought of as another dimension of sensory spaces was channeled out through movements, whereby we could read it out of the nervous systems' continuous performance. To begin testing this idea, I needed empirical data from unconstrained systems mastering a new motor skill in real time. The trouble for me was that under the computational models guiding the empirical research in neural control of movement, there was no room to explore any of these ideas. I had to go *"solo"* on this and that was a bit hard in a world where (1) peer-review was as constrained as the prevalent ideas of the time and (2) largely determined one's academic success. I simply could not publish any of it.

Notwithstanding the high risk of likely not even being able to do science at some point, I decided to initiate a different path whereby we could eventually begin the objective characterization of nervous systems' functions in a way that was independent of how we thought the brain was doing it. Even if I did not have a complete model of motion generation involving all joints, masses, and general dynamics, I could still build a model that enabled me to explore these questions in the context of unconstrained behaviors, so as to gain a window into possible strategies the mind may use to solve such complex problems, or to cope with the aftermath of brain injury, neurodevelopmental stunting or neurodegeneration. I was also interested in learning from athletes and performing artists, the *"geniuses"* of motor control. How did they acquire their exquisite timing and coordination?

TWO VANTAGE POINTS: THE EXTERNAL OBSERVER VS. THE INTERNAL NERVOUS SYSTEM

I reasoned that one of the main problems in the field I was trying to understand was that it was easy to fall in the trap of *the external observer*. This is to look at a problem exclusively from one vantage point, external to the problem and miss other sides not immediately apparent but that could nonetheless help us rethink the issues at hand. To illustrate this notion, consider the following example. Every weekday my dad used to leave the house to go to a senior center at 9:00am and come back around 6:00pm. By 6:00pm, every day of the week, my dog would go to the front door and wait for my dad to arrive. Nobody ever encouraged the dog to do this. It just happened. At times, we joked that it must be around 6:00pm now because Yuzu (my dog's name) was by the door strategically positioned anticipating my dad's arrival to salute him.

To an external observer, this exactness in the prediction of time of arrival may be somewhat puzzling. It would seem as though my dog had a "ticking clock" and knew exactly when 6:00pm was going to be each day of the week (Figure 4.3A). That is one way to look at it, that is, from the external observer perspective. Yet, it is also possible to take the internal vantage point of my dog's nervous system. Its nervous system does not need to have a clock that marks 6:00pm as the time to go and wait by the door. Its nervous system can build that prediction of the time of arrival from probabilistic estimation and inference (Figure 4.3B). By sampling cyclically the fluctuations in the occurrences of sensory events that his systems register, his brain can build a sort of density function that predicts the likelihood of the event signaling my dad's arrival. Specifically, given that my dog possesses an exquisite sense of smell, his nervous systems may rely on my dad's scent as an external source to guide internal processes of my dog's nervous systems while building this predictive code. At 9:00am, my dad's sent in the house is at its major concentration, but after he leaves, such level of concentration may decay at some rate. All things being more or less equal each day, by 6:00pm the concentration has decayed to a minimum, and

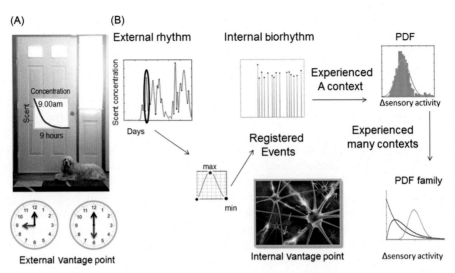

FIGURE 4.3   **External vs. Internal Vantage Points.** (A) The external observer may prescribe a pre-defined time duration and time course to the actions of my dog based on its seemingly accurate anticipation of the time of my dad's arrival. In this case a centralized ticking clock dictating the orderly occurrence of such behavior would be assumed to generate the action trajectory. (B) In contrast, the internal vantage point, taken from the perspective of the neurons registering events due to changes in sensory processing, provides a probability density function (PDF). This reflects the likelihood of the event of my dad entering through the door when the concentration of his scent in the house is ready to increase to a local maximum. In such cases experiencing the events as registered by the neurons (red dots) leads to a probabilistic predictive code and eventually to a family of such PDFs for which the same geometrically optimal (length minimizing) global action path can be prospectively traveled with different profiles of sensory-based changes in activity. Here a time-dependent optimal trajectory for the dog's action is not needed.

is ready for replenishment. Given that my dad's return does increase the concentration of the scent that my dog has associated with him, my dad's cyclically coming and going within that interval (9am–6pm) produces a rhythm that my dog's nervous systems can follow probabilistically: an external rhythm becomes an internal nervous systems biorhythm. This is so because that rhythm has peaks and valleys created by the sensory processing of my dog's nervous systems. The fluctuations in the occurrences of the external event (e.g., my dad may not go to the senior center some days or may go elsewhere and return to the house sooner) give rise to fluctuations in the amplitudes of the peaks and the valleys of the internal biorhythm, along with fluctuations in the inter-peak-distances (or inter-valley distances). These are sources of information that can provide a temporal code based on the *distances* between the internal registrations of such sensory fluctuations by my dog's nervous system.

These distances in sensory space can be internally coded in the nervous systems' language of transduction delays, action potentials and transmission delays linked to the processing of electro-chemical signals, that is, the language of the *internal observer*. That internal observer is trying to predict the next event leading to replenishing the concentration of my dad's scent in the house (a local computation each day) but over time, the global effect is giving the impression to the external observer that the dog has a ticking clock and knows when 6:00pm is going to be. More generally, looked at from the vantage point of internal sensory spaces, one can use these infinitesimal fluctuations and build a line element to represent a variety of distance metrics. Such distance metrics would be of use to the sensory systems in different situations according to sensory goals (e.g., get the reward—in the touch-sensory domain—of being petted by my dad when saluting him). Globally one can build a path leading to a sensory goal that will produce some reward as outcome. In other words, we can build a notion of time from a notion of distance in sensory space.

This line of thought would lead me to a geometric formulation of the motion problem that did not depend in any way on time per se; one that would rather arrive to the notion of time via the orderly probabilistic predictive code one could build from fluctuations in sensory events that the nervous system registered. Registration of sensory events in this case is critical because if the fluctuation is not registered by the system; then it did not happen as far as the system is concerned. And this is very important because the cumulative information from those locally internally registered sensory fluctuations in association with external events would lead to a global solution, one that stood uniquely in spite of all other possible variations. As such, my conceptualization of this problem depended on the registration of sensory input above and beyond the noise levels inherently present in (or characteristic of) that individual's nervous system. Notably, I thought of actively self-generated motions (kinesthetic reafference) as the conduit channel to identify and broadcast to the rest of the nervous system the most adequate (informative) source of sensory guidance, that is, the one with the lowest signal to noise ratio and the highest anticipatory content. The reason why I thought of motion as the conduit channel is because of its inherently dynamic nature whereby there are changes in the state of the system with the passage of time: position, velocity, acceleration, etc. Each one of these higher order derivatives of position contain the notion of distance and as such afford a geometric formulation, one that makes sense when the underlying sensory processes do not explicitly depend on time. For instance, physiologically, the reaching movements my brain produces at 9:00am do not have to be any different than the movements my brain produces at 6:00pm. In this sense, the generation of movements my brain does is time independent.

Of course, this is only an ongoing hypothesis of my lab, but as we saw in Chapter 1, when the nascent nervous system is "polluted" with noisiness and randomness, the mere transitions from spontaneous random noise to systematically well-structured signal becomes challenged. And as we saw in Chapter 3 for the case of autism spectrum disorders (ASD) and the related ADHD, such transitions in the probabilistic code are critical to form proper temporal dynamics and help scaffold predictive decision making. As such, I imagined that the type of mechanical formulation based on time-dependent forces to produce goal-directed motion were not going to lead me anywhere to think properly about the problems the internal observer faced. The formulations in the extant literature at the time I was searching for an idea for a PhD in Cognitive Science were based only on observations by an external observer; one holding a ticking clock to prescribe a priori the duration of the motion along with its time course. If I adopted that route, the question of how the baby's nervous system arrived at its own temporal dynamics for prospective action would forever remain elusive.

JUSTIFYING A GEOMETRIC APPROACH
TO MOVEMENT MODELING

I decided to build a geometric model of action simulation. One of the challenges was on deciding how to choose suitable maps and coordinate functions that allowed for the recruitment of internally—or externally defined task constraints and goals—on demand. For example, how would I build a solution that allowed automatically freezing some DoF in the arm while releasing others to move freely? (As when we open the door to our house recruiting the hand's and forearm's joints, while holding up a grocery bag, with the upper arm stabilizing the bag against our trunk). The model should also allow for a change in priority while in route to a target? (e.g., as when transiently prioritizing obstacle avoidance with a sudden bend of the hand path along an avoidance curve, momentarily moving away from the direct path to the target.)

In all these hypothetical situations, the temporal-dynamic profiles of the arm-hand system would carry an element of uncertainty because the movement duration was not predefined (as it was in the optimal control framework.) In contrast, the geometric path could be computed on demand if we posed the problem independent of the dynamics and addressed different issues that the developing brain would have to deal with. One of these issues was how to build the notion of an egocentric frame of reference to compute relative distances between targets and other objects (or people) in the environment; and eventually be able to adopt an allocentric (disembodied) perspective. Such a disembodied perspective would allow a person to imagine what it would be like to do something from the vantage point of another person or object (as if standing outside the body.)

Research of such complex problems always poses a trade-off between how much to reduce or simplify the problem to gain insights into a possible solution, and reducing the complexity so much that we may miss the opportunity of ever finding first principles governing the systems functioning. In my case, I would clearly disclose that *this is not the way the brain works*. This is just a model that enables exploration of what it would be like for an unconstrained system with sensors to learn to move, guide itself while learning to self-supervise its self-generated motions. By developing geometric invariants that could *self-emerge* from the systems' own performance, the controller of that system could sample, keep and encode large amounts of information in compressed forms. Under these assumptions, I set to design a geometric equation for trajectory generation amenable to explore multiple problems of the reaching system. These included global path determination, online local corrections, and automatic satisfaction of multiple constraints embedded in both internal and external sensory spaces. I would treat time as another adaptable dimension of sensory spaces.

Using such a model I could generate hypotheses and empirically test ideas without explicitly constraining the participants to move in a certain way or complete the action within a given time. By allowing the sensory-motor systems adopt the natural solution within a given context, and recording those spontaneous responses, as they naturally unfolded, I could learn to identify self-emerging solutions of the system, rather than imposing my own solution. It was a different way to address an old problem and possibly pose new questions.

THE GEOMETRIZATION OF THE DOF PROBLEM TO MOVE ALONG THE SHORTEST PATHS IN SPACE AND TIME

If you ask a 4—5 years old child to draw the shortest distance path between two points in space, very likely the child will draw a straight line. That child has no formal knowledge of Euclidean geometry, no notion of what a distance metric is and no mathematical knowledge of what a geodesic curve is. And yet, intuitively his brain learned somehow that the straight line gets him along the shortest path between the two points in his visual peripersonal space. It is a *unique solution* to a problem that affords many possible solution paths between the same two points (Figure 4.1). It is also the fastest solution: the shortest path in space and in time. How does the child's brain arrive to this conclusion?

My first question with regard to the reaching problem was how can we draw a general unique geometric solution between two points in space? Clearly, we had multiple spaces to deal with (e.g., bones, muscles, joint angles, and hand). In the absence of knowledge on what spaces the brain may use to solve these problems, I settled for (at least) two co-existing spaces, Q representing the space of joint angles and X representing the space of hand positions. My model would have to simultaneously operate in these two spaces under multiple constraints and goals. As such, adopting different coordinates and transformation maps to represent joints and task-contexts would play a role in generalizing the model to handle different situations. I started out by learning how roboticists would do this mapping for anthropomorphic robots.[21,22]

My second question was how to produce incremental changes that were informed by the past performance, considering as well the local present situation and projecting ahead a solution a bit into the future. I wanted the model to do all of that while aiming toward the overall global goal. Ultimately, the model had to bring the hand to the target from a given initial position, while flexibly accommodating possible transient changes such that both spaces co-participated and co-contributed to the final solution. This iterative unfolding would have to be driven by a known goal that we could actively update. A simplified version of the problem closely following the computational planar design prevalent in the field in the late nineties (i.e., a planar iterative solution, without redundant DoF) is shown in Figure 4.2. The model also had to make room for a recursive solution—one that knew how to automatically stop at the goal while spontaneously steering the hand without explicit guidance, if faced with additional constraints (e.g., an obstacle.) I reasoned that these types of spontaneous acting had to be implicitly embedded in the spaces of interest. This could be done though a coordinate transformation that adjusted the notion of distance accordingly, or one that kept the old coordinate system while changing the metric distance (Figure 4.4)[1] and Appendix 5.

A two-dimensional "toy model" of reaching a target can help us visualize this. In Figure 4.4, I show motions that unfold along geodesic paths in the Cartesian and joint angle spaces represented as Cost surfaces. Here (as in the case explained in Figure 4.2) the Cost is measured as the remaining distance to the target. Figure 4.4A contrasts that case with Figure 4.4B, where the same problem of reaching the target now involves an obstacle along the way. A change in the distance metric, which changes the Cost expression also produces a change in the Cost surface and leads to a different path. The paths in

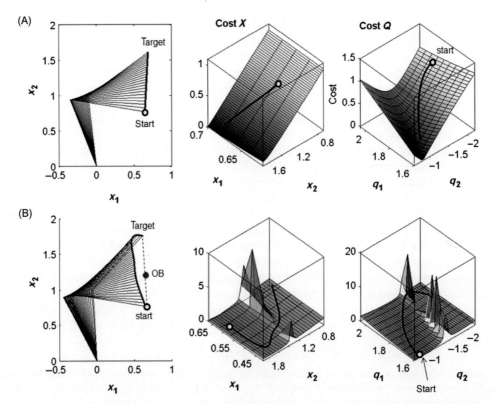

FIGURE 4.4 **Different tasks lead to different cost forms but similar idea: follow the shortest distance path.** (A) The shortest path in a Euclidean sense takes the hand straight to the target in *X*. (B) The shortest path in a non-Euclidean sense takes the hand along a curved path in *X*, one that optimally avoids the obstacle along the straight path. Both paths are geodesics with respect to different notions of distance. The corresponding cost surfaces in *X* and *Q* are also shown.

Figure 4.4B (blue curves) are also geodesics, as are the paths in Figure 4.4A, yet they are the *"straight lines"* or shortest distance paths with respect to a different distance metric (representing the Cost as well in this example). This type of simplified modeling allowed me to explore geometric aspects of this problem before scaling up the model to a more realistic anthropomorphic design.

My third question for the scaled up version of this model was then, how to deal with the excess DoF problem at any given moment, that is, how to manage DoF that changed in relation to the present goal and DoF that were incidental to that goal at some point, but recruited to change in response to other priorities at some another point.

Overall, I wanted to generate some predictions concerning the speed invariance feature that Atkeson and Hollerbach had discovered in their empirical work of 1985. Specifically, I wanted to simultaneously generate unique goal-compliant paths in both spaces of interest: joint angles' rotational displacements and hand positions and orientations. These paths would have to remain conserved despite changes in speed, so as to mimic what humans did when moving the arm-hand linkage in unconstrained ways.

In Figure 4.5, I elaborate on the questions I had with a simple example contrasting the role of vision for perception of a static object *vs.* the role of the visuomotor systems when the same object affords a goal-directed action. In the first case, the spatial affordances could define the problem geometrically in terms of spatial distances and orientations. In the second case, however, the unfolding of the proper orientations and displacements of the hand to successfully grasp the object calls for a dynamic formulation involving time-dependent parameters to model the time-dependent forces. This was where the field was stuck. But we could instead model the geometric path of the forces by adding that sensory-dependent temporal dimension and treating it independent of the actual movement duration. I called this extra dimension tau $t(\tau)$ and made it dependent upon distance (δ) information in sensory spaces. This was now an adaptable quantity whereby a simple coordinate transformation would make it a function of the distance and its adaptive rate of change (Figure 4.6).

This formulation of the DoF problem invoked the *geometrization* of the dynamics. Instead of computing the time-dependent forces to move the arm and produce a where-and-when solution, that is, with a fully pre-specified time-dependent course; here I would answer the question of where the solution was in *sensory space* and at what rate it was changing. In this formulation "motor" was sensory (reafferent) input defined by sensory-based rates of change. As such, time emerged from the rates of change in the amplitude of a sensory signal (which I could then tie to the line element of sensory space). In this way, time would enter the distance metric as another coordinate dimension and could be linked to the notion of *change* in sensory spaces. Possible modification of the distance metric and/or the coordinate functions through coordinate transformations (see Appendices 1, 5) would then allow

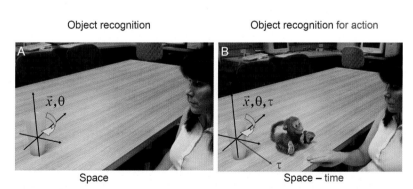

Object recognition   Object recognition for action

Space   Space – time

FIGURE 4.5 **Static perception vs. perception for action requires different sensory-based dimensions.** (A) A static object has different perceptual properties that can be geometrically represented (e.g., shape, orientation, color, and brightness) but temporal information for the static object is not relevant. (B) When that same static object affords an action (e.g., reaching to grasp and hit it down) temporal dynamics comes into play because the arm must move in a certain way such that the hand reaches for and grasps the object, particularly around the obstacle (that is, the stuffed animal) obstructing the path straight to it. How do we represent the temporal information for action? Do we need to know the duration of the impending motion ahead of time? Do we need to know its full profile a priori? The τ-dimension added to the object's geometry is proposed here to be related to the *registered change* in sensory information relevant to the motion plan and its impending execution, such as the rates of change in sensory transduction and transmission delays necessary to integrate sensory information from multiple-sources and experience the multimodal percept simultaneously across all senses.

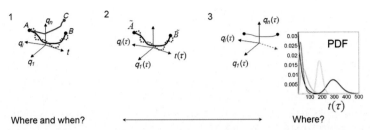

FIGURE 4.6   Geometrization of the dynamics problem: A coordinate transformation to represent time as another sensory spatial dimension. (1) A time duration and time course of the dynamic trajectory can entirely specify the motion from A to B to accomplish the end goal along the least action path (black line) chosen among neighboring variations of this path (e.g., the dotted blue path). Yet another pre-specified duration leads to a different dynamic path from A to C (red path), while B is still the goal. How do we go from A to B along the least action path, independent of the time duration? (2) A different representation of the notion of time attained through a coordinate transformation whereby time emerges from sensory processing distances (e.g., the registered *rates of change* in the amplitudes of internally generated biorhythms from sensory processes, denoted τ) provides a new way to represent the dynamical path of the system. Whereas the original representation tells us *where and when* the action is completed, the new transformed one expresses *where* in sensory space the system attains the action and registers its occurrence above and beyond baseline levels of sensory-motor noise. (3) Given a context then, temporal dynamics are experienced through active action (kinesthetic reafference) integrated with other sensory inputs to form a family of probability density functions. In this sense all we need to know is the geometric path in sensory space attainable from sensory distances, rather than the exact time to travel along that path. The exact time emerges from probabilistic changes in our internally experienced sensory-motor processing activities.

me to introduce later on additional elements involving the probabilistic sensory consequences that the "affordances" of an impending action would create. These could exist independent of the actual physical laws governing motion generation. In a mental space, such laws could be violated or adapted to create physically impossible scenarios (we do this all the time in our dreams.) I did not have to set rigid bounds on my way of thinking about this problem to conform to the physical constraints of a mechanical system. I was exploring ideas to propose a PhD thesis in Cognitive Science, the science of the mind, so I had some room to wiggle out-of-the-box ideas. And I thought to myself, why not? Geometry seemed an appropriate language to describe mental laws and their representations.

The geometrization of dynamics was a formulation of this problem difficult to imagine for the motor control engineers developing computational models, but it was not unknown to physicists using contemporary (geometric) approaches to classical dynamics.[23] More specifically, solving the DoF problem in this way invoked Maupertuis principle of least path length,[24,25] explained by Cornelius Lanczos[26] and Richard Feynman[27] in various text books. Figure 4.6 shows in schematic form the basic idea behind the path-length minimization framework that I adopted to address the DoF problem as a geometrization of the dynamics of the nervous system. The model determines a path that could represent the full history of the dynamics up to a given point and serve to predict ahead the subsequent geodesic path as well. This would be a path that was prospectively updated online (based on current feedback and past performance) rather than computed a priori. This path was the *shortest in space* but because time entered into the model as an adaptable sensory-space dimension to become aligned within sensory-based distances, the proposed solution would be (interchangeably) the *shortest in time* as well.

GENERATING UNIQUE GEOMETRIC SOLUTIONS

The problem of path determination requires the use of at least two spaces to represent sensory information from *internal* and *external* sensory sources, so that the solution path simultaneously unfolds in both. Owing to the anthropomorphic robotic literature, I adopted the space of joint configurations leading to arm postures as the internal space, and the space where the hand changes positions and orientations in route to the target as the external space. For simplicity, I used two manifolds to represent them.

To model Bernstein's DoF problem, there is a dimensional disparity between the internal and external spaces (Figure 4.7). Task features expressed in a target for action define a point in m-dimensions of the external space, denoted X, while the points used to characterize the corresponding postural configurations have n dimensions in the internal space, denoted Q. In the context of voluntary arm movements, $n \gg m$. (Additional spaces may be added to monitor the evolution of these parameters as the motion unfolds, for example, muscle-related spaces, such as a space representing the lines of action of the muscles but for simplicity, we will restrict the model to these two spaces only.)

Composition of differentiable maps linking these spaces is possible, as the governing operation I chose to capture iterative changes along each dimension is the chain rule. The chain rule operates on smooth-differentiable manifolds, where the parameters of interest take values through continuous differentiable coordinate functions. The chain rule is used to obtain the *Jacobian* transformation matrix from space to space. Since we are interested in studying the case of redundant DoF, we use seven joint angles, so the dimensionality of Q, $n = 7$. Figure 4.8A shows one possible way (out of many) to parameterize the internal Q-space of joint angles. The number of goals and constraints describing the target of a task may vary, and span the dimensions of X. Figure 4.8B shows the position target for a pointing motion while Figure 4.8C expands the problem to match the orientation of an object with the orientation of the hand. Figure 4.8D provides one way to represent a postural goal in an economical manner (as defined by the plane of the arm spanned by the

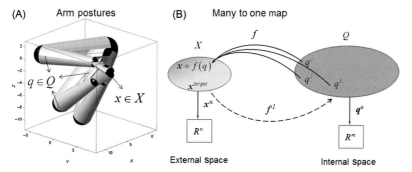

FIGURE 4.7 The many to one map in Bernstein DoF problem. (A) Schematic representation of the arm in different postural configurations in Q-space that map to the same hand position in X-space. (B) Schematic representation of the problem at hand: how to determine the inverse map from hand to postural configurations? Given a hand position, which postural configuration describes that hand position? Note here that specifying other external positions along the arm (e.g., shoulder, elbow, and wrist) besides their orientations helps reduce the number of possible postures to describe a given hand position.

shoulder-elbow and shoulder-wrist vectors; the normal vector to this plane and the angle θ they form with the horizontal plane). Human participants seem to have awareness of this "plane of the arm".[28] More formally, these three hypothetical examples with different m dimensions are:

1. Reaching for a target object in space, where each of the coordinates representing the object's center of mass location spans a dimension, so $m = 3$ Figures 4.8B.

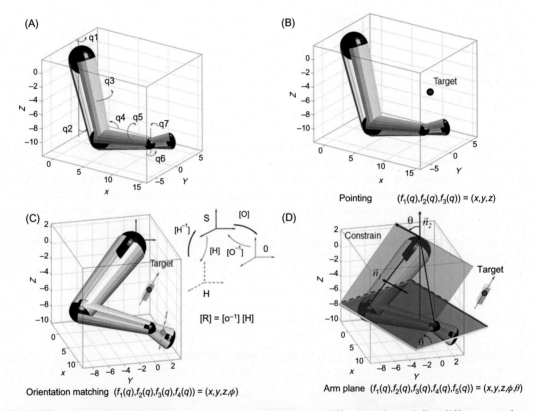

FIGURE 4.8 **A given parameterization of the posture space and different tasks to define different cost functions.** (A) Seven joint angles define a posture whereby the shoulder rotates about three axes, the elbow around 2 and the wrist also around 2 axes. (B) Given a starting postural configuration of the arm and a target, move the hand to the target along a straight path. (C) Given an object located and oriented in space, bring the hand to it. This requires rotation and translation of the hand in relation to the body and the external object. These relations can be represented through the special group of rotation matrices SO3 to express the hand and the object relative to a common reference frame anchored on the body (e.g., at the shoulder). Three axes are used to represent each node and to transform from one frame of reference to another. The body could also be referenced based on the external frame anchored on the object (or on another person) so as to *"imagine what it would be like to move the body"* from the outside. (C) A different task to reach to grasp the object located and oriented in space subject to the constraint: the plane spanned by the upper arm and the forearm must be oriented in a certain way in relation to the horizontal plane. For example, as when performing an experiment where the participant is instructed to return the arm to a posture where the hand is vertically at face level, with the palm facing the ear and the vector normal to the plane of the arm parallel to the horizontal plane.

2. Suppose that besides bringing the hand to the object's location, we also want to match the objects' principal orientation axis, then $m = 4$, where the new dimension gives the angular discrepancy between the hand and the object's orientation Figures 4.8C (see details in ref. [29]). Thus the target is position (giving the CM location) and object's spatial orientation. This problem involves different frames of reference to specify the target in relation to the body (egocentric formulation); or to describe the hand (anchored at the body) in relation to the external target (allocentric formulation).

3. If besides the endpoint-related goals in (2) there is a postural goal, then we may have $m = 5$. The new dimension could give the inclination of the plane spanned by the forearm–upper arm segment in relation to the shoulder as explained in Figures 4.8D and.[28] Thus the target has dimensions spanned by (the object's location (3 DoF), the orientation axis of the object (1 DoF), and the tilt of the plane of the arm (1 DoF)). All dimensions depend on the seven joint angles defined by a point in Q and five DoF are recruited to comply with the goals and constraints of the task.

In each case, there is a map $f:Q \subset R^n \rightarrow X \subset R^m$ that takes a posture $q \in Q$ and expresses it as a point $x \in X$. Each coordinate component of the current point $f_i(\mathbf{q})$ relates to a component of the goals, and has to be a differentiable function of the arm posture, so that changes in the joint angles that change the current posture also change the hand configuration via f. The dimensional disparity between m and n makes f a many-to-one map (Figure 4.7A and 4.7B), so there is no analytical inverse $f^{-1}(\mathbf{q}) \in X$. This problem can be solved locally with a locally linear isometric embedding of X into Q. Before explaining it, we define a scalar function to measure the distance (remaining at any time step) from the hand to the target in such a way that the coupled arm–joints–hand system simultaneously changes in relation to the target features and the joint restrictions mentioned above in the suggested examples 1–3.

We first define a positive scalar function $r:X \rightarrow R^+$ that expresses the difference between the current configuration of the hand and the current set of goals (e.g., the target for action). We choose this function as the distance in X. As the hand changes configuration on its way to the goals, we can monitor the remaining distance at any given step.

If we were to operate on the X space alone, to reach the target we would intuitively have to explore all directions with point of application at x^{init} and build a set of directional derivatives to choose the one that would change the value of r at a maximal rate. Recall here that the directional derivative $\nabla_u f(x_0, y_0, z_0)$ is the rate at which the function $f(x, y, z)$ changes at a point (x_0, y_0, z_0) in the direction u, with $\hat{u} = (u_x, u_y, u_z)$ a unit vector. It can be defined as:

$$\nabla_u f \equiv \nabla f \cdot \frac{u}{|u|} = \lim_{h \to 0} \frac{f(x + h\hat{u}) - f(x)}{h}$$ for $\hat{u}$ and the "*nabla*" ∇ operator. This notation gives a vector form of the usual derivative, $\nabla_{\hat{u}} f = \frac{\partial f}{\partial x} u_x + \frac{\partial f}{\partial y} u_y + \frac{\partial f}{\partial z} u_z$.

The direction ∇f along which the directional derivative has the largest value, as measured by a norm $|\nabla f|$ is called the gradient[30] (Figure 4.9 gives a visual example along a cost surface). Thus, which norm we measure the directional derivative with so as to define the gradient, will play an important role in our formulation. We want the gradient that points the hand along the path-length minimizing direction, that is, the shortest distance path.

In a Euclidean (flat) world this is the straight line the 4–5 years old child in our example would intuitively describe. But in a more general, non-Euclidean sense there are other

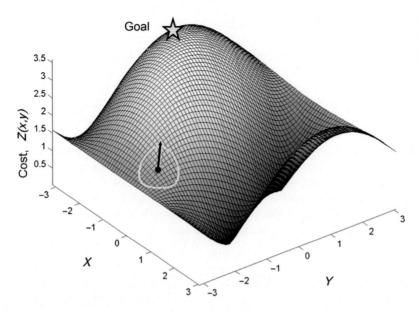

FIGURE 4.9 Schematic of the gradient vector on a cost surface pointing towards the goal.

"straight lines" called geodesics (e.g., Figure 4.4A *vs.* 4B show different geodesic curves connecting two points in two spaces related by some map *f*). They are the shortest distance paths, as measured by non-Euclidean distance metrics that determine as well the gradient direction we choose to update our iterative (optimization) process.

In our specific reaching problem, the hand is the end effector of the arm linkage with seven joint angles. As such, we want to minimize the hand distance to the target in a general sense. By that, I mean minimization in a broader sense than Euclidean distance, whereby the *norm* of choice to select the gradient in Q will change on demand, as a function of the context of the task and the goals ahead. These would be defined by sensory information emerging from the spaces of internal and external sensory sources. The joints of the arm will iteratively rotate and translate as dictated by the *r*-distance of choice in X, which will involve as well the norm of choice in Q. We achieve this by composing *r* with the map *f*.

The composition of *r* with *f* builds a map $(r{\circ}f){:}Q{\rightarrow}X{\rightarrow}R^{+}$, which is a function on Q. This construction is the *pullback* action of *r* by *f*, and it is denoted by $f^{*}r = (r{\circ}f)$ (Figure 4.10). As the name indicates, f^{*} pulls back the function *r* from X into Q, so we can *simultaneously* monitor and control the rate of change of *r* due to changes in the set of joint angles representing a posture in Q, which in turn cause changes in X.

A path in Q is a line directed from the initial posture q^{init} to the final posture q^{final} at the target, smoothly connecting these two postures with intermediate postural configurations that map to points in X and move the hand in X. The path traced by the hand in X is the image by *f* of the path in Q.

The (negative) gradient chosen according to the distance metric compatible with the task iteratively brings the hand to the target along a length-minimizing path until the value of *r* is *zero* at the target (Figure 4.10). This amounts to incremental gradient descent,

$$f^* r = r \circ f(q)$$

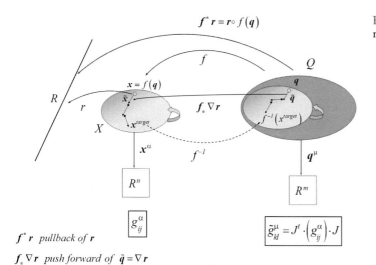

FIGURE 4.10 Locally linear isometric embedding.

$f^* r$ pullback of r

$f_* \nabla r$ push forward of $\tilde{q} = \nabla r$

a well-known optimization method to minimize an objective function (our cost) expressed as a sum of differentiable functions.[31]

The function r is thus used as a cost function representing the distance in X between $f(q)$ and x^{target}. The function r has a unique global minimum *zero*, which occurs when the hand is at the target. This feature is important because the optimization process we follow to search for the optimal path will not stop until the cost reaches this unique *zero* value (i.e., undesirable local minima can be avoided). To move the hand away from its initial position, we need to wiggle the arm joints a bit by some Δq and check that the image changes $x^{new} = x^{init} + \Delta x = f(q^{init} + \Delta q)$, that is, ensure that a change in Q corresponds to a change in X. This is tricky. Because of the arm's redundant DoF (Figure 4.7), we could hit a series of q-postures that did not move the hand at all. They correspond to the so-called *self-motion manifold* from the excess DoF. These are joint rotations that change the arm posture, but do so in such a way that the arm motions result in no displacements of the end-effector (i.e., the hand does not change position.)

How do we chose the Δq such that the hand changes position? And more to the point of modeling the goal-directed reaching motion, how do we do so in such a way that the hand actually moves toward the target and eventually lands at the target?

The solution to selecting the appropriate Δq resulting in appropriate Δx, such that the hand moves toward the target, comes from selecting the gradient of $(r \circ f)$ under the proper distance metric. The partial differential equation that accomplishes this and governs the path motion determination in both Q and X spaces in an iterative and recursive mode is:

$$dq = G_q^{-1} \cdot \nabla r^{task} \circ f(q) \cdot \Delta \tau \tag{4.1}$$

described in refs.[29,32] Here the inverse of $G_q^\mu = J^T G_{f(q)}^\alpha J$ provides the proper gradient to move along the path-length minimizing direction in Q that maps to the corresponding path-length minimizing direction in X, such that the distance metric in X is preserved under pullback of Q (Figure 4.10). This equation generates unit-speed paths (parameterized by arc-length) that minimize the r-distance defined by the task. Here $\Delta\tau$ is adaptable and refers to the step size for the unit-length gradient direction, and not to the time parameter (as in Figure 4.6). In what follows, I describe how to obtain the geodesic direction through this local linear isometric embedding.

LOCALLY LINEAR ISOMETRIC EMBEDDING OF X INTO Q

The negative gradient of the pullback yields an adequate direction $\tilde{\mathbf{q}} = \Delta\mathbf{q} \in T_q\mathbf{Q}$, an element in the tangent space to the Q-manifold that allows reducing the hand-target distance related to the X-manifold.

A vector transformation action under a change of coordinates that operates on the tangent spaces yields the desired direction $\tilde{\mathbf{x}} = \Delta\mathbf{x} \in T_{f(q)}\mathbf{X}$. This is a **push forward** action, denoted by $(f_*\nabla r)$ (Figure 4.10). This action builds the (differential) map $df : T_q Q \to T_{f(q)}\mathbf{X}$ (Figure 4.11). It works as follows: using bases $\left\{\partial_\mu = \frac{\partial}{\partial q^\mu}\right\}$ in Q and $\left\{\partial_\alpha = \frac{\partial}{\partial f(q)^\alpha}\right\}$ in X, one can relate the components of the gradient $\nabla r = \nabla r^\mu(\partial_\mu)$ in Q, to those of its push forward $(f_*\nabla r)^\alpha\partial_\alpha$ in X: $(f_*\nabla r)^\alpha\partial_\alpha = \nabla^\mu\partial_\mu(f * r) = \nabla^\mu\partial_\mu(r{\circ}f)$ but by the chain rule, this gradient with respect to the ∂_μ -basis is $\nabla^\mu\frac{\partial f(q)^\alpha}{\partial q^\mu}\partial_\alpha r$ which is $\frac{\partial r}{\partial f(q)^\alpha} \cdot \frac{\partial f(q)^\alpha}{\partial q^\mu}$, whereby the first term is the gradient with respect to the ∂_α -basis and the second term $(f_*)^\alpha_\mu = \frac{\partial f^\alpha(q)}{\partial q^\mu}$ is the Jacobian, the

coordinate transformation matrix, for example, $J(f(q)) = \begin{pmatrix} \dfrac{\partial f_1}{\partial q_1} & \cdots & \dfrac{\partial f_1}{\partial q_7} \\ \vdots & \ddots & \vdots \\ \dfrac{\partial f_3}{\partial q_1} & \cdots & \dfrac{\partial f_3}{\partial q_7} \end{pmatrix}$ for the case

where X has $m = 3$ dimensions and Q has $n = 7$ dimensions.

Later we will see that the *Jacobian* plays a key role on converting the analytical expressions of the partial derivatives into numerical expressions useful to test the model using empirical data.[33] Each term serves as a source of error-correction at each level of representation (i.e., based on internal and external sensory sources). Figure 4.11 provides the schematics of this locally linear decomposition into the tangential and normal projections on the tangent spaces to Q and X. They separate the DoF in $T_{f^{-1}(x)}Q$ related to the *goals* in $T_{f(q)}X$ and into the remaining (incidental) q-DoF in $T_q Q^\perp$.

We can visualize path determination in X through various examples in 3-space (Figure 4.12A). Different hand paths emerge under different r distance metrics, depending on the task definition 1–3 above. Examples of $(r{\circ}f)$ used to generate the paths in Figure 4.12A are:

$$(r{\circ}f)^{\text{reach}} = \sqrt{\sum_{i=1}^{3}(x_i^{\text{target}} - f_i(q))^2}, \text{ pointing}$$

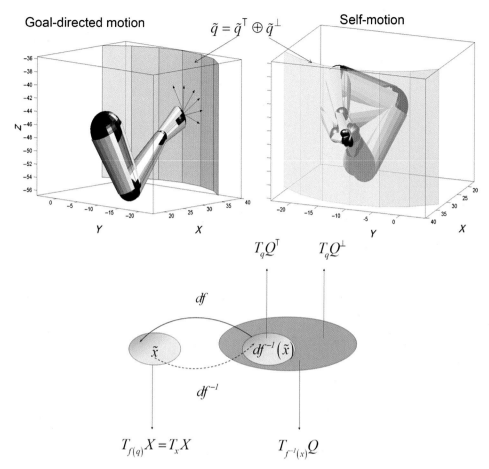

FIGURE 4.11 **DoF Decomposition using the locally linear map** into the rank and null subspaces of the tangent space representing the task-relevant and task-incidental DoF to study their variability patterns.

$$\left(r \circ f\right)^{\text{or}} = \sqrt{\sum_{i=1}^{3}\left(x_i^{\text{target}} - f_i(q)\right)^2 + a\left(\phi^{\text{target}} - \phi(q)\right)^2}, \text{ orientation matching}$$

$$\left(r \circ f\right)^{\text{tilt}} = \sqrt{\sum_{i=1}^{3}\left(x_i^{\text{target}} - f_i(q)\right)^2 + a\left(\phi^{\text{target}} - \phi(q)\right)^2 + b\left(\theta^{\text{target}} - \theta(q)\right)^2} \text{ and matching the tilt}$$

of the plane of the arm (all defined in Figure 4.8B–D, respectively).

In each of these cases the coefficient of the term (e.g., a, b are scalars to provide a proper uniform scale across all physical units, cm, and degrees).

These simulated sample paths from each one of these task-dependent distance metrics described in Figure 4.8 and examples 1–3 above produce different curvature profiles for different goals starting at an initial hand position in space, as defined by an initial arm posture. These changes in curvature can be measured relative to the Euclidean geodesic

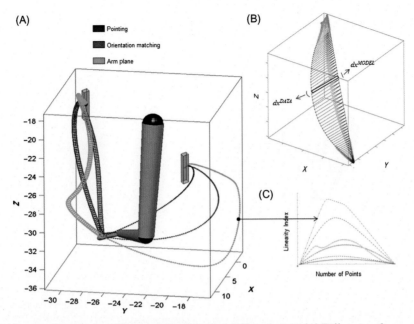

FIGURE 4.12 **Simulated tasks of the reach family using different costs.** (A) Different paths corresponding to different costs (distance metrics) functions. (B) Linearity metric to compare non-Euclidean geodesic curves to the Euclidean (straight line) geodesic. First the data trajectory is resampled at equal intervals, then each point on the data curve is projected on the line and each normal distance recorded. Further the incremental dx^{MODEL} from the model Euclidean straight line is obtained for each point as well as for the corresponding dx^{DATA} and the coordinate transformation matrix computed across all space (as in Appendix 6) subject to the positive-definite constraint. (C) Profiles of the linearity index obtained for each of the curves in (A) corresponding to different model-generated geodesics.

path connecting the hand initial location and the target. To that end, we can project each point on the non-Euclidean geodesic to the Euclidean geodesic and measure the normal-distance deviation (bending), then divide the quantity by the total path length and obtain an index of linearity. The step to obtain the linearity index is shown in Figure 4.12B with the resulting profiles of this index in Figure 4.12C for each of the simulated paths in Figure 4.12A. We can also use the curvature and torsion measures from the differential geometry of curves and implicit (e.g., cost) surfaces in general[34] (Appendix 2), but I will stick to this linearity index for now to show in the next section how to measure in the empirical data a geometric invariant that emerges from the hand trajectories of goal-directed motions.

The iterative process of computing $\tilde{q} = \Delta q$ at the current posture and pushing it forward to obtain the corresponding $\tilde{x} = \Delta f(q)$ at the current hand configuration builds vector flows on the tangent spaces to both manifolds. Since r is 0 if and only if the goals are achieved, the paths thus generated (in Q and X) autonomously move the hand to stop at the goals. The point at the end of the path in Q is the posture corresponding to the task-target point in X. Locally; this solves the problem of finding a point in the pre-image of X, that is, the inverse map. Both paths are continuous and first-order differentiable by virtue of their construction, but in order for them to be local geodesic directions, we needed to conserve in Q the notion of distance defined in X.

As we have seen our objective function for optimal path determination is precisely r, the scalar cost defining as well the norm in the task space. In general we defined r to have the form,

$$r^{\text{task}} = \sqrt{\sum_{i=1}^{n} \sum_{j=1}^{n} g_{ij}^{\alpha} \left(f_i^{\text{target}} - f_i(q) \right) \left(f_j^{\text{target}} - f_j(q) \right)}$$

A gradient pointing along a geodesic direction is thus obtained by pulling back into Q the geometry of X via its metric tensor and the *Jacobian*, $G_q^{\mu} = J^{\mathrm{T}} G_{f(q)}^{\alpha} J$.[35] In addition, it is also possible to modify the metric in Q to add other joint angle constraints as in.[29] We also showed how to do this in Figure 4.4 for a simpler map with no redundant DoF, while Figure 4.12 scales up to more realistic scenarios in examples 1–3 defined graphically in Figures 4.8B–D.

It is important to keep in mind that although we have built this model to represent what it would be like to follow this proposed geometric solution for the arm-hand system, the solution extends to other redundancy cases in general. As such, geodesics on other surfaces can be generated using the same distance-gradient idea. Examples of different surfaces are shown in Figure 4.13. Choosing the shortest geodesic enables us to traverse the path between two points on the surfaces in free-fall on "one step". In this sense, that initial direction that the gradient dictates points the system straight to the global goals and serves as a *global solution* while allowing room for other iterative modifications along the way, as the task may demand in real time, gives rise to a *local solution*.

Preserving the notion of distance through the pullback action on the X metric makes the linear map $d\mathbf{f} : T_q Q \rightarrow T_{f(q)} X$ an isometry. The corresponding inverse map $d\mathbf{f}^{-1}$ starting at a given posture is locally injective, and the path thus obtained gives a continuous map onto its image $d\mathbf{f}^{-1} \left(T_{f(q)} X \right)$ that is, an embedding. This local operation preserving in Q the notion of distance in X makes $d\mathbf{f}^{-1}$ a local isometric embedding. The local linearization of this map separating the DoF into task-dependent and task-incidental (Figure 4.11) makes our model a local linear isometric embedding.

DOF DECOMPOSITION, RECRUITMENT, AND RELEASE ACCORDING TO TASK DEMANDS

Figure 4.11 illustrates the decomposition of the arm's DoF using the projection of the gradient vector $\tilde{q} \in T_q Q$ onto the orthonormal basis from the singular value decomposition (SVD) of $G_q^{\mu} = J^{\mathrm{T}} G_{f(q)}^{\alpha} J$ providing the tangential and normal projections: $\tilde{q} = \tilde{q}^{\top} \oplus \tilde{q}^{\perp}$. The tangential projection has a one-to-one correspondence (up to a coordinate transformation) with the elements of the tangent space to $X, \tilde{q}^{\top} \in T_{f^{-1}(\tilde{x})} Q \doteq \tilde{x} \in T_x X$, where the differential $d\mathbf{f} : T_q Q \rightarrow T_{f(q)} X$ is a linear map (Figure 4.11).

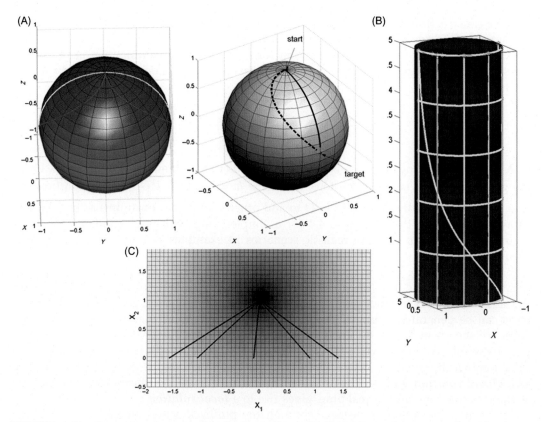

FIGURE 4.13 **Computing the geodesics of other surfaces.** (A) Using the model to move along the shortest distance path (continuous line) on the unit sphere using the map $f: \theta \subset R^2 \to S^2 \subset E^3$ from θ-configuration defining $f_1(\theta) = \sin(\theta_1) \cdot \sin(\theta_2)$, $f_2(\theta) = -\cos(\theta_1) \cdot \sin(\theta_2)$, $f_3(\theta) = \cos(\theta_2)$ and proper metric $\tilde{G}^{\alpha}_{f(\theta)} = \begin{pmatrix} 1 & 0 \\ 0 & \cos^2(\theta_2) \end{pmatrix}$ in contrast to dashed path using $G^{\alpha}_{f(\theta)} = \begin{pmatrix} 1 & 0 \\ 0 & 1 \end{pmatrix}$. (B) Geodesics on the surface of the unit cylinder (the folded Cartesian plane) with $G^{\alpha}_{f(\theta)} = \begin{pmatrix} 1 & 0 \\ 0 & 1 \end{pmatrix}$ and $\theta \in [0, 2\pi)$ with coordinates $f_1(\theta) = \cos(\theta_1)$, $f_2(\theta) = \sin(\theta_2)$, $f_3(\theta) = \theta_2$. (C) Multiple geodesics to one target point on the cost surface in Figure 4.2, with $(r \circ f)$ taking values across X and $r^{surface} = \sqrt{\sum_{i=1}^{n} \sum_{j=1}^{n} g^{\alpha}_{ij} \left(f_i^{target} - f_i(\theta) \right) \left(f_j^{target} - f_j(\theta) \right)}$

Recall here that we are addressing with this model Bernstein's DoF problem (Figure 4.1): to guarantee that the direction $\tilde{q} \in T_q Q$ with $q \in f^{-1}(x)$ (q in the lift) maps uniquely to the desired direction $\tilde{x} \in T_{f(q)} X$ that the task demands, two extra conditions must be met:

1. $df(\tilde{q}) = \tilde{x}$, that is, the intermediate direction $\tilde{q}$ for action simulation maps linearly through the differential df of the map f to the goal-related direction $\tilde{x}$, and

2. $\langle \tilde{q}, q + \tilde{q} \rangle = 0 \ \forall \tilde{q} \in \mathrm{Ker}(df) = \{\tilde{q} | df(\tilde{q}) = 0\}$, that is, $\tilde{q} \perp q + \tilde{q}$ for all changes in postures that do not change the configuration of the hand (i.e., the self-motion subspace Figure 4.11).

Under these two conditions one can restrict the solution to be the unique local geodesic direction in posture space with a corresponding geodesic direction in the task-goals space that iteratively builds the geodesic path. Figure 4.14 shows the DoF decomposition for the examples of Figure 4.12 whereby each different task recruits 3, 4, and 5 dimensions, respectively using the projection of $\tilde{q} \in T_q Q$ onto the unitary (orthonormal) bases V of the SVD of G_q^μ subject to $G_q^\mu = J^T G_{f(q)}^\alpha J$ and conditions (1−2) above.

We have therefore addressed Bernstein DoF problem: that is, we have solved the many-to-one map problem by locally finding the inverse map and also solved the path determination problem by finding a unique global solution path that simultaneously unfolds in both spaces. The local changes of the joints lead to changes of the hand in such a way that given a task with a set of goals and constraints, we can always find the global shortest path to the goals, while obeying the additional constraints.

EMPIRICAL EVIDENCE FOR SPEED INVARIANCE

When this model was completed in 1998, we immediately thought of many different applications, ranging from artificial neural networks that could be trained to produce complex goal-directed paths using this generative solution, to simulations of different behaviors involving movements of the reaching family. I eventually moved to CALTECH and trained under the guidance of Professor Richard Andersen to address other empirical questions predicted by simulated neural solutions from artificial neural networks that I designed as part of my PhD thesis. Along those lines, Richard Andersen and David Zipser had collaborated in the mid-eighties and produced very interesting work involving

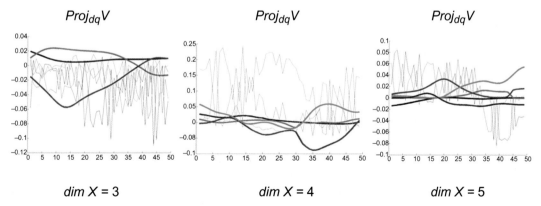

FIGURE 4.14 DoF decomposition using the locally linear embedding to separate rank and null subspaces for the sample tasks represented in Figure 4.11. We use 3, 4, and 5 goals/constraints to illustrate the separation of the task-relevant from the task-incidental DoF. Smooth curves are the task-relevant dimensions while variable jerky curves are the task-incidental dimensions.

coordinate transformations of the kinds the brain must resolve to produce goal-directed behavior.[36]

My main interest on exploring the model in the empirical arena was still on the issue of speed invariance. I wanted to avoid having to assume that the duration and time course of the movement had to be known a priori in order to be able to determine the solution trajectory of the arm-hand system. Having a more realistic model of the arm-hand system had already allowed me to explore many questions that were not part of the motor control discourse in 1998. I was looking forward to examining other issues in the area of cortical activity and potentially extending the knowledge I had acquired to define and resolve new problems in the clinical domain.

I moved to CALTECH toward the end of 2001, but 2 years prior to that I regularly visited the Gastby unit of computational neuroscience in London. At the time, Professor Geoffrey Hinton was the director of the center. There was a tea-time meeting every afternoon around 3:00 pm, where very interesting problems were discussed. It reminded me a bit of the Z-lab meetings, so I felt at home at that place. One day, the late Professor Sam Roweis, who was at the time a graduate student of Hinton and was working on the local linear embedding (LLE) algorithm[37] saw my arm-simulations running in Mathematica and asked me how I had resolved the redundant DoF of the reaching problem. I explained it to him and he liked the solution very much. He suggested I gave a talk at the tea-time, which I eventually did encouraged by Professor Zoubin Gaharmani. During the talk, I learned from Hinton himself that he had a 1984 paper with similar intuitive idea for the use of a gradient-descent like solution to the reaching problem.[38] He stepped out to his office and brought back a copy of the paper. I was thrilled to see this because I was not familiar with the work and joked "oh well, I am over a decade late!"

During the eighties, the Parallel Distributing Processing (PDP) group at UCSD (specifically at the Department of Cognitive Science, where I was completing my PhD) had popularized backpropagation and developed the gradient descent-based learning algorithm.[39,40] Hinton and my advisor Zipser were part of that group. The gradient descent algorithm is still in use today in many learning settings of artificial neural networks. As it turned out, in the process of trying to solve Bernstein's DoF problem I had—unintentionally—made a nontrivial modification that could improve the use of gradient descent simultaneously in two spaces. In the context of artificial neural networks and learning algorithms I learned a year later at the neural information processing systems (NIPS) meeting that Professor Sun Shi Amari of Japan had developed a similar idea that sped up the learning during the training of artificial neural networks. He coined it the *"natural gradient learning"*.[41] I had the chance to attend his workshop and see firsthand a different application of the geometric method. Though we all had different problems in mind, a geometrically effective solution served all three different purposes.

We discussed at that tea-time in great length various ways to simulate new problems in motor control and other areas of research that up until that point were not possible to do or visualize. It was a very exciting tea-time meeting for me as I found a group that spoke similar language and did not mind the fact that I was the only female in the room—and the speaker that day. In fact, we joked about that. Since I was staying for over a month, they kindly agreed to turn the restroom on the top floor of the building into unisex option, so I did not have to go two flats down to the ladies room in Psychology. I must say that this has been the only time in my scientific career when I actually felt I was a colleague among a group of people with similar interests and level of curiosity. That sense of belonging somehow made the day magic. We ended dining at *"Hare and Tortoise"* which I adopted as one of my favorite spots to dine in the vicinity of the Gatsby unit.

Upon my return to the States, I thought the model would be also well-received by the neural motor control community. I was dead wrong. I was not able to publish the work until 2002. Even after that publication came out, I had a tremendous hard time trying to convince my colleagues in the neural motor control field that there are many different possible ways to ask empirical questions with a computational model. I simply could not share any of those questions with that community. It was an impenetrable field, so I shifted gears to translate my geometric solution to the clinical arena. A great ally in this new endeavor was Professor Howard Poizner. Howard had a very rigorous and ingenious way to collect kinematics data from very special patients, dating back as well to the mid-to-late eighties. He was way ahead of his time! I will get back to some of that data later, when I explore invariants predicted by the model present as well in the empirical data that I received from him.

I moved to CALTECH in the winter of 2001 to begin my training in electrophysiology and computational neural systems. In parallel, while I was training and learning at the Andersen lab, I continued to work on the model's simulations. I explored various avenues to further test the question of speed invariance since I had remained very interested in that problem.

Back in 1999, a paper from the lab of Professor Marta Flanders at the University of Minnesota had extended the findings of the 1985 paper by Atkeson and Hollerbach to the full arm. Flanders and her colleagues had found that in reaching movements the postural trajectories were also conserved and remained robust to changes in speed.[42] This was a very important empirical result that encouraged me to explore the postural space across different tasks that demanded speed changes.

As a graduate student, the Minnesota group was always in my radar. Their work was my favorite in the field. From motor psychophysics to neural systems level, I followed very closely what they were doing. In particular, the Soechting and Flanders labs were among the few sources I could rely on to gather information on empirical findings that were relevant to our model. Most (if not all) of their experiments took place in unconstrained ways. They were the real thing! Owing to their results, I explored the idea of speed invariance for different postures of the arm that resulted in similar hand position but different hand orientations. For example, Figure 4.15 shows a set up I built to that end.

It was challenging to have a target for which I could systematically change the orientation while maintaining the position. But one day, I mentioned this during class to Marty Sereno who had built a bite piece to stabilize the head of participants during his brain-imaging experiments. By biting the piece the head did not move as much, so the motor artifacts would not contaminate the imaging data as much. The "concoction" Marty came up with to that end had at least 6 degrees of freedom provided by a couple of spherical joints and hinges. It was perfect for me to maintain the spatial target location while probing different orientation axis for the same target position. I asked him to let me borrow a prototype and made the three-dimensional drawing that adapted Marty's design to my problem's needs. Then, I took the three-dimensional drawing to a shop in Chulavista (very far from UCSD), where they had something like a three-dimensional printer and a

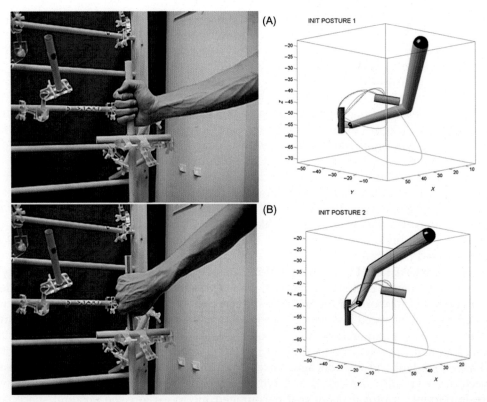

FIGURE 4.15 **Experimental setup and simulations of the many-to-one map problem in the reach-to-grasp task.** Different hand configurations requiring different initial arm postures at the same starting hand location in space can be attained by holding a cylinder and orienting it with palm towards or away from the board of the targets. Each target was held in place by an apparatus that permitted changing the orientation of the principal axis of the cylinder while maintaining the location of the center of mass. Positional and orientation coordinates were registered using the Polhemus Fastrak electromagnetic system (120Hz). Simulated paths generated with the model to mimic the postures of the participant's picture in (A) (red path) and (B) (blue path). Green paths of more pronounced curvature are due to a different cost used to simulate avoiding obstacles placed along the way while aiming at the oriented target.

machine that would reproduce the shape that came out of the printer. The resulting apparatus, made out of Plexiglas was held together with plastic screws. This is shown holding the targets in Figure 4.15. The entire set up had to be made out of wood and glued together without screws or nails because I was using an electromagnetic system (the Polhemus Fastrak 120Hz, Colchester, VT) to capture the motions and anything ferrous would introduce distortions.

It took me a while to put it all together but USCD had an amazing machine shop, so I spent many hours in that shop all by myself building this thing. It stood on a wooden platform and looked quite impressive when it was all ready. It was my very first experimental setup and I was well on my way to become an experimentalist. In my mind, the way to learn about all the hidden mysteries of the motor control system was by gathering empirical data in an as unconstrained as possible way. When it was all ready, I stopped by Marty's lab to thank him once more and to show him the end product of his design adapted to my experiment. To my dismay, he was ecstatic with joy to see it, and asked me to let him borrow it. On his way to the magnet, he had driven off while leaving the box with all the bite pieces on the roof of his car, so they were probably scattered somewhere in the freeway. We laughed at the idea of *the absent minded Professor did it again!* I gave him the piece, since I had made quite a few extras.

Using this target-holding device, I could have the participant match the orientation of a cylinder while coming from different postural configurations, all starting at the same initial position of the hand. They would hold a cylinder in their hand (as the participant did in Figure 4.15) and be instructed to match the given target orientation with the principal axis of the hand-held cylinder (the red stripe on the hand-held cylinder of the picture) aligning to the target's orientation. In this way, the reach-to-grasp action was not an actual grasping action, where the fingers would have to wrap around the target cylinder. This task was simpler and it was something the model could do (Figure 4.15A–B). In this sense, I could explore the joint angle paths empirically while varying the target's orientation, the initial arm posture and the speed of the movement, all without constraining the arm. I could exploit various configurations of the DoF and study the DoF decomposition problem as well. Then, I could compare what was geometrically conserved in the empirical data to what was predicted by the model. As I left out the finger component of the grasp, this was a major simplification and I risked losing information at the expense of studying the DoF behaviors across multiple scenarios. Yet, I set to explore these questions knowing that this could not possibly be the brain solution to the problem. This was rather a way to get at possible cognitive strategies in the precise sense that they were impervious to variations in the dynamics and would help me further ask if they broke down in compromised nervous systems.

To that end, I started to play with forward-and-back motions in the model; I tried to learn about the effects of intermediate targets and intermediate obstacles across space. In particular I could simulate obstacles that were avoided while coming from different initial postural configurations (e.g., the green curves in Figure 4.15A–B). To test orientation-matching at variable speeds, I built a simple computer interface that instructed people to move at different randomly called speeds and orientations while they repeated the movements from an initial location to a randomly given target. This enabled me to map the cost values across space. The speeds I set as a goal were slow, normal and fast. To determine their ranges for each person, I had to test the comfortable pace of each participant and personalize the experiment accordingly.

The experimental task was cognitively loaded because the person had to match a position in space, an orientation in space, vary the initial posture and vary the speed, and do all of it with minimal practice since my interest was to discover how people moved when they had not rehearsed and acquired a notion of the motion's tempo. The hand-path conservation despite speed changes can be appreciated in the averaged paths. Figure 4.16 A and 16B shows the corresponding speed profiles grouped by maximal speed taken across random trials. Virtually the same path in space could be traveled by the hand at different speeds with the remarkable property that across movements to a given target performed at random speeds, the distance from the initial location of the hand to the point where the

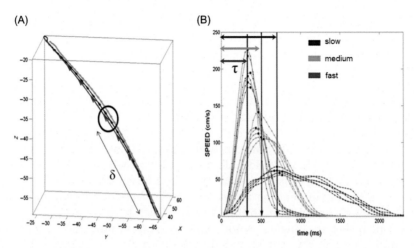

FIGURE 4.16 **Empirical data from human subjects performing the orientation-matching reach task at different speeds.** (A) Hand path conservation across multiple speeds. The hand travels a consistent distance δ up to the velocity peak marked with a circle at slow, medium and fast speeds that each subject determined based on his/her most comfortable pace. (B) The speed profiles corresponding to each speed type with the τ-variations and the peak speed marked with a dot.

velocity reached its peak was similar. That is, to modulate the speed, the system varied the time τ to reach the velocity peak in space but conserved the distance it traveled up to that point in time. Interestingly, the profiles were not always symmetric and as such, the duration of the deceleration phase was more variable than the initial temporal segments. This caught my attention. It was something I decided to explore later on when I studied a different type of speed variation. In that later work I did not explicitly instruct the changes in speed, but rather elicited those changes form variations in hand path curvature required to successfully avoid obstacles.

It was remarkable that despite the changes in speed, all participants conserved the path across all locations that I probed in the external space. Recall here this is the space where the target orientation and the target position were visually available. This task was a bit more challenging than the pointing task of Atkeson and Hollerbach, and a bit more challenging as well than the Flanders et al task. The extra-requirement of a target orientation to be matched was cognitively more demanding. As such, it evoked—for the same target—a family of paths that depended on the expected orientation. Figure 4.17A shows the simulated paths for two different expected orientations (coming to the target from above *vs.* coming to it from below). Figure 4.17B shows the corresponding joint angle excursions the model generated for the two different expected target orientations. Figure 4.17C shows empirical data from one participant in trajectories that reached the target using a family of hand orientations landing at the target differently while moving at different speeds. Figure 4.17D shows the actual empirical paths of seven joint angles. Even the most variable joint at the wrist had no statistically significant difference across the paths under different speeds for a fixed orientation and position of the target. I am

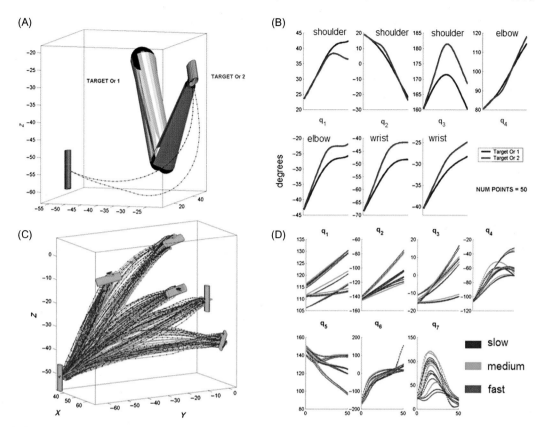

FIGURE 4.17 **Model simulations and empirical data.** (A) Model simulating the task of matching two different target orientations while coming from the same initial hand position. (B) Modulations of the modeled joint angular excursions corresponding to each target orientation. (C) Actual empirical data from a subject performing the task with multiple target orientations and multiple speeds. Note the family of curves and the modulation of the path curvature with multiple target orientations. (D) Different postural excursions recovered from the positional data generated for different target orientations. The average excursions of the joint angles maintain their course across the joints despite changes in speed. Even the most variable joint at the wrist maintains the variance independent of speed level.

convinced that the system has limits regarding path conservation in the face of dynamic manipulations. The hand path probably ceases to be conserved for extreme speeds. Yet, within the ranges we worked, which were typical of activities of daily living, this path conservation in visual space was intriguing.

An important element of the model is that the postural excursions from the simulations can be matched to the veridical postural excursions of the arm using a simple coordinate transformation (Appendix 5), a point that we developed in detail to convey the notion that the model shall not be equated with what the brain may be doing. Rather, the important results are from coordinate-invariant features that remain robust to dynamics. Yet the expectation in the modeling community was different. The assumption of most modelers in the field of neural motor control was that their model must reproduce with exactness the actual behavior, at all cost. This involved a number of hidden parameters one could adjust to produce veridical trajectories in space and time. Surely one could fit an elephant with so many free parameters to get the two-dimensional or three-dimensional trajectories; but what was the point of that exercise? I was not interested in mimicking exactly the behavior but rather capturing nontrivial, hidden aspects of it. These were features that were not apparent in the data but remained invariant to the coordinate functions we chose, to the dynamics and to the context in which the experiment took place. These law-like relations predicted by the model could then be found in the empirical data regardless of how that data were obtained.

My dream-idea since graduate school was to build a new way to do motor neuroscience, one where we did not need to be in a lab performing rather artificial tasks to understand what the brain may be doing. Instead, I wanted *a lab on the go*, where no matter how the action was performed; the behavioral outcome had certain invariant features that we could examine at a clinic in nervous systems with pathological conditions.

During these early experiments challenging the model, it was also very interesting that all participants conserved the postural paths despite the speed changes. These paths were variable—as were the hand paths; but their variability in the joint angle excursion from beginning to end was negligible with respect to the changes in speed dynamics that the task required. I searched for other two geometric relations that the model predicted and found that they held as well despite the changes in dynamics.

One was the conservation of the path described by the changes in orientation and displacement in relation to the expected position and expected orientation at the target. To solve this task one could adjust the orientation of the hand and then independently translate the hand, then rotate the hand once more, etc. But subject after subject simultaneously rotated and translated the hand and to that end the joints of the arm co-articulated in a speed invariant manner. More specifically, these forms of rotational and translational changes recruited four DoF in the arm, such that in the local linear decomposition $\tilde{q} = \tilde{q}^\top \oplus \tilde{q}^\perp$, the $\tilde{q}^\top \in T_{df(\tilde{q})}\mathbf{Q}^\mathrm{T} \equiv T_{df^{-1}(\tilde{x})}\mathbf{X}$ as in Figures 4.11 and 4.14. The path profiles of those four DoF were conserved as were the four dimensions spanned by the hand position and orientation differences as they were updated toward the target (Figure 4.18). That is, despite the speed changes (**18A**), the changes in the initial posture of the arm (**18B**) and the changes in the target orientation (**18C**), the system co-articulated the arm's DoF recruited for the task goals and it did so similarly across speeds, postures and expected target orientations.[43] The course of the change in cost was conserved, suggesting that the geometric solution was a good descriptor of the phenomena under study. This particular speed invariant relation prompted me to address a new aspect of the Bernstein's DoF problem. In the process of exploring obstacles and generating avoidance paths with the model (Figure 4.18D), I came across another geometric relation that also held in the empirical arena. This geometric invariant turned out to be very important to help me later understand Howard Poizner's data across different patient populations with neurological disorders, and to extend it as well to patients with neuropsychiatric conditions like schizophrenia.

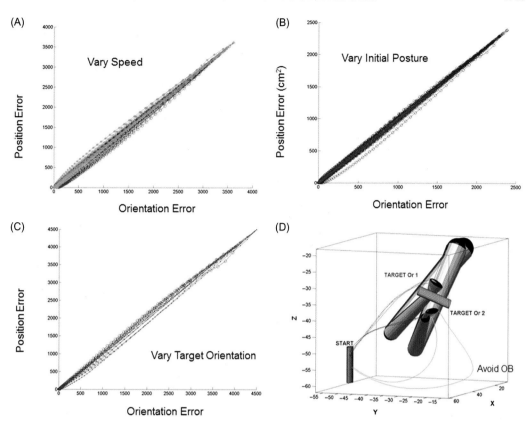

FIGURE 4.18 Empirically registered speed invariance of the co-articulation in the task-relevant DoF of the paths in Q maintaining the simultaneous rates of change in hand rotations and displacements along the paths in X. (A) The co-articulation remains invariant to variations in speed for the same initial posture for movements to different locations of targets with a given orientation. (B) The co-articulation remains invariant to variations in the initial posture of the arm (as in Figure 4.15) corresponding to the same starting location and aimed at the same target location and orientation. (C) The co-articulation remains invariant to variations in the required target orientation for the same target location, starting from the same initial position and orientation of the hand. (D) Simulated paths for different desired orientations change the final posture and the path in X (green curves contrast the model's geodesics to avoid obstacles located at different places along the way).

INVARIANTS OF THE GEODESIC-GENERATING PDE

By the end of my PhD work in 2001 I had explored a number of questions related to the model's predictions as they pertained to the behavior of unconstrained arm movements required to generate goal-directed paths that were geometrically compliant and robust to changes in dynamics. Besides the evidence provided by work in motor psychophysics on postural data, I also gained some confidence for the geometrization of dynamics as a framework to study human motor performance because of an extremely interesting result published in 1999 by the group of Professor Ferdinando Mussa-Ivaldi of Northwestern

University.[44] Sandro (as we call him) and his student at the time Michael Conditt (Senior Director of Clinical Research at MAKO Surgical Corp.) discovered a lack of evidence for a centrally controlled representation of time. According to force field experiments they designed specifically to adapt the arm system to perturbations induced explicitly by time-dependent forces, there does not seem to be a time-dependent representation of forces in the neural structures of the CNS related to motor learning. To understand the significance of their findings in relation to the *time-independent* formulation I was proposing, I need to elaborate a bit on the force-field adaptation paradigm.

The field that studies motor learning and adaptation through computational modeling often uses state dependent force fields whereby as the subject attempts to move the hand to a target, there are forces that perturb the arm. Such a perturbation (initially) results in highly curved movement trajectories in the general direction of the perturbation flow field. As the arm-hand system adapts to such forces, the nervous systems learn to overcompensate and cancel out the perturbing forces. Eventually, the hand trajectories to the target become straight. However, this is a transient phenomenon for if one turns off the forces, the trajectories are curved again, but they tend to curve in the opposite direction, overcompensating when there is no force field present. This is not something that one can actually consciously control, but eventually voluntary repetitions of the movement lead to de-adaptation from those force-field *"after-effects"*. Up to then all the adaptation studies had been carried on using force fields that depended on the state of the arm (e.g., position, velocity and acceleration). But these states of the arm are time varying and time dependent. They vary in time, so it was tricky to come up with the idea of a purely time-dependent force field. Tricky indeed and ingenious! But this group was always coming up with such interesting questions that as the Minnesota group for three-dimensional unconstrained motions, I kept their work in my radar as a graduate student. This was the group I would follow to learn about motor learning.

The subtle yet important difference between strictly time dependent and time varying force fields is beautifully explained in their paper. The problem lies in the ambiguity of the co-dependencies between the state of the arm and the notion of time during the movement. When the force field is *state dependent*, this means that it depends on the position and velocity of the arm, as the arm moves. As such, for a same time instance (across repeated trials of the motion to the target) the arm may produce larger forces if it moves faster than if it moves slower. If the force field is *time dependent*, there is a similar ambiguity. As the arm moves, a time-dependent force field varies with the arm's state because the arm changes position and velocity with the passage of time. In order to avoid such ambiguities, these researchers assigned a *specific unique force value at a specific time* of the movement each trial. In this sense, these were not "time-varying" forces but rather strictly "time-dependent" forces. They established a unique functional dependency between time and force to be able to probe the system's adaptive capabilities under these conditions. As such, the temporal variation in this case was not inherent from the time-dependence of the state variable on the arm motion. Such an experiment had never been done, yet it was so important to better understand what the system was using to learn these weird force fields.

I parenthetically experienced twice the planar robotic manipulandum used to perform these experiments. One was around 1999 when Professor Emo Todorov (at the time a PhD student of Professor Michael Jordan) hosted me at their MIT lab. The other occasion was over 10 years later, when in 2011 I was invited by Dr. Stewart Mostofsky to give a talk at his lab in Johns Hopkins University and Professor Reza Shadmehr, a leading expert on motor learning, let me test their force-field set up. In both occasions I thought it was really cool to do that type of research, but nonetheless really weird to experience the adaptation /de-adaptation routine. It was sort of like when you walk on a treadmill for some time and then stop all of the sudden and try to walk on the ground. You feel some "ghost" force pulling you back. The sensation is similar to that, but restricted to the arm system. The effect is transient but quantifying its evolution can be really informative of the nervous systems adaptation strategies.

My whole issue with this paradigm had been the assumption of time-dependency that I have discussed throughout the chapter. Thus, the 1999 PNAS paper by Conditt and Mussa-Ivaldi was reassuring that the point I was trying to make about the assumptions on time-dependent forces was not a trivial one. Their result showed that when asked to compensate for these repeated and highly predictable strictly time-dependent force fields, the sensory-motor system learned to counter the field by producing forces that did not depend on time, but rather on the state of the arm, that is, on the position and velocity of the arm.

Now, think about that briefly: Position and velocity are both distance dependent, that is, we can use the distance component in them to geometrically construct a motion path that can be traveled with multiple temporal profiles, independent of time. Their result implied that *"time and time-dependent dynamics are not explicitly represented within the neural structures that are responsible for motor adaptation"*.[44] In other words, the centralized ticking clock that had been assumed all along to model and test the motor control phenomena in the primate system was not present during basic motor learning. I was thrilled to see this, yet the paper remains under people's radar in that field. The same type of external observer assumption and time-dependent forces continues to prevail in the computational models driving the research.

MY TIME AT CALTECH: IN SEARCH FOR EVIDENCE WITHIN THE NEURAL CODE

During my PhD thesis I had also explored artificial neural networks trained under the rubric of supervised learning using simple back-propagation algorithms. If neurons in the brain had to resolve the problem of redundant DoF, there were many aspects of this problem that were amenable to model using the geometric equation. Among those was the problem of transforming back and forth between frames of reference; for example, from a retinocentric to an ego-centric (body-anchored) frame. Richard Andersen of CALTECH was one of the leading researchers in the field working in such problems. And as mentioned, he had worked with David Zipser on a version of this problem concerning the posterior parietal regions responsible for the generation of saccades (eye movements to spatial targets).[45,46] In 1999, his lab had discovered the parietal reach region (PRR) with neurons that responded to the planning of target-reaching behavior in various ways.[47] Richard hired me as a postdoc and gave me the problem of recording and analyzing the PRR activity while participants planned obstacle avoidance reaches.

The problem I saw with my new job was that I was totally and utterly ignorant of all aspects of electrophysiology, from training monkeys to recording from their brain to analyzing spike data. Thus, the learning curve was going to be steep beyond belief. Luckily, the lab was full of experts on all of the above who helped me learn the skills I needed to independently train the animals, upkeep their recording chambers and perform the recordings and analyses of the data.

I was interested in the learning component of the reaching problem. I wanted to know how an animal that was awake, behaving and learning a new motor skill would do so in real time. To that end, I decided that I would not train the animals to perform the obstacle (OB) avoidance task. Instead, I would train them to perform the straight reaching task and then place a variety of OBs along the way to targets so as to evoke paths with variable degrees of curvature around the OB. One aspect of the problem I brought along was the expertise in the analyses and modeling of the reaching behavior. Most of the literature from systems neuroscience would describe the behavior without in-detail analyses or modeling of it. In this regard, the spiking activity was interpreted in a somewhat disembodied way. I realized that behavioral neuroscience really meant description of behavior by observation, so there was a paucity of detailed kinematics analyses and modeling to help interpret the spike data. Most spike data even today (2017) is described without considerations of the movements underlying the behaviors the subjects produce while their spikes are being registered.[48]

I could immediately see why this was the case. It is extremely hard to do motion caption on the animals and continuously record from their body in nonobtrusive ways. Less invasive camera-based motion caption systems are available, but they suffer from considerable data loss due to occlusions of body parts that interpolation methods cannot fix. As such, three-dimensional analyses of natural movement trajectories of the arm were uncommon. I decided to adapt a primate jacket (**Notes** [2]) by adding long sleeves and using adhesive Velcro strips to hold the sensors in place along the arm. I had brought along the Polhemus Fastrak electromagnetic sensor system (120Hz) that I had used at UCSD for the orientation-matching experiments with human subjects and decided to give it a try with the monkeys. I realized they were very curious and would chew away on the cables thus destroying the sensors, so I built fake sensors that had similar texture and weight as the real ones. The animals got used to them and eventually I could use the real sensors without having to worry about their destruction. This took some patience, some singing to the monkeys and some discovery of their food preferences. I spent many hours doing this, and day by day, I learned to build a relationship with these animals and to appreciate their high intelligence and exquisite motor control. I was very afraid of them and they knew that but somehow, they did allow my proximity and mutual trust emerged during a co-adaptation process that took months. In the end, both animals wore the jackets and the sensors affixed to the jackets so I could get the behavioral data I needed to help me interpret the planning activity prior to the motions and then the neural activity after movement onset.

The delayed reach paradigm I used had been developed long ago to study the planning activity of the cells in the PRR, so I used it with a subtle yet very important adaptation. I did not impose a priori a fixed duration of the movement epoch. In this way, the animals had enough time to complete the much longer curved reaches that the OB avoidance task

[2]https://www.sai-infusion.com/products/primate-jackets

was going to require. I had modeled the task already and had a very good notion of several aspects of the hand kinematics and the geometry of this task in relation to the direct reaching task. Figure 4.19 shows the different epochs of the task before (**19A**) and after (**19B-C**) the experimental block initiated. The figure essentially illustrates what the animals were seeing in the dark. The example shows this for one target location (up and to the right). I used an additional experiment (**19C**) where I abducted the posture of the arm on purpose to see if the activity of the same cell was similar to the activity the cell had during the spontaneous abduction of the arm posture that the animals performed prior to the OB-avoidance task. I reasoned that this anticipatory activity prior to the actual movement onset could inform me of the nature of the cells that were spatially tuned to the external locations of the board where the targets were placed. *Were they also tuned to postures?*

We have seen that the hand is attached to the rest of the arm and the posture space contains information that is not apparent from the kinematics analyses of hand trajectories. Analyzing the postural kinematics and the relations of that internal space (hidden to the naked eye) with the neural activities that we evoked using external objects could help us infer some aspects of the spatial tuning these cells manifested in relation to the board located external to the body. The analyses of the hand kinematics were going to be helpful but incomplete when the rest of the arm is also "behaving". I wanted to know how postural activity of the arm modulated neural activity in PRR, a region that was visually modulated.

Prior work suggested a retinotopic, eye centered code in this region.[47] However, activity in response to postural changes had never been examined, thus leaving open the question of a possible hybrid representation to generate trajectory plans in both internal configuration spaces (e.g., to position the arm posture) and external configuration spaces (e.g., to code the hand-target vector[49]). If besides the hand-target vector representation, this region also contained modulatory postural activity, it would have all the elements necessary to produce a geometric transformation (e.g., a change of coordinates) to transform from one sensory modality (vision) to another (proprioception and kinesthetic reafference) in order to produce the plan of motion trajectories simultaneously in the posture and the hand spaces. As it turned out, it did. By 2007, I had systematically confirmed this hybrid representation in cell after cell, over the span of several years of recordings in the PRR of two monkeys, but I was not able to publish the empirical results until 2013.[50]

Here is how the PRR posture-encoding story unfolded: During my mapping of the recording chamber to locate the PRR patch, distinguish it from the LIP region (**Notes** [3]) and later record from it, I had noticed that the cells were modulated by posture. Since I was continuously recording the arm posture I could estimate the arm configurations the animals had during the anticipatory changes that modulated the cell's activity. I decided to explore that aspect of the problem and designed an apparatus to evoke a family of postural configurations similar to those I recorded during the anticipatory activity. In this way, I could compare the modulation of the cells' activity in response to postures that were spontaneously self-generated by the animal in anticipation of the impending OB-avoidance trajectory. Then, I could also record the same cell during similar postures that the apparatus evoked, when no OBs were present.

---

[3]Lateral Intra Parietal region (LIP) known for the planning and execution of saccades.

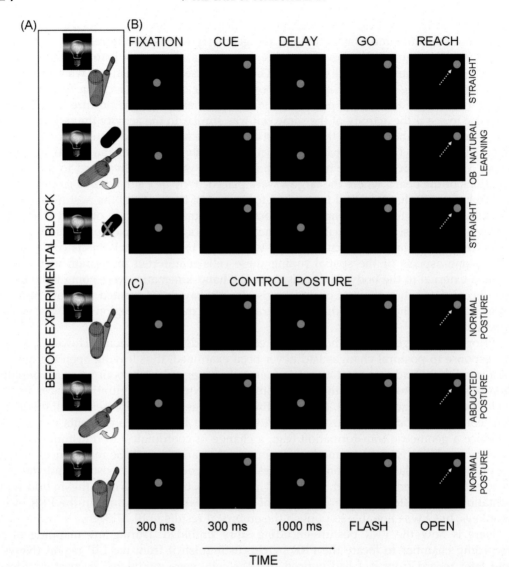

FIGURE 4.19 **Experimental paradigm to study the coding of postures by PRR cells.** (A) Information available to the animals and experimenter in the fully lit setup, before the room turned dark and the epochs of the experiment started. They would see the board (see next figure) in full light and the experimenter would have access to postural information from the sensors. They would see the placement of the obstacles and their removal, so they knew the location of the obstacle to be avoided but also they knew when the OB was no longer there in the second set of straight reaches. Lastly they would sense the apparatus to evoke an abducted posture similar to the anticipatory posture we recorded during the OB positioning, prior to the planning period. (B) The epochs of the delayed reach paradigm: fixation light comes up for 300Ms; then a flash cues the target at a random location on the board (e.g., a target located up and to the right in this case). There is a delay period randomly varying between 700-1000Ms and then the central light goes off to indicate the "GO" signal to the animal. At that point the movement begins and the duration is left open until the animal completes the reach. The sensors continuously record from the arm so we can examine the posture at all times, pre- and post-movement.

I brought my drawings to the CALTECH shop and Rick Paniagua built this apparatus. Rick was someone we all thought had the most amazing three-dimensional visual ability to build whatever it was we needed at the lab. I was then able to study in great detail the kinematics of postural movements in relation to the activity preceding the actual reach. These were the activities spanning 1.6 seconds, from the fixation epoch to the end of the delay period. At the "GO" signal, when the central fixation point went off, the monkeys performed the reach and had no time restrictions to complete it. Figure 4.20 shows the vertically oriented board with a configuration of obstacles (2 on each side in this case) used to obstruct the straight path from the central (fixation) location to the targets. The red light emitting diode (LEDs) in the center of the buttons were for delayed saccades (to map the LIP region) while the adjacent green LEDs were for reaches, to map the PRR. I used both LED's in different tasks. I had to train the animals to perform delayed saccades, delayed reaches straight to the target and hope they would generalize the direct reaching task to perform the delayed reaches to the targets while navigating around OB. I did not want to pre-train them to do that, so I could capture *in vivo* the evolution of the learning each day I recorded cells. Once this posterior parietal patch of cortex was mapped out, I could proceed with the reaching experiments (straight *vs.* OB-avoidance) and record from the identified PRR.

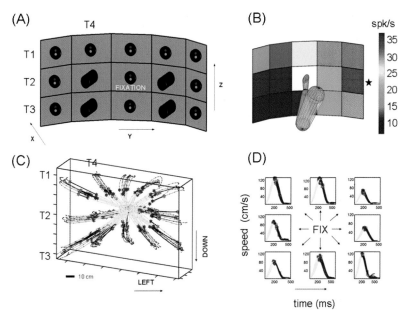

FIGURE 4.20 **Mapping the spatial PRR cell's preference on the reaching board.** (A) Schematic of the experimental board of push buttons. Each button is a target cued through a central LED (red for saccades and green for reaches). (B) Corresponding response field of a PRR neuron during the *planning* period of straight reaches. The initial arm posture is recorded with the sensors. This cell responds maximally towards the right side of the board with the maximal location marked by a star (i.e., the preferred target location of the cell on this board). Color bar represents the range of planning spike activity measured in spikes per second across the board. (C) Hand trajectories to all 14 targets plotted in blue with kinematics landmarks: δ in yellow, red star is the peak velocity and gray star is the distance traveled to the peak curvature, black line marks the path segment from the peak velocity to the peak curvature. (D) Speed profiles from 8 targets around the central location using similar landmarks as in (C).

If I recorded from the more superficial Parietal Cortex area 5, cells were more active as the animals changed postures, but deep in the sulcus, in the PRR cells, I had to wait for the planning period to actually be able to map out the spatial-reaching "receptive fields" of the neurons prior to the movement execution. This is so because I was listening to the spikes through the amplified signal the speakers broadcasted and in the PRR the neurons were silent unless the animal engaged in planning, then the neurons would "talk". Such cells were really active in this period of the experiment. An example of a planning epoch map registered in a PRR cell during the planning of straight reaches to the board is shown in Figure 4.20B. This is the board in Figure 4.20A color-coded with the average spiking activity of this PRR cell during the planning period. The color bar shows the range of spiking activity per second. The spatial board-tuning was maximal to the right, where the highest firing rate of that cell is marked with a star. We call this the preferred direction of the cell at this board setup. I would then use that straight-reach planning activity as the baseline to measure against during the planning epoch of the obstacle-avoidance reach task. In this way, I could examine the difference in planning activity for the planning of the highly curved OB-avoidance reach relative to the planning of the automatic straight reach. Importantly, I underscore that I did not train the animals to perform the OB-avoidance task. Since I was very curious to examine the adaptive behavior and the adaptive cell's responses to a naïve situation, I let the animals gradually learn, one day at a time.

In the meantime I studied the kinematics in great detail.[51,52] Figure 4.20C shows the actual straight reaching trajectories to the board locations. The red star marks the peak velocity while the yellow segment is the δ distance traveled to the peak velocity. The gray stars are the points of maximal deviation from the Euclidean straight line using the linearity measure in Figure 4.12B. Figure 4.20D shows the speed profiles from the trajectories in **20C** aimed at a subset of 8 targets when no obstacle was present. These targets are the ones located around the fixation point (starting with the T4 location and going around counter-clock wise, including the locations in **20A** where the OBs are plotted). I plot on the speed curves the same kinematics landmarks as those I plotted along the corresponding hand trajectories in **20C**.

The work concerning the planning activities of the PRR neurons resulted in the discovery of different populations of cells that had different spike widths and changed their responses dramatically with the presence of the obstacle.[50] I captured the result in Figure 4.21 for a small sub-sample of PRR neurons. About two-third of the neurons I recorded in PRR would become silent for several seconds (which seemed like eternity in the time scale of spikes), but then these neurons would recover their original spiking rates for the straight reaches. Those neurons had wide spikes. They are the ones with the blue waveform. Then, about one-third of the PRR neurons I was able to hold throughout the entire experiment (111 cells total) had narrow spike width and doubled and tripled their firing rates for several seconds before returning back close to their original rates registered for straight reaches to targets (those are the ones with red waveforms).

The waveforms of a subset of the cells are plotted in Figure 4.21 along the surface derived from the mean difference in spiking rate at the cell's preferred location registered on the reaching board. Importantly some of these cells with large negative and positive gains in firing patterns maintained the preferred board-spatial location throughout the entire adaptive task of learning to avoid the OB and performing the direct reaches again.

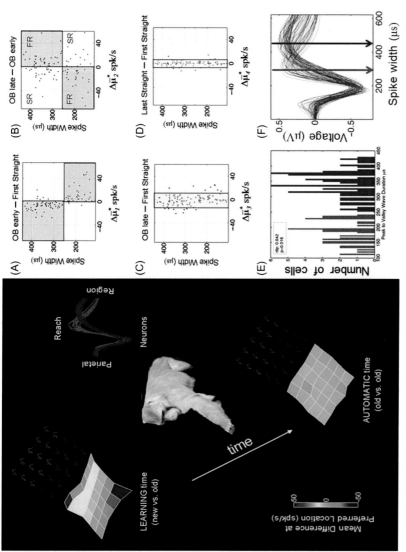

FIGURE 4.21 The real time *in vivo* evolution of PRR cells *planning* activity preceding the movement as monkeys transitioned from naive to skillful obstacle (OB) avoidance in relation to an automatic straight reaching task. (A) Parameter plane of the cells' waveform spike width *vs.* the cells's average change in activity at the preferred direction in space (the target direction with the maximal tuning across 14 targets of board in Figure 4.20A). Each dot represents a cell (111 cells total) tracked each day across different blocks of three tasks (automatic direct reaches, OB-avoidance reaches and direct reaches de-adaptation after OB-avoidance task) The colored quadrants are blue for slow rates (SR) and red for fast rates (FR). The novel task of OB avoidance was divided into early learning (multimodal speed profiles which occurred earlier) and automatic avoidance (unimodal speed profiles which occurred later). The OB-block was interleaved between two blocks of straight reaches (labeled First Straight and Last Straight, respectively.) The Last Straight required de-adaptation from the former curved reaches that were avoiding the obstacles. Blue dots represent cells that decreased their firing rates from straight to curved reaches (2/3), while red dots are cells that increased the firing rates from straight to curved reaches (1/3). (B-D) Evolution of the activity of each cell across the tasks registered in real time each day. (E) Histograms of cell type count based on the spike width (peak-to-valley time) was significantly non-unimodal according to the Hartigan's dip test.[53] Two Gaussian distributions fitted to the histograms crossed at 250µs. We used this cutoff to color code the waveforms in (F) Waveforms of the 111 cells color-coded by width (red narrow less than or equal to 250µs and blue broad greater than 250 µs).

In marked contrast, others shifted the preferred location in complex ways during the OB avoidance task, but returned close to their original spatial tuning registered in the first block of straight reaches.[50] This can be seen as well in the spike rates of the preferred location. The surface in the lower area of the figure shows the change in activity is nearly zero, when comparing the planning activities preceding the automatic portions of the performance in the OB task and the spiking activity from the last 10 trials of the second block of automatic straight reaches. When learning had resumed, the cell went back to its "baseline" planning activity for straight reaches. This automated state of the cell was what we were used to seeing in Neuroscience. In the particular case of PRR, we had never seen such dramatic changes and much less identified different classes of posture-sensitive neurons engaged in temporal dynamics learning. *Was this a neural characterization of embodied intent?*

The spike activity indicated that these cells, which had tuning for the visual targets flashed in the dark (i.e., had spatially tuned visual responses in the absence of overt goal-directed movement) also had tuning for the postural changes. The cells had modulations in their pre-movement responses when the postures were induced by the anticipation of obstacle avoidance (i.e., during the planning period, in the absence of overt movement). They also had such responses when I evoked similar postural changes with the apparatus that forced the arm to abduct. These were not cells with externally evoked visual responses or internally evoked postural responses. They were cells that had both types of responses. They contained multiple representations of the postures and the visual targets, all of which were modulated by the obstacles in very systematic ways during the fixation, visual and *planning epochs*, that is, up to 1,600Ms before any actual movement took place.[50]

The PRR had all the ingredients in place to perform geometric transformations of the type required to plan full trajectories simultaneously in spaces of internal and/or external sensory sources. My guess in relation to implementing a robotic arm-brain interface using the PRR was that those cells with gain fields that maintained the spatial direction across different postural fields would be likely useful to control the task-relevant DoF subspace of a multidimensional anthropomorphic robotic arm. Those cells with complex gains that rotated the preferred direction with different impending postures were likely useful to control the task-incidental DoF. This is now a testable hypothesis, yet regardless of the outcome, the population of PRR cells I recorded and analyzed could blindly decode the impending postural path over a second ahead of the actual arm movement.[50] Their code without a doubt could be used to guide and update trajectory generation of a robotic arm.

Even during the presentation of the target in complete darkness, when the animals were not seeing the obstacle, the arm already made anticipatory postural adjustments that the Polhemus system recorded. The cells responses during this fixation period could already blindly decode which impending postural path would be the most appropriate to avoid the obstacle.[50] In the context of the model, this activity would correspond to the gradient vector pointing in the direction of posture space along the geodesic for obstacle avoidance in route to a target. This vector would change in posture space to reach the target straight. But Appendices 4 and 5 explain how a simple geometric transformation can achieve the change of the geodesic direction that provides a global solution to the reaching problem. We can change the distance metric (our cost) or perform a coordinate transformation under the old metric. In either case, the gradient of our local linear isometric embedding can provide both the global and the local solutions to this problem. And the message from this work in electrophysiology was that cells in this region have the ingredients necessary to do the job.

Figure 4.21A shows the dramatic changes in firing rates registered in the cells once the OB task is introduced. The cells were color-coded by their spike width whereby the Hartigan's dip test of unimodality[53] yielded significance ($p < 0.016$) for nonunimodal distribution. A mixture of two Gaussians yielded a cutoff value of 250 µs (the cross point of the Gaussian PDFs) separating neurons with a narrower from a broader waveform. These are shown in Figure 4.21E−F. The evolution of these cells planning activity in real time, as the animals acquired the OB-avoidance motor skill, was registered every day. In the initial months the cells' changes were huge but they decreased as the animals were acquiring the new motor program. I could say this with confidence because I was recording the arm activity in tandem with the cell activity every experiment. I could track the evolution of this behavior each day and across months. Figure 4.22 shows a parameter plane I built using relative indices. Along the horizontal axis is the change in the coefficient of variation (ΔC_V) during the OB task in relation to the straight reach task. Here the mean and variance to obtain the CV-ratio refer to the maximal change in *planning* activity at the preferred direction obtained in the straight reaches. Along the vertical axis of this parameter plane is the change in mean firing rates at the preferred location of the cell when comparing OB *vs.* straight reaches ($\Delta \bar{\mu}_1^*$). This subset of the cells encompasses the largest and the lowest changes in the early and late days (respectively) spanning several months. Some of the behavioral analyses I did were eventually published many years after the work was completed.[51,52]

As with the model and all other work I did, it was very difficult to publish the results concerning the neuronal activities. I had not trained these monkeys (Corky and Wally were their names) to perform the obstacle avoidance task. They had learned the direct reaching task, but were naïve to the obstacle. The neurons had to learn on the spot as their arm-hand systems adapted and de-adapted from the task (each day there was a little bit of learning). The field was not used to seeing detailed modeling and analyses of postural behavior in tandem with the neural analyses. I think that all these factors played a role in the delays I experienced to be able to have the editor send the paper to reviewers. I simply could not get even past an editor and I tried all sorts of journals to that end. This was a very depressing time for me and I almost quite science altogether.

It was not until 2013 that I was able to publish work that was ready since 2003. I continued the recordings and systematically confirmed what I had already learned but it was sad that scientists in that field remained skeptical or set on their old ways (or both.) In the span of time from 2003 to 2008 many people in the field of computational neural systems became interested in the work. I was able to present talks at the COSYNE meeting and at their workshops (**Notes** [4]), at the Annual Meetings of the Society for Neuroscience[54] and at invited colloquia in multiple venues (e.g., the Salk Institute, UCSD, College de France, IBM research in San José, Northwestern University, Wellesley College, among others). The results were very well received by the audiences owing to their novelty; but when it takes 10 years to be able to publish work in a peer-reviewed system, one simply gives up.

---

[4]http://www.cosyne.org/c/index.php?title = Cosyne08_The_cortical_microcircuit_and_cognitive_function

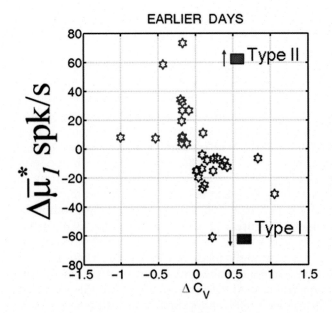

FIGURE 4.22 **Parameter plane of longitudinal learning effects on the PRR cells planning activities** captured in cross-sectional data from a subsample of the 111 cells in Figure 4.21. Type I cells are broad spiking (above 250μs peak-to-valley spike width) decreasing rates with the learning of the obstacle avoidance task. Type II cells are narrow spiking cells (below 250μs peak-to-valley spike width) increasing their rates with the learning of the obstacle avoidance task. Early days show very large effects on the coefficient of variation (mean to variance ratio) of the changes in spike rates while the changes in spike rate differences from obstacle to straight reaches are plotted on the vertical axis. In the later days these differences are not as dramatic, but continue to have statistical significance.

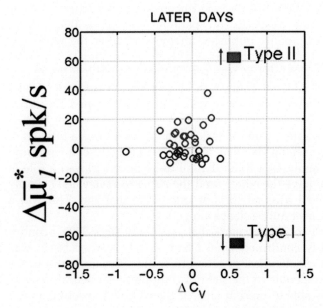

I gave up and put the work to rest for years. In 2012, I was already on a tenure-track position at Rutgers University and had shifted gears to autism research when I came in touch with the new open access platform of Frontiers. I then submitted the work to be judged beyond the confines of the small community of monkey electrophysiology and neural motor control in the US. As with the talks and colloquia, the work was received with great interest. This encouraging response has prompted me to re-analyze the PRR neural data concerning the actual reaching motion, as the activities unfold from trial to trial. To that end, I have developed new methods to better understand these fascinating phenomena from these two classes of PRR cells that engaged during the planning of a complex reach. I wonder what their reaching activities may reveal. Without the *"publish or perish"* pressure and having already reached a point in my career where it matters very little what the neural control of movement field does for basic research, I look forward to these new analyses and statistical inferences on my old data from postdoctoral times. I have sabbatical time severely overdue. I know that I will model some of these activities and will finally characterize the postural motor behavior in relation to its planning stage.

Without a doubt, these cells had complementary functions that will become more apparent when we pair their planning to their movement activities. However, what surprised us all when we examined their spike widths was the fact that their spikes' waveforms were of two different classes, narrow (increasing rates) and broad (decreasing rates). I came across this information thanks to a colleague, Jude Mitchell (now Professor at the University of Rochester), who had discovered cells in area V4[55] with different spike widths during a spatial attention task. His adviser at the time, Professor John Reynolds of the Salk Institute organized a workshop at the COSYNE meeting of 2008, where among other speakers I also presented our results from the PRR area (**Notes 4**). Others discussed criteria derived from intracellular recordings to differentiate pyramidal neurons from interneurons. In appearance, the extracellular recordings criteria of spike width, functional differences in spike rates and cell yield (2/3 broad spikes *vs.* 1/3 narrow spikes) across different areas of cortex were mapping well to the intracellular criteria. The COSYNE workshop organized by Reynolds in 2008 was incredibly interesting and brought together many well renowned electrophysiologists to discuss these new findings across the field and in different animal models (including rabbits, rats and monkeys).

At our end, we had discovered at the neural level two different types PRR cells that contained postural and visual information in their planning activities. At the behavioral level, how the arm-hand reaching unfolded over a second and a half later, after the target was flashed in the dark will soon become apparent.

GEOMETRIC INVARIANTS OF UNCONSTRAINED ACTIONS

The analyses of the obstacle avoidance movement trajectories across the posture space and the hand space brought me back to the question of speed invariance. In particular, they helped me assess the question of geometric invariants using the linearity index that I introduced in Figure 4.12B. This measure also relates to the notion of geodesic curvature of the manifolds I had chosen to model this problem. All trajectories I had examined thus far across the pointing and orientation matching tasks had an interesting invariant. How to obtain this invariant is represented in Figure 4.23.

If one takes the measure of linearity of Figure 4.12B and chooses the point of maximal deviation from the Euclidean straight line (named in the picture *"max bending"*), one can obtain the area enclosed between the two curves up to that point (yellow area in the picture). Then, dividing that quantity by the total area enclosed between the two curves

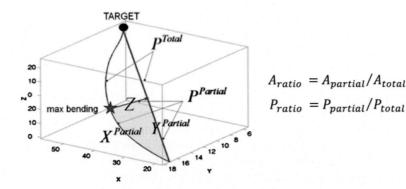

FIGURE 4.23 Area-Perimeter symmetry from hand trajectories of the reaching family.

$$A_{ratio} = A_{partial}/A_{total}$$

$$P_{ratio} = P_{partial}/P_{total}$$

amounts to $1/2$. If one does the same with the perimeter, the ratio also amounts to $1/2$. This symmetry remains invariant to the changes in dynamics that the arm-hand system undergoes to produce and control highly curved paths. This is shown in Figure 4.24 for representative obstacle-avoidance trajectories to a target. As the trajectories evolve from earlier to later trials, in the order in which the animal executed the movements, their speed profiles change in very specific ways.

We must keep in mind here that there were 14 targets distributed across the workspace in front of the animal (Figure 4.20A) and the target locations for the reach were flashed in random order. As such, owing to the arm's DoF, there was no obvious reason to have consistency across all the trajectories to a give target. Surprisingly, the spatial path (i.e., the length and shape) was conserved for each target, even for those maximally affected by the presence of the obstacle. The speed profiles did change from trial to trial, and their evolution was very informative of the learning process.

Early on, the peak speed seems very variable, but a closer look reveals that the time to reach that first peak (τ) is very consistent. The quantity that changes dramatically is rather the distance traveled by the hand during that initial segment, before the hand decelerates and accelerates several times along the stable spatial path that conserves the path length (Figure 4.24A). I will call this distance δ. At some point, the speed profiles become stable and δ also stabilizes (Figure 4.24B). At that point, when the time duration of the entire trajectory has also become stable, it would be comparable to interchangeably speak of τ or δ as another space dimension providing feedback to prospectively plan ahead impending movement dynamics.

If we were to build a model of the trajectory dynamics, under the general rubric of the Maupertuis' least action principle[24,25], as proposed in Figure 4.6, the empirical equivalence of τ and δ that we find for automatic motions would make them good candidates to represent the temporal axis as a sensory-space dimension bridging internal and external information. In this case, δ would provide the information to build metric distances for each task (i.e., to learn about geodesic and normal curvatures of many trajectory solutions across families of reaching movements), while τ would be tied to the internal sensory processing (transduction and transmission) delays. The model would help us understand the empirical data and simulate how the system learns to transform information in the space of external sensory sources (e.g., visual cues) into the space of internal sensory sources

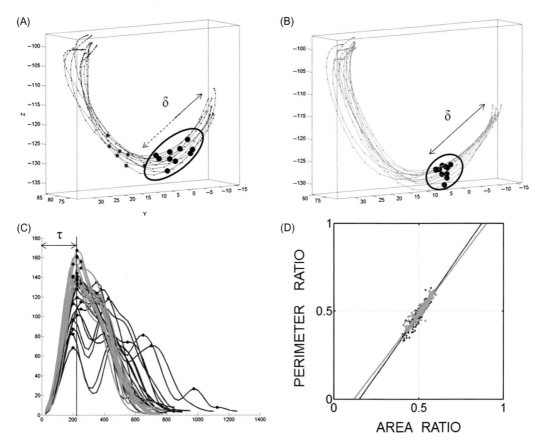

FIGURE 4.24 **Obstacle avoidance learning behavior registered in tandem with cell activity** one of the recording days for one of the cells in Figure 4.21. Each day the behavioral activities were recorded and tracked over months. In any given day the hand trajectories and the postural trajectories evolved as the animals acquired the obstacle avoidance motor program. (A) Early trials were characterized by variations in δ (distance to first velocity peak) under a stable τ (time to first velocity peak). Red and green dots indicate second and third velocity peaks. Path length conservation and conservation of path shape prevailed over temporal variability to traverse the paths with variable speeds and acceleration-deceleration profiles. (B) Later trials of obstacle avoidance learning with stable δ, a single velocity peak and stabilized deceleration phase and overall movement duration. (C) Speed profiles for the early (blue) trials in (A) and late (green) trials in (B). Note the stability of τ and the variability of the overall time duration. (D) Scatter of the area-perimeter ratio symmetry across all learning trials remains invariant to changes in dynamics.

(e.g., movement-based kinesthetic reafference related to the internal dynamics). In this sense, in view that δ was the most variable and adaptable parameter of this experimental setup and in view that tracking its evolution as it reached the consistency of τ for the already-stable path length, I proposed that these PRR cells were participating in a geometric (coordinate) transformation to derive internal temporal information from external spatial information (i.e. deriving time from space). The learning process that we registered in the neurons during the planning activity of these cells and the ensuing behavior of the arm-hand systems supported this notion of spatiotemporal adaptive alignment. Indeed, the geodesic path in space preceded the stability of the temporal course. In fact we registered time-varying dynamics, whereby for each given instances of time we had different speeds across trials (i.e., different force profiles) as the arm and the hand traveled along consistent paths in the posture and hand spaces, respectively. Such different dynamics along a similar path produced different time durations under conservation of several geometric invariants. These empirical data were not consistent with the optimal control formulation of the models of planar motion dynamics that we previously mentioned. The evidence suggested a different model for motor learning and adaptation, one where the path of the dynamics could be geometrically modeled. Given the involvement of the posterior region of the parietal cortex in the nonhuman primates, I wondered if in the human primates, the δ-τ relations would be disrupted after damage to the posterior parietal cortex. And more important yet, I wondered if such relation could be restored with some form of sensory guidance.

At the behavioral level, it was remarkable that despite the changes in speed, the geometry of the path and its length remained invariant from trial one, until the system learned the skillful movement and turned automatic at it. This is captured in Figure 4.24D which plots the scatter of points centered at ($^1\!/_2$, $^1\!/_2$) corresponding to the values of the area ratio $vs.$ the values of the perimeter ratio. Further, the conservation of the postural paths despite the changes in speed was reflected in the arm trajectories at various levels. Figure 4.25A shows them for the sensors' located on the hand, the forearm and the upper arm. Figure 4.25B shows them for the seven joint angles that I used to represent the postural configurations. Tracking δ, the full path length and τ was also important because of the path length conservation. This conservation of the path geometry stood in marked contrast to the large variations in the time duration of each trajectory during the learning period. Once again, this empirical result pointed at a degree of independence between the generation of the geometric reach plan and the implementation of the time-dependent forces (the movement dynamics) to execute that plan in a geometrically robust way. Figure 4.26A shows a surface estimated from the 14 target locations across the reachable workspace of the board in Figure 4.20A. The surface figure highlights the large variability of the adaptable δ for 6 targets at the extremes of the board (T1−T3 on the left hand side and their counterparts on the right hand side), 3 of which on the left were affected by the presence of the obstacle. These are the targets with the largest error bars representing the variability of the δ parameter across 30 trials. The "T" in the figure represents the target location for which representative trajectories in Figures 4.24−4.25 were plotted for reaches to T1 in Figure 4.20A.

Figure 4.26B shows the low variability of the path length across the trials, also in line with the low variability of τ depicted in Figure 4.26C, suggesting a geometric link independent of the varying dynamics. The panel in Figure 4.26D underscores the consistency of the area perimeter ratios despite the large changes in speed and timing along these trajectories. Because of their spatial consistency, these trajectories can be modeled as the geodesic curves representing the path solution of different task costs. The locally linear isometric embedding procedure explained above can be used to model these geodesic solution paths.[51] Further, the geodesic direction determining the unique solution path for each task can also be obtained from the implicit surface representing the cost across multiple tasks. This is explained in the Appendices 2 and 3. Appendix 4 explains how to switch

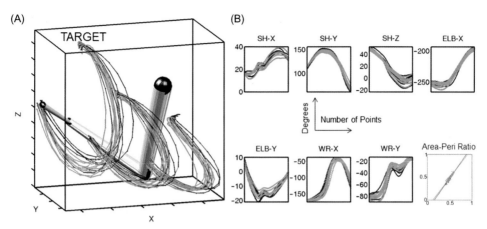

FIGURE 4.25 **Speed invariance in postural excursions recovered from the sensor position trajectories.** (A) Polhemus sensors were affixed to the shoulder, upper arm, fore arm, and hand. Trajectories from the initial position of the hand to the target remain consistent in shape and had similar path length at each arm location. This was despite different speeds reflecting the learning (blue) *vs.* automatic (green) phases of the adaptation process while the subjects acquired a new motor skill. (B) The corresponding postural excursions for seven joint angles of the arm reconstructing the positions of the hand, wrist, elbow and shoulder from the sensor data.

directly between different tasks using different coordinate maps and distance metrics representing the cost functions to build corresponding implicit surfaces across the reachable peripersonal space.

PART II   SWITCHING RESEARCH PATHS FOR CLINICAL APPLICATIONS

One of the extraordinary properties of the nervous systems is the power to find a path of self-healing. In various instances the system learns on its own how to cope with injury and develop strategies to replace the damaged (or altogether missing) source of sensory input with a new sensory source the injury may have spared. This is called sensory-substitution, or sensory augmentation in the cases where an artificial source is provided to augment the faulty sensory feedback. An example of such a technique was introduced in Chapter 3 for the children with autism. One can think of this technique as an accommodation to boost the systems access to the sensory world and help the nervous systems better integrate external with internal sensory information during the self-generation of actions and their self-supervision and correction by the person's nervous systems.

Most times, however, it is not obvious at all what source or sources of sensory guidance may be spared in an injured system, or in a system with neurodegeneration. Acquiring this information is crucial to help a patient's nervous systems process and integrate sensory information with ongoing movements during therapy. Most of the information we can gather from visual inspection during a patient's visit to the clinic is apparent to the naked eye, but information that is not apparent to the naked eye, that is, occurring largely

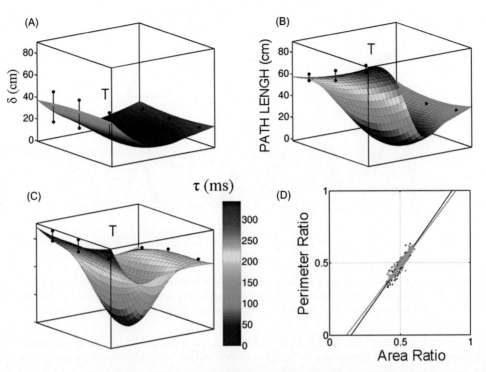

FIGURE 4.26 **The geometric nature of τ and its map across the reachable board space.** (A) The parameter δ with adaptable variability from straight reaches to OB avoidance reaches. The "T" represents the target for which the trajectories in Figures 4.24–4.25 are shown. (B) The stability of the path length across all locations corresponds to the stability of τ. The larger paths are due to the presence of the obstacle increasing the curvature of the trajectories. (C) The τ parameter map and its stability despite the changes in dynamics induced by the OB avoidance task. (D) The area perimeter ratio symmetry scatter across the learning (blue) and automatic (green) trials of different path curvatures, different postural excursions and different speeds across the space used to build these maps in A-C.

beneath awareness, is also important and very relevant to the question of choosing the appropriate source of sensory-guidance to adopt sensory substitution or sensory augmentation techniques. As such, it is critical to have methods that reliably inform of optimal sources of sensory guidance while using activity harnessed from the nervous systems during action generation, in the presence of different sources of sensory guidance. In this sense, the theoretical model presented in Part I provides several geometric invariants that in the typical system remain impervious to the movement type or to the context. We can use them to probe the injured system, investigate their disturbance, and utilize the outcomes of our biometrics to help us identify which source of sensory guidance (externally *vs.* internally generated) would be most adequate to help the person' system carry on sensory-motor integration in such a way that the invariants would return to their typical signatures.

More precisely, the original motivation to uncover geometric invariants in a theoretical anthropomorphic model of the arm-hand system was to identify law-like relations inherent in the movement trajectories that remained robust to dynamical or contextual changes across different tasks. The idea was to try and understand if these invariants were merely the byproduct of the system's biomechanics (e.g., present in a robotic system as well), or if they depended on the brain's voluntary control. If they broke down in patients with nervous systems injury, the question then was whether they could be restored back to typical levels. If such cases existed, the source of guidance restoring the invariant could identify the source of the enabling sensory modality (i.e., external *vs.* internal) spared by the injury.

When I was wondering about these questions, I was already at CALTECH learning about the electrophysiology of monkey neural recordings and had no access to patient data. Yet, by a lucky turn of events, I was contacted by Professor Howard Poizner who had been at Rutgers University up to then. He had moved to UCSD and opened an amazing motion studio lab with state of the art motion caption systems and plenty of space to carry on movement science. Poizner had led an effort to develop motion caption systems since the eighties. The recording techniques and experimental paradigms he had developed back then to study patients are part of today's cutting edge methods for clinical research across many disorders of the nervous systems.

Poizner studied my PhD thesis work and decided to give the theoretical ideas a try. He drove up to Pasadena from La Jolla to ask me if I could train a postdoctoral fellow from Professor Terry Sejnowski's group at the Salk Institute to perform the types of analyses I did on the human subjects I had recorded in the speed-orientation matching task I described above. I agreed to this and provided this fellow with all my Mathematica notebooks and Matlab scripts to carry on these types of analyses on Poizner's data. In a fortuitous turn of events, the postdoc had to return back home to Japan, so the project dissolved. Yet, later on Poizner asked me if I would be interested in analyzing the data. I jumped of joy! It was precisely what I had hoped for, so I did.

THE EXPERIMENTAL PARADIGM

The experimental paradigm that Poizner developed was exactly what I was looking for to address the question of identification of an appropriate source of sensory guidance for sensory substitution/augmentation therapies. He had published some work involving these data[56] but none of the questions I had concerning geometric transformation invariants and postural analyses had been addressed. As such, there was room for new analyses using the model's invariants. Figure 4.27 shows the set up (Figure published in[33,57,58]). The experiment involved pointing to a target using unconstrained three-dimensional motions that naturally engaged all DoF of the arm. Notably, unlike all other pointing paradigms, which only registered and analyzed the goal-directed portion ending at the target, Poizner's data contained the full kinematics of the trajectories described by the hand. They also had the positional trajectories of the entire arm movement loop while the participant retracted away from the target and ended at another position. This was to my advantage since I was interested in the automatic transition the arm-hand linkage would make when reaching the target and retracting away from it. This type of transition is uninstructed. It is something the participant spontaneously does. Given the DoF problem, I was interested in the interplay of these movement epochs during the execution of these segments that fall

(A)

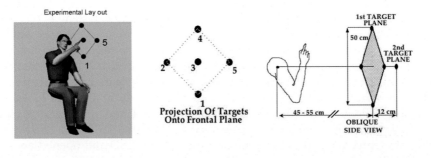

(B)

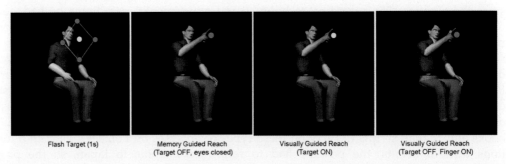

FIGURE 4.27   The experimental paradigm to study different sources of sensory guidance and their potential use in therapies for different injuries of the nervous systems. (A) Experimental lay out where the person seats comfortably and points at a target using a forward and back loop from an initial location of the hand to a posture shown in schematic form. Five targets and their location and physical dimensions are also shown. The central target is in depth and within the fovea at the body midline level. (B) The pointing motions in the dark are performed to reach a target flashed at random for 1 second on one of the 5 target locations and then bring the hand near the face area as in (A). The memory guided reach is performed with eyes closed; the visually guided task with the target continuously ON provides visual guidance from the external target; the visually guided reach where the target is OFF but the LED on the moving finger is continuously ON, guiding the reach forward to the target and backwards to the instructed posture.

largely beneath the person's awareness—unless one specifically instructs the participant to attend to the retraction. There was no literature addressing these questions since motor control studies were restricted to goal-directed movements. In 2005, when Poizner brought me these data, questions involving spontaneous behaviors of the nervous systems were not part of the narrative of the neural motor control field. Today (2017) they are still not mainstream questions in that field. But spontaneous aspects of behavior hold the key in many problems the brain has to resolve to control the body in motion.

In some settings Poizner had asked people to actually attend to this retraction and return to a particular posture, but in others he let people do this on their own. The retraction was a totally unexplored motion. How would the invariants behave for cases where the movement segment was spontaneous *vs.* cases when it was deliberately performed?

Further, he had used three conditions where the visual sensory feedback was systematically manipulated. As subjects moved in the dark, he had a programmed robot presenting the targets to the subjects.

The experimental lay out is represented in schematic form in Figure 4.27A. There were five targets in front of the participant located at a comfortable distance, four targets on a plane surrounding a middle one in depth (see dimensions in the figure). The experiment had three blocks with different experimental conditions. In one block the target was flashed in the dark and the light went on to eliminate the retinal memory of the target. In one case the pointing movement to the target was performed in complete darkness. The participant had to remember the visual location while performing the reach with eyes closed (this was termed memory-guided reach). In another condition, the participant could guide the hand movement using continuous visual feedback from the target which was lit from the beginning to the end of the movement. In the third condition, the target went off after it was flashed and on the movement onset a light emitting diode (LED) was lit on the subject's pointing finger. In this case, the visual information of the moving finger was continuously available in tandem with the kinesthetic reafferent information coming from the moving arm.

Poizner gave me data from a patient with a stroke in the left posterior parietal lobe, from typical controls of similar age and from patients with Parkinson's disease that had different levels of severity. The results of my query regarding the geometric symmetry are shown in Figure 4.28. The figure shows that the geometric invariant relating the hand trajectories and the postural trajectories held for normal controls in the forward and backward segments of the reach across all conditions of visual feedback (first column of the figure). The patients, however, had a very different pattern that violated the symmetry ratios and the disruption pattern changed with the type of injury. More important yet, a specific type of sensory guidance selectively restored the symmetry to typical levels for each patient type.

The patient with the stroke in the left PPC (third column of the figure) did not benefit from the continuous feedback of the finger LED. Instead, the continuous visual feedback from the target LED restored the symmetry. This is interesting in light of the major disruption this patient had with the τ and δ parameters during his movements to the target in the absence of visual feedback (in the dark). Figure 4.29A shows the irregularities in the hand trajectories. These are marked by random trial to trial variations in the distance δ traveled to the point of peak velocity (yellow segment) and the distance traveled from the point of peak velocity (red star) to the point of maximal deviation from the straight line (black segment ending in a gray star). Typical performance showed straighter trajectories (Figure 4.29B) and more regularity in the $\delta-\tau$ parameters (e.g., on average, the peak velocity was reached near the peak curvature of the hand path) and from trial to trial they systematically maintained these trajectory landmarks (i.e., low variability in the parameters). To appreciate the irregularity of these kinematic features in the patient's movements we show the temporal speed profiles derived from the trajectories of the top panel in Figure 4.30 across all five targets. Unlike all typical controls I had studied (including as well the nonhuman primates) who had manifested a consistent map of τ across the reachable space, this patient had a profound disruption in the stability of the time to reach the peak velocity. This can be appreciated in Figure 4.31A where a surface across the

(A)

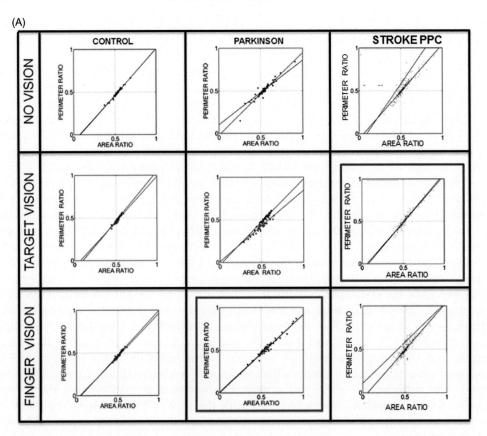

FIGURE 4.28   Identifying the source of sensory guidance that restores the symmetry from the geometric transformation between internal configuration space and external visual space. (A) Conservation of the symmetry across conditions for the forward (red) and backward (blue) motions in controls. (B) Symmetry break and lack of correspondence between forward and backward segments in patients with Parkinson's disease. Continuous vision of the moving finger helps their system re-align the forward (to target) and backward (to posture) segments. (C) The continuous feedback of the visual target helps the patient with a stroke in the left PPC restore the forward and back symmetry correspondence.

reachable workspace was estimated and at each target location the estimated mean τ and the standard deviation are plotted with error bars. The representative controls data summarizes the typical performance whereby τ is stable across space. In marked contrast, the patient has large τ-variability across locations. This τ-parameter normally shifts smoothly across space according to the type of visual guidance provided to the participant (as shown by the color coded surface). Yet the variability typically remains steady. However, in this patient this is not the case, and as it can be seen in Figure 4.31B, the continuous visual feedback of the target shifts the values closer to typical ranges and the variability moves to steadier regimes. As such, this parameter is a good outcome measure of performance contributing to our objective quantification and understanding of the systems' responses to sensory guidance.

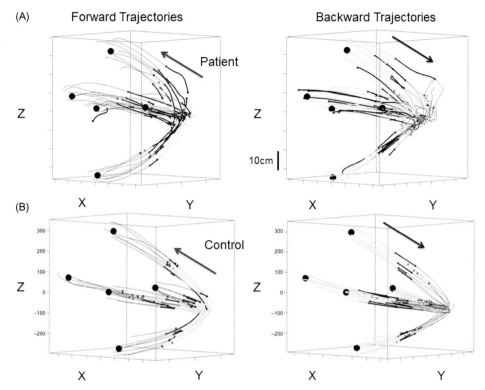

FIGURE 4.29 Hand trajectories to the target and back to a posture from patient with a left PPC lesion due to a stroke and age-matched control. (A) Forward trajectories to the five targets in Figure 4.27A from the patient show higher curvature than control in (D) and highly irregular kinematics landmarks (i.e., δ in yellow, peak velocity red star, peak curvature gray star, distance from peak velocity to peak curvature black segment) and larger endpoint errors at the targets represented by black circles. (B) Patient's backwards trajectories to the instructed posture with the hand near the face and the palm facing the ear. Kinematics landmarks as in (A) and (D). Notice the irregular variability of the patient at the end posture when retracting the hand in marked contrast to the controlled landing of the age-matched control. (Arrows mark the directional flow of the hand motion.)

As seen in Figure 4.29A, the patient's three-dimensional hand trajectories to the memorized targets were much irregular throughout, including and particularly at the end of the loop in the backward segment; yet τ (and δ) regained the consistency that typical subjects showed when the patient was provided with continuous visual feedback of the lit target. This was the case across all five positions and extended to the maximal curvature landmark.[58] Consequently the variability in the scatters of the area-perimeter symmetry returned closer to typical levels for both the forward and backward movement segments, a featured highlighted by the enclosing red square in Figure 4.28 (third column).

The left PPC patient did not manifest the type of hemispatial neglect that patients with a stroke in the right PPC often manifest. His deficits were more evident in the temporal parameter τ that we tied to the distance δ, suggesting that some sort of asymmetry in spatiotemporal coding may be present in the human PPC. In both the stroke patient and the parkinson's disease (PD) group it was possible to individually probe which source of

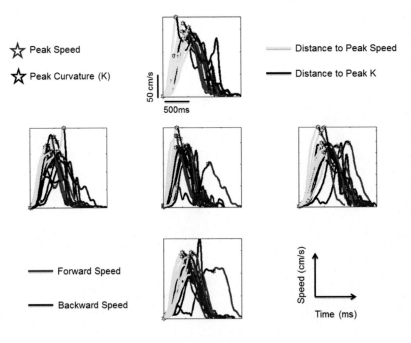

FIGURE 4.30 **Speed profiles for the patient's trajectories during the memory-guided reach with eyes closed** shown in Figure 4.29(A) for the forward segments and 29 (B) for the backwards segment using similar kinematics landmarks. Notice the irregularities in δ, τ and the deceleration phase of the motion (blue) to land at the desired posture.

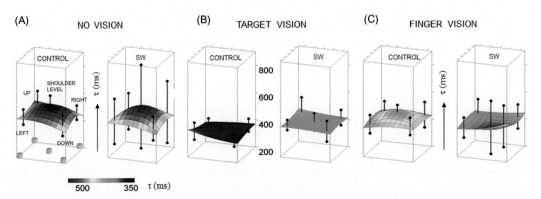

FIGURE 4.31 **Susceptibility of τ to the lesion in the left PPC manifests across all 5 targets with the worst variability in the midline,** (A) reaching in depth with eyes closed. (B) The map of τ is closer to the normal controls with continuous target vision. (C) Continuous vision of the finger does not help the patient.

sensory guidance was the most adequate to restore τ, δ and more generally the trajectory ratios reflecting aspects of the geometric transformations between postural and hand spaces. Years later, this early work informed me of ways to uncover useful forms of sensory-guidance to help children with ASD habilitate and re-habilitate volition by boosting their nervous systems access to somatic-motor information from their self-generated motions and consequently helping them restore sensory-motor integration processes in real time.[59]

PARSING DOF REVEALS EXCESS DELIBERATENESS IN AUTOMATIC MOVEMENT SEGMENTS OF PD PATIENTS

The model explained in Part I is easily implemented with real data.[33] To that end we need the recordings of time series of arm positions. The Poizner data sets contained the time series of positions from the shoulder, the upper arm, the forearm, and the hand. These data enabled me to recover the postural trajectories of seven of the joint angles of the arm. They also allowed for the identification of the parameters defining the task goals and constraints to then use the model to study the decomposition of the DoF.

The data we need to make use of the model are shown in Figure 4.32A, where the positional trajectories from each sensor location are shown for the forward (bottom plot with the start posture) and backward segments (top plot with the final posture) of the reaching movement to the target. These positional trajectories are then used to uncover the initial posture of the arm in Figure 4.32A and recover the joint angle excursions in (Figure 4.32B). The joint angular excursions obtainable from the positional trajectories of the arm sensors will then be used to separate the task-relevant from the task-incidental DoF for each forward and backward segment (Figure 4.32C). The position and orientations of the hand and arm as well as the endpoint error at the target are used to define the forward and backward goals as well as to track the physical distance from the start to the end of the segment. This physical distance reflects the evolution in the values of our cost function, with a unique global minimum occurring when the arm is configured with the hand at the target.

Defining Forward Reach Goals

Each forward and back task segment spans different goal-related dimensions. The dimensions defining the *pointing goals* in this task were three, corresponding to the x, y, z positional coordinates of the hand at the target. The decomposition of the seven DOF joint angular velocity vector in this case spans three DOF for the intended-goal (task-relevant) components and four DOF for the automated (task-incidental) components.

Defining Goals for Reaching Back to a Sensed Posture

In this particular experiment, the participants were asked to return to a posture of the arm near the face, similar to that shown in schematic form in Figure 4.27A (right). The dimensions defining the goal in this portion of the task were five. They corresponded to the spatial position of the hand (3 dimensions that depend on the arm posture, $(f_x(q), f_y(q), f_z(q))$) near the face; the orientation of the palm of the hand (1 dimension, the angle ϕ defined by the Euler-Rodrigues' parameter[60]) with the palm of the hand facing the ear—which also depends on the arm posture $f_\phi(q)$; and the orientation of the plane of the arm θ (1 dimension which also depends on the arm posture $f_\theta(q)$) (defined by[61]) and used in Figures 4.8 D and 4.12 A to provide examples of different criteria to define different distance-metric costs). This is the angle that a vector normal to the plane spanned by the upper arm and the forearm makes with the horizontal plane. In this case the vector normal

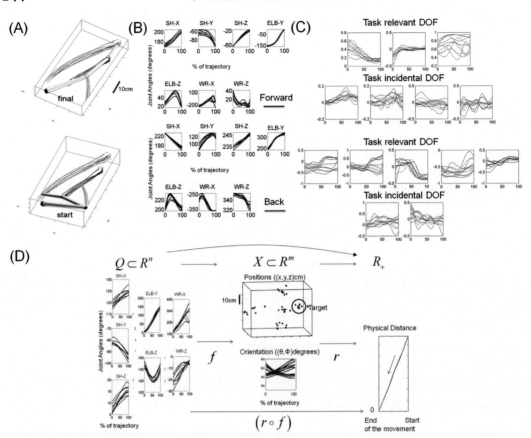

FIGURE 4.32   Implementation of the model using empirical data. (A) Positional trajectories at the shoulder, elbow, wrist and hand for the forward and backward motions are used to recover the initial posture and reconstruct the postural paths. (B) The postural paths reconstructed from the positional trajectories in (A). (C) The DoF decomposition into the task-relevant (rank) and the task-incidental (null) subspaces generated by the locally linear isometric embedding for the forward and backward segments. (D) The locally linear isometric embedding model implemented in the empirical data.

to the plane of the arm was aimed at being approximately parallel to the horizontal plane, when the arm returned to its initial configuration. Whereas the forward motion decomposition is three DoF for the target and four DoF for the self-motion dimensions, in the backward segment the seven DoF joint velocity vector spans five DoF for the intended-goals. They are the coordinates of retracting the hand to a certain position with the desired inclination of the plane of the arm and the hand oriented toward the ear. This defines a total of five DoF for the task-relevant subspace and two DoF for the remaining task-incidental subspace.

To obtain the orientation of the hand $f_\phi(q)$ we used the Euler-Rodrigues angle-vector parameterization of rotations, $\phi = \arccos\left((A_{11} + A_{22} + A_{33} - 1)/2\right)$ from A the rotation matrix at the hand. We obtained A from the cosine angles associated with the three axes defining the three-dimensional positions of the hand sensors. Then we obtained the unit vector $e = [e_1, e_2, e_3]^T$ defined as $e_1 = (A_{32} - A_{23})/(2\sin\theta)$, $e_2 = (A_{13} - A_{31})/(2\sin\theta)$, $e_3 = (A_{21} - A_{12})/(2\sin\theta)$.

To obtain the angle of the plane of the arm $f_\theta(q)$ we computed $\cos(\theta) = \langle n^\rightarrow, (0, 0, 1)\rangle$ and $\vec{n} = \vec{u} \times \vec{v}$ for $\vec{u}$ and $\vec{v}$ unit vectors from the shoulder to the elbow and from the shoulder to the wrist, respectively (Figure 4.8D). These quantities are all obtainable from the positional data.

These parameters of the hand positional trajectories in X space serve to build the position-dependent matrix G_x^α providing the coefficients necessary to transform the model dx to the data dx. Recall here that in Part I we have generally characterized these motions geometrically with a *"natural"* gradient vector with respect to (w.r.t.) a non-Euclidean (Riemannian) norm and approximated the g_{ij}'s corresponding to this metric using the actual data velocity vector flow from the veridical three-dimensional movement trajectory. We can approximate by linear regression the linear transformation matrix conditioned to be positive definite that takes the model dx into the data dx. This is explained in Appendix 6 and also Appendix in ref.[1]

The dx^{MODEL} is a vector along the Euclidean straight line representing the 'ideal' hand positional displacement that builds the shortest path (w.r.t. the Euclidean norm) between the initial position of the hand and the target. Our job is to approximate the metric that goes best with the true path from the data, which is curved. Each point on the straight line defines the starting and the ending three-dimensional coordinates of the dx^{MODEL} vector which we normalize to obtain a unit-speed curve. This is shown in Figure 4.12B where we plot the dx^{MODEL} as the black straight line segment from the projection of the curve to the Euclidean line and the dx^{DATA} as the red curved segment representing the line element from the participant's trajectory.

To change from the ideal model path to the veridical data path we set $dx^{\text{DATA}} = G_x^\alpha dx^{\text{MODEL}}$ where dx^{DATA} comes from the actual hand displacements recorded by the sensors and dx^{MODEL} comes from the projection of the veridical displacement onto the straight line connecting the initial hand position to the target (see Appendix 6 at the end of the chapter for details on the implementation of the algorithm and numerical approximation of the G_x^α metric coefficients). We approximate the G_x^α matrix to define the distance metric of the space of goals, and then pull it back into the joint angle velocity space. This is necessary to later obtain from the arm data movement trajectories the range-null decompositions of the joint velocity movement segments described below.

To measure the contributions of both the range and the null components of the joint angle rotations as the movement unfolds we projected each component on the unitary basis from the singular value decomposition of the metric in the tangent space to the joint angle space (the postural configuration manifold). The elements of the tangent space are the joint velocities. Recall that in our formulation of the inverse solution we preserve the metric from the hand-goal configuration space (hand configuration-goal manifold) under coordinate transformation[35]: $G_q^\mu = J^T G_x^\alpha J$ where G_x^α is *3x3* in the forward pointing case and *5x5* in the case of reaching back to the initial arm posture. The *Jacobian* matrix of partial derivatives is *3x7* in the former case and *5x7* in the latter.

We can estimate the *Jacobian* matrix directly from the data using the physical displacements quantified in the hand and in the arm spaces. Indeed, the data provides us directly with what we need to solve for J in $dq = \nabla r_x \cdot J$ where ∇r_x is directly measured by the sensors, dq is also provided by our reconstruction algorithm from the sensors, with minimal error to reconstruct the three-dimensional displacement-trajectories of the shoulder, elbow, wrist and the tip of the hand using seven joint angles (Figure 4.32A−B). Thus we can obtain $J = \nabla_x^T \cdot dq$ where the transposed of the gradient in X is $mx1$ and the dq is $1xn$. In this way we obtain directly from the physical positional displacements of the joint angles Δq and of the hand Δx, the relative contributions of each joint to the decrements in the physical three-dimensional distances from each component of the hand's current position to the target position.

The paths of relative joint displacements in time (to build the joint velocity vectors) Δq that best sub serve the hand's displacements Δx over the course of the reach can provide a geometric description of how the variability of the movement dynamics unfolds. They can be helpful to distinguish aspects of the movement dynamics that remain conserved from aspects of the movement dynamics that change along the motion path in agreement with the natural trial-to-trial fluctuations in movement dynamics. Critical to this decomposition step is the manipulation of the source of sensory guidance across blocks (of experimental conditions) and the effects that different sensory sources have on the redundant DoF. While the task-relevant DoF are known to remain with low variability, perhaps to keep the intended course of the action on track, the redundant incidental DoF co-vary with the changes in the source of sensory guidance.[58] They serve to channel out through movement variations the most effective form of sensory guidance in a compromised system. In our experience with studies of redundancy in movements, the task-incidental have been the most informative DoF with regard to the automated/spontaneous aspects of behavior.

The resultant matrix $G_q^\mu = J^T G_x^\alpha J$ is $7x7$ positive definite and its coefficients, the g_{ij} 's, define the new metric under change of coordinates (from hand to arm-postural configurations in this case).

We use the singular value decomposition to factorize the $G_q^\mu = U \sum V*$ where U is an $m \times m$ unitary matrix in the real field K, the matrix Σ is an $m \times m$ diagonal matrix with nonnegative real numbers on the diagonal, and $V*$, an $m \times m$ unitary matrix over K (*) denotes the conjugate transpose of V. A common convention—which, for example, MATLAB follows—is to order the diagonal entries $\Sigma_{i,i}$ in descending order. In this case, the diagonal matrix Σ is uniquely determined by G_q^μ (though the matrices U and V are not). The diagonal entries of Σ are the singular values of G_q^μ. In each version of the task we project the joint angular velocity unit vector from the data trajectory onto U and extract the range and null components (3 vs.4, or 5 vs. 2).

In the model of Part I, the angular velocity vector is the gradient (path length minimizing) direction with respect to the metric G_q^μ in the tangent space to the manifold of postural configurations. The numerical estimate of the goal-directed metric for G_x^α uses the scaling factors related to the translational distance components and the rotational components (ϕ) for the palm orientation and (θ) for the inclination of the plane of the arm. These scaling factors are important to characterize how the system drives the rates of change of rotations (e.g., degrees or radians) and linear translations (e.g., cm), whose interplay change from task to task. We then obtain the lengths of the projections properly normalized by the number of dimensions in the rank and null components, respectively, and

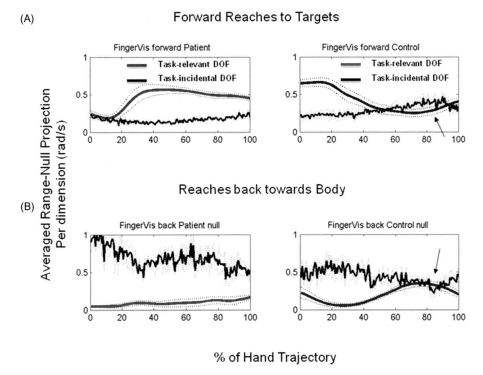

FIGURE 4.33 **The modulation of the DoF in task-relevant and task-incidental subspaces during forward and backward pointing loops.** (A) In forward reaches towards the target, patients with PD cannot modulate the interplay between the arm-joint angles recruited for the pointing task and the remaining joint angles. On average their motions tend to have a dominance of the task-relevant DoF while controls modulate the recruit-release course of the joint angles to balance the deliberate and automated components of the reach. Towards the end of the reach the task-incidental DoF take over the task-relevant DoF in contrast to the PD patients who maintain throughout the reach the same goal-directness DoF dominance. (B) In the retraction segment towards the body patients also have dominance of one subspace, the task-incidental DoF turn more variable and take over the task-relevant subspace. In contrast the controls do modulate the dominance of the goal-directed DoF as the hand approaches the face area.

assess the variability of the DoF across targets (1–5), for all trials and subject type (normal *vs.* PD patient). The resulting summary data is shown in Figure 4.33 for both the forward and backward segments of the PD patients and controls.[33]

We learned that controls modulated the task-relevant DoF whereby in the initial stages of the forward movement to the target these DoF dominated over the task-incidental DoF. However, midway to the target the task-incidental DoF dominated over the task-relevant DoF. In the backward segment this pattern was complementary, suggesting a much automated motion in the early segment of the backward reach but a more controlled ending whereby the task-relevant DoF once again dominated. This made sense, since the hand was coming back near to the ear location, toward the face area. Without properly controlling the arm motion, the hand could hit the person's face. In the case of the PD patients the patterns revealed a very different performance. There was no modulation of the DoF.

In the forward reach, the task-relevant DoF dominated the full motion, whereas in the backward reach, the incidental DoF took over with higher variability and poor control of the task-relevant DoF.

From this study we concluded that the kinesthetic reafferent feedback from the reaching motion of PD patients was much more different than that of controls. We later confirmed this hypotheses in various experiments that examined the stochastic properties of their arm-hand system[57] in relation to those of Ian Waterman, the special subject without proprioception that we saw in Chapter 3. That result explained why the lit LED on the finger was the most adequate source of sensory guidance to help restore the geometric symmetry in these PD patients. In the presence of persistent noisy reafference[57] (i.e., comparable to the signatures of an actually deafferented participant), the continuous visual feedback of the finger motion helped these PD patients realign proprioceptive input with visual input. This realignment was reflected in the recovery of the motions toward typical levels, as measured by various parameters, including the area-perimeter ratio in Figure 4.28. These ratios expressing the posture-hand geometric relations returned to the typical congruency that controls manifested between the forward and backward segments of the full, continuous reach.

In the case of Ian Waterman, he had found a way to sensory-substitute proprioception with vision of his moving hand to restore his motor control. As Ian did, these PD patients also used the visual feedback anchored on their moving finger to replace kinesthetic reafference with dynamically moving visual reafference and motor imagery. Using vision of their moving finger in route to the target, the PD patients were able to dampen the excess motor noise and dynamically boost their proprioceptive input with continuous visual guidance of their moving limb. The biometrics we used to quantify this dynamically updating process captured the changes in variability and signaled which source of sensory guidance worked best.

This form of dynamic sensory-augmentation helped me years later design interventions for ASD whereby noise-cancellation in the reafferent signals from the self-generated motions could be attained with dynamically updated external sources of sensory guidance. These included visual input from movies and associated sounds. In the presence of such guiding signals and encouraged by their self-discovery of cause and effect, their nervous systems were able to "tame" the randomness and noisiness of their reaching movements and shift their stochastic signatures toward typical regimes. One of the pressing questions we have today is whether we can retain these gains permanently, beyond the confines of the lab, to help the person cope better with activities of daily life.

SCHIZOPHRENIA PATIENTS: DID I MEAN TO DO WHAT I JUST DID?

The studies of reaching movements with various patient populations led me to interact with engineers and clinicians during my time at Rutgers to extend the use of these methods to clinical cases. Along these lines, the labs of Professors Steven Silverstein (Psychiatry) and Thomas Papathomas (Biomedical Engineering and Perceptual Sciences) were investigating the effects of a depth-inversion visual illusion in patients with schizophrenia. They had found that such patients were generally less susceptible to these visual illusions than typical controls.[62]

Jillian Nguyen, a graduate student in the lab completing her PhD at the time decided to pursue this question in controls first, and then compare her results to the performance of patients with schizophrenia. To that end, Jill teamed up with students from the biomedical engineering program at Rutgers. These were students that worked with Papathomas and I designing and building the experimental set up.[63] Ushma Majmudar (today at the Albert Einstein School of Medicine) and Jay Ravaliya (today software engineer at Apple) built and programmed the controllers to automate the task. Jill (today an analyst at a finance firm in NYC) helped design the experiment and piloted the first group of participants. From that study, we learned a great deal on the crosstalk between the visual and sensory-motor systems in the intact nervous systems.[64] However, after piloting six patients with schizophrenia, Jill noticed that the task was much more challenging to them than we had anticipated. The patients with schizophrenia were not experiencing the illusion to the same extent than controls, but more troubling was that their movement patterns were very different from what we had seen, even in other patient populations. This was surprising because as a mental illness, schizophrenia is not associated with disorders of the sensory-motor systems. However, their motions were visibly disorganized and uncoordinated, clearly and systematically (patient after patient) showing signs of movement disorders. Further, we could see a type of avolition that prevented some of them from finishing the experiment altogether. We decided to simplify the task and have the patients perform a biomechanical, forward and backward pointing motion with full and continuous visual feedback of the target. The task is shown in Figure 4.34 in schematic form. Under those conditions, we asked if visuomotor transformations involving the arm-hand systems were impeded in some special way detectable through our geometric measures.

In the process of setting up this new experiment and recruiting participants, I decided to do some detective work on the possible relationships between schizophrenia and motor dysfunction. I found very little in the literature but then asked a colleague and friend Professor Anne Donnellan about the issue. Anne Donnellan and Martha Leary had introduced the notion of sensory-motor differences in autism[65] with the purpose of creating accommodations to help affected individuals with their activities of daily living. Their expertise on mental illnesses, developmental disabilities and their ties to motor disorders was impressive. Anne pointed me to a wonderful book by the Neurologist Dr. Daniel Rogers, "*Motor disorder in psychiatry: toward a neurological psychiatry*". It was a great read that taught me a great deal of the historical context that led to the emergence of psychiatry as we know it today, preceded by very different beginnings.[66] As it turned out, the early definitions of mental illnesses were based on motor descriptions and criteria expressed in terms of motor disorders. The birth of Freudian psychoanalysis did away with that, as it was somehow seemingly easier to provide a psychological explanation as the cause of a disorder than to dig deep into the underlying neurophysiology.

Schizophrenia and related disorders of mental function were among those initially defined in terms of motor phenomena, so it made total sense that these patients with a diagnosis of schizophrenia who came to the lab and could not complete the experiments were visibly affected in the motor domain. Historically speaking, prior to Freud, these patients would have been textbook cases of motor description of mental illness. As such,

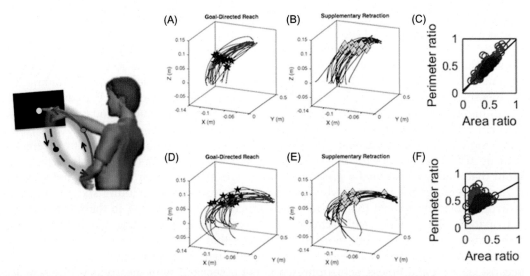

FIGURE 4.34   **The forward and back pointing task in the schizophrenia group.** Schematics of the task performed in full light with continuous vision of the target. Movement forward (red) are continuously followed by uninstructed retractions where the hand travels away from the board and back towards the body. These spontaneous segments were contrasted with the deliberate ones directed to the target. (A) Hand trajectories from representative control participant (red denoted goal-directed and blue denoted supplementary). Kinematic landmark of the peak velocity plotted as black stars and yellow diamonds in each trajectory. (C) Scatter of the area-perimeter ratios across all trajectories similarly color-coded and linear polynomial fit congruent for both segments. (D-E) Hand trajectories from a representative patient with schizophrenia. (F) Disruption of the area-perimeter ratio and the relationship between the deliberate and spontaneous aspects of the task.

given that motor output is kinesthetic reafferent input, they were likely affected in the movement sensory domain as well. How would this be manifested in the stochastic rhythms of their arm-hand motions and the interplays between deliberate and spontaneous segments of the action?

The lab members trained to see the relevance of motor control as a source of sensory feedback to the brain could immediately see the relationship to avolition and motor delusions.[67] The implications for lack of agency and bodily action ownership were clear to all of us. Sadly, the majority of those diagnosing and treating these disorders exclusively as mental illness could not possibly see this because there is very scarce scientific evidence collected with the necessary rigor and mathematical formalism.

As of today in 2017, the National Institute of Mental Health is still trying to figure out whether or not to include motor issues in the Research Domain Criteria (RDoC)[68] that its former director, Dr. Thomas Insel created.[69,70] The RDoC was supposed to do away with the pervasive presence of the Diagnosis Statistical Manual[71] in research that involved mental illness and perhaps steer research away from the tremendous conflicts of interest between the American Psychiatric Association and the leading pharmaceutical companies that ultimately drive DSM criteria, medication prescription and taxpayers coverage for all of that.[72]

Despite these socio-economic issues and their profoundly negatively impact on the science, my lab continues to move forward with a research program that is driven by personalized somatic-motor driven criteria to characterize mental illness, so here we are, writing a book on that. The idea is to provide an objective characterization of these motor issues across disorders of the nervous systems that are rendered as neuropsychiatric and/or within the realm of mental illnesses today. As such, Silverstein's team helped Jill assemble 26 patients with schizophrenia and 26 age- and sex-matched controls to examine the somatic-motor patterns in the forward and back pointing task. She registered motion from all positions of the arm and hand, so we were able to use the model in Part I to characterize the signatures of variability extracted from both the area-perimeter symmetry and the DoF decomposition of the joint angles. The hand trajectories of representative controls are shown in Figure 4.34A while those of representative patient are shown in Figure 4.34 D−E for the forward and backward segments of the pointing movement depicted in schematic form in the figure.

The first thing to note is that across all controls the area perimeter ratio held the symmetry for both movement segments, so the linear fits nearly overlapped (Figure 4.34C). This was in marked contrast to the patients' performance, whereby the symmetry did not hold for both segments and the linear fits had different slopes and intercepts (Figure 4.34F).

Given this disparity in the performance, and the fact that this ratio encompasses the posture-hand geometric relation, we tried to understand the relationship between the variability of the scatter from the area-perimeter symmetry manifested in the hand space and that of the DoF decomposition in the postural space. To that end, we first characterized the stochastic signatures of the former and then recovered the joint angles of the arm, performed the DoF decomposition and studied the variability of the task-relevant DoF *vs.* that of the task-incidental DoF. The idea was to address the distinctions between the variability from the goal-directed (forward) segment and that from the supplementary (backward) segment of the pointing movements these participants performed.

The goal-directed segments in this experimental paradigm are deliberately intended to the target. In fact they were explicitly instructed. However, the retracting segments were not instructed in this experiment. As such, they occurred spontaneously and pursued no instructed goals. The result of our research thus far was that these two segments had very different stochastic signatures in neurotypical controls. As such by blindly examining the variability of their trajectory parameters we could predict the type of movement the participant had most likely performed, that is, whether it was intended or spontaneous. For example, in a complex boxing routine we could blindly tell the difference between the intended jab deliberately aimed at the opponent and the retracting jab, automatically performed beneath the athlete's awareness.[73] We had done this extensively within the reaching and pointing paradigms; but the extension to naturalistic complex movements of sports gave us more confidence that deliberate and spontaneous movements are in different classes of motor control.[73,74] *Would these differences manifest in the patients with schizophrenia?*

There are two main striking features that we consistently noted from the patients. One was the avolition given by an apparent lack of will in the forward motions that were supposed to be deliberately intended to point to the target. The other one was the totally visible lack of coordination across body parts (e.g., head, trunk and arms-hand linkages). How would these odd synergies we were seeing impact the automated retractions that the patients spontaneously performed without instructions?

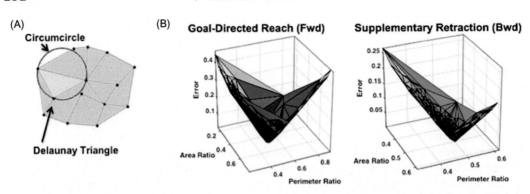

FIGURE 4.35   Geometric analyses of the three dimensional scatter generated by the area-perimeter ratio symmetries and their errors from the linear fit. (A) Delaunay triangulation of the scatter shown for a 2-dimensional scatter ensuring that the circumcircle associated with each triangle contains no other points in its interior. (B) The Delaunay triangulation of the three dimensional scatter generated by the goal-directed (red) and supplementary retraction (blue) hand trajectories.

   To address the variability in the scatter of the area-perimeter ratio we used a number of geometric methods. Recall that in the ideal case when the hand path is modeled as a geodesic (using equation (4.1) under some distance metric cost), the geometric transformation yielding the area-perimeter ratio relationship should fall on a straight line with a slope of 1, passing through the ($1/2$, $1/2$) point. In the model, this means that there is a scaling factor relating the curvature of different geodesic curves of the reach family. While the implicit (cost) surface where these solution paths change curvature, the changes are such that they globally scale the curvature value while preserving the symmetry. From the empirical assessment of these ratios we infer that the balance between both spatial trajectory parameters is equal ($P_{ratio} = A_{ratio}$) and there is a congruent map between the local postural increments and the local hand increments that results in a smoothly connected path between the hand's initial location and the target. We also know from the empirical data of other patient populations that this ratio falls apart when the geometric sensory-motor transformations are affected. To assess the extent to which they are affected in this cohort of schizophrenia patients, we calculated the error of each area-perimeter ratio value by determining the shortest normal distance to the line $y = mx$, in which y is the P_{ratio}, x is the A_{ratio}, and $m = 1$. Since we now introduce a third dimensions to our data, the error E, we can generate a three-dimensional surface representation of all three variables to perform topological and geometric queries (Figure 4.35A). To this end, we built a matrix with values for all $A_{ratio}, P_{ratio},$ and E for each subject and movement class, and created a Delaunay triangulation surface on the three-dimensional scatter.

   Delaunay triangulation creates a surface representation of a matrix P out of triangles connected by points $[x,y,z]$ in each row of matrix P, each point serving as a vertex, such that the circumcircle associated with each triangle thus formed contains no other points in matrix P in its interior.[75] Figure 4.35A demonstrates an example of how

Delaunay triangles are constructed in 2D, and Figure 4.35B–C illustrate the surfaces formed using the sample data of Figure 4.34C generated by forward goal-directed reaches (left) and the surface generated by the spontaneous retractions of the hand (right).

We then calculate the areas of the triangles that make up each three-dimensional surface, and find the parameters that most likely determine the underlying probability distribution of these geometric parameters in each movement class. To that end, we apply the statistical platform for individualized behavioral analyses (SPIBA) framework described in Chapter 1 where we use a Gamma process to empirically estimate the shape and scale parameters of the Gamma probability distribution functions to perform statistical inference in these data. We use maximum likelihood estimation with 95% confidence intervals for the parameter determining the shape of the distribution (a) and for scale parameter (b) describing the dispersion. Recall here that this parameter is also the noise to signal ratio (NSR) that we used to study transitions from spontaneous random noise to well-structured signal in the acceleration data from the babies and the pregnant coma patient in Chapter 1. In Physics they call it the Fano Factor, the variance-to-mean ratio that in the Gamma case amounts to (b), $b = \frac{a \cdot b^2}{a \cdot b}$.

We then examine the log (a) $vs.$ the log of (b) scatter to find the first-degree polynomial function that best fits the grouped data in a least-squares sense. We calculated delta (Δ), the error of the linear fit for each group and mapped Δ values against the Fano Factor calculations. They are shown in Figure 4.36, presented as scatter-box plots for visualization purposes. A striking result from these analyses was that the patterns of variability between deliberate and spontaneous segments of the reach were inverted between controls and schizophrenia patients.

The patients' backward patterns aligned with the controls' forward patterns and the patients' forward patterns aligned with the controls' backward patterns. This individualized analyses consistently showed that for each patient the goal-directed segment of the pointing act, that is, the segment deliberately intended to reach the target, was performed in spontaneous mode. And yet, the retraction segment which was not instructed and as such was supposed to be automatically performed (beneath awareness) had the typical signatures of deliberateness that the control subjects manifested. How the nervous systems of these patients would then "know" they intended to do something on purpose? Furthermore, how action ownership and agency could be characterized in the forward movements of these patients, when these patterns had signatures of spontaneous performance without intent?

Owing to the consistency of the inversion in intent $vs.$ spontaneity across different movement parameters and levels of analyses, I decided to analyze the individualized decomposition of the DoF of the arm across the seven joint angles used to represent a posture in Q and the three spatial dimensions used to represent the hand position in X. As with the Figure 4.32, I recovered the joint angles from the positional trajectories of the shoulder, upper arm, forearm and hand and estimated the metric G_q^μ to perform the SVD and compute the projections $q^{\mathrm{T}} \oplus q^{\perp}$ of the gradient vector dq—as we did with the PD

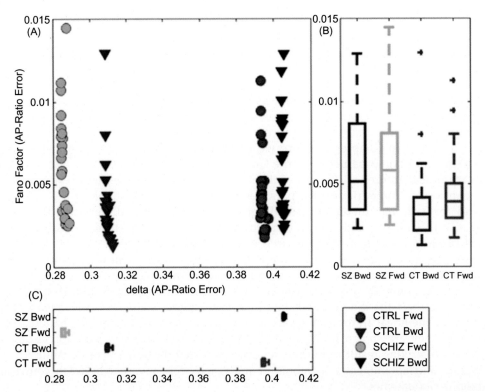

FIGURE 4.36 Inversion of deliberate and spontaneous nature of motion in patients with schizophrenia as measured by the Fano Factor or noise to signal ratio (NSR). (A) Scatterplot of the NSR of the area-perimeter error *vs.* the actual error (delta) obtained from the polynomial fit. Around 0.4 the forward (deliberate) goal-directed segments of the controls align with the retraction (spontaneous) supplementary segments of the schizophrenia patients, while the opposite is true for the errors around 0.3; that is, the spontaneous segments of the controls align with the deliberate segments of the patients with schizophrenia. (B) Box-plot version of the scatter shows the estimated means and error bars for the NSR and (C) the box plots with error bars of the delta values measuring the normal distance from the point in the scatter to the polynomial linear fit.

patients study using the model translation (of Figure 4.14 for the $m = 3$ case) to the empirical data of the pointing action (as in Figures 4.32–4.34).

I reasoned that if these patterns were inverted across the visuomotor system involving the posture-hand map, and also according to the area-perimeter geometric transformation measures related to the local linear isometric embedding, they may also be inverted at the level of internal configurations. If this was the case, then we could safely conclude that the patients with schizophrenia have a different representation of the map connecting the sensory spaces where the sources of input are internally vs. externally anchored. This presumed difference would then impede certainty about

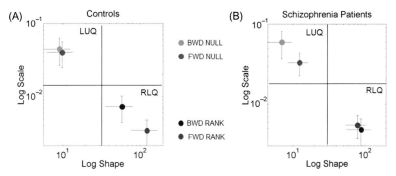

FIGURE 4.37 Stochastic signatures of the DoF locally linear decomposition confirm the reverse trends between controls and patients with schizophrenia regarding the deliberate and spontaneous character of the goal-directed and uninstructed retractions of the pointing task. (A) The motions of control participants distinguish between the forward (deliberate) and backwards (spontaneous) segments of the motions within the rank subspace but not within the null subspace. The task-incidental DoF have higher noise along the scale component in the LUQ and tend towards the exponential ranges of the shape axis. The task-relevant components fall in the RLQ towards symmetric distributions with lower dispersion. (B) The motions of participants with schizophrenia do not distinguish within the rank subspace the forward and backwards segments of the motions. These are rather distinguished in the null subspace, where the values are in the LUQ (higher skewness and dispersion in the PDFs of the task-incidental DoF).

action intent. In this sense, action ownership and self-control-at-will might also be affected. In other words, the assessment would provide a bodily foundation to characterize a type of ownership-confusion possibly scaffolding mental delusions: *Did I just do that or did someone else do it?*

I found that as a group, the patients also had inverted Gamma plane signatures of variability in the linear decomposition of the arm's DoF for the rank and null subspaces corresponding to each of the forward and backward movement segments. This is shown in Figure 4.37 using the Gamma plane visualization tool. Recall here that in this plane, each point represents a probability distribution function from the continuous Gamma family. We plot the empirically estimated points with 95% confidence intervals for each parameter indicating the shape and the scale (dispersion or NSR) of the PDFs. Points on the Left Upper Quadrant of the gamma plane (LUQ) have higher noise and are closer to the exponential distribution. Points in the Right Lower Quadrant of the gamma plane (RLQ) have lower noise and shapes that are closer to the Gaussian distribution. We also delineate these quadrants in the figure to highlight the different signatures between the variability of the rank and null subspaces.

In the case of the controls, the variability in the rank subspace distinguished between the deliberate and spontaneous segments of the full reaching movement. However, the variability of the null subspace did not. When the motions were spontaneously performed the joint angles recruited in the self-motion manifold were noisier and had distributions with higher skewness, tending toward the exponential ranges of the Gamma plane (to the left on the shape axis), but their PDFs were no different between forward and back segments. In the intended segments the PDFs were different with higher variability in the backward case. The probability distributions of these motion parameters could distinguish between levels of intent in the task-relevant subspace of typical controls.

In the case of the patients, the variability of the rank subspace—where task-relevant dimensions recruit joint angles to purposely sub-serve a set of goals—could not distinguish between deliberate and spontaneous segments. Oddly, the null subspace had signatures of variability that could distinguish between the forward segments deliberately intended toward the goal and the spontaneously retracting segments. Recall here that the null subspace corresponds to the self-motion manifold, where changes in posture do not produce displacements of the hand. Typically the variability in this subspace supplements the rank space variability relevant to the realization of the goals. As such, the distinction between the PDF's corresponding to the deliberate and spontaneous segments existed in a space that is incidental to the goals of the reach, beyond any typical level of awareness for explicit agency or intended purpose.

As in the controls, the variability in the null subspace was higher than that of the rank subspace. It manifested in the LUQ with higher noise and more skewed distributions toward the exponential range. Yet this inversion between what was intended and what was spontaneous in this action systematically pointed to a system that lacked purposefulness. This was congruent with the systematic type of avolition we observed in the visual illusion pilot experiment demanding too much of the system. In that experiment the inverted depth illusion and the reach to grasp action had many more goals to handle than this simpler pointing task. Yet, even after simplifying the task we could quantify many motor disturbances in these patients.

In a last set of analyses we included the temperature data from accelerometers we placed on the participants' wrist. The temperature data provides a sense for the amount of spontaneous fluctuations in movements that are not goal-directed. These are spontaneous involuntary micro-motions that the sensors register and since they are small, the battery consumption is lower than the battery consumption for larger self-directed motions. The changes in temperature values for such overt motions tend to be higher than for the involuntary ones because they reflect a higher rate of battery consumption. As such, I decided to (1) examine the individual patterns of null-rank decomposition across the cohort, (2) plot the summary statistics derived from the individually estimated Gamma parameters and (3) color code the scatter according to the temperature output by the accelerometers.

The results of these analyses are shown in Figure 4.38. They confirm fundamental differences between the statistical patterns of the typical controls and those of the schizophrenia patients. The figure is a five-dimensional plot summarizing the empirically estimated statistical signatures and the levels of spontaneous involuntary fluctuations in acceleration reflected in the temperature values. Along the x-axis I plot the mean value of the micro-movements extracted from the DoF decomposition waveform (e.g., in Figure 4.14A) corresponding to the rank (top panels) and null (bottom panels)

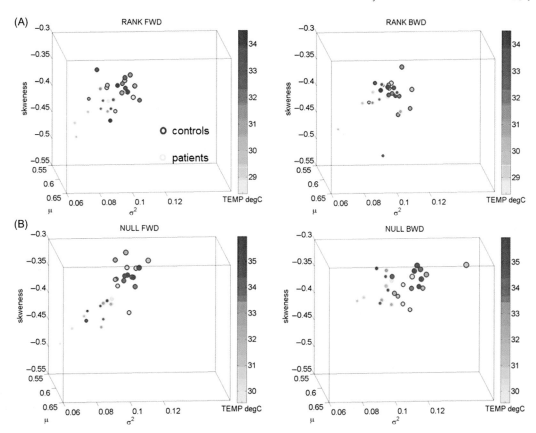

FIGURE 4.38 Summary statistics of the individualized local-linear decomposition of the arm's DoF using the gradient-based method separate controls from patients with schizophrenia in the null subspace. (A) Rank subspaces forward and backward segments show more of a mixture between patients and controls with higher kurtosis in the controls. Each point represents a participant. The x-axis is the mean, the y-axis is the variance, the z-axis is the skewness (higher values towards symmetric shape) and the size of the circle is the kurtosis. Controls have higher kurtosis indicating peakier distributions with lower dispersion. The face color of the circle indicates the average temperature of the sensor across the trials. The edge color of the circle denotes the mode (the most frequent value) of the temperature across the trials. Lower values (towards yellow) denote a prevalence of small fluctuations in the involuntary micro-movements while larger values (towards red) indicate larger overt voluntary motions with higher volition. (B) Null subspaces forward and backward segments. Note the separation of the scatter in the null subspace as color coded by the sensor temperature with a prevalence of lower values in the patients (lower volition likely from excess spontaneous involuntary motion).

Gamma signatures individually estimated for the each participant. Recall that these are the minute moment by moment fluctuations in the amplitude—as registered by the peaks of the time series waveform. The y-axis is the variance while the z-axis is the skewness of the distribution. The size of the marker is the kurtosis and the color conveys the temperature information also reflected in the color bar. The edge of the marker is the mode of the temperature (the most frequent value of the experimental session). The face of the marker is the average temperature across the session. Here the patients had lower ranges suggesting lower rates of temperature change. This indicates excess spontaneous involuntary fluctuations in marked contrast to the controls with higher temperature values. Further the controls have peakier distributions (higher kurtosis) with lower dispersion (lower noise) and higher range of skewness than the patients (here a value of 3 is perfectly symmetric and negative value indicates skewed to the right). The patterns of the controls have higher variance and cluster apart from the patients in the null subspace, a pattern that shows high statistical significance ($p < 0.01$, given that each point is a different Gamma PDF).

The overall analyses of these data sets provide a series of biometrics that can be used to assess patients with schizophrenia above and beyond descriptive inventories. They clearly show that schizophrenia is not just a mental illness but rather a disorder that involves as well the neuromotor and reafferent sensing components of the nervous systems. As such, the disorder can be objectively quantified and dynamically tracked over time.

When we discussed our results with psychiatrists they argued that there are confounding factors between motor disturbances that may be core features of the disorder and motor disturbances that may be a byproduct of the psychotropic medications' side effects. Psychotropic meds have a profound negative effect on the somatic-motor systems because they tend to increase the levels of NSR of covert involuntary micro-movements.[76] They manifest in many other overt ways, including tremors, ticks, dyskinesia, avolition, catatonia, among others. My reply is: *Why not measure it? Why not characterize it?*

If we aim for an objective unbiased science of mental illness, we should measure such disorders of the nervous systems and do so in a personalized manner. We should track somatic-motor dysfunction over time with and without medication intake. We should carefully examine the outcomes of treatments to help the patient and the physicians regulate and adjust the treatments as a function of the nervous systems responses. It is not as though the "mind" with a mental illness exists in a vacuum. Its functioning depends on the well-functioning of the brain, which in turn depends on the well-functioning of the PNS, including the autonomic nervous systems (ANS) as well. A persistent noisy and random feedback from the PNS-ANS, including that from the internally generated kinesthetic reafference that dynamically emerges from self-generated motions, is bound to affect the brain functioning and with it the interpretation and inference the mind may create in order to make sense of the self, the others and the social world where all these entities co-exist.

Combining mathematical tools with data-driven analyses can bring basic science a step closer to implement Precision Psychiatry.

APPENDIX 1

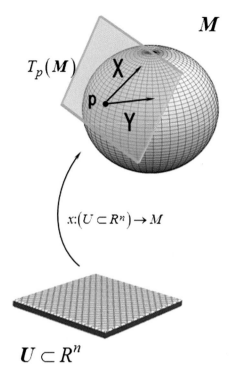

$$T_p(M)$$

$$M$$

$$x:(U \subset R^n) \to M$$

$$U \subset R^n$$

COORDINATE CHARTS AND THE INNER PRODUCT

Given a coordinate chart $x:(U \subset R^n) \to M$ the inner product may be represented by a symmetric matrix $\left(g_{ij}(x^1,\ldots,x^n)\right)$ of smooth functions.

If X and Y are tangent vectors at $p = x(u^1,\ldots,u^n)$ then $X = \sum_i a^i \frac{\partial}{\partial x^i}$, $Y = \sum_i b^i \frac{\partial}{\partial x^i}$, and

$$\langle X, Y \rangle_p = \left(a^1,\ldots,a^n\right)\left(g_{ij}(u^1,\ldots,u^n)\right)\begin{pmatrix} b^1 \\ \vdots \\ b^n \end{pmatrix}$$ is the inner product.

A coordinate system is a set of functions that map points into scalars: $x_{1\ldots N}:p \to R$, for example, the natural coordinate functions of R^n are the set of functions $\{x_1,\ldots,x_n\}$ defined by $x_i(p) = p_i$ for $p = (p_1,\ldots,p_N)$

APPENDIX 2

The differential of the gauss map is self-adjoint

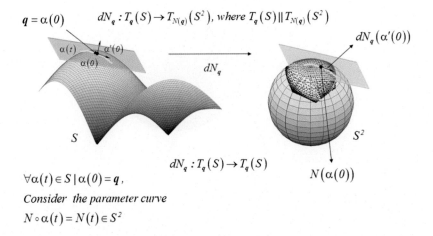

$q = \alpha(0)$ $dN_q : T_q(S) \to T_{N(q)}(S^2)$, where $T_q(S) \| T_{N(q)}(S^2)$

$dN_q(\alpha'(0))$

$dN_q : T_q(S) \to T_q(S)$ $N(\alpha(0))$

$\forall \alpha(t) \in S \mid \alpha(0) = q$,

Consider the parameter curve

$N \circ \alpha(t) = N(t) \in S^2$

THE GAUSS MAP AND ITS DIFFERENTIAL

Let $S \subset R^3$ be a surface that admits a differentiable unit normal vector field $N^{\to}(q) = \frac{x_v \wedge x_w}{|x_v \wedge x_w|}(q)$ defined on the whole surface, that is, an orientable surface. The map $N:S \to R^3$ takes values in the unit sphere S^2 is called the Gauss Map.[77] Its differential dN_q is a linear map from $T_q S$ to $T_{N(q)}(S^2)$. Since these are parallel planes dN_q as a linear map on $T_q S$. This linear self-adjoint map $dN_p:T_p S \to T_p S$ operates as follows: For each curve C parameterized by $\alpha(t)$ in S, take $\alpha(0) = q$ a point on the manifold. Consider the parameterized curve $N \circ \alpha(t)$ as $N(t)$ in the unit sphere S^2 by restricting the normal vector $\vec{N}$ to the curve $\alpha(t)$. The tangent vector $N^{\to}{}'(t) = dN_p(\alpha'(0))$ is a vector in $T_q S$, which measures the rate of change of the normal vector $N^{\to}$, restricted to $\alpha(t) = q$, at $t = 0, dN_q$ measures how $N^{\to}$ pulls away from $N(q)$ in a neighborhood of q. For curves this measure is the curvature, for surfaces, it is this linear self-adjoint map dN_q.

To say that dN_q is self-adjoint means that $\langle dN_q v, w \rangle = \langle v, dN_q w \rangle$ $\forall v, w$ in $T_q S$. If $\{e_1, \ldots, e_n\}$ is an orthonormal basis for $T_q S$ and (h_{ij}) a matrix of dN_q relative to that basis, then (h_{ij}) is symmetric: $\langle dN_q e_i, e_j \rangle = h_{ij} = \langle e_i, dN_q e_j \rangle = \langle e_j, dN_q e_i \rangle = h_{ji}$.[78] To this self-adjoint map dN_q one can associate a map $H:T_q S \times T_q S \to R$ defined by $H(v, w) = \langle dN_q v, w \rangle$, H is bilinear (linear in v and w) and the fact that dN_q is self-adjoint implies that $H(v, w) = H(w, v)$, that is, H is a symmetric-bilinear form in $T_q S$. Conversely, if H is a bilinear symmetric form in $T_q S$, one can define a linear map such as $dN_q:T_q S \to T_q S$ by $\langle dN_q v, w \rangle = H(v, w)$ and the symmetry of H implies that dN_q is self-adjoint.[77,78]

To each symmetric-bilinear form in T_qS, there corresponds a quadratic form Q in T_qS given by $Q(v) = H(v, v)$ with v in T_qS. Thus there is a one to one correspondence between quadratic forms and self-adjoint linear maps on T_qS. The Hessian $\nabla^2 r^{task} \circ f(q)$ at the point q is the quadratic form in this example and the gradient $\nabla r^{task} \circ f(q)$ its leading eigenvector.

APPENDIX 3

(A)

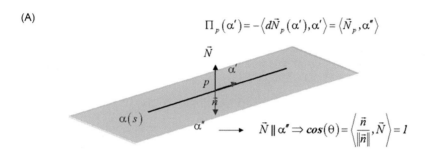

(B)

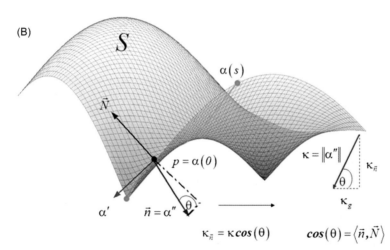

TESTING GEODESIC PROPERTY AND OBTAINING CURVATURE MEASURES WITH THE MODEL

The cost surface of the form $S = \left(q_1, \ldots, q_n, r\left(f(q), x^{target}\right)\right)$, where r is the distance metric representing the task, is an implicit surface whereby the gradient and the Hessian of the pullback $r \circ f$ can be used to obtain the Mean and Gaussian curvatures. For example, the formula for the Mean curvature is the average of the principal curvatures obtainable from the diagonalization of the Hessian, $K_M = \dfrac{\nabla(r \circ f) \cdot H(r \circ f) \cdot \nabla(r \circ f)^{\mathrm{T}} - \left|\nabla(r \circ f)\right|^2 TraceH(r \circ f)}{n \cdot \left|\nabla(r \circ f)\right|^3}$

(where n denotes the number of dimensions). This formula involves the Second Fundamental Form (the shape operator[35]) of the implicit cost surface, denoted Π_p.

In our model, as the iterative optimization unfolds, we track this quantity along the curve to enforce the curve's geodesic property. Specifically, when the curve is a geodesic, the surface normal at the point parallels the normal to the gradient (the curve's acceleration) at the point, that is, the geodesic curvature is 0 (**Appendix** Figure 4.3A). $\Pi_p(q), q \in T_pS$ has associated a symmetric, bilinear quadratic form, the Hessian $H(r \circ f)$, which relates to the gradient, given in our construction by $\tilde{q} = G_q^{-1} \nabla(r \circ f)$, $\Pi_p(\tilde{q}) = -\langle H_{r \circ f}(\tilde{q}), \tilde{q} \rangle = -\langle dN\tilde{q}, \tilde{q} \rangle, \tilde{q} \in T_qS.^{79}$

The gradient vector $\tilde{q} \in T_qS$ is the leading eigenvector of this quadratic. Its corresponding eigenvalue is one of the principal curvatures (the maximal). In practice, this can be iteratively measured (**Appendix** Figure 4.3B) through the cosine of the angle between the normal vector to the surface at the point, $\vec{N_q}$ and the normal vector to the gradient $\alpha' = \tilde{q}$ at the point, that is, the acceleration of the curve at the point, α'' expressed as $\Pi_q(\alpha') = \langle \vec{N_q}, \alpha'' \rangle$, where we have used the notation $\alpha(s)$ to represent our unfolding unit-speed curve on the cost surface.

The diagonal form of the Hessian, $D = diag(\lambda_1, \ldots, \lambda_n)$ with $D = C^t H C$, where C is an orthogonal matrix such that $y = \tilde{q} \cdot C$ reduces the quadratic from $Q(\tilde{q}) = \tilde{q} \cdot H \cdot \tilde{q}^t$ to its diagonal expression $Q(y) = y \cdot D \cdot y^t$, whereby providing a more compact representation of the task solution and as such, a solution scheme simpler to encode and retrieve for a system using it. In this sense, besides providing a solution that can be self-supervised and (deliberately) deployed in flight, when faced with a sudden perturbation; this scheme is also amenable to encode (automatic) pre-programmed, path-length minimizing solution curves in compact form for a wide family of problems. In the context of human self-generated movement, the model can capture different levels of control (deliberate and automatic), which may be occurring within disparate slower *vs.* faster time scales, respectively. Given the uniqueness of this solution, one can then empirically study temporal aspects of the actions without having to predefine them a priori.

As we have seen in this part of the chapter, it is possible to model multiple tasks within the reaching family, each performed within a different context. The question is then how to readily transition from one version of the same task to another version and produce the solution path appropriate to a given situation. Within the present schema, is it is also possible to study the transitions from one task to another using the analytical expression for the gradient solution expressed in terms of different coordinate functions (see next **Appendix** Figure 4.4).

APPENDIX 4

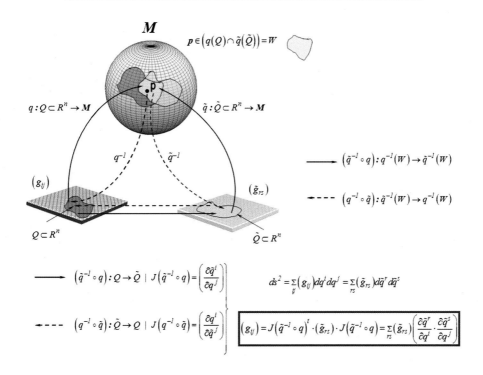

SWITCHING BETWEEN TASKS

Coordinate transformations paired with the use of different distance metrics allow us to study the transitions between different tasks. For example, suppose that you have two different tasks involving different families of arm postures, for example, those for curved reaches avoiding obstacles and those for direct reaches straight to the target. We would like to know how to represent both tasks simultaneously and readily switch from one representation to another. This is possible to model with the present framework.

Denote M the manifold to represent sensory information internally generated such as that given by postures. Denote N the manifold to represent sensory information externally generated, such as that related to end goals and additional external constraints. They are further related by some map f. It is possible to use two overlapping coordinate charts on two open sets of the manifold M to provide two different task representations selectively recruiting DoF in the spaces of internal and external sensory sources (**Appendix Figure 4.4**)

Denote $q:M \rightarrow (Q \subset R^7)$ and $\tilde{q}:M \rightarrow (\tilde{Q} \subset R^7)$ the two overlapping injective coordinate charts on M with elements in the intersection, $W = q(Q) \cap \tilde{q}(\tilde{Q}) \neq \varnothing$, $p \in W$ and let $q^{-1}, \tilde{q}^{-1}$ be the corresponding inverse maps.

The compatibility condition $W \neq \emptyset$ makes it possible to construct the continuous injective maps: $(\tilde{q}^{-1} \circ q): q^{-1}(W) \to \tilde{q}^{-1}(W)$ and $(q^{-1} \circ \tilde{q}): \tilde{q}^{-1}(W) \to q^{-1}(W)$. They are the transition functions (or change of coordinates) $(q \circ \tilde{q}^{-1}): Q \to \tilde{Q}$ and $(\tilde{q} \circ q^{-1}): \tilde{Q} \to Q$ from one task to another (**Appendix** Figure 4.4).

In general, the posture manifold M can be covered by a collection of coordinate charts (an atlas). Each chart may define a task defined by multiple goals. For instance, the distances for the two tasks of interest are defined as:

$$
r^{\text{OB-avoidance}} = \sqrt{
\begin{aligned}
& \sum_{i=1}^{3} \sum_{j=1}^{3} g_{ij}^{\alpha} \left(f_i^{\text{safe}} - f_i^{\text{current}}(q) \right) \left(f_j^{\text{safe}} - f_j^{\text{current}}(q) \right) \\
& + \sum_{i=1}^{3} \sum_{j=1}^{3} g_{ij}^{\alpha} \left(f_i^{\text{target}} - f_i^{\text{current}}(q) \right) \left(f_j^{\text{target}} - f_j^{\text{current}}(q) \right)
\end{aligned}
}
$$

where in the OB-avoidance task, f^{safe} represents a safety region around the obstacle that the visual system can detect to avoid hitting it and $r^{\text{pointing}} =$

$$
\sqrt{\sum_{r=1}^{3} \sum_{s=1}^{3} \tilde{g}_{rs}^{\alpha} \left(f_r^{\text{target}} - f_r(\tilde{q}) \right) \left(f_s^{\text{target}} - f_s(\tilde{q}) \right)}.
$$

As explained above, these distance functions are pullbacks that also serve as costs to define the task and take as inputs points on the posture manifold regions (the domains) of their corresponding coordinate chart.

The metric tensors $G^{\mu} = J(f(\mathbf{q}))^{T} \cdot G^{\alpha} \cdot J(f(\mathbf{q}))$ and $\tilde{G}^{\mu} = J(f(\tilde{\mathbf{q}}))^{T} \cdot \tilde{G}^{\alpha} \cdot J(f(\tilde{\mathbf{q}}))$ lend structure to their M regions. They preserve metric information from N in G^{α} and $\tilde{G}^{\alpha}$, and space information in the *Jacobian* coordinate-transformation matrix. The coefficients of these metrics smoothly change with position along their corresponding regions on M and N.

The transformation from one task to another can be obtained via the intrinsic geometry expressed in the metric coefficients, $g_{ij}^{\mu} = J(\tilde{q}^{-1} \circ q)^{T} \cdot \tilde{g}_{rs}^{\mu} \cdot J(\tilde{q}^{-1} \circ q)$, and the *Jacobian*s $J(\tilde{q}^{-1} \circ q) = \left(\frac{\partial \tilde{q}^i}{\partial q^j} \right)$ and $J(q^{-1} \circ \tilde{q}) = \left(\frac{\partial q^i}{\partial \tilde{q}^j} \right)$ associated to the transition functions $(q \circ \tilde{q}^{-1}): Q \to \tilde{Q}$ and $(\tilde{q} \circ q^{-1}): \tilde{Q} \to Q$, respectively.

The gradient solution works in general regardless of the task representation because the line element on the manifold is an invariant. For our case of seven dimensions:

$$
ds^2 = \sum_{ij}^{7} \left(g_{ij}^{\mu} \right) dq^i dq^j = \sum_{rs}^{7} \left(\tilde{g}_{rs}^{\mu} \right) dq^r dq^s
$$

Thus, the gradient will always point in the "straight-line" geodesic direction with respect to the task distance. A simple planar-arm example[29] (**Appendix** Figure 4.5) shows that computing the gradient of r with metric tensor $\tilde{G}$ associated to the original q-chart is equivalent to doing so after a coordinate transformation $\tilde{q}(q) = \tilde{G} \cdot q$ with metric G: that is, $\tilde{\nabla} r(q) = \tilde{G} \cdot \nabla r(q) = G \cdot \nabla r(\tilde{q})$.

The compatibility condition makes $J(\tilde{q}^{-1} \circ q) = \left(\frac{\partial \tilde{q}^i}{\partial q^j}\right)$ invertible with inverse defined as $J(\tilde{q}^{-1} \circ q)^{-1} = J(q^{-1} \circ \tilde{q}) = \left(\frac{\partial q^i}{\partial \tilde{q}^j}\right).$

When computing the gradient for pointing, the transformation rule (the chain rule) yields $\frac{\partial r}{\partial \tilde{q}^i} = \frac{\partial r}{\partial f(\tilde{q})} \cdot \frac{\partial f(\tilde{q})}{\partial \tilde{q}^j} \cdot \frac{\partial \tilde{q}^j}{\partial \tilde{q}^i}$ and multiplication by the *inverse Jacobian* $\left(\frac{\partial q^i}{\partial \tilde{q}^j}\right)$ on both sides gives $\frac{\partial r}{\partial q^i} \cdot \frac{\partial q^i}{\partial \tilde{q}^j} = \frac{\partial r}{\partial f(\tilde{q})} \cdot \frac{\partial f(\tilde{q})}{\partial \tilde{q}^j} \cdot \frac{\partial \tilde{q}^j}{\partial q^i} \cdot \frac{\partial q^i}{\partial \tilde{q}^j} = \frac{\partial r}{\partial \tilde{q}^j}$, which is the gradient of r with respect to the obstacle-avoidance $\tilde{q}$—coordinates, that is, $G^{-1} \cdot \nabla r(\tilde{q})$ for the metric G.

This framework opens the possibility of exploring different tasks in search for geometric transformations at the behavioral level that can inform us as well of geometric computations the neocortical cells in the posterior parietal cortex are thought to perform.[46,80–82]

APPENDIX 5

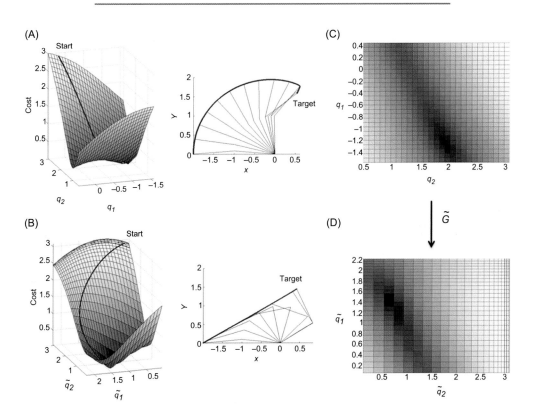

COORDINATE TRANSFORMATIONS AND NEW DISTANCE METRICS

The paths in posture space depend on the set of coordinate functions we use. The paths may not be the same as the veridical paths but a coordinate transformation can get veridical gradient paths. For example, the two-dimensional toy model presented in Figures 4.2 and 4.4 served to illustrate the outcome of the gradient using different coordinate representations or keeping the old representation and changing the metric distance to select the gradient. We model the arm with two unit-length segments and two joints q_1 at the shoulder and q_2 at the elbow. The map in Cartesian configuration space is:

$$f_{x_1}(q) = \cos(q_1) + \cos(q_1 + q_2), f_{x_2}(q) = \sin(q_1) + \sin(q_1 + q_2)$$

where the $\left[f_{x_1}(q), f_{x_2}(q)\right]$ represent the coordinates of the arm's end-effector (the hand) and the cost function to translate the hand from its initial position to the target is:

$$r(x^{\text{target}}, \mathbf{q}) = \sqrt{\left(x_1^{\text{target}} - f_{x_1}(\mathbf{q})\right)^2 + \left(x_2^{\text{target}} - f_{x_2}(\mathbf{q})\right)^2}$$

A representative curved path produced by the gradient under Euclidean metric is shown in **Appendix** Figure 4.5A with the corresponding cost surface path.

To achieve the straight line path we can find a coordinate transformation, $\tilde{\mathbf{q}} = G \cdot \mathbf{q}$ via a matrix G with coefficients transforming from the original to the new coordinates such that the new gradient-generated path is straight. The gradient with respect to the new coordinates is:

$$\tilde{\nabla}r(x^{\text{target}}, \mathbf{q}) = \tilde{G}^{-1} \cdot \nabla r(x^{\text{target}}, \mathbf{q})$$

This gradient is obtained by pre-multiplying the original gradient by the inverse of the transformation matrix. This matrix must be positive definite, but it is not restricted otherwise. If the $\tilde{G}^{-1}$ is symmetric, the transformation matrix is the metric tensor. In the present example G can be found analytically[35] as:

$$\tilde{G}(\mathbf{q}) = J(f(\mathbf{q}))^T \cdot G(f(\mathbf{q})) \cdot J(f(\mathbf{q}))$$
$$= \begin{pmatrix} 2(1 + \cos(q_2)) & 1 + \cos(q_2) \\ 1 + \cos(q_2) & 1 \end{pmatrix}, G(f(q)) = \begin{pmatrix} 1 & 0 \\ 0 & 1 \end{pmatrix}$$

The **Appendix** Figure 4.5B gives an example of the straight line path and corresponding cost surface path. The **Appendix** Figure 4.5C−D illustrates the effects of the coordinate transformations on the Q-space representation that leads to end-effector straight-line paths, the shortest in a Euclidean sense.

APPENDIX 6 NUMERICAL ESTIMATION OF POSITIVE DEFINITE COORDINATE TRANSFORMATION MATRIX

Let $Y = M \cdot X$, where $Y = dq^{DATA}$ and $X = dq^{MODEL}$.

We define the error $E = \sum_{i=1}^{n} (y_i - M \cdot x_i)^2$ and want to solve for M for all n, the number of points in the data set involving all points in all paths from all trials. To that end we need to minimize E over M by solving $(\partial E / \partial M) = 0$. Taking the derivative with respect to M and using he chain rule amounts to solving

$$\frac{\partial E}{\partial M} = 2 \sum_{i-1}^{n} (y_i - Mx_i)(-x_i^T) = 2 \sum_{i-1}^{n} (-y_i x_i^T + Mx_i x_i^T) = 0, \text{ where}$$

$$M = \left(\sum_i^n y_i x_i^T \right) \left(\sum_i^n x_i x_i^T \right)^{-1} \text{ and we set } G^{-1} = M$$

When we want the matrix M to be symmetric positive-definite, $M = B \cdot B^T$, we solve the same least square problem subject to the constraint

$$
\begin{aligned}
E &= Trace\left\{ (Y - BB^T X)(Y - BB^T X)^T \right\} \\
&= Trace\left\{ (Y - BB^T X)(Y^T - X^T BB^T)^T \right\} \\
&= Trace\left\{ -YX^T BB^T - BB^T XY^T + BB^T XX^T BB^T \right\} \\
&= Trace\left\{ -2YX^T BB^T + XX^T BB^T BB^T \right\}
\end{aligned}
$$

Set $(\partial E / \partial B) = 0$ and minimize E with respect to B to solve for B and obtain $G^{-1} = B \cdot B^T$.

References

1. Torres EB, Zipser D. Reaching to grasp with a multi-jointed arm. I. Computational model. *J Neurophysiol.* 2002;88(5):2355−2367.
2. Flash T, Hogan N. The coordination of arm movements: an experimentally confirmed mathematical model. *J Neurosci.* 1985;5(7):1688−1703.
3. Uno Y, Kawato M, Suzuki R. Formation and control of optimal trajectory in human multijoint arm movement. Minimum torque-change model. *Biol Cybern.* 1989;61(2):89−101.
4. Pontriagin LS. *Optimal control and differential games : collection of papers. Proceedings of the Steklov Institute of Mathematics.* Providence, RI: American Mathematical Society; 1990:278. vii.
5. Bernshteĭn NA. *The co-ordination and regulation of movements.* 1st English ed. Oxford, New York: Pergamon Press; 1967:196. xii.
6. Georgopoulos AP, Kalaska JF, Caminiti R, Massey JT. On the relations between the direction of two-dimensional arm movements and cell discharge in primate motor cortex. *J Neurosci.* 1982;2:1527−1537.
7. Kalaska JF, Cohen DA, Prud'homme M, Hyde ML. Parietal area 5 neuronal activity encodes movement kinematics, not movement dynamics. *Exp Brain Res.* 1990;80:351−364.
8. Kalaska JF, Hyde ML. Area 4 and area 5: differences between the load direction-dependent discharge variability of cells during active postural fixation. *Exp Brain Res.* 1985;59(1):197−202.
9. Kalaska JF, Caminiti R, Georgopoulos AP. Cortical mechanisms related to the direction of two-dimensional arm movements: relations in parietal area 5 and comparison with motor cortex. *Exp Brain Res.* 1983;51 (2):247−260.
10. Hamel-Paquet C, Sergio LE, Kalaska JF. Parietal area 5 activity does not reflect the differential time-course of motor output kinetics during arm-reaching and isometric-force tasks. *J Neurophysiol.* 2006;95(6):3353−3370.
11. Kalaska JF, Cohen DA, Hyde ML, Prud'homme M. A comparison of movement direction-related versus load direction-related activity in primate motor cortex, using a two-dimensional reaching task. *J Neurosci.* 1989;9:2080−2102.
12. Scott SH, Kalaska JF. Changes in motor cortex activity during reaching movements with similar hand paths but different arm postures. *J Neurophysiol.* 1995;73(6):2563−2567.
13. Scott SH, Sergio LE, Kalaska JF. Reaching movements with similar hand paths but different arm orientations. II. Activity of individual cells in dorsal premotor cortex and parietal area 5. *J Neurophysiol.* 1997;78 (5):2413−2426.
14. Georgopoulos AP, Ashe J. One motor cortex, two different views. *Nat Neurosci.* 2000;3(10):963. author reply 964-5.

15. Scott SH. Reply to 'One motor cortex, two different views'. *Nat Neurosci*. 2000;3(10):964–965.

16. Todorov E. Reply to 'One motor cortex, two different views'. *Nat Neurosci*. 2000;3(10):963–964.

17. Atkeson CG, Hollerbach JM. Kinematic features of unrestrained vertical arm movements. *J Neurosci*. 1985;5 (9):2318–2330.

18. Torres EB, Smith B, Mistry S, Brincker M, Whyatt C. Neonatal Diagnostics: Toward Dynamic Growth Charts of Neuromotor Control. *Front Pediatr*. 2016;4:121.

19. Torres EB, Lande B. Objective and personalized longitudinal assessment of a pregnant patient with post severe brain trauma. *Front Hum Neurosci*. 2015;9:128.

20. Von Holst E, Mittelstaedt H. In: Dodwell PC, ed. *Perceptual Processing: Stimulus equivalence and pattern recognition*. New York: Appleton-Century-Crofts; 1950:41–72.

21. Benati M, Gaglio S, Morasso P, Tagliasco V, Zaccaria R. Anthropomorphic rRobotics (I). Representing mechanical complexity. *Biological Cybernetics*. 1980;38:125–140.

22. Benati M, Gaglio S, Morasso P, Tagliasco V, Zaccaria R. Anthropomorphic Robotics (II): Analysis of manipulator dynamics and the output motor impedance. *Biological Cybernetics*. 1980;38:141–150.

23. Jose JV, Saletan EJ. *Classical Dynamics: A Contemporary Approach*. New York: Cambridge University Press; 1998:670. xxv.

24. de Maupertuis, P.L., *Accord de différentes lois de la nature qui avaient jusqu'ici paru incompatibles.Mémoires de l'Académie Royale des Sciences*, in *Oeuvres,4, 1-23 Reprografischer Nachdruck der Ausg*. Paris; 1744:417–426.

25. de Maupertuis PL. *Les Loix du mouvement et du repos déduites d'un principe métaphysique*, in *Mémoire Académie Berlin* 1746:267.

26. Lanczos C. *The Variational Principles of Mechanics*. Fourth ed New York: Dover Publications Inc; 1970.

27. Feynman RP, Leighton RB, Sands ML. *The Feynman Lectures on Physics.*. Reading, MA:: Addison-Wesley Pub. Co; 1963.

28. Soechting J, Buneo C, Flanders M. Moving effortless in three dimensions: does Donders' law apply to arm movement? *J Neurosci*. 1995;15(9):6271–6280.

29. Torres EB, Zipser D. Reaching to grasp with a multi-jointed arm (I): a computational model. *J Neurophysiol*. 2002;88:1–13.

30. Spivak M. *A Comprehensive Introduction to Differential Geometry*. 3rd ed. Houston, TX: Publish or Perish, Inc; 1999.

31. Bertsekas DP. *Dynamic Programming: Deterministic and Stochastic Models.*. Englewood Cliffs, NJ: Prentice-Hall; 1987:376. viii.

32. Torres EB, Zipser D. Simultaneous control of hand displacements and rotations in orientation-matching experiments. *J Appl Physiol*. 2004;96:1978–1987. Highlighted Topic Neural Control of Movement.

33. Torres EB, Heilman KM, Poizner H. Impaired endogenously evoked automated reaching in Parkinson's disease. *J Neurosci*. 2011;31(49):17848–17863.

34. Carmo MPOD. Differential Geometry of Curves and Surfaces. Englewood Cliffs, NJ: Prentice-Hall; 1976:503. viii.

35. Gray A, Gray A. *Modern Differential Geometry of Curves and Surfaces with Mathematica*. 2nd ed Boca Raton: CRC Press; 1998:1053. xxiv.

36. Zipser D, Andersen RA. A back-propagation programmed network that simulates response properties of a subset of posterior parietal neurons. *Nature*. 1988;331(6158):679–684.

37. Roweis ST, Saul LK. Nonlinear dimensionality reduction by locally linear embedding. *Science*. 2000;290 (5500):2323–2326.

38. Hinton G. Parallel computations for controlling the arm. *Jf Motor Behavior*. 1984;16(2):171–194.

39. McClelland JL, Rumelhart DE. *Explorations in Parallel Distributed Processing: A Handbook of Models, Programs, and Exercises. Computational models of cognition and perception.*. Cambridge, MA: MIT Press; 1988:344. ix.

40. Rumelhart, D.E., J.L. McClelland, and University of California San Diego. PDP Research Group., *Parallel distributed processing : explorations in the microstructure of cognition*. Computational models of cognition and perception. 1986, Cambridge, Mass.: MIT Press.

41. Amari S. Natural gradient learning for over- and under-complete bases in ICA. *Neural Comput*. 1999;11 (8):1875–1883.

42. Nishikawa KC, Murray ST, Flanders M. Do arm postures vary with the speed of reaching? *J Neurophysiol*. 1999;81(5):2582–2586.

43. Torres EB, Zipser D. Simultaneous control of hand displacements and rotations in orientation-matching experiments. *J Appl Physiol*. 1985;96(5):1978−1987. 2004.

44. Conditt MA, Mussa-Ivaldi FA. Central representation of time during motor learning. *Proc Natl Acad Sci U S A*. 1999;96(20):11625−11630.

45. Andersen RA, Zipser D. The role of the posterior parietal cortex in coordinate transformations for visual-motor integration. *Can J Physiol Pharmacol*. 1988;66(4):488−501.

46. Zipser D, Andersen R. A back-propagation programmed network that simulates response properties of a sub-set of posterior parietal neurons. *Nature*. 1988;331(25):679−684.

47. Batista P, Buneo CA, Snyder LH, Andersen RA. Reach plans in eye-centered coordinates. *Science*. 1999;285:257−260.

48. Krakauer JW, Ghazanfar AA, Gomez-Marin A, MacIver MA, Poeppel D. Neuroscience Needs Behavior: Correcting a Reductionist Bias. *Neuron*. 2017;93:480−490.

49. Buneo CA, Jarvis MR, Batista AP, Andersen RA. Direct visuomotor transformations for reaching. *Nature*. 2002;416:632−636.

50. Torres EB, Quian Quiroga R, Cui H, Buneo CA. Neural correlates of learning and trajectory planning in the posterior parietal cortex. *Front Integr Neurosci*. 2013;7:39.

51. Torres E, Andersen R. Space-time separation during obstacle-avoidance learning in monkeys. *J Neurophysiol*. 2006;96(5):2613−2632.

52. Torres EB. New symmetry of intended curved reaches. *Behav Brain Funct*. 2010;6:21.

53. Hartigan JA, Hartigan PM. The Dip Test of Unimodality. *Annals of Statistics*. 1985;13(1):70−84.

54. E.B. Torres, C.A. Buneo, H. Cui, R. Andersen, in *Annual Meeting of the Society for Neuroscience*. (Washington DC, 2008), vol. Slide.

55. Mitchell JF, Sundberg KA, Reynolds JH. Differential attention-dependent response modulation across cell classes in macaque visual area V4. *Neuron*. 2007;55(1):131−141.

56. Adamovich SV, Berkinblit MB, Hening W, Sage J, Poizner H. The interaction of visual and proprioceptive inputs in pointing to actual and remembered targets in Parkinson's disease. *Neuroscience*. 2001;104:1027−1041.

57. Torres EB, Cole J, Poizner H. Motor output variability, deafferentation, and putative deficits in kinesthetic reafference in Parkinson's disease. *Front Hum Neurosci*. 2014;8:823.

58. Torres EB, Raymer A, Gonzalez Rothi LJ, Heilman KM, Poizner H. Sensory-spatial transformations in the left posterior parietal cortex may contribute to reach timing. *J Neurophysiol*. 2010;104:2375−2388.

59. Torres EB, Yanovich P, Metaxas DN. Give spontaneity and self-discovery a chance in ASD: spontaneous peripheral limb variability as a proxy to evoke centrally driven intentional acts. *Front Integr Neurosci*. 2013;7:46.

60. Altmann SL. *Rotations, quaternions, and double groups. Oxford science publications*. Oxford Oxfordshire New York: Clarendon Press; Oxford University Press; 1986:317. xiv.

61. Soechting JF, Buneo CA, Herrmann U, Flanders M. Moving effortlessly in three dimensions: Does Donders' law apply to arm movement? *J Neurosci*. 1995;15:6271−6280.

62. Keane BP, Silverstein SM, Wang Y, Papathomas TV. Reduced depth inversion illusions in schizophrenia are state-specific and occur for multiple object types and viewing conditions. *J Abnorm Psychol*. 2013;122:506−512.

63. Nguyen J, Papathomas TV, Ravaliya JH, Torres EB. Methods to explore the influence of top-down visual processes on motor behavior. *J Vis Exp*. 2014.

64. Nguyen J, Majmudar UV, Ravaliya JH, Papathomas TV, Torres EB. Automatically characterizing sensory-motor patterns underlying reach-to-grasp movements on a physical depth inversion illusion. *Front Hum Neurosci*. 2015;9:694.

65. Donnellan AM, Leary MR. *Movement Differences and Diversity in Autism/Mental Retardation: Appreciating and Accommodating People with Communication and Behavior Challenges. Movin on series.*. Madison, WI: DRI Press; 1995:107.

66. Rogers DM. *Motor Disorder in Psychiatry: Towards a Neurological Psychiatry*. New York: Wiley; 1992:159. viii.

67. Nguyen J, Majmudar U, Papathomas TV, Silverstein SM, Torres EB. Schizophrenia: The micro-movements perspective. *Neuropsychologia*. 2016;85:310−326.

68. Bernard JA, Mittal VA. Updating the research domain criteria: the utility of a motor dimension. *Psychol Med*. 2015;1−5.

69. Insel T, Cuthbert B, Garvey M, Heinssen R, Pine DS, Quinn K, Sanislow C, Wang P. Research domain criteria (RDoC): toward a new classification framework for research on mental disorders. *Am J Psychiatry*. 2010;167:748−751.

70. Insel TR. The NIMH Research Domain Criteria (RDoC) Project: precision medicine for psychiatry. *Am J Psychiatry*. 2014;171(4):395−397.

71. American Psychiatric Association and American Psychiatric Association. DSM-5 Task Force., *Diagnostic and Statistical Manual of Mental Disorders : DSM-5*. 5th ed. Washington, D.C.: American Psychiatric Association; 2013:xliv, 947 p.

72. Cosgrove L, Wheeler EE. Industry's colonization of psychiatry: ethical and practical implications of financial conflicts of interest in the DSM-5. *Femin Psychol*. 2013;23(1):93−106.

73. Torres EB. Two classes of movements in motor control. *Exp Brain Res*. 2011;215(3-4):269−283.

74. Torres EB. Signatures of movement variability anticipate hand speed according to levels of intent. *Behav Brain Funct*. 2013;9:10.

75. Delaunay B. Sur la sphere vide. *Izv. Akad. Nauk SSSR, Otdelenie Matematicheskii i Estestvennyka Nauk*. 1934;7 (793−800):1−2.

76. Torres EB, Denisova K. Motor noise is rich signal in autism research and pharmacological treatments. *Sci Rep*. 2016;6:37422.

77. Do Carmo M. *Differential Geometry of Curves and Surfaces.*. Englewood Cliffs, New Jersey: Prentice Hall; 1976.

78. Apostol TM. Linear Algebra, A First Course with Applications to Differential Equations. New York: Wiley; 1997.

79. Berger M. *A Panoramic View of Riemannian Geometry*. Tokyo: Springer; 2003.

80. Andersen RA, Snyder LH, Bradley DC, Xing J. Multimodal representation of space in the posterior parietal cortex and its use in planning movements. *Annu Rev Neurosci*. 1997;20:303−330.

81. Andersen RA, Buneo CA. Intentional maps in posterior parietal cortex. *Annu. Rev Neuroscience*. 2002;25:189−220.

82. Andersen RA, Buneo CA. Sensorimotor integration in posterior parietal cortex. *Adv Neurol*. 2003;93:159−177.

CHAPTER 5

Learning to Detect Expertise in Sports Aided by the Gift of Our Students

Here's to the crazy ones, the misfits, the rebels, the troublemakers, the round pegs in the square holes... the ones who see things differently—they're not fond of rules... You can quote them, disagree with them, glorify or vilify them, but the only thing you can't do is ignore them because they change things... they push the human race forward, and while some may see them as the crazy ones, we see genius, because the ones who are crazy enough to think that they can change the world, are the ones who do.

Steve Jobs.

UNCERTAIN TIMES

There have been major technological advancements in our lifetime. These advances have changed the way we organize information, the way we connect and communicate with each other, the way we make decisions and ultimately, the way our lives are legislated. The control we may think we have over our lives somehow dissipates when you think of how much others know about us now, and how much we are being watched by others. One can go insane thinking about these things. I nearly did sometime under communist Cuba.

I left the island precisely out of the terror of being under surveillance by others. I had no control over even the most immediate mundane decisions I wanted to make. Others made those decisions for me. The lack of freedom to even plan one's immediate future was blatant in Cuba. There was no ambiguity about that. There was no uncertainty about what we could not do and in a sense, looking back now, this was comforting to know. At least we all knew it. But there was not much we could do about it. Musicians wrote songs about it, artists painted about it, dancers and choreographers captured it, filmmakers made movies about it, so that somehow, we were fully aware of this lack of freedom and free will.

Many in my generation decided to leave Cuba at some point. It is a diaspora across the globe that transiently reunited through the magic of the internet, when Facebook came

DOI: https://doi.org/10.1016/B978-0-12-804082-9.00005-8

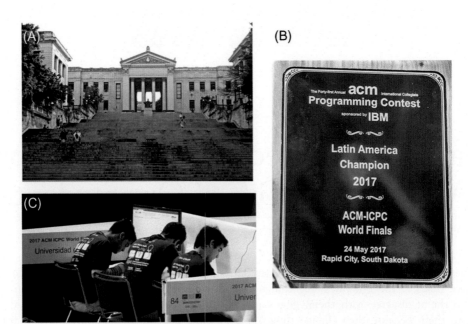

FIGURE 5.1 **A place that values Math.** (A) The University of Havana (UH), funded in 1728 is the oldest in Cuba and one of the first in the Americas. It sits on top of a hill and has a long staircase that one has to climb to end at the feet of the Alma Mater monument, welcoming all to learn with her open arms. (B) The Cuban team won the Latin American Champion title at the ACM contest 2017. *(http://www.cubadebate.cu/noticias/2017/05/30/ uh-los-codigos-de-una-universidad-historica-fotos-y-video/#.Wcr8dWhSzyQ)* and took this trophy home. (C) Solving the problems as a team. *From ICPC News Photo by Bob Smith, Photo: L Eduardo Domínguez / Cubadebate.*

alive. For the most part, we are professionally connected to something related to our original studies at the University of Havana (UH). We had an incredible group there; a stellar program that even today shines amidst embargoed trading and many hardships (Fig. 5.1).

Our own group self-emerged in the late 80s, after the first final exam in the first semester of *"Analisis Matematico"* (Mathematical Analysis) at the *"Facultad de Matematica y Computación."* This class was the nemesis of our original group of about 120 students. The students that had made the cut to enter the relatively recent program of *"Matematica-Cibernetica"* at the school of Math (Fig. 5.2A,B) were among the best in their schools. But they had to pass additional placement tests to be able to join this new and highly competitive program. Everyone was intrigued by the new computers the University had acquired (allegedly bypassing the United States embargo through secret trades with Panama's government.) In the mid-to-late 80s we were already mastering and reverse engineering the PC's and Apple computers of the United States. So, they created an official program to that end. And they did so at the school of Math, because we needed to master the math behind all of that to be able to *reinvent the wheel*—without having proper access to basic resources.

FIGURE 5.2 The school of math and computer science at the UH. (A) Our favorite spot to hang out between classes, sitting by the stairs and chatting away. (B) The classroom was located where the arrow points. The beautiful inner gardens and corridors at the Facultad de Matematicas. (C) Right after my arrival to the United States in the Fall of 1990. (D) My dear friend, my brother, Edilio Hernandez Jr. at the time he helped me get the official Cuban transcripts on time to start the Fall semester of 1991 at SJSU. (E) Present times in Miami where he lives and teaches college math nowadays.

The first exam of Analyses was dreadful. We were allowed open notes, open books and no one in the room was needed to supervise us for 4 hours. Every student kept to himself and solved problems on his/her own. The pride in it all was to Ace that exam with no help whatsoever. You were on your own and whatever you did was whatever you were able to do. It was all very self-evident, very self-corrective, with no need to cheat and no room for it either. There were some 15–20 questions and if you could solve some 5 of them you were good to go. The trick though, was that these were not questions on the book, or exercises we had studied prior to the exam. They were demonstrations that required having previously mastered the demonstrations of theorems that we had studied in class. Yet these class-exercises were not the direct answers of what those problems were asking from us in the exam questions. In those exams, you had to be creative and bring on your own solution to arrive at the correct result. We knew what the answer would have to be at some level—as someone else had already demonstrated the base theorem in the 1800s or the 1900s, etc. The trick was how to solve the given demonstration using our own methods. This class taught us to think in such a way that after passing it, we could pose problems and solve them in any general domain. It prepared us to *learn how to learn* and how to solve problems in general.

Out of approximately 120 students, some 50 remained in the group after the first semester of Analysis was over. We were then split into two groups, and after the second final exam of Analysis in the second semester (same fashion as above), we lost others. The group was about 25 or so and then some foreign students joined the class. We had a steady number after that. The series of Analysis comprised five semesters. After going through such hard exams as a group, there was a sense of comradery and solidarity that self-emerged and remained all the way until we re-connected in Facebook. We came to know each other rather well in that class. I left Cuba around the time I had barely started the fourth year of this degree. I had survived the first three years and proudly completed the full series of Analysis. It is somewhat puzzling to me that not even my PhD completion, gave me as much satisfaction as going through the Analysis series in the UH. Perhaps it was all in the magic of being so young and innocent (Fig. 5.2C).

I moved to the United States in pursuit of more personal freedom but was lucky enough that I could finish the degree here. They had something equivalent to my degree in Havana at San José State University (SJSU), which was affordable to me as I could live-in with my family and bike to the campus in 40 min. A bus ride took even less time, and I could take the bike on the bus (all buses had a rack to that end), so it was all doable. During the nights, I could work as a janitor in the family-owned business and help the family in exchange for room-and-board. My family was very supportive of my idea to finish my studies in the United States. During the day, all I had to do was attend the public schools that taught English as a Second Language (ESL), while I figured out how to continue with my college degree at that nearby University. I was here under a tourist visa for three months, issued in Mexico City, where I was also visiting. As such, I could not work and receive a salary, so I decided to learn English. To switch to a foreign student visa and be able to attend college here, I had to manage to learn enough English in those three months to pass the TOEFL exam, get back to Mexico City and petition the foreign student visa. I could do that provided that my TOEFL scores were outstanding and a college here in the United States accepted me. The question we had then was if the Math-CS department in SJSU (the closest college to my family's home) even knew there was a university in Havana. As it turned out, they did. Several of the professors at the SJSU-Math department knew my professors in Havana University. I came to know this in a somewhat serendipitous way.

My family in Santa Clara, CA knew other Cuban families in the Bay Area. Cubans have this feature whereby, everyone seems to know everyone else and for the most part, everyone wants to help everyone else. As such, as soon as I arrived in the United States and my family learned of my desire to continue attending college and finishing up my degree here, they called a friend who happened to be a Math Professor. He was not just any Math Professor (though my family at the time had no idea of the significance of what he had done for the community). He was José Antonio Valdés (Fig. 5.3), *"the Math Professor who truly believed that all students can learn mathematics and worked to prove this true."* (http://www.josevaldesmath.org/cms/page_view?d = x&piid = &vpid = 1458202011630)

To us he was simply Valdés. Valdés, as we called him, talked with me over the phone and after our conversation, he set up an interview with the Counselor of foreign students at SJSU (Mr. Loui Barozzi). Valdés and Barozzi helped set up an interview with the head of the Math-CS, program, so I could be interviewed there and they could determine the

FIGURE 5.3 Our beloved Professor Valdes his legacy lives on at the Math Institute created to honor his extraordinary service as a citizen of CA. https://valdes-math.squarespace.com/history-of-jvmi/ and http://www.josevaldesmath.org/.

placement level. Valdés was rather ill at the time. He in fact passed away on December 12, 1991, a few months after I started school in the Fall Semester. I owe it to him to have been able to continue my studies on Math-CS at SJSU. The conversation we had, and his legacy have left a tremendous impression on me. It was in the tradition of our roots, whereby education is first, and education that involves mathematics is top. *"Tienes que terminar tu carrera,"* (you must finish your degree) he said to me.

I could transfer in full, all the credits, from my first three years completed at the University of Havana because our textbooks were the same as those in the program at SJSU, and the structure of the program mirrored the program here. All I needed was my official transcripts, the Chair told me, and I could start exactly where I left; and I needed to complete all the General Education requirements for a BSc degree in the United States. Easier said than done though. With the embargo and without diplomatic relations between Cuba and the United States, it was not possible to request my transcripts and get an official copy all the way here on time for the Fall semester. I left the interview hopeful yet, it was highly uncertain that we could do it. My uncle said to me *"Mi'ja si se quiere, se puede. Dale, vamos a hacer lo que hay que hacer!"*. (Mi child, where there is a will, there is a way. Let's go and do what we need to do!)

In another stroke of luck, my best friend's father Edilio Hernadez, a top singer of the Bell Canto in the Havana Opera (and former pupil of Luciano Pavarotti) was coming to Miami for a work/family visit. He could then bring the official transcripts from Cuba and hand them in person to someone trustworthy, on the way to CA. As it so happened, my aunt and uncle (also by chance) were stopping in Miami that very weekend on their way back from their vacation in the Bahamas. All I had to do was coordinate with my friend Edilio Jr. in Havana, so his dad could bring the transcripts to Miami on time for me to hand them to SJSU. Mr. Barozzi needed them to set up my registration at SJSU on time for the Fall semester, or I would miss that window of opportunity.

A minor detail was that the Cuban government would not release an official copy of any college transcripts unless the processing fees were paid for in US dollars. But no one in Cuba was even allowed to carry dollars. In fact, it was against the law back then to own dollars. One could go to jail if caught with a single US dollar. A type of *catch 22* situation that a dear friend from Mexico resolved. He should remain anonymous as he is a very private person, but it suffices to say that a phone call and the mentioning of what I needed, placed him in Havana the next day. He met my friend Edilio Jr. at the Dean's office in Havana University. My friend was also part of that reduced group that had finished the Analysis series and loved Math (he is in Fig. 5.2D at that time when he helped me out while taking some risks.) (It was forbidden in Cuba at the time to talk to foreigners, or to carry dollars. By accompanying my Mexican friend to the Dean's office in search for my transcripts, my friend risked to be detained by the authorities). He knew how much I wanted this. As a matter of fact, he came to the States eventually and has been awarded several distinctions for teaching excellence at the Miami Dade College—distinctions he holds dear. Teaching Math is his life's passion since we were young (Fig. 5.2E, present time).

They collected the official transcripts at the Ministry of Foreign Affairs properly stamped by the University and bearing the Cuban government seal. They paid the dues and my friend's father brought them just in the nick of time for my family to bring the sealed envelope back from Miami to San José, CA. What a trip! It was all lined up and as my late mother would say *"estaba escrito"* (It was meant to be, it was fate). To this day, all parties who were part of this sort of marathonic, frantic race against the clock remain in awe that it all happened on time.

I could finish the BSc degree here in the United States, and I was ever so lucky once again that within the second semester in the program I received a letter from the office of the registrar offering me a scholarship. The letter came directed to me and had identified me as a 4.0 GPA student with Hispanic roots eligible for the *Minority Access to Research Careers (MARC)* scholarship. The letter described the benefits of the scholarship and what was expected from me should I accept the offer. The benefits were that it paid tuition for two full years and a stipend to help defray the cost of living (so high in the Bay Area) as an incentive to help the student focus on classwork and excel. The expectation was to do research with a Professor of our choice and to present the research results at a yearly conference (all expenses covered). Lastly, the recipient of the scholarship was expected to pursue a PhD or an MD/PhD (doing a PhD was my dream).

You may not believe this, but I thought this was some form of advertisement and too good to be true, and I threw the letter away. My uncle had been complaining (around those days) of all the junk mail he always received. He had shown me some adverts on this Publishers Clearing House marketing company and told me "look at it, too good to be true; you ought of work hard and disregard such nonsense." When I saw this letter, I thought this was a joke of bad taste someone was trying to play on me and threw the letter away without even considering it twice "too good to be true, I thought, as my uncle would say." At some point, about a week later, I told my family about the "letter-joke." To my dismay, they said this was likely real and *"Muchacha, tu estás loca? Esta niña es tonta!"* which in plain Cuban, which is not a politically correct culture, it translates into *"you are an idiot!"*

Luckily for me, Professor Helbert Silber (Fig. 5.4) https://www2.calstate.edu/csu-system/news/Pages/stellar-roster-csus-white-house-honorees-for-science-guidance.aspx, who in 1998 became the recipient of the Presidential Award for Excellence in Science, Mathematics and Engineering Mentoring (http://paesmem.net/node/1497), did not give up.

Herbert Silber - Professor of Inorganic Chemistry and Nuclear Science

Dr. Silber has a distinguished 25-year record of seeking-out and mentoring minority and disadvantaged students at the high school, undergraduate and graduate levels. The focal point of Dr. Silber's mentoring activity is the laboratory. As evidenced by 90 publications co-authored by students, Dr. Silber involves his students in significant research and supports their efforts. At the national level, Dr. Silber has had a 20-year involvement with the American Chemical Society's SEED program (Summer Educational Experiences for the Disadvantaged), including a very successful three-year appointment as chair of the National SEED Committee.

FIGURE 5.4   The Presidential Award presented to our dear Professor Silber.

"Mr. Silber" as we called him then, came to my class (I recall I was in the Combinatorics class, which was as challenging in the SJSU math program as was our Analysis series in Havana—to the extent that someone literally had a seizure during the final exam) and asked, *"who is Elizabeth Torres?"*. I responded, *"I am"* to what he replied, *"See me at my office in Chemistry (next building over) ASAP."* I did. It was about that letter…his face lit up when I apologetically (for my lateness) accepted. But in view of my own's family's reaction to my throwing the letter away and thinking of it as being a bad joke, I did not mention any of that.

During my time at CALTECH, Prof. Silber located me and invited me to SJSU to join other seven former students from the MARC program to celebrate our accomplishments in science. We were to share some time with the members of Congress behind these programs and the President of SJSU and talk about our research. We had all made it through a full PhD or MD/PhD program, were at a Postdoctoral training position, and had published our work in peer-reviewed journals. I was a Sloan-Swartz Postdoctoral Fellow at CALTECH in their Computational Neuroscience program; had already published part of the work from my PhD thesis, and was planning on an academic tenure-track career path. Another former MARC student who had started in the program with me, was at Mayo Clinic developing biomarkers for cancer research. And a former MARC student who started the program a year after we did, was at the CDC heading a special group of extremely rare infectious diseases. There was even someone who had created a company related to a cancer drug with international reach. It was amazing to see there the fruit of these programs Congress supported. Today such programs are endangered because of the current Trump administration. Yet, these programs planted many seeds that not only helped me and others like me gain access to science careers. They helped me pay it forward and help many others along the way advance in science, math and engineering.

Professor Silber was very proud of our career path and of the programs he directed at SJSU, which made it all possible for us. The NIH-NIGMS keeps track of our career paths, and mine is here (https://www.nigms.nih.gov/training/alumniupdates/pages/torres.aspx) (NIH stands for the National Institute of Health. NIGMS stands for the National Institute of General Medical Sciences). Professor Valdés, Professor Silber and the SJSU Counselor Mr. Barozzi opened the door to the path that brought me here today. They are in that special class of people that move the world forward and help anyone, regardless of race, creed or religion. In their honor I try to do the same, every day, in my lab.

When the Facebook opened, people from that small group in Havana suddenly realized that only very few of us had remained in Cuba by choice. We were so happy to momentarily re-connect and we found out that most of us had professions related to our original degree. We had found applications for our knowledge base, and whether in the United States, Spain, Switzerland, Canada, Australia, etc. or back home in Cuba; we were working on something we really liked. Ironically, even though we had no freedom in Cuba, the foundation that the Cuban Math-CS program gave us all at the UH, led us to find the path towards the freedom to work on something we liked. At least in my case, I love what I do so much that I do not have enough hours in the day to do it all. I found this to be the case for those I interacted with from our group on Facebook. The professors and the program we had in Havana taught us how to think independently, how to define problems and how to arrive at a solution from many different angles. They planted the seed so we could build our future.

Each person in that group was *absolutely* unique. It has been some time now since I was confronted with and passed those challenging exams we all had to take to master the Analysis series at UH. Along the way, since I came to the United States and took the academic path, I have been fortunate enough to be part of the Computational Neuroscience group at the UCSD/Salk Institute, visit MIT labs, spend time at the Gatsby Unit of Computational Neuroscience in London during my PhD training and train as a Sloan-Swartz fellow at CALTECH during my postdoctoral years. I have met very unique people in those circles, innovators, inventors of new technologies, people who are creating the knowledge base that is transforming the world into a better place. My interactions with them very much measure up to what we experienced as a group in those Analysis exams at the Math-CS program of the University of Havana.

The pride we felt to be able to solve something on your own was very empowering then. Today, it is also very telling of what we would have been capable of, had we been given the chance back home. I know we would have changed Cuba for the better if given the chance. We had the brain power to do it, but lacked the free will to act on it. Most of us left, utterly heartbroken, somehow forced to find our own path elsewhere.

The United States and Cuba have no normal channels of communication. The historical steps President Obama took were for the most part undone by the Trump administration. As such, I cannot be in touch with the folks in Havana; but through the magic of the internet, I know that new generations of students from that Cuban Math-CS program are excelling and showing what a wonderful tradition in math the UH has

(Fig. 5.1 and see **Notes** [1]). They will be the generation that moves that beautiful island forward someday. I have no doubt of that, as I have no doubt that those from our original class who did not leave Cuba, did the right thing, that is, teaching these young fellows how to think independently. I only hope—politics aside—that I can someday help keep that tradition and pay back my dues to the land where I was born and raised.

In Cuba, it was obvious to all of us that freedom was a commodity we did not have. It was unambiguous to all of us what we could or could not do. It is subtler here in the United States. Freewill tends to get lost in the fine print lawyers write on the back of everything we sign. *Who reads that stuff*? So, 30 years later it is beginning to feel the same here as I felt back in Cuba during my adolescence and early youth in college. It can be asphyxiating at times. Who would have thought that some of the technological advances of our times would ultimately entrap us so much? I now keep the webcam of my computer covered at all times—unless I skype with someone; but then cover it back right away. And I deactivate the mic of my computer, or remove it altogether when I am not skyping or doing some face-to-face remote communication.

Surviving Immigration

By the time I arrived in the United States in 1990, endowed with coding skills from the CS-Cuban program, and as soon as I was allowed to work, I could survive by traveling all over the Bay Area, taking programming jobs here and there. While I took classes towards the completion of my undergraduate degree, I witnessed from afar the rise of several companies that are well known today, and of others we no longer hear about. As a computer science major at SJSU, in the heart of the Silicon Valley, I necessarily kept track of such developments because several of my professors were entrepreneurs in the Bay Area and most of my classmates opted to take a job at those companies.

I recall recruiters from the Valley (IBM, INTEL, Atari, Microsoft, Apple and many others) lining up outside our Department of Mathematics and Computer Science, handling out brochures and trying to interview us even before we had finished our degrees. Indeed, computer programming was a great commodity that certainly helped me survive my early days as an immigrant in this country. I rode my bike from San José to Palo Alto, providing my services as a programmer in small businesses and local offices during the off-school hours. I would prepare their taxes and quarterly reports, do some of their accounting and occasionally, even do some of their services (e.g., I once served as a security guard for a fashion convention working for one of the companies I did the quarterly taxes for. That weekend-pay covered several of my textbook expenses.) I did not like having a boss telling me what to do, so the idea of taking up a job in any of those big companies was not my

[1]https://www.facebook.com/matcomuh/
¡MUCHÍSIMAS FELICIDADES al equipo de UH++, el profe Somoza, Marcelo Fornet Fornes, Ariel Cruz Cruz y Eloy Pérez Torres por el espectacular resultado en la ICPC!
¡Campeones de América Latina! #Cuba #ICPC2017 #UH
http://www.cubadebate.cu/noticias/2017/05/30/uh-los-codigos-de-una-universidad-historica-fotos-y-video/#.WcU2jWhSzyQ

FIGURE 5.5 At the Faculty recognition during the half- time of a football game between Rutgers University and Washington University. Too much sensory stimulation transiently cut off my language comprehension.

cup of tea. I rather enjoyed having the freedom to solve problems that I chose to work on, or choosing the type of work I wanted to delve into.

Seeking this type of independence is mostly because all my life, I have had hard time following instructions. It is a handicap I have never been able to overcome. In special occasions, I have been known to freeze and lose my ability to understand spoken language altogether (as when I recently had to face a large football crowd at the Rutgers Stadium and for all practical purposes, went *mute and deaf*, Fig. 5.5). My husband had to help me out with the instructions they were giving me to be on TV during the half-time of the game. In a way, I can completely relate to some people in the spectrum of autism who cannot talk. Well, I know for a fact that I am *"one of them."* Whereas my condition is transient and may be triggered by certain circumstances surrounding excess audio-visual stimulation, and excess motion from large crowds, theirs (seemingly permanent) must not be everlasting. In our SMIL lab, we have witnessed—more than once—some of the children who cannot usually talk, articulate full phrases when their body is supported and their biorhythms are brought to a general and sustained state of self-regulation and self-control. For this reason, we are exploring new avenues to evoke and sustain that magical state of the nervous system that enables vocalization and in some cases, even entire speech phrases.

SPOTTING GENIUS

The nervous system works in mysterious ways. It can at times freeze and leave you inert and speechless; or it can be hyperactive and unleash such a fast-multi-tasking regime that it becomes difficult to (socially) interact with others. At our SMIL group we are all much too familiar with these extremes and have somehow endured both *"maladies"* one way or another. Most of the time, we are in multi-tasking mode and get a lot accomplished in any given day; but the cost of it is that to be able to function socially it does take some effort outside our medium. We must force ourselves to slow down a bit, so every now and then we do that and connect to the social world out there. Our group, as many other labs, is very happy in this world of science, computers and math. This is our comfort zone, where what others might consider as "work" is actually "play." We do enjoy what we do immensely and somehow the lab ends up naturally attracting people with a similar frame of mind.

Joe Vero

Although we are located at the Psychology Department of the Busch campus at Rutgers, we somehow attract students from other more technical departments. This happened with Joe Vero, who came from the Biomedical Engineering Department to complete a Senior Design project in the lab (Fig. 5.6). Joe continues with us while completing his Master's Degree and debates between industry and a CS-PhD degree. I jokingly keep reminding Joe that the founders of Apple, Microsoft and Facebook did not need even a college degree to do what they did and revolutionize the world. In many ways, Joe reminds us of them. We have the feeling that in due time he might end up doing something transformative out there. Just give it time and you will see.

FIGURE 5.6 Joe Vero at work in the lab.

Joe started to teach himself Python to help me translate all my MATLAB code for analyses into Python. I decided to do this so that our biometrics for this book could be easily implemented in a language that is readily available, and openly accessible to everyone. MATLAB, the language we initially used to implement our analyses is very nice, but restricted to academic license. Our aim is to reach beyond academics and provide open access to our work and data to anyone who would be interested in taking what we have done, test it, improve it and perhaps even make it into something more practical. Opening these findings to the world at large may even result into building a device that already does some of these analyses up front, when the biosensor is harnessing the data from the nervous systems.

While teaching himself Python and querying Google about a few things he needed to be able to process all these time series data, do signal processing and perform various stochastic analyses on the micro-movements parameterization of the data, something rather interesting happened. The Google foo bar dropped and asked Joe a question (Fig. 5.7.) *"You're speaking our language. Up for a challenge?"*. Joe was given six challenging problems in increasing levels of complexity. He passed them all and went on to the next level of interview, which he seems to have passed as well.

Essentially Google may hire him away and the prospect of this possible hiring has the lab very happy; particularly, given that we had spotted his incredible problem-solving abilities the moment he showed up in the lab.

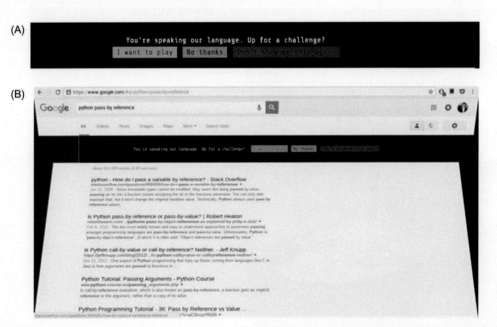

FIGURE 5.7   The Google interview that surprised Joe Vero amidst his programming spree. (A) The google foobar that Joe got all of the sudden while searching for some terms related to our statistical platform and Gamma process. (B) The drop in the google search bar adopting some 3D form we never see in the interface, unless google asks you a series of questions as in (A). It is an eerie feeling to know someone is 'watching' our search patterns at the other end.

Joe can talk to you while doing multiple things at once, including googling his cell phone for random facts he thinks about, video gaming and solving a homework problem simultaneously, and he will still give you the answer you need.

He managed to set up Lab Streaming Layer (LSL) for the lab and now we can synchronize streaming data from multiple instruments (e.g., EEG, IMUs, EKG) and study the patterns of the nervous systems' biorhythms as they unfold in tandem, in real time (as we saw in Chapter 6, Fig. 6.26). Since our methods enable rapid parameterization of these synchronous biophysical signals, we can use the parameterized output, update it (i.e., re-parameterize it) and provide instantaneous feedback to the end user in various forms. For example, we can feed it back as audio-visual input combined with movements in the virtual form of an artificial agent or avatar.

Joe's LSL code facilitated several other projects in the lab, where full-body motions are harnessed, re-parameterized, and updated in real time. If he decides not to go to Google and remains in the lab, we will be very happy to have our very first male doctorate student. Regardless of what the future brings to Joe, we will be grateful to share various samples of his Python code implementing our biometrics in the companion site of this book.

Ushma Majmudar

"*Ushmita*," as we dearly call her, blew us all away. She designed and built an Arduino-controller based system for Jill's experiment that could automatically turn on and off the lights in the room, move some apparatus to time the presentation of a visual illusion stimulus to the participants and retract the illusion as needed. While she coded these, and got the hole concoction to work properly [1], she decided to help us out with another project. We were tracking Vilmi's sleep cycles and exercise routines as she trained for the Nut Cracker show (Vilmi is our cover photo dancer) (Fig. 5.8). Ushma, just for fun, built a user interface to process the data daily (Fig. 5.9A) and to give Vilmi daily feedback on what the sensors were reading out of her body in motion during the training, and then during her sleep [2] (Fig. 5.9B). Amidst these hacking adventures, she completed pre-med and got into Albert Einstein Medical School in NYC, while receiving many honors and prizes of academic excellence during her graduation.

The interface she built helped us use the IMUs to profile the nervous system of the person longitudinally, in response to exercises and sleep states. That corpus of data served to build the empirical foundation that later helped other projects, including the Qualcomm competition for the innovative fellowship. Vilmi and Ji were among the selected presenters in the final cut and were flown to beautiful San Diego to present their proposal at Qualcomm (Fig. 5.10). The experience was magical for all of us.

Neha Tadimeti

She knocked at our SMIL door one day. "*Professor Torres, I am interested in working with you for my Senior Project.*" Neha was a student from Electrical Computer Engineering and came to us when she found our lab on the web. She was interested in analyzing EEG data, and she came to the right place because we had just completed a very interesting project

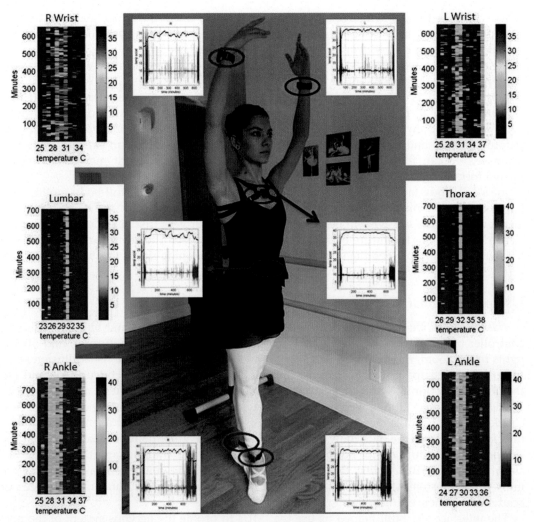

FIGURE 5.8 Tracking Vilelmini with wearable biosensors as she prepared for her Nut Cracker performance. We followed her bodily acceleration, gyroscopic rotations and temperature.

comparing the brain activities of folks in the social sciences, mathematicians and people with applied math background, as they learned to control a cursor in the screen by mere will [3]. In the midst of this project we had received the surprising visit of Ian Waterman, the man without proprioception [4], and Dr. Jonathan Cole, his neurologist.

Ian and Jonathan had visited the lab on occasion of Dr. Oliver Sacks 80th birthday (taking place in NYC at the time) and since I had studied Ian's movements in the past and characterized his stochastic signatures [5], I took the occasion to study his brain EEG waveforms. We discussed part of those results in Chapter 3 (Figures 3.5–3.8) but of particular interest to use were the results in Figure 3.6 of Chapter 3. There we compared Ian's

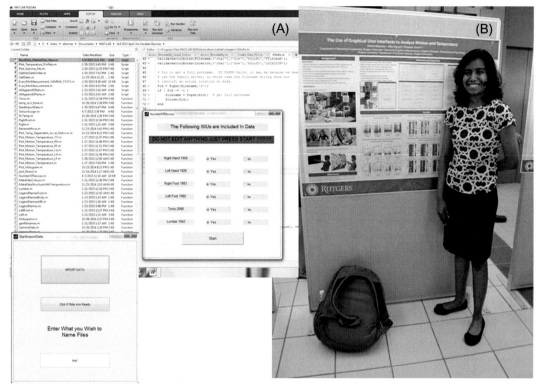

FIGURE 5.9 Ushma Majmudar and her interfaces (A) helping us process Big Data in no time. (B) Here she is presenting her poster at a local event in our Perceptual Science program at Rutgers (usually happening in May). Here we prepared for the Society for Neuroscience 2015 taking place in Chicago that year.

brain waveforms with those of a fellow from the social sciences (Chapter 3, Figure 3.6D) and a mathematician (Chapter 3 Figure 3.6E).

Neha performed those and other analyses while mastering new techniques we developed in the lab to parameterize the EEG signals and track the electrode leads across the 64 nodes of our EEG system. We had adapted old Phase Locking Value metrics of synchronization (Fig. 5.11) to visualize the ongoing scalp leads' activities as the person learning to control the direction of an external cursor on the screen was merely thinking of shifting the cursor left or right (Figs. 5.12–5.13). The novelty of our methods was we could use the micro-movements parameterization as a first step to automatically detect the electrode leads with the maximal amount of signal to noise ratio and reconfigure the grid based on that statistically optimal reference each trial. Within a learning paradigm this was very advantageous, as the brain signal was being trained to control this cursor using a closed loop paradigm whereby we were re-parameterizing the brain's self-generated output and feeding it back as input. This was not a *"black box"* approach to the Brain Computer Interface paradigm. In our case, we knew exactly the stochastic nature of the

FIGURE 5.10 (A) Ji, Vilmi and I at Qualcomm, very happy they were among the finalists' presenters competing for the Qualcomm Innovative Fellowship. (B) Vilmi and Ji after their presentation.

ongoing signal being co-adapted with the evolution of the person's accuracy to control external stimuli; because we were empirically estimating it and updating it, as the signal changed.

The idea was to explore and compare the differences between a brain that relied on afferent feedback from the body versus one that was deafferented, such as that of Ian. While Ian's online re-afferent feedback was from audio-visual input and motor imagery, the other participants relied as well on additional feedback from their bodies in motion. Even when seemingly still during the experiment, the mere visualization of a directional signal and the instruction to think about moving the cursor in that direction, would evoke a type of motor imagery across the body and with it, the activation of intentional maps, known to exist in the Posterior Parietal Cortex (PPC) [6].

We had an interesting result, namely Ian's PPC leads activation from both hemispheres remained stronger than those of the prefrontal electrode leads throughout the whole learning period. This was interesting because each trial we changed the reference lead electrode to the area of minimal noise. Yet the robustness of the fronto-parietal network remained. And within that module, the PPC stood out throughout all the trials for Ian Waterman. Typically, afferent connections project to the PPC and are thought to serve to update information necessary to maintain internal models for action [7–9]. But in the case of Ian Waterman, those connections were severed by the infection when he was 9 years old. As such, it was rather surprising that electrode leads over this area had sustained strongest activation in his case.

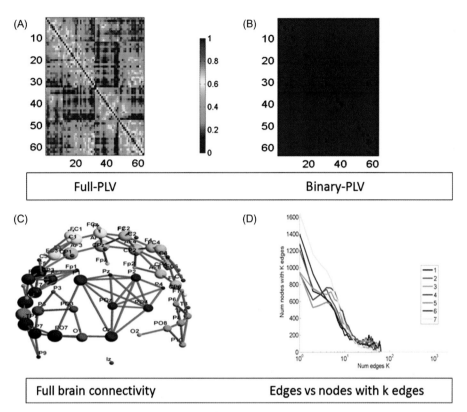

FIGURE 5.11 **Visualization tools to study EEG activity during Brain Computer Interface adaptation.** (A) Phase Locking Value (PLV) representation of a 64x64 grid showing pairwise sample activity across the brain nodes. (B) The binary matrix (upon thresholding at 0.85). (C) The BrainNet viewer representation of the nodes and links of a state of the connectivity network analyses showing various modules and connectivity patterns. (D) The degree distribution showing the evolution of the network connectivity across seven blocks comprising 10 seconds each.

We could monitor all 64 leads *"listening"* to all brain regions (Fig. 5.13A) as they evolved in time over seven blocks of training. Each block had 30 trials, each lasting 10 seconds. Each trial was well-structured into three main epochs (2 seconds of fixation, 6 seconds of the imaginary task trying to move the cursor in the required left vs right direction, and 4 seconds of resting state upon feedback from the performance.)

Fig. 5.13A shows the evolution of a representative 64x64 matrix of PLV values during the seven blocks of learning to control the cursor direction. Each participant's evolution was different, but when examining in detail the noise to signal ratios and the connectivity patterns we could find clusters of participants that transitioned to the state of brain control over the cursor direction in similar fashion. For example, Fig. 5.12 shows the two main types of participants we found with fundamentally different evolution in connectivity patterns, as illustrated by the degree distribution graph. This graph quantifies the evolution

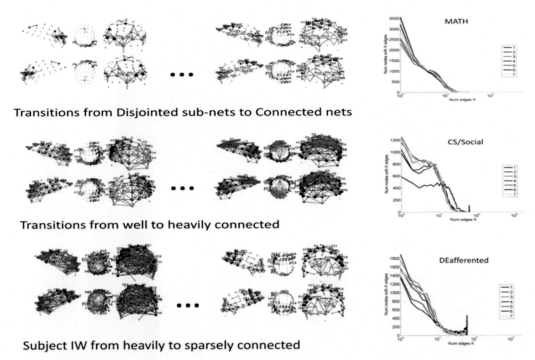

Transitions from Disjointed sub-nets to Connected nets

Transitions from well to heavily connected

Subject IW from heavily to sparsely connected

FIGURE 5.12   Examples of visualization tools capturing personalized patterns of EEG-based network connectivity for three different participants representing different patterns: a representative mathematician, a representative social scientist and Ian Waterman (IW), the deafferented "man who lost his proprioception." Patterns are shown for the first and last block of learning (spanning 70 seconds.).

of the edges of the graph (the links in the network). In the first case (the mathematicians), they started out with sparse front and back disjointed subnetworks and evolved towards a more connected brain pattern. The other example is from a representative pattern of the other cluster of participants. This person (a social science PhD student from Psychology), starts out with more distributed activity across the brain and as time and learning progresses, acquires more links, for a fully connected brain, Both cases contrast dramatically with that of Ian Waterman, who showed the opposite dynamics in connectivity evolution. His brain starts out heavily interconnected and ends up upon he learns to control the cursor on the screen by mere thought, with a far sparser pattern.

To build these graphs we adapted a publicly available software tool, the BrainNet Viewer [10] using binary PLV matrices from the original matrices in Fig. 5.13A. Fig. 5.13B shows the binary matrices of values with a threshold above 0.85. Fig. 5.13C shows the sequence of all participants during the learning. This plot is a parameterization of the signal using the empirically estimated Gamma moments (mean, variance, skewness and kurtosis) of the pre-frontal and parietal regions (the most engaged regions across the task for all participants [3]). The idea was then to automatically determine across the 30 trials, which of the pre-frontal or PPC groups of lead electrodes sustained the highest signal to

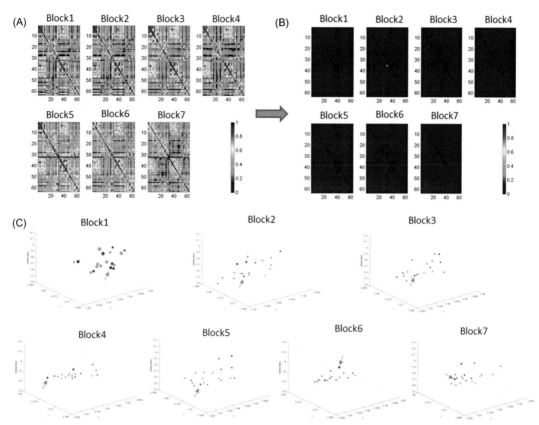

FIGURE 5.13 Tracking the stochastic signatures of learning across seven blocks of 10 seconds each. (A) PLV 64x64 matrices showing pairwise synchronous activity across the electrode leads on a standardized index spanning values from 0 to 1 (0 representing no synchronicity and 1 representing synchronous activity). (B) Binary PLV matrices after thresholding activity above 0.85. (C) Gamma moments estimated from the balance between prefrontal and PPC recruitment, with green dots representing dominance of prefrontal nodes of the network and red representing dominance of the PPC. Evolution of all 20 participants across the seven blocks of learning. Arrow points at Ian Waterman's patterns, which maintained dominance of the PPC throughout.

noise ratio. Those colored in red had a prevalence of the sustained activation in the PPC with the highest signal to noise ratio; while those colored in green had a prevalence of such sustained activation in the prefrontal regions. Ian's activity (marked by the arrow) throughout the seven blocks had a prevalence of activity in the PPC regions.

In the absence of kinesthetic reafferent feedback, this result concerning the adult brain suggests that audio-visual reafferent activity (i.e., from external stimuli) may be sufficient to close the biofeedback loops in fronto-parietal networks and provide an intentional code to volitionally direct impending motions. The adult brain has already matured the sensory motor maps, so cortically adapting them ought to be possible. Even Ian had such maps formed when he lost his proprioception at the age of 19 years old. In the absence of proprioception and continuous afferent inputs updating the cortical signal, Ian managed to have a robust code that reflected his intentions to direct motion. But how would this be for a nascent nervous system that has yet to mature such cortical maps and establish the bridge between the intention to move and the physical realization of that intention? Would such continuous updating feedback from the periphery be readily available during neurodevelopment, when the nascent nervous systems are still shaping maps reflecting the sensory consequences emerging from internal self-generated dynamics (i.e., from ongoing kinesthetic reafference)?

The work we did in this area helped us shape several of the ongoing closed-loop paradigms for sensory-motor augmentation and substitution in ASD.

Neha participated in other research in the lab while she finished her Masters' degree in ECE with a specialization in Deep Learning. She went on to NVIDIA and is now in the Deep Learning team creating technology that will help us advance the very science she was once part of in our SMIL. There she is in Fig. 5.14A, helping us harness data and sharing the fun at the Vision Science Society Conference in Saint Pete, Fla, Fig. 5.14B. The entire lab is traveling to India for her wedding in March 2018. That will be outrageously fun!

FIGURE 5.14 Neha Tadimeti (shown with an arrow) (A) working with the SMIL group as in any given regular day at the lab; (B) how many Math-CS wizards does it take to get a selfie? We did it in *take 3* after several mysteriously black photos.

Jillian Nguyen

With teary eyes and a big disappointment about a Behavioral Neuroscience Professor in our Psychology Dept., Jill knocked at our SMIL door. She came recommended by Prof. Thomas Papathomas, our undergraduate Dean for the Busch campus and a dear collaborator of the lab. Thomas is part of the Biomedical Engineering Dept. where Jill had converged to from NJIT. She had graduated with honors at NJIT and was a McNair scholar when she came to us. After six months volunteering in a lab upstairs, day in and day out, mastering the techniques of a wet lab and under the impression that this was going to be her doctorate-degree lab, she had just found out that she was dead wrong to think so. The PI of that lab told her that after all that time and effort, he had no funding to support her and in any case, *she would never amount to anything in Behavioral Neuroscience.* It is the kind of stuff that makes me mad beyond belief.

Thomas came to the rescue and Jill, he and I worked on an application for the highly competitive NSF Graduate Research Fellowship Program. With our help, she put together the package, connected all the work she had done at NJIT with the work she did at our SMIL, and provided proof of concept for the NSF proposal. In no time, nearly magically, Jill managed to set up our Eye Link Eye tracker system, build the proper setup for reaching experiments and collect pilot data for the proposal. She got it! She became a recipient of the prestigious NSF GRFP. I had to repress my *Jewban* roots and the fact that I was raised in the matriarchal tradition (Cuban heritage and Jewish descent. As many in Cuba, my family has Jewish roots and preserves many Jewish traditions; but as they were part of the wave of conversos from the Iberian Peninsula, Judaism was not part of my upbringing https://www.youtube.com/watch?v = 4-Poh3TmKtU). I cannot stand injustice, much less unjust treatment of such a brilliant female student. I wanted to go upstairs and say, *"See this? She is funded for her entire doctorate program and she does have what it takes to do that and much more. Get a clue!"* But that would have been socially inappropriate in this culture.

Jill completed several projects in the lab, in addition to what she proposed for her NSF fellowship. She authored several peer-reviewed and conference papers [1,2,11−17] and before she left to become an analyst in one of those Finance companies of Wall Street, she made sure to teach the undergraduates (including Ushmita and Sejal) everything she learned and worked on. Before leaving, she wrote all the manuals we still have in the lab and organized all her data in ways I have made sure to preserve for upcoming projects. Her legacy has already payed it forward to others in our lab (here she is the day she was being interviewed for her JOVE paper [15] Fig. 5.15). And Figs. 5.16 and 5.17 summarize the results from her work bridging perception and action through our new biometrics. For the first time, we could blindly predict the mental state of a visual illusion through the motor-output performance.

But it was during our NSF-I Corps adventure that Jill and I grew very fond of each other's love for the cause of autism. There I learned to appreciate her genius and her infinite kindness. She took the entrepreneurial lead while we conducted 120 interviews to try and discover the *"pain point"* of the autism community: their critical need. She was outstanding in the ways she communicated with people and connected with those in need along the broad spectrum. After 18 interviews, where we failed miserably to connect with the critical need for our biometrics, she stayed up all night and figured it out. I remember

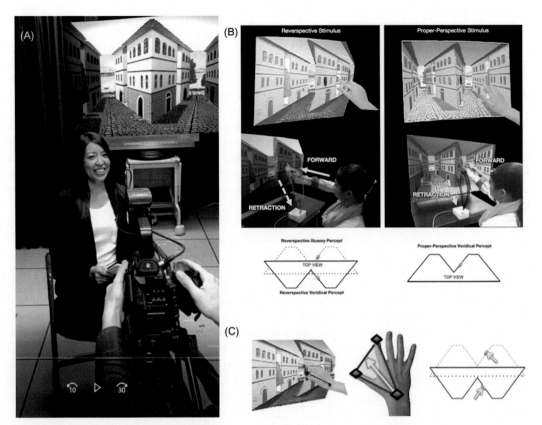

FIGURE 5.15 Jill Nguyen explaining to the camera her work on visual illusions and neuromotor control. (A) Jill with the Journal of Visual Experiments (JOVE) cameraman with the visual illusion we used in our collaboration with Thomas in the background. Our engineer Tom Grace built the bi-stable illusion which one experiences as popping out, or as caving in. (B) Sejal serving as model participant to show the hand configurations. Using these configurations, Jill managed to blindly predict the mental state of the illusion by examining the shapes of the trajectories of the hand and the fingers' aperture as they performed a reach-to-grasp action while experiencing the illusion. (C) The motor output gives the illusion away as the motion unfolds in route to the target but even more surprisingly, the accuracy of the prediction sharpens as the hand spontaneously (without any instructions) retracts to rest.

falling sleep around 3:00 a.m., out of exhaustion, after a very long day driving around in Michigan and interviewing people to no avail. Jill woke me up and said, *"I have it Liz! We need to interview the parents of the children with autism. Our biometrics will help them track their kids' neurodevelopment."*

This was not exactly the pain point in the end, but it was the entry point to the path leading us to the discovery of the market fit for our technology. By interviewing the parents of children in the spectrum, we found out that ABA was not fulfilling the daily needs of the children receiving it. Parents relayed to us that their children could not develop independence and autonomous control to perform simple activities of daily

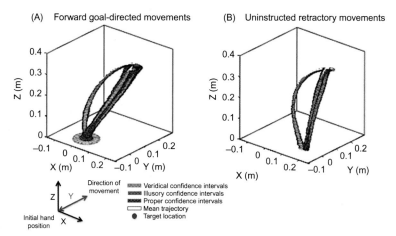

FIGURE 5.16 Automatically reading out the illusion from the motor output of the hand movement trajectories. (A) Representative hand trajectories departing from the resting position (large variance on the bottom of the path) to the target located on the illusory or proper perception of the surface in Figure 5.15. The illusory trajectories contrast with those from the veridical case. The cylinder of uncertainty computed from the empirically estimated trial by trial variance did not overlap in any section of the hand path when comparing between the illusory and proper cases, or when comparing those to the veridical case. This result was consistent in both the deliberate segment intended to the target, and (B) the spontaneous (uninstructed) retraction back to rest.

living. To achieve that, they would go to the Occupational Therapists (OT) who were specialized on sensory-motor integration. But unlike ABA, OT based on sensory-motor integration had no insurance coverage. The hourly rate and overall cost per session was prohibitive to most. And yet, it was the only thing helping the children gain self-confidence and do simple things like get dressed on their own, take a shower on their own, eat on their own and navigate on their own. Getting insurance coverage for OT that was specifically directed to alleviate sensory-motor issues, was a major pain point of that ecosystem. Our biometrics could get the parents and OTs what they needed. All we had to do was work hard to give the insurance companies the age-dependent biometrics they needed as outcome measures of neurodevelopment. These biometrics could provide objective evidence on the effectiveness (or lack thereof) of sensory-motor based OT, so they could decide on codes for continuous coverage.

This finding is the reason why I wrote this book in the first place. I owed it to the families Jill and I interviewed together.

MEASURING MOTOR LEARNING IN SPORTS

The first 20 interviews Jill and I conducted included in-person meetings with numerous athletic programs from the University of Michigan and other programs of the Big 10 group. We also interviewed the owners and trainers of some Gyms specialized on the

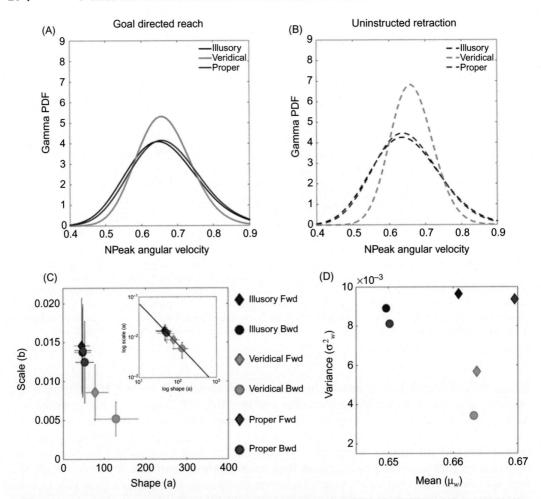

FIGURE 5.17   The variability patterns of the speed of the hand trajectories also revealed the distinct patterns in the empirically estimated PDFs (A) forward and (B) retraction and (C) in the signatures of the parameters locating each condition (pooled across participants) on the Gamma parameter plane (D) estimated Gamma mean and variance provide separation across conditions for the deliberate (forward) and spontaneous (backwards) segments of the reach to grasp action.

training of top athletes that at some point suffered an injury and required special retraining of specific muscle groups.

During those interviews, we discovered that our biometrics, while desirable to some, were not attractive to others. The notion of being able to predict the consequences of a training session on the athletes' performance (including the possibility of injury) or even unveiling information on sleep patterns, was not welcomed by several of the coaches and individual athletes we interviewed in person and over the phone. Phone interviews were also included in our data collection because sometimes it was too late in the day to set up an in-person interview and we had to for example call the West Coast and interview folks there. The goal we had of achieving a minimum of 100 interviews to discover the market fit for our technology within a few weeks was very difficult to attain, but somehow doable if we expanded the data acquisition to other states and countries. To discover the market fit for our biometrics, we were working against the clock.

Many of the coaches and athletes that we interviewed did not feel comfortable with technology dictating scores of performances to advance decisions on their place within a team; not even as mere complementary information on what the coach was already detecting by the naked eye. Besides potential legal implication (e.g., lawsuits because of infringement on privacy matters by surveilling the athletes 24-7), the prospect of for example losing the top player in a team the day of the competition, *"just because some smart gadget predicted imminent injury"* was not going to cut it. *"This technology could never replace us"*—said the coaches and trainers we interviewed. Clearly, the role of the technology would be merely to monitor performance and indicate the rate of change of adaptive learning (e.g., the degree of plasticity of the nervous systems of the athlete when training within new contexts); but it was consistently taken as a form of intrusion by all those people we interviewed across the US college athletics system.

I decided nonetheless to pursue the use of the biometrics in sports research, as a springboard to generate ideas for clinical research. Given the complexity of the sports routines, the data generated from the athletes in motion, along with their narratives on how they learned the routines and what they felt throughout that adaptive process, proved to be a very rich testbed for ideas that I could later use in the development of clinical motor biomarkers.

Uri Yarmush

Genius is not confined to academic performance or intellectual capacity to solve written problems or to formalize logical reasoning. The bodily performance is something to take very seriously as it entails the solution of many difficult computational problems. One of them, which we saw in Chapter 4, is the optimal recruitment-release and coordinated control of the many DoF of the human body to produce timely responses to the sensory flow of the environment, as well as to control the internally generated sensory flow. As I was planning on extending the studies of the motions of the reaching family to the full body, a new undergraduate student knocked at the door of our SMIL. He was Uri Yarmush, a martial arts expert who set up the Jab-Cross-Hook-Uppercut (J-C-H-U) (and other combination) boxing routines to study in the lab. Fig. 5.18 shows the digitization of the (J-C-H-U) using the Motion Monitor software (from SportsInn, Chicago).

Uri was an undergraduate student in Psychology who rotated through the SMIL and taught us a great deal about sports from the vantage point of a highly skilled athlete. His family owned a gym and taught martial arts. Among their pupils were children with neurodevelopmental disabilities, such as those in the spectrum of autism. Uri had been practicing since he was 4 years of age. By the time he came to work with us, he held a black belt second degree level in karate and was a competitive athlete in kick boxing. In his free

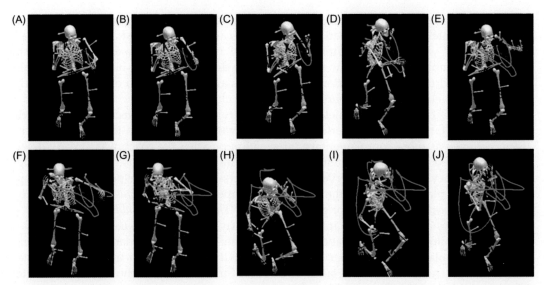

FIGURE 5.18   Three-dimensional digital rendering frames from the expert's performance of one trial of J-C-H-U beginner's white belt technique using the real-time sensor outputs. (A−J) Arrows mark the locations of 15 electromagnetic sensors recording at 240 Hz. The first isolated technique is called a Jab. The Jab starts with the front hand extending toward the imaginary opponent's nose (J1, forward Jab segment), keeping the hand in a tight fist, making sure that the elbow does not hyperextend; the hand should be retracted while it is still slightly bent. At the same time, that the Jab is being retracted (J2, Jab backwards), the Cross is being extended forward (C1, Cross forward). Again, the imaginary target is used and the Cross is directed toward the nose. Simultaneously, the body is twisting, beginning with the back foot, then the torso and ending with the back of the hand extending forward. Because the body is already twisted, this motion naturally sets up the staged portion of the Hook (H1) aimed at the opponent. As the Cross (right hand) reverts back to its original position (C2, Cross backwards), the left forearm is made into a C-shape with the hand in a fist and the palm facing down, and the body untwists itself, using the momentum of the body rather than the force of the hand to achieve the intended goal (to reach the opponent's face) in the H1 Hook forward segment. As the body untwists itself in a supplemental motion (H2) Hook backwards segment, the knees bend slightly in preparation for the intended Uppercut (U1, Uppercut forward segment which goes upwards to the chin of the opponent). After the knees are bent and the left hand is returning to its original positioning to protect the face, the right-hand fist shoots up in a motion that resembles throwing a bowling ball in U1; but the hand is kept tighter aligned to the body and the palm is facing the body. The supplemental portion (U2, Uppercut backwards) brings the hand back, and the body adopts the defense position again. It is important to note that all routines where done in the presence of an expert instructor in order to minimize risk of injury [27,28].

time, he taught children and because my lab was interested in autism, he dropped by and offered his help.

We were extremely impressed by his exquisite sense of time and coordination. If I asked Uri to cut the time of a routine by half and measured it with millisecond time precision, his performance indeed complied with the request with exactness. His motions readily adapted to any of the manipulations I used (darkness, mirror, avatars, sound and others) and within a matter of seconds, his movement trajectories would be as efficient as they were in full trained mode. We had some fun with Uri's teachings in the lab. He coached novices to martial arts and Rutgers athletes who were part of some other sports.

One of the questions we were both curious to answer was whether being an expert on other sports would help or interfere with the learning of the boxing routines. From the few people we tested in the sports category, we felt they varied widely. They were all really good, but those who played in a group (e.g., lacrosse, volleyball and basketball) asked very different questions during the instructions than those who played in dyads (e.g., tennis) or alone (e.g., swimmers and runners). It seemed to me that the most important thing all athletes were trying to learn was how to situate the opponent relative to their own body and reference everything egocentrically. Interestingly, we tested a couple of performing artists too. They took longer to learn these routines and were extremely well-aware of their bodies. They were also very aware of their external surroundings. It was as though they were "acting" for an audience and the allocentric vantage point of that audience did matter to them. Further, the form of the routines was rapidly acquired but the effectiveness of the punch took some time to master.

We discovered a few interesting things during the experimental sessions regarding how people may learn these routines in general. Uri was a very good instructor and explained the logic behind each configuration of the body postures, so it was possible to extract the precise goals of each segment. He would explain as well why one should move in a certain way to attain effectiveness in the punches, etc. And when we got our first adolescent with autism, he certainly excelled. He had a very good sense on how to engage the participant and keep his interest in the structuring of the routines. Years after my experience with Uri and the participants in the spectrum, I traveled to Buenos Aires and met a couple of physical education instructors who had created a program specially tailored to teach children in the spectrum of autism how to play in groups [18]. They did so using sports as well. It has been the best program I have ever seen and it reminded me of how Uri taught these fellows with autism, here in the lab.

I would say that by far, the biggest discovery in this project was the empirical finding that the fluctuations in the hand's speed amplitude of individuals with poor proprioceptive feedback was well characterized by the exponential distribution. Among the participants showing this signature, was our first individual with autism [19], a result that we then reproduced in many others across ages, sex, levels of severity and spoken language abilities [12,20,21]. This feature, present in neurotypical infants would transition into skewed and towards symmetric distributions as children passed the 4 years of age; but it would remain exponential in autism [19,20].

I decided to further explore the issue of speed variations and find out how it would help me track learning in a personalized manner and across sections of a given cohort of individuals.

DESIGNING A MOTOR CONTROL EXPERIMENT USING SPORTS

One of the aspects of speed sensing as the movements unfold is the conservation of the path in the motion trajectory of the hands in route to a goal. We had seen this conservation in many examples in the family of reaching, pointing and reach-to-grasp actions that we modeled and empirically studied in Chapter 4. But how would this phenomenon manifest when the hand reaching action was embedded in more complex motions engaging the full body? Usually we had studied very simplified versions of hand motions pointing to a visual target, but how would this translate into full-body actions?

Unlike other studies in the field, our research on pointing behavior also included the movements that spontaneously retracted the hand to rest. We saw in Chapter 4 that such motions were highly informative in Jill's experiments involving patients with schizophrenia as it signaled the lack of coordination across their arm joints and the inability of their motor systems to differentiate intended versus spontaneous behaviors. Further, Jill's work on the visual illusion experiments revealed, in neurotypical controls, how the variability inherently present in the spontaneous retraction movements, clearly broadcasted the state of the illusion, as it was being experienced in the subject's mind. Here, in the context of these complex boxing motions, we wanted to learn if manipulations in speed still had the feature of path conservation for intended movement segments versus large (speed-dependent) departures from the path geometry in spontaneous segments unfolding beneath the person's awareness. The idea was to try and understand how the brain and bodily joints coordinated the co-occurrence of complex synergistic movements of the limbs so that the temporal dynamics of full body motions remained compliant with the high-level goals of a task. How would spontaneous segments of the motions funnel out the speed variability in cases when the speed was instructed in blocks (expected) versus cases when the targeted speed was called at random?

To investigate this question, it was important to chart out the speed profiles of the full routine for both the fast speed case and the slow speed case. Fig. 5.19 shows the left and right hands' trajectories and speed profiles for the slow case and the fast cases. The alternating patterns of the J-C-H-U are broken down into the deliberate segments (J1-C1-H1-U1), intended to the opponent, and the spontaneous segments (J2-C2-H2-U2) automatically retracting the hand towards the body (beneath the person's awareness). Indeed, when I showed Uri his own performance and explained the meaning of each curve, he was surprised that I was capturing the retractions because he did not realize, or ever thought about them, while learning or perfecting this routine. At his family-own Gym, he would instruct his pupils about the deliberate segments aimed at the opponent. Further, his martial arts instructors always taught him about those as well. These spontaneous retractions were a *wild card*. But now that he learned about them, and could capture them with high grade instrumentation, he could (perhaps?) use these segments somehow *to surprise his opponent*.

One would think that when repeating these routines over and over, the changes in speed—as called by a computer program at random, would have an impact on the hand paths of the movements' trajectories. However, as shown in Fig. 5.20, the intended motions (J1-C1-H1-U1) manifested similar paths across different randomly called speeds. In this sense, as with the goal-directed hand motions pointing to a visual target (e.g., as in the tasks we studied in Chapter 4), this more complex behaviors requiring full-body joint coordination, did preserve the intended geometric path of the hand. But, what about the spontaneous retractions?

Here was where things got a bit different. The retractions changed the path according to the speed. When I pulled all the trajectories corresponding to the fast speed, plotted the paths superimposed across trials, and compared them to those from the slow speed trials, the geometry of these paths was completely different. The right panels of Fig. 5.20C,D

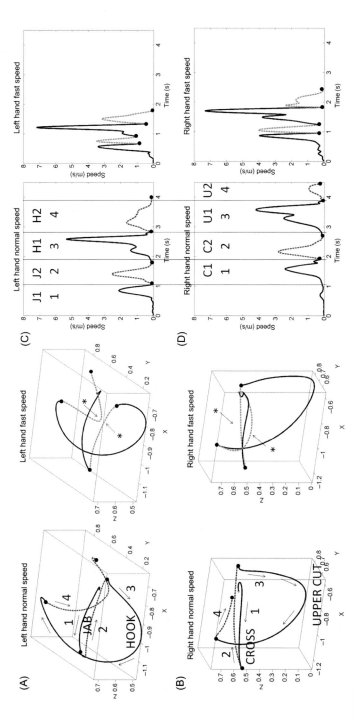

FIGURE 5.19 **Movement trajectory decomposition and speed profiles.** (A) Left-hand trajectories in a continuous, sequenced routine with each dot marking the beginning and the end of a segment. Solid segments (1 Jab1 and 3 Hook1) mark the goal-directed segments (forward strikes) intended towards a goal located on an imaginary opponent. Arrows mark the directional flow of the movement. Dashed traces mark the supplemental transitional segments at normal or fast speed. Arrows and asterisks mark the differences evoked by the speed in the trajectories of the supplemental segments of each technique. (B) Same as in (A) for the Cross and Uppercut techniques. (C) Instantaneous left hand speed profiles from the fluid sequence labeled in correspondence with the hand trajectories (dashed curves are the supplemental segments). (D) The speed profiles from the right hand.

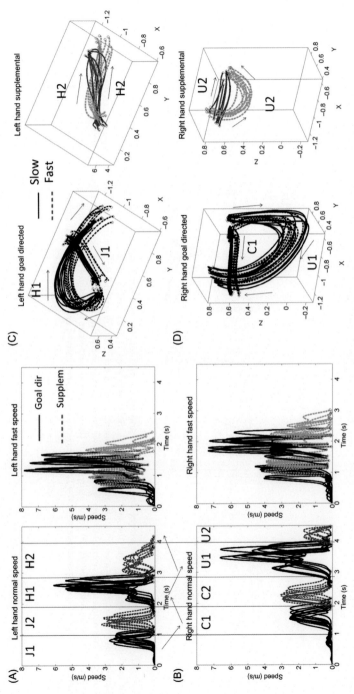

FIGURE 5.20 **Different effects of speed changes on the supplemental and staged movement trajectories:** Expert's performance of a fluid sequential full set of techniques. (A, B) The instantaneous speed profiles from all 10 trials performed at the slow instructed speed from the left-hand alternating between staged and supplemental segments of each technique. The technique segment is indicated for each hand speed. Arrows show the alternating order and mark the simultaneous performance of an intended (staged) and a supplemental segment. (C) All trajectories from the staged segments of the left hand at the fast and the slow speeds grouped, thus showing similarity in space. Contrast the left-hand trajectories from the supplemental segment of the Hook back, which changed their spatial geometric properties (e.g., curvature and torsion) as a function of speed. Faster movements were far more curved on the way back towards the body and changed the curve space orientation. (D) Staged trajectories for the deliberate segments in the Cross and Hook techniques performed by the right hand also maintained the spatial properties despite the randomly instructed speed. The returning supplemental trajectories from the Uppercut changed in space with the speed. Notice that fast U2 occupies a different region of space (different curvatures, orientations and lengths) than slow U2 despite the fact that the preceding U1 segments were statistically invariant to speed changes. Arrows indicate the flow of the motion.

show the different geometric features of these paths, changing as a function of the speed for the H2 and U2 segments (shown for simplicity, thought the J2 and C2 segments also changed). Interestingly, this was the case regardless of the way in which the speed was sensed back through kinesthetic reafference from the movements. That is, whether the participant experienced the fast and slow speed in random order (Fig. 5.21A,B), that is, without the certainty of knowing prior trials to accurately predict impending trials; or under a

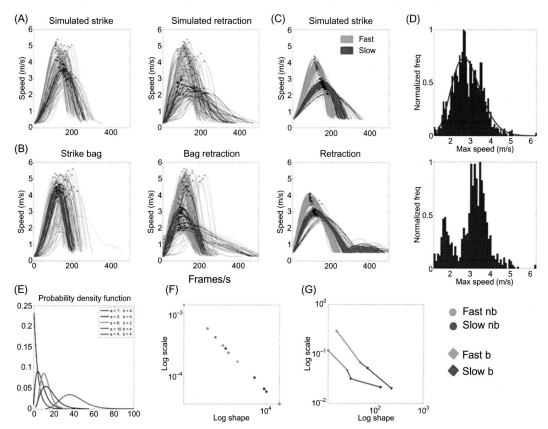

FIGURE 5.21 **Analytical methods.** (A) Representative hand's instantaneous speed profiles during the Jab-strike (left) and retracting-Jab (right) in the block where the participants mentally simulated an opponent under randomly instructed speeds. Sampling resolution of 240 Hz, movements lasting between 0.8 and 2.1 seconds). (B) Same as (A) but striking against a real opponent in the physical form of a punching bag, and retracting from it. (C) Same as (A), but with instructed speeds within a block design. (D) Empirical frequency distributions of the ensemble data from (A) (randomly instructed speeds on top) and speeds from the block design (bottom). (E) The continuous Gamma family of probability density function curves across a subset of values for the shape (a) and scale (b) parameters in the legend. (F) The plots of some subjects for fast and slow speeds (simulated and punching-bag intermixed) using the normalized maximum velocity and estimating the stochastic signatures of each condition. The log-log plot of the shape and scale plane aligns the points along the line of unity. (G) The stochastic signatures dynamically measured in real time: stochastic trajectories of intended movements for two subjects across different training contexts with 110 trials each (fast-bag, fast-no-bag, slow-bag, slow-no-bag) measuring predictability towards the right extreme (Gaussian range of the Gamma plane) and randomness towards the left extreme (Exponential range of the Gamma plane).

blocked condition (Fig. 5.21C), whereby only slow speeds or only fast speeds would be experienced in one block. In both cases (random and blocked), the person's nervous system maintained the deliberate route intended to the target; but manifested geometric changes in the retractions that were automatically occurring beneath awareness. The fluctuations in peak speed amplitude provided us with a different frequency histogram for the random case (Fig. 5.21D top) when the speed was called at random versus when it was called within a pre-defined block (Fig. 5.21D bottom). Further, we contrasted cases where the participant was aiming at an imaginary opponent versus cases where the opponent was an actual punching bag.

This distinction between virtual and physical opponents was important because in one case the person is simulating the punch and receives no physical feedback from the physical opponent (the bag), whereas in the other case, the punching bag does provide haptic feedback. Haptic feedback serves to modify the outgoing motor command to achieve the desired goal. The speed distinction in the randomly called case was less obvious for the intended motions (Fig. 5.21B left) than for the retractions (Fig. 5.21B right). In contrast, (and as expected) the speed differentiation was more obvious in the blocked condition (Fig. 5.21C), yet once again, the retracting motions were different as they had a long tail for the simulated-strike condition (Fig. 5.21C bottom panel) that was absent from the intended segment under the same cutoff motion criterion to group the trials.

When we examined the frequency histograms and fit different Gamma probability distributions across different shape and scale parameters (Fig. 5.21E) to the peak speed data, we discovered that different families of probability distributions could be used to characterize these various conditions. In this sense, we found a way to plot the evolution of the stochastic signatures in speed variability of the athlete's performance across different contexts. This map shown in Fig. 5.21F for the micro-movements parameterization and in the Fig. 5.21G for the actual speed peaks, provided a tool to assess the empirical meaning of various regimes of the Gamma parameters. For the first time, we could start charting out the motor performance of a person in a rather dynamic way, as it was impacted by different contexts and different, varying stimuli.

Clearly, we could now extend these methods to assess the performance of the novices that Uri was teaching in the lab (Fig. 5.22) and compare them with the signatures of other experts. When we did so, we realized that (1) typical novices and experts had skewed distributions of speed peaks (e.g., inset in Fig. 5.22A) and (2) experts stood apart from novices in that their distributions were more symmetric (towards the Gaussian range of the Gamma parameter plane, increasing in value to the right along the horizontal axis) and the scale parameter was at the lowest end, decreasing in value along the vertical axis of the plane. This can be appreciated on Fig. 5.22B (using the micro-movements parameterization of the speed peak amplitudes), where the two experts lined up at the right lower quadrant of the Gamma parameter plane (the RLQ). To obtain these quadrants, we can use the median of the shape and scale taken across all members of the cohort for the empirically estimated values of the two Gamma parameters. Since experts were the best-case scenario, this empirical result provided comparative criteria to know when the novice person was performing at the level of the expert. Further, since these values changed dynamically on the Gamma parameter plane, it was possible to use this dynamic map to monitor the performance of the members of a given cohort. Given the values of the best

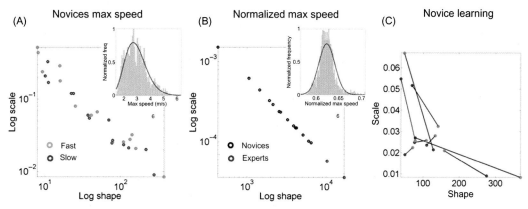

FIGURE 5.22   The velocity-dependent parameters reveal learning according to the subject's somatosensory stochastic signatures in each training context. (A) The empirical frequency distribution of speed maxima extracted from velocity trajectories across subjects (inset) and the MLE of shape and scale Gamma parameters for each subject, for the fast and slow instructed speed condition. (B) The micro-movements parameterization using the deviations of the peaks from the mean and normalizing it from 0 to 1 (to be invariant to possible allometric effects due to individual differences in limb sizes). This parameterization aligns participants across the line of unity, locating experts at the far-right symmetric range of the Gamma. Empirically, this location at the right lower quadrant of the Gamma parameter plane turned out to hold higher predictive power than locations on the left upper quadrant. Inset shows the empirical frequency distribution across subjects for the micro-movements' parameterization of the speed peaks. (C) The individual learning progression for novices as they performed slow and fast versions of the jab. Notice that the stochastic signatures of their speed maxima shifts towards the right for the fast condition (more predictive) in some cases, whereas in other cases it is instead the slow condition which has this effect. Notice also that given the same number of repetitions, the rate of change of the stochastic signatures is very small for some subjects and very large for others. This plot captures the individual's learning progression and unveils which training context is most adequate to make the subject's motions more predictable (i.e., to shift down and to the right, towards lower noise and higher symmetric shapes).

performers, we could estimate not only how close or how far a given person was from the *ideal expert zone* of the RLQ; but also, we could learn about the rate at which the person's signatures shifted values along the stochastic trajectory of its own Gamma PDF values. This was a new way of profiling stochastic shifts relative to the person's own baseline state, rather than doing so relative to a population assumed grand average. Such ideal mean was a theoretical assumption the field of behavioral neuroscience was making a priori, without any empirical justification. Our interest here was on a personalized profiling of the physical motions of the athlete as captured objectively by our analytics; not on how the athlete's nervous system would ideally behave. The latter was a subjective supposition without empirical grounds to support it.

The variations in individual learning rates underscoring the need for a personalized approach can be appreciated in Fig. 5.22C for a group of novice individuals that learned the J-C-H-U routine. This figure plots the starting and ending points of the empirically estimated Gamma trajectory of their dominant hand as they performed the JAB segment in intended mode (forward towards the opponent) at various speeds. The signatures of learning this segment of the routine are distinct for the slow and fast versions and so is the individual performance of each person.

One of the advantages of these methods is that they facilitate comparison of these personalized signatures across the population. While we can uncover self-emerging clusters across the population, we can also use the profiles of an expert athlete as an anchor to track the novice's learning trajectories and ranges of values of motion parameters. Indeed, Fig. 5.23 provides an example of fundamental statistical differences between an expert and

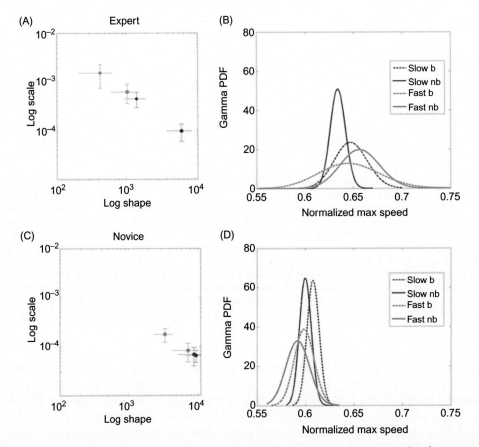

FIGURE 5.23 **Statistics of the normalized maximum speed labeling subjects on the Gamma plots for representative novice and expert.** (A) Expert case, MLE for each speed condition and training context (bag vs no-bag) with 95% confidence intervals. (B) The corresponding Gamma probability density function (PDF) curves reveal in the expert a broad bandwidth of parameter values across training contexts. It also shows an unambiguous distinction between bag and no-bag conditions for each speed level. Speed levels are not confused by the expert's kinesthetic data. (C–D) The novice however shows a narrow bandwidth of parameter values with no clear distinction between slow motions that are against the bag or towards a simulated opponent. The novice's kinesthetic data does distinguish between the fast-bag condition and the other training contexts. Notice the degree of dispersion of the probability distributions measured through the Fano Factor (noise to signal ratio), the ratio of variance to mean taken within the time window (the time in ms to reach the peak velocity, on the order of 200 ms in this case) is indistinguishable in the novice for the slow case (6.47×10^{-5} slow-bag vs 6.30×10^{-5} slow-no-bag) and for the fast case (1.76×10^{-4} fast-bag vs 2.47×10^{-4} fast-no-bag). The novice can however differentiate between fast and slow (Wilcoxon ranksum test of equal medians $P < 10^{-3}$). Compare to the expert with Fano factors that distinguished speed within each training context (slow-bag 4.4×10^{-4} vs fast-bag 0.0015; slow-no-bag 9.7×10^{-5} vs fast-no-bag 4.08×10^{-4}).

a novice across the various conditions discussed above, namely varying the speed and examining the punches against a virtual (imaginary opponent) and a physical one in the form of a punching bag.

The plots in Fig. 5.23A show the empirically estimated Gamma parameters for the expert across all these conditions on the left-hand side. On the right-hand side, the corresponding PDFs are shown. Notice the spread of the expert on the probability distribution function space and the broader range of data values his motions span. In contrast, the novice has a very narrow bandwidth of range in data values. Further, the PDFs are all constrained to a narrow location on the Gamma plane with very small range in probability space. Using these maps, we could assess these fundamental differences between expert and novice performances. Further, we could track that outcome as a function of various training contexts and situate the person's skill relative to the expert's personalized regime. In essence, we had empirically discovered how to set expert criteria to optimally guide outcome measures of human performance towards an ideal case scenario.

These stochastic methods are usually employed to predict how much the past events may contribute to the accurate prediction of future events, for example, as when incoming phone calls on a switch board are tracked in the order in which they arrive under the different conditions. For example, the calls could be related to each other because different groups of callers may be inter-related, and as such the probabilities of such events be conditioned upon receiving certain calls in certain order. They could also arrive independently from unrelated people with not apparent relation to past calls predictive of future calls, etc. Here, the requirements imposed by the very nature of the nervous systems' functioning were different. Events in the future may or may not be independent of events in the past. This is so, because the nervous systems are self-supervising and self-correcting their own self-generated motions in very non-linear ways with very complex dynamics. As such, the probability distributions governing the random processes that emerge from interactions across the body and brain, while they cooperate with each other in a closed loop, had to be empirically determined. There was no data available to us in textbooks or peer-reviewed publications to acquire this kind of information. This was uncharted territory, particularly in the face of such non-linear complex dynamics emerging from excess DoF simultaneously changing and forming synergies as the movements unfolded. I went back to my original question in Chapter 4, how would the parameter manifolds behave in the face of changing dynamics? To that end, I studied the DoF decomposition we talked about in Chapter 4, that is, guided by the Torres-Zipser PDE that I described there.

BERNSTEIN'S DOF PROBLEM IN THE BOXING ROUTINES

The complexity of the arm's motions that we saw in Chapter 4, when the movements were performed in isolation to reach a visual target (i.e., without having to simultaneously coordinate this movement with the rest of the body in motion), posed the question of how such arm motions would be performed, when embedded in highly complex routines. In the context of the J-C-H-U movements, we retake this question and study the DoF decomposition, while considering a subset of goals the martial arts expert identified to effectively Jab the imaginary opponent (i.e., without haptic feedback from contact forces hitting a physical opponent.) Fig. 5.24 unfolds the motions. For simplicity, we zoom in the upper body joints. We use the helical axes representing the simultaneous rotations and translations of each joint relative to immediate joints [22]. Such interactions are very important in the types of complex motions demanded by these boxing routines. They depend on the number of instantaneous axes of rotations a joint may afford and in turn, we refer to these as the DoF.

Joints rotate and bones translate. For example, the ball-and-socket structure of the shoulder joint generates rotations and translations that we can study using a helical axis

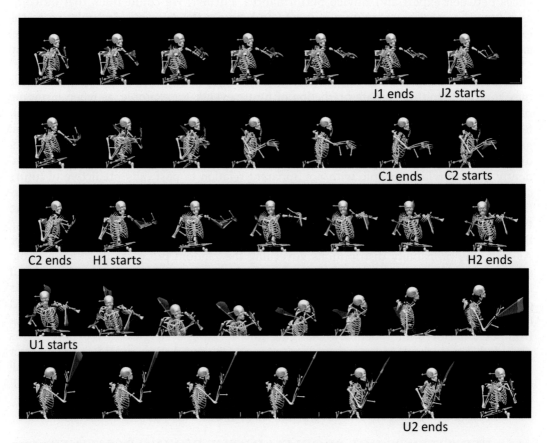

FIGURE 5.24 **Full routine breakdown according to upper limbs' motions.** Rendering of a subject's upper body and extremities with x-y-z axes measuring changes in position and orientation of the limbs, head and trunk, relative to a global frame of reference. The Jab forward (J1) ends as the retraction (J2) starts simultaneously with the Cross forward (C1). This is followed by the retraction of the Cross (C2), which rotates the body and simultaneously initiates the Hook forward (H1). The helical axes in light yellow (spanning a fan of vectors) show relative rotations between two coupled body parts. The length of that vector is proportional to the net coupled rotation. They are evident during the Upper Cut.

and the relative translation of the bone along the instantaneous axis of rotation while attached to the rotating joint can be quantified using these representations. In Fig. 5.24 we use the Motion Monitor's graphs to highlight with a yellow vector the helical axes of selected joints and provide a sense for their excursions. This form of visualization, as the movement unfolds, helps us appreciate the amount of relative joint to joint interactions that takes place as the J-C-H-U routines unfold. Note the changing size of the vectors and the changes in the spanning area of their excursions. They help us understand major synergistic interactions across the upper body representation in this figure.

As in Chapter 4, to simultaneously study parameters in the end effectors' space (e.g., the hand) and the internal joint space, we start out with a toy model of the problem we are interested in. The problem at hand is once again the DoF decomposition during the generation of geodesics curves that move the arm from an initial configuration to a set of goals along a geodesic path. In the case of deliberate motions intended towards a goal (or set of goals), we learned across a variety of reaching tasks that the empirically measured hand trajectories remained robust to changes in speed. Further we learned that the geometric path of this trajectories could be well characterized as geodesics along a task-relevant manifold, and that they could be set in correspondence with joint angles rotating and translating the arm, also along geodesic paths in joint angle space. We in fact learned to build a locally linear one-to-one correspondence between the points along these two geodesic paths and could examine the joint angle space decomposition into the task-relevant DoF and the task-incidental DoF. Of course, all of that was for simpler pointing motions, but here the problem is far more complex because the arm motions are embedded in highly complex movements that co-exist with those of other body parts. It is possible that the hand trajectories remain robust to speed, but the trajectories of the joint angles would not be independent of the speed changes; or would be only partially independent for some portion of the path. Further, we wondered now about the spontaneous movements' segments that the athlete was unaware of and that we certainly did not instruct, or prompt in any way.

To simulate different scenarios of potentially evoking change in the geometric properties of the path with different speed dynamics, we studied the differential of the Gauss Map (as in Chapter 4) and plotted the projection of the geodesic direction (gradient vector) on the unitary sphere of the various surfaces generated by a two-dimensional joint angle arm model. The endpoint (sometimes referred to as the end effector) of this arm model was then set to move on a cylindric surface. The cylinder is a Euclidean plane folded (as in the Fig. 5.25) and the main geodesics run horizontally (Fig. 5.25C), vertically (not shown) and diagonally (Fig. 5.25D) with their corresponding joint angle cost surfaces in Figs. 5.25A and 5.25B respectively. On these cost surfaces we plot the geodesic path unfolding as the joints rotate and translate also along a geodesic path. The endpoint (modeling the hand or end effector of the two-jointed linked arm in Fig. 4.2 of Chapter 4) approaches the target (i.e., the hand reaches the target when the cost is at 0 value).

We could gain a sense using these simulations of the regions on the unit sphere that the Gauss Map would span when we examined the Second Fundamental Form (as in Chapter 4) and probed the Hessian operator to confirm the locally geodesic (length minimizing) property of the gradient vector spanned by our PDE (i.e., the principal eigenvector of the Hessian corresponding to its leading largest eigenvalue). The projections of the

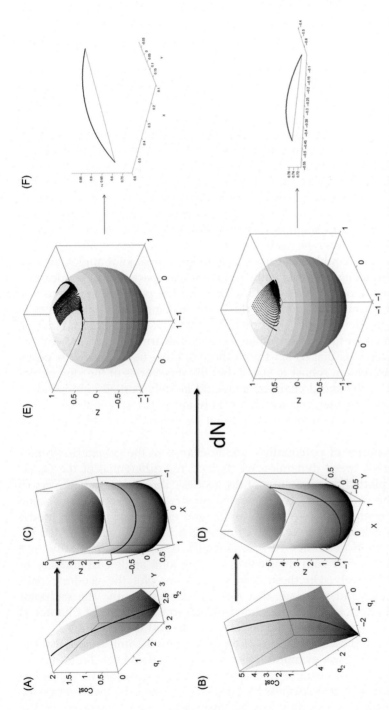

FIGURE 5.25 **Toy model of the two link arm motions along geodesics of the joint angle and the end-effector spaces.** (A–B) Along the cost surface spanned by the joint angles space and minimizing the distance to the target, mapped onto the cylinder's horizontal geodesic (C) or diagonal geodesic (D) and then via the differential of the Gauss map projected to the unitary sphere in (E). The three-dimensional paths in (F) are compared to the Euclidean geodesics for each case.

fields from the cost surface and the geodesic path on the unitary sphere are shown on Fig. 5.25E, while their corresponding paths in the Euclidean three-dimensional space are compared to the Euclidean geodesics in Fig. 5.25F.

Using this Torres-Zipser PDE approach to decompose the bodily DoF into task-relevant and task-incidental submanifolds during the Jab, we proceeded to scale up the model and find out the empirical decomposition of the Jab motions, while embedded in the motions of other 15 joints. This was a far larger space than the arm's joint angles space; whereby the general geometric solution guiding the numeric quest, would inform us on the decomposition of the DoF manifold and the speed-independence of deliberate motions in the empirically generated data set.

We measured the angular speed profiles across the 15 joints (including the hands) and defined several goals the Jab required. These were defined by the martial arts expert according to how pupils are trained to produce an effective Jab punch. Fig. 5.26 shows the angular speed profiles of the 15 joints we tracked over the span of 5 seconds per trial. Fig. 5.27 shows our empirical exploration of several goals defined by the instantaneous displacements and rotations of the hands to achieve an effective punch. These included decreasing the cumulative distance along the joint excursions to converge to 0-rotation

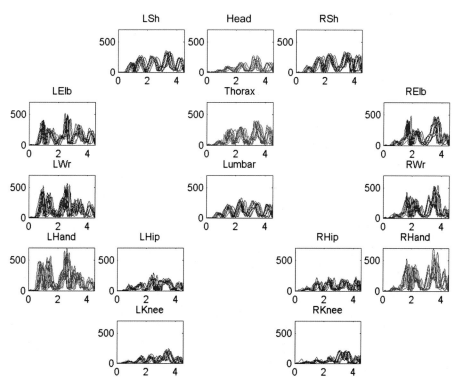

FIGURE 5.26 Angular speed temporal profiles across 15 joints of the body spanning 5 seconds of the J-C-H-U routine.

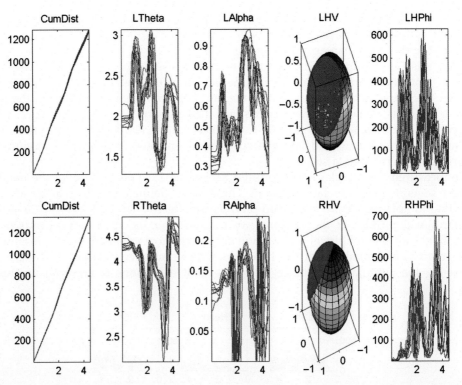

FIGURE 5.27 **Goal identification by the expert to model the full forward and back Jab routine.** Representative elements to build the analytics of the goal representation. These include (for both the left and the right hands) the distance to target; the orientation of the plane of the arm; the orientation of the palm of the hand (parameterized by the Euler-Rodrigues angle-vector and projected on the unitary sphere) and the orientation of the hand relative to the shoulder.

and displacement at the target (shown in the first panel of Fig. 5.27.) It also included in the second panel of Fig. 5.27 how to orient the plane of the arm in an appropriate manner. The plane of the arm was defined in Fig. 4.8D of Chapter 4. The orientation of the plane of the arm (Theta angles of the left and right arms) is critical to direct the punch straight to the nose area of the opponent without over extending the elbow and hurting that joint. Goals also included how to modulate the angle associated to the instantaneous rotational vector of the hands using the Euler-Rodrigues parameterization described in Chapter 4 (denoted Alpha and represented on the third panel of Fig. 5.27). The hand orientation path while delivering the Jab's punch is critical. The expert explained that a common injury of beginners was hurting the knuckles of their closed fist landing on the punching bag. If the hand came from a path leading to the wrong fist orientation at contact, the impact could fracture the small bones on the surface of the hand.

Once we parameterize the hand's orientation by the Euler-Rodrigues angle-vector, we can also track the projection of the unit gradient on the unitary sphere across the manifold of possible hand configurations spanning these values in Theta and Alpha. This is represented for the left hand (top panel) and the right hand (bottom panel) on the fourth column of Fig. 5.27. Further, another goal included the actual orientation of the palm of the hands (denoted Phi and represented on the last column of Fig. 5.27). This orientation, computed relative to the shoulder orientation can be attained by chain-multiplying the orientation matrices of the shoulder, elbow, write and hand sensors, and then attaining the angle as a rotation relative to the shoulder. All these goals plotted in Fig. 5.27 as derived from the empirical data while using elements of screw theory [22] and SO3 groups [23].

The schematics of the implementation of our PDE orbits as extracted from empirical data in numeric form appears in Fig. 5.28. This version of our model used the 15 joints in the joint angles space, the multiple goals in the goal space defined in Fig. 5.27, the maps relating these configurations and the distance notions we need to track in order to optimally converge to the desired outcome. Using these schematics (which we explained in Chapter 4 within the context of forward-and-back pointing motions for the case of PD patients), enabled us to track the DoF decomposition for different speed conditions. Fig. 5.29 shows the results of tracking the evolution of the path along the two submanifolds, the task-relevant (Rank projection) and the task-incidental (Null projection) for the deliberate and supplemental segments of the Jab motion. In this figure, we name staged Jab the forward punch with the dominant hand, which precedes the supplemental retracting motion of the Jab (spontaneously occurring beneath the athlete's awareness.)

The main message of this decomposition is that the initiation of the path of the DoF in the task-relevant submanifold is the same along the first 250 ms, despite the changes in the speed of the motion. Clearly, the slow-motion path lasts longer and must incur in corrective feedback after the first 250 ms of the first half of the motion. The fast motion lasts 250 ms total and during that time, the task-relevant DoF corresponding to the deliberate punch travel the exact same shape in both speed regimes. In contrast, the DoF decomposition of the task-relevant submanifold during the supplemental Jab retraction movement has complementary shapes for slow and fast speeds.

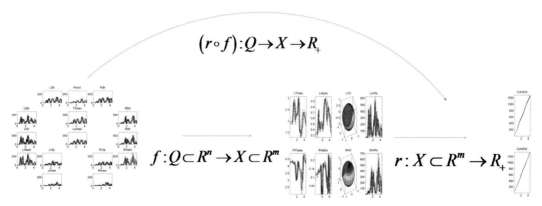

FIGURE 5.28 The map from the Torres-Zipser PDE explained in Chapter 4 expressing the numerical formulation of the problem of partitioning the DoF into the task-relevant (rank) and task-incidental (null) projections from the local linearization of the isometric embedding of the task manifold (of goals) in the joint angles manifold.

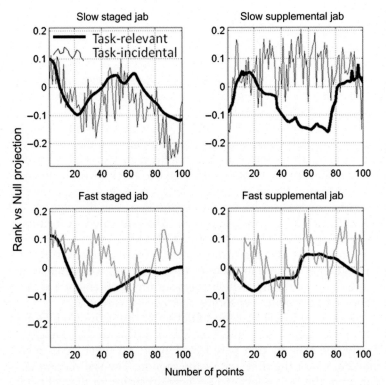

FIGURE 5.29 The conservation of the path from the DoF decomposition into the rank and null locally linearized subspaces for the deliberate (forward) and spontaneous (retraction) of the Jab routine. Conservation of the initiation of the Jab across the first half of the movement (first 250 ms) in the forward segment regardless of speed level and changes in arm dynamics.

The resulting profiles of the unfolding DoF decomposition revealed that the initial portion of the Jab remains invariant to the speed of the motion for those DoF intervening in the achievement of the goals defined in Fig. 5.27. These goals recruit the DoF spanning the dimensions of the task-relevant subspace in the 15-dimensional space of joint angles that we probed. The time axis here refers to the number of frames to generate a unit speed path (100 frames) upon resampling of the original temporal trajectory to create a curve of equally-spaced points (also called a time-normalized curve in some circles).

The original timing for the slow-speed case of these motions is 500 ms; but in the fast-speed case, the movements last 250 ms. This original timing can be appreciated in Fig. 5.19C,D where the slow cases are on the left-panels and the fast cases on the right-panels. The unit speed length parameterization of the Jab trajectory under consideration is necessary to be able to partition the geometric path equally and estimate the empirical mean using the same number of points taken across all trials. In this way, we can present the average curve describing the DoF partition into the task-relevant and the task-incidental submanifolds. Then, we can study the staged Jab and the supplemental Jab separately and learn about the effects of speed on the DoF decomposition.

To examine the issue of conservation of the three-dimensional paths in relation to changes in movement dynamics, once the resampling of these paths is attained, we refer the reader to our work [1,15,24] involving humans. There, we extended methods I developed for my postdoctoral work involving non-human primates to study these geometric issues with proper statistical techniques [25,26]. The companion site will contain code and explanation to implement these graphs and analyses of three dimensional paths.

The recruitment and coordination of the task-relevant DoF during the initiation of the staged Jab turned out to be insensitive to the speed level, at least within the speed levels we prompted the athlete to do. In the slow condition, during the second half of the staged Jab, over 250 ms of the motion had passed already. As such, the peripheral nerves across the body would have enough time to transmit feedback to the brain and enable corrective changes at the end effectors based on sensed speed levels.

To better understand speed influences and their sensing along with the anticipation of the action's consequences, I examined the shifts in stochastic signatures for the case whereby the target speed level was called at random by a computer program. Along those lines, I later explored the contributions of the trial-by-trial amplitude fluctuations in speed and acceleration from previous trial to the forward prediction of impending speed in the next trial. But, before we delve in the derivation of stochastic rules that could possibly anticipate impending levels of speed within the general rubric of random processes, we would like to take the speed adaptation issue to another level of discussion, one that serendipitously, ended up being useful to the athlete's competition. In fact, it did help him defeat his opponent in a rather unexpected way.

De-adaptation from New Dynamics can Surprise your Opponent

During the first set of experiments probing the reaction of the system to its self-generated distinct levels of speed, the speed was instructed. We instructed the targeted level within a block, so the performer knew which force level he needed to effectively deliver his punches with. We also instructed the targeted speed level at random, so from trial to trial, the required forces to effectively punch the opponent were more uncertain to predict amidst such fast occurring sequence. Notwithstanding different levels of variability quantified in Fig. 5.21D, the main result we had found in the experiments of Chapter 4 (within simpler pointing experiments) extended to the case where the movement is deliberate. The changes in movements' dynamics do not alter the geometry of the intended hand path to the target. Thus, whether the hand's goal-directed movements are performed in isolation (i.e., comfortably sitting and pointing to a visual target), or standing up and performing the outreach embedded in a complex routine that recruits all DoF in the body, the path conservation is still strong. This path conservation extends to the recruitment of task-relevant DoF during the launching phase of the punch, that is, the phase occurring within the first 250 ms of motion. This is important, since such changes in dynamics require the self-generation of very precisely timed synergies to attain a geometric path that is common across a wide range of speeds. Further, the conservation of the path geometry in the task-relevant submanifold of the staged Jab, was accompanied by the non-conservation of the DoF path of the task-relevant submanifold in the spontaneous Jab retractions. The sensory-motor systems seem to have different strategies to implement the dynamics of deliberate and spontaneous segments of complex movements engaging the whole body when the speed is deliberately targeted by top down instructions. But how about situations where the speed changes are evoked from bottom-up adaptive mechanisms, evoked by changes in the mass of the arms?

To address this question, we attached loads to both arms of the participants and let them adapt to the change in the arms' weight. This condition would alter the movement dynamics, so the athlete had to recalibrate the output forces to be able to adjust the speed of the punch and be effective to knock out the opponent.

The idea here was to contrast the (bottom-up) loads condition evoking a broad range of speeds with the (top-down) explicitly instructed speed condition. Fig. 5.30A shows the top-down speed condition. Fig. 5.30B shows the bottom-up loads condition. The movies contrasting the top-down speed case with the bottom-up adaptation case and

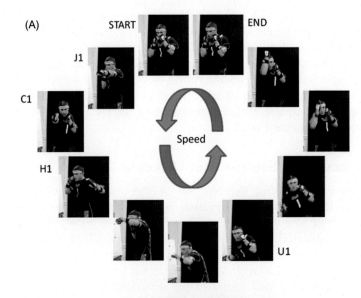

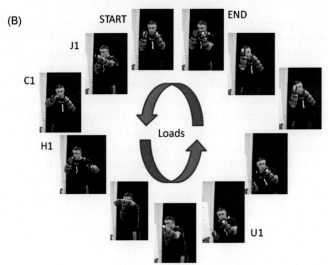

FIGURE 5.30 Uri Yarmush performing the J-C-H-U during the speed condition (A) and the loads condition (B) with the loads attached to his arms.

then the de-adaptation block can be found in this https://www.youtube.com/watch? v = c1fVZ-3YbaQ.

When I plotted the hand trajectories, I found that once again the staged segments of the routine, those performed with more deliberateness than the spontaneous retractions, were far more conserved despite the dramatic changes in dynamics and speed that the loads induced. The retracting trajectories that brought the arm back towards the body and helped the athlete transition from one segment of the routine to the other (e.g., transition from the Jab to the Cross) were however different for different speed levels. To determine speed levels, I took the median value of the peak velocity across all trials and partitioned the set into the trials above the median (fast) and the trials below the median (slow).

I measured the deviations from the straight line over the trials and found it to be far more variable and overall distinct for the changes in speed during the retractions. Fig. 5.31 shows the examples of the Uppercut and Hook motions (forward intended to the targeted opponent and backwards retracting from the staged punch in transition to the next staged segment.)

An unexpected reaction from the athlete occurred when I removed the loads from the arms. As the athlete performed the J-C-H-U without loads this time, he surprised himself of how fast he had become. We captured his reaction on video. It was truly remarkable to see this. The speed of his arms was such that at 60 Hz the camera could not capture the frames clearly. The hands and the arms came out hazy owing this to the speed he had acquired. This de-adaptation *after effect* lasted a few minutes before his motions went back to a normal speed. He also acquired much more effectiveness in his punch. The amount of energy his body stored up from the adaptive process in response to the alteration in the arm dynamics that the weights produced was tremendous. When he fully adapted and performed the punches at full speed under the weighted arms, he had to recalibrate his entire body. He took this result and used it during competition. Right before the contest, he practiced and adapted his speed to the weights. Then he removed them and the opponent was surprised by the unexpected speed and the effectiveness of the punches. The effect lasted several minutes and this was enough to buy him time to win the contest.

As with the top-down prompted speeds, the bottom-up evoked speeds, as induced by the loads, did not alter the intended path of the hands as much as it did the spontaneous retractions in transition to the next staged segment of the routine. The nervous system seems to understand this dichotomy between deliberate and spontaneous motions rather well. I suspect that neocortical structures are involved in this form of deliberate control of the motions. I also suspect that subcortical brain structures and structures in the spinal cord control the spontaneous segments that are so automated that the person does not even realize they are being performed. Further, I believe that the peripheral networks innervating the limbs and trunk are not merely serving local transduction and transmission of sensory motor signals. I think that it is possible that nodes along these networks gate information through local feedback loops and actively modulate the flow of information, thus contributing to the autonomous control of muscle synergies to output the necessary forces that modulate the movements' acceleration. In other words, I believe that *motor intelligence is distributed across the PNS*. The control strategy consists of achieving proper balance between central and peripheral centers to produce this flow of coexisting deliberate and spontaneous motions.

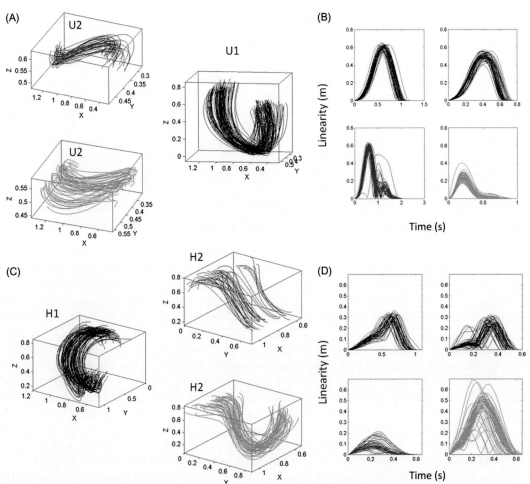

FIGURE 5.31 Sample hand kinematics of the complex sequences in which the Jab was embedded: Conservation vs non-conservation of trajectories according to changes in body dynamics (speeds and loads). (A) The intended Uppercut motions U1 performed at different speeds maintain the curvature of the trajectories despite the changes in body dynamics. (B) In marked contrast, the speed changes separate the curvatures of the spontaneous retracting segments, measured through a simple linearity metric capturing changes in curvature. (C–D) Similar behavior was registered for the Hook under speed and loads condition. Notice that the addition of loads makes the linearity more variable.

The spontaneous segments of our behaviors have not been systematically studied, but they hold many clues as to how the brain is controlling the peripheral limbs as the limbs transition through such complex sequences in a fluid manner. Their evolution in the learning and adaptation processes that we studied during these martial arts sessions provided important information on how the balance between controlled and free-flowing motions manifested across the various contexts we tried. To better capture this dichotomy within the framework of stochastic processes, whereby the prior information may be predictive of the future information, we derived stochastic rules and tested them within different conditions that altered the sensory input to the system. Thus, we tested the same routines under different conditions and measured the systematic differences between deliberate and spontaneous segments.

SENSING THE FUTURE SPEED IN DIFFERENT CONTEXTS

We used different conditions to evoke different movement-gated sensory inputs [27].

1. *Simulation:* The participants performed the routines from memory, without receiving guidance from the instructor, pretending as well that an opponent was present.
2. *Mirror feedback:* The participants performed the routines in front of a mirror and were instructed to use the feedback from the reflection of their body on the mirror to help practice.
3. *Dark with eyes closed:* The lights in the room were turned off and the participants performed the routines with eyes closed. We placed a bandana over the eyes to help them keep the eyes closed.
4. *Loads:* The participants carried training loads on both of their arms (12 lbs in each forearm). The loads were distributed along the forearm and consisted of 3 lbs sand bags attached with Velcro that secured the loads to the forearms. These small sand bags are commonly sold at sports stores for training.
5. *Mirror with body lights:* We turned off all lights in the room and participants performed in front of the mirror with glowing sticks attached to the body (using Velcro). They were instructed to use visual feedback from the reflexion of the body lights on the mirror. These resembled a simplified version of the body (as a stick figure with point-lights).
6. *Body lights:* The participants were instructed to perform the routine in the dark with glowing sticks placed on the body but no mirror. They were instructed to use visual feedback from the glowing sticks as much as they could while they moved.

These conditions were performed in blocks of 10 trials each. When speed was prompted we programmed a computer to call the speed level (fast or slow) at random. Fatigue in motor control experiments is common. For this reason, providing resting breaks to the participants is important. However, inevitably some fatigue will influence the motor noise during training, so we thought of these experiments as analogous to a training session an athlete would commonly undergo. We hypothesized that the differences between deliberate and spontaneous segments of the motions would still be present at some level. As such, I decided to study the noise patterns of these routines for different randomly prompted speeds.

I considered the trial by trial fluctuations in speed and acceleration maxima as a stochastic process. As before, I used the continuous Gamma family of probability distributions to fit the frequency histograms of the speed and acceleration peaks because it had proven a good-enough family to characterize human behaviors. I wanted to know the extent to which the system would have correctly updated the slow versus fast velocity in an impending trial, based on the kinesthetic sensing of its self-generated changes from a previous trial—despite the random instruction. Was the system sensing its own self-generated acceleration? To this end, I examined the noise of the scatter of points according to a stochastic rule. This rule helped me understand the evolution of the motor noise under various contextual influences. It was not a rule of *"how the brain updates the velocity signal."* But it would help me characterize the adaptation process taking place under subtle variations of the sensory stimuli the person self-produces and the sensory stimuli that it receives from the external environment. I wanted to study these interactions between self-generated speed and acceleration both in natural training situations and in artificially constructed scenarios.

Velocity and acceleration are co-dependent parameters. Thus, their noise is expected to co-vary. Any split in this process which is systematically modulated by training context or instructed speed could inform us of possible anticipatory strategies. I was particularly interested in the spontaneous segments the subjects did not even notice were performing, because of the above described finding. These segments changed the geometry with changes in speed. Would these changes be in any way systematic with the different conditions, or would they have a unique pattern across conditions?

Since the Jab (which we focus on here) was performed embedded in the J-C-H-U sequence and in isolation, I was interested in the noise profiles of the deliberate and spontaneous segments for these two conditions.

The trials were taken in the order in which they were acquired and plotted according to the rule below. To derive the rule, I used the relationship between velocity and acceleration, and from trial to trial, I set the velocity in the next trial proportional to the acceleration and the velocity of the current trial:

$$\left(A_{\max}^t + v \cdot V_{\max}^t\right) = \left(V_{\max}^{t+1} - V_{\max}^t + v \cdot V_{\max}^t\right) = \left(V_{\max}^{t+1} + \nu \cdot V_{\max}^t\right) \tag{5.1}$$

Here I derived the constant of proportionality $v = 10$ with $\nu = 1 - v$ for the entire data set comprising all 15 participants (six females and nine males). I did this to spread the fitting error across all participants and avoid biasing it towards one participant or a reduced group. In this way the scale I built to assess these interdependencies was common to all these participants and allowed me to study the noise patterns of this cohort. Clearly, for a different cohort, I would have to adjust the constant of proportionality and consider additional age-dependent factors. Here the cohort was quite homogeneous in age, as most were college students in their early twenties.

I gathered all speed peaks into a frequency histogram, which turned out to be skewed. The log transformation adjusted the distribution from skewed to normal and was well fit by the lognormal distribution. Yet, the result on the autistic adolescent we tested (see Chapter 3) doing these routines revealed an exponential distribution representing an additive rather than a multiplicative process. In this sense, I chose an additive family and using maximum likelihood estimation confirmed the Gamma family as our best fit. Since our scatter on the log-log Gamma parameter plane was well fit by a linear relation, we here used the linear model $f(x) = mx + b + \varepsilon$ to characterize the log relations of the noise present in velocity-dependent measurements of the movement positional trajectories from trial to trial. Here m is the slope of the line, b is the intercept and ε refers to the fit-error.

In the case of the Jab, I consider the training context as the task and replacing the above approximation on the equation of the line gives,

$$\left(V_{\max}^t(task) + v \cdot (task)\right) = m\left(A_{\max}^t(task)\right) + b(task) + \varepsilon(task) \tag{5.2}$$

I then take the natural logarithm of the parameters of interest, and assume that the properties of the noise will change with the task context. This is a reasonable assumption that required empirical validation. As it turned out, this assumption was correct:

$$\ln\left(V_{\max}^t(task) + v \cdot (task)\right) = m\ln\left(A_{\max}^t(task)\right) + b(task) + \varepsilon(task) \tag{5.3}$$

Here the parameters of interest are the velocity and acceleration maximum in each trial, which recall here, I empirically approximated as a Gamma process. The time scale of these motions range from 0.8 to 1.7 seconds per trial. The experimental session under a given context has 100 trials and the parameters t, $t+1$ in equations (5.1) and (5.2) refer to the order of the trial number. As explained above, v is approximated from the cohort and m and b are also empirically taken by fitting the line to the scatter, while ε is the residual error attained from the fit. The residual error distributed normally for each participant and the dispersion served to track the rate of learning, as the variance reduced over practice at a unique rate for each person. Intuitively this made sense because some of the participants were athletes in other sports, while others were not.

Exponentiation of Equation (5.3) gives:

$$e^{\ln\left(V_{\max}^t(task)+v\cdot(task)\right)} = e^{m\ln\left(A_{\max}^t(task)\right)+b(task)+\varepsilon(task)} = e^{m\ln\left(A_{\max}^t(task)\right)}e^{b(task)+\varepsilon(task)} \tag{5.4}$$

Here I use the logarithmic and exponent rules $a^{m+n} = a^m \cdot a^n$ and $a = e^{\ln a}$, $a^x = \left(e^{\ln a}\right)^x = e^{x\ln a}$ for each real x and if a^x is to preserve the logarithmic and exponent rules,

$$e^{m\ln\left(A_{\max}^t(task)\right)}e^{b(task)+\varepsilon(task)} = e^{\ln\left(A_{\max}^t(task)\right)^m}e^{b(task)+\varepsilon(task)} \quad \text{such that}$$

$$V_{\max}^{t+1}(task) + v \cdot V_{\max}^t(task) = \left(A_{\max}^t(task)\right)^m e^{b(task)+\varepsilon(task)} \tag{5.5}$$

One example comes from the fitting parameters of the novice in $f(x) = mx + b + \varepsilon$, $m = 1.03$, $b = 0.87$ with correlation coefficient 0.98 for the intended Jab segment. From the slope value, which we can write as $m = 1 - \delta$ we take Equation (5.5) to leading order and obtain:

$$V_{\max}^{t+1}(task) = A_{\max}^t(task)[1 - \delta\ln A_{\max}^t(task) + O(\delta^2)] \cdot e^{b(task)+\varepsilon(task)} - v \cdot V_{\max}^t(task) \tag{5.6}$$

With $\delta = -0.03$ implying a stochastic updating-rule that anticipates the peak velocity of the next trial using the combination of the current trial's velocity maximum and acceleration maximum with multiplicative error. I use this rule to characterize the movement across different contexts and expect changes in the slope m (the exponent of the power relation given by the linear fit to the log-log of the values), the intercept and the error (scatter) as a function of context, effort, fatigue etc. I then ask if given each task context, the scatter maintained this first order stochastic rule. Given that the range of speed maxima falls between 0.25 and 9 m/s the squared absolute value $\delta^2 = (-0.03)^2 = .0009$ affecting the slope of the scatter can be considered negligible.

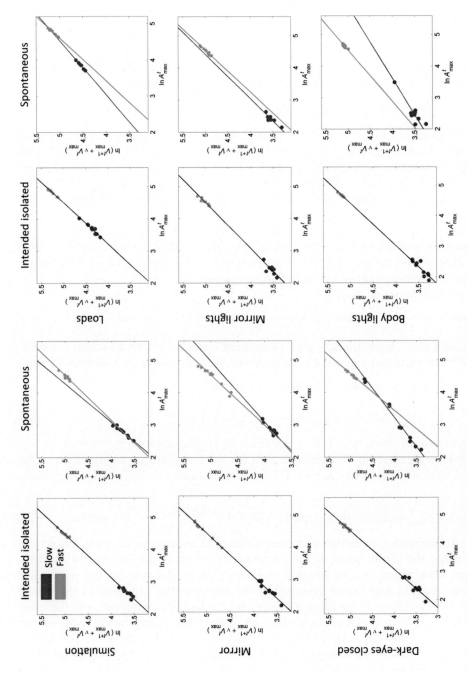

FIGURE 5.32 Systematic effects of speed level and training context on the noise properties of the spontaneous retractions in the expert system as captured by the first order stochastic rule predicting the speed in a trial ahead, based on the prior speed and acceleration.

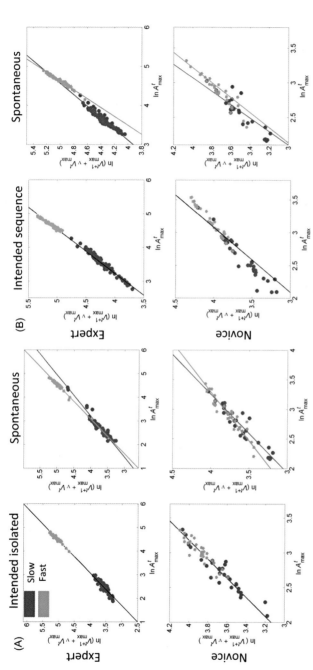

FIGURE 5.33 **Anticipatory performance of the expert vs representative novice participant across different training sessions.** The scatter is comprised by the trials from fast and slow speed according to the first order stochastic rule used to parameterize the relation between the maximum velocity and maximum acceleration from trial to trial. (A) Isolated Jab trials performed at different speeds for intended and spontaneous segments form self-aggregates. Top is from the expert and bottom from the representative novice. (B) Performance from a subsequent session, where the participants executed the Jab embedded in the full fluid sequence. Notice the improvement in the novice upon training whereby the Jab embedded in the complex sequence begins to cluster correctly as a function of instructed speed. Notice also that spontaneous movements "channeled" out through a different slope the type of instructed speed. Both segments had slopes and intercepts that differentiated the sequenced from the isolated routines.

Fitting errors from the scatter taken across each condition are reported elsewhere [27] for both intended and incidental segments. The goodness of fit parameters yielded adjusted R-square values within the ranges of 0.97–0.99 and root mean square errors in the range of 0.002–0.5. Fig. 5.32 provides the representative expert's performance as quantified by this rule for each of the conditions and each of the Jab segments (deliberate and spontaneous). In all cases, the prediction error of the speed level in the next trial using the speed maximum and the acceleration maximum from the prior trial resulted in a common slope for each of the speed scatters in each condition. In other words, for a similar slope and similar intercepts, the condition shifted the scatter along the same slope when the Jab was staged. In contrast, the slopes and intercepts of the spontaneous retraction of the Jab changed for each speed level in different ways for each condition.

The spontaneous motions differentiated the condition and speed. This distinction between speeds during the spontaneous retractions manifested for both cases, the performance of the Jab in isolation (i.e., repeating just the full forward and back Jab) and for the Jab embedded in the full sequence. This can be appreciated in Fig. 5.33.

The study of the predictive properties of speed—and acceleration-dependent signals self-generated and output by the nervous systems during the adaptive process of learning (or practicing) these routines provides a good example of new personalized approaches to sports with the potential to give athletes and coaches more than meets the eye.

The road to biometrics leading us to the discovery of various sensory-motor biomarkers of neurological disorders started with these studies of top athletes, but their translation to the clinical arena may connect technological advances with the research and medical fields to help improve the quality of life of individuals affected by such disorders. We may see these types of analytics someday embedded in our smart watches.

References

1. Nguyen J, et al. Automatically characterizing sensory-motor patterns underlying reach-to-grasp movements on a physical depth inversion illusion. *Front Hum Neurosci*. 2015;9:694.
2. Majmudar U, Nguyen J, Torres E. The use of graphical user interfaces (GUIs) to analyze motion and temperature. *J Vis*. 2015;15(12):491.
3. Choi K, Torres EB. Intentional signal in prefrontal cortex generalizes across different sensory modalities. *Journal of Neurophysiology*. 2014;112(1):61–80.
4. Cole J. *Pride and a daily marathon*. 1st MIT Press ed. Cambridge, Mass: MIT Press; 1995:194. xx.
5. Torres EB, Cole J, Poizner H. Motor output variability, deafferentation, and putative deficits in kinesthetic reafference in Parkinson's disease. *Front Hum Neurosci*. 2014;8:823.
6. Andersen RA, Buneo CA. Intentional maps in posterior parietal cortex. *Annu Rev Neurosci*. 2002;25:189–220.
7. Torres EB, et al. Neural correlates of learning and trajectory planning in the posterior parietal cortex. *Front Integr Neurosci*. 2013;7:39.
8. Andersen RA, Cui H. Intention, action planning, and decision making in parietal-frontal circuits. *Neuron*. 2009;63(5):568–583.
9. Mulliken GH, Musallam S, Andersen RA. Forward estimation of movement state in posterior parietal cortex. *Proc Natl Acad Sci U S A*. 2008;105(24):8170–8177.
10. Xia M, Wang J, He Y. BrainNet Viewer: a network visualization tool for human brain connectomics. *PLoS One*. 2013;8(7):e68910.
11. Torres EB, et al. Characterization of the statistical signatures of micro-movements underlying natural gait patterns in children with Phelan McDermid Syndrome: towards precision-phenotyping of behavior in ASD. *Front Integr Neurosci*. 2016;10:22.

12. Torres EB, et al. Toward precision psychiatry: statistical platform for the personalized characterization of natural behaviors. *Front Neurol*. 2016;7:8.

13. Nguyen J, et al. Schizophrenia: the micro-movements perspective. *Neuropsychologia*. 2016;85:310–326.

14. Nguyen, J., et al. *Characterization of visuomotor behavior in patients with schizophrenia under a 3D-depth inversion illusion*. in *The Annual Meeting of the Society for Neuroscience*. 2014. Washington DC.

15. Nguyen J, et al. Methods to explore the influence of top-down visual processes on motor behavior. *J of Vis Exp*. 2014;(86).

16. Torres, E.B., et al. *Noise from the periphery in autism spectrum disorders of idiopathic origins and of known etiology*. in *The Society for Neuroscience*. 2013. San Diego, CA.

17. Nguyen J, et al. *Quantifying changes in the kinesthetic percept under a 3D perspective visual illusion. Vision Science Society*. Naples: Fla; 2013.

18. Biasatti M, Lombardo MV. Autism sports and educational model for inclusion (ASEMI). In: Torres EB, Whyatt CP, eds. *Autism: The Movement Sensing Perspective*. CRC Press Taylor and Francis; 2017:271–280.

19. Torres EB. Atypical signatures of motor variability found in an individual with ASD. *Neurocase: The Neural Basis of Cognition*. 2012;1:1–16.

20. Torres EB, et al. Autism: the micro-movement perspective. *Front Integr Neurosci*. 2013;7:32.

21. Torres EB, et al. Stochastic signatures of involuntary head micro-movements can be used to classify females of ABIDE into different subtypes of 3 neurodevelopmental disorders. *Frontiers in Integrative Neuroscience*. 2017;11(10):1–17.

22. Zatsiorsky VM. *Kinematics of human motion.*. Champaign, IL: Human Kinetics; 1998:419. xi.

23. Altmann SL. *Rotations, quaternions, and double groups*. Oxford Oxfordshire New York: Oxford science publications; 1986:317. Clarendon Press; Oxford University Press. xiv.

24. Torres EB, et al. Sensory-spatial transformations in the left posterior parietal cortex may contribute to reach timing. *J Neurophysiol*. 2010;104(5):2375–2388.

25. Torres E, Andersen R. Space-time separation during obstacle-avoidance learning in monkeys. *J Neurophysiol*. 2006;96(5):2613–2632.

26. Torres EB. New symmetry of intended curved reaches. *Behav Brain Funct*. 2010;6:21.

27. Torres EB. Signatures of movement variability anticipate hand speed according to levels of intent. *Behavioral and Brain Functions*. 2013;9:10.

28. Torres EB. Two classes of movements in motor control. *Exp Brain Res*. 2011;215(3-4):269–283.

Rethinking Diagnoses and Treatments of Disorders: The Third (Objective) Neutral Observer Assessing the Interactions between the Examiner and the Examinee or the Therapist and the Client

"…I danced in my mind. Blinded, motionless, flat on my back, I taught myself to dance Giselle" *Alicia Alonso (1921-)*.

FROM BALLET TO PERSONALIZED PRECISION PSYCHIATRY

For many years during the cold war, the United States barred the Ballet Nacional de Cuba from appearing on its soil. Reportedly the New York Times dance and theater critic, Clive Barnes, after seeing the company perform in Canada in 1971, wrote, "*We may be so struck by the way they dance 'Swan Lake' that as a nation we may spontaneously demand Fidel Castro as president.*" Clearly, as a regular citizen I would not go that far; but as someone who appreciates what goes into producing and evoking such motions and emotions, I would think twice about it. Indeed, in Cuba they managed to create a school of ballet with a unique seal and then made it rather accessible to all, from the illiterate peasant to the well-educated scholar (Figure 6.1). Growing up in Havana, I was a regular at the *"Gran Teatro de La Habana, Federico Garcia Lorca"*. There, I watched how their *prima ballerina assoluta* Alicia Alonso redefined *Giselle* and created *Carmen* in rather unique ways. But more importantly, I saw the legacy she was already leaving behind in the newer generations of dancers trained by the Cuban school of ballet she created.

FIGURE 6.1   Alicia Alonso, once a blind ballerina, who danced Giselle in her mind. Her legacy lives across several generations. She made ballet accessible to all in my native Cuba.

I thought Alicia Alonso's movements during her performances were extraordinary. She could float so lightly it seemed as though there was no gravity when she moved across the scenario. Whether dancing soloist or in partnership, she transmitted that sensation of effortless motion in every performance I watched. I knew she was nearly blind and that alone made the performance extraordinary. Yet, watching a documentary about her persona and legacy, I learned in her own words something that seemed even more interesting to me today. When she was merely 20 years old, she recalled having to remain motionless in bed for a whole year so as to avoid entirely losing her eyesight from retinal detachment. I cannot imagine what that must have been like for a dancer. During that time, she recalls playing the role of the public in her mind, by taking the vantage point of the audience and mentally watching herself dance with her partner and the rest of the ensemble on the scenario. From that perspective, she could mentally rehearse every small detail of her performance but also position herself in many different configurations relative to her partner and the rest of the ensemble. In the absence of physical movement, mental motion imagery helped her build each one of the characters in her mind. Then, later on, during the actual physical performance while nearly blind, she used lights positioned in strategic places across the scenario as reference points for guidance. She could see light at a specific angle, so by cleverly positioning lights throughout the scenario and theater, she could rehearse and then instruct her partner and ensemble precisely what to expect during her actual performance, that is, where and when she would be during the performance. And sure enough, there she was as a tangible physical extension of her representational mental space. Perhaps because each person's mental space is unique, her performances too were so uniquely irreproducible. Choreographers created characters for her to interpret, but then the characters became *Alicia*, each one with a distinctive seal. She somehow managed to transmit that unique seal to newer generations of dancers.

SOME THOUGHTS ON OUR SOCIAL MENTAL SPACES

One of the most intriguing aspects of the mental rehearsal Alicia Alonso described as a dancer was the integration of the bodily rhythms of the soloist, the dyad in partnering and then the entrainments of this core dyadic unit with the rest of the dancing ensemble as a whole. This form of ensemble-dance acting components, so obvious to the spectator at a describable macro-level of overt behaviors, carries as well hidden coupled rhythms difficult to verbalize or even consciously perceive. They seem to exist on a different dynamic realm conveyed through gestures and body language continuously unfolding within temporal and frequency scales of coupled rhythms at a micro-level that instrumentation can capture. In turn, the audience's responses contribute to the closed feedback loops guiding the dancer. They are part of what we call in the lab *"the Dark Matter of Science"* (Figure 6.2); the part that sneaks up on us while we are busy concerned with the more obvious phenomena that we can consciously decode.

Taking different perspectives and zooming in and out of phenomena through varying scales in the time- and frequency- domains allows us to uncover relative relations; relations that remain conserved across different layers of the nervous systems despite the disparity of control levels and classes of processes involved in the performance of complex actions. Along those lines, we have been able to analyze human-generated data in a personalized manner. Taking such data analyses across the general population, we have identified power laws relating key aspects of probabilistic behavior across ages. Further, using this approach, we have been able to infer various consequential relations for the development and maintenance of neuro-motor control. Our hope has been to uncover emerging properties of the multilayered nervous systems whereby phenomena we identify at the hidden micro-level enable us to better understand co-existing aspects of overt and covert motions embedded in natural behaviors the person performs. A case in point has been the

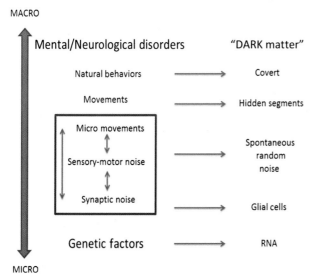

FIGURE 6.2 The dark matter of science across many disciplines. Each field of scientific pursuit has neglected some information while busy studying the obvious. In our case in point, hidden, covert segments of behavior at the macro-level of description remain unexplored while we focus on goal-directed ones. Likewise, spontaneous variations in our motions at the micro-level remain under characterized while grand averages taken under Gaussian mean assumption smooth out as noise the minute fluctuations in our motions—that is, the information that our platform renders as the very signal of interest.

personalized characterization of neurological and neuropsychiatric disorders that we have provided in previous chapters of this book. While up to now the aim was to develop analytics for personalized approaches to medicine (Figure 6.3), the purpose of the present chapter is to extend such ideas to the case of dynamically coupled behavior taking place within the social scene, including the clinical settings as well, the laboratory and the classroom environments.

For example, in Chapter 7 we saw that boxing routines and the tennis serve provided methods and clear-cut criteria to identify and automatically differentiate deliberate and spontaneous co-existing segments of complex behaviors executed by one person. Yet, how to extend those processes to the social realm, where two or more people participate in the subtle social dance, is far more challenging. Here we will first explore the dyad in the realm of more overt motions of performing artists, as they partner in a dance. Then, we will show how to extract hidden layers of information from the coupled biorhythms that scape the naked eye.

The tracking of multiple biorhythms unfolding in tandem across the nervous systems of the interacting dancers as they produce coordinated behavior requires methods that allow for the continuous measurements, analyses and inferences of multiparty processes unfolding in real time. Such a fast dynamic coupled exchange would not explicitly occur in the actual social scene. However, the partnering in staged-dancing contains key elements common to actual social exchange that can help us design visualization tools and outcome measures amenable for the personalized analyses of more subtle elements unfolding in the social dance of a dyadic exchange. For example, given the simplest of cases of a complex social scene such as that of a diagnosis setting; we would aim at tracking each individual person within that social dyad. We would also track a hidden third element we call

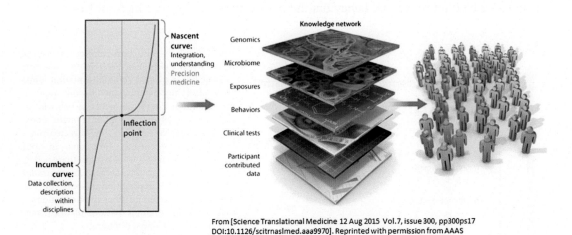

From [Science Translational Medicine 12 Aug 2015 Vol.7, issue 300, pp300ps17
DOI:10.1126/scitrnaslmed.aaa9970]. Reprinted with permission from AAAS

FIGURE 6.3   Poised for accelerated change in medical research and patient care using the Precision Medicine platform[19] (From (*Science Translational Medicine* 12 Aug 2015: Vol. 7, Issue 300, pp. 300 ps17 DOI: 10.1126/scitranslmed.aaa9970) Reprinted with permission from AAAS). Yet we are missing the behavior link to bridge lower layers of the knowledge network with higher layers of genomic data. Without solving that problem we will not make much progress in the development of target treatments disorders of the nervous systems.

"the neutral observer". This neutral observer comes in the form of self-emerging, dynamically coupled biorhythms, unfolding largely beneath awareness. *How are we going to track this dark matter of behavioral science?*

NOBODY PRINTS!

A little detour is in order (before we delve deep into the methods) to see how it all began to unfold in my lab: By 2012, the time period of our NSF grant was nearing its end we had about $140,000.00 left in our budget because the lab had been extremely careful to save every precious penny during those 2 years of autism research. I had to set up a mobile lab at the Christian Sarkine Autism Center at Bloomington, Indiana and provide funds to support Dr. Isenhower during the time he spent there setting up the equipment and collecting the data. To that end, Dr. Nurnberger and Dr. José funded the clinical support personnel and covered the cost of the space at the hospital. Thus, through synergistic collaboration we were able to gather data from a large patient population and boost the number of subjects we needed for our publications.[1,2] Amidst these efforts, I put in for a no-cost extension (NCE) of the NSF grant. The NCE is some procedure that would enable us to use the remaining funds past the 3-year period of the grant, provided that we set a hard deadline to spend it and complete the project. While trying to submit the paperwork to extend our research time under the grant, we learned that this was an ARRA grant. (ARRA stands for American Recovery Reconstruction Act.) What that meant at the heart of a re-election year for then President Obama was that all funds had to be spent or returned. There was no NCE. Imagine my shock.

The ARRA NSF grant was shared with co-PI Professor Dimitri Metaxas from the Computer Science Department at Rutgers University, where I had also a graduate Professorship appointment. As such, Metaxas provided me with access to a graduate student and suggested that we rebudget and request funds for equipment. These two pieces of advice were invaluable since they allowed for the development of the methods I am about to present in this **chapter**.

I did request a change in the budget and the NSF approved it. I was able to acquire (upon much research on state-of-the-art motion caption systems) the Phase Space (**Notes** [1]) and install it in the lab to initiate the path of developing real-time co-adaptive interfaces. The main idea was to use one's own motor output in tandem with those of an artificial agent to learn and relearn various nonapparent aspects of a given behavior. The second part of Metaxas suggestion was to advice a graduate student from CS toward the completion of a PhD thesis. The student who became interested in the lab's work is Vilelmini Kalampratsiduo, or Vilmi as we dearly call her today. Vilmi is featured on the cover of this book and as you can see ballet is her passion and inspiration for her thesis. I was incredibly lucky to have an ARRA grant expire! Both solutions to the one spending problem came hand in hand to become a key piece of the final implementation of the research program from that very proposal.

---

[1]http://phasespace.com/portfolio-item/bot-and-dolly/

Vilmi had the skills to understand the beauty and complexity of the partnering dance in ballet, and to fully appreciate the social dyadic exchange that is staged in front of a large audience. Trained since she was 4 years of age, and having ample experience performing as a soloist, in dyadic partnership and as part of the choir (Figure 6.4**A**), she could properly appreciate the type of mental rehearsal that would place her as the audience to see herself performing, her partner and the group dancing on the scenario; or the mental rehearsal that would switch vantage points and look at all of that from the perspective of her own nervous systems. As Alicia Alonso did in her mind, Vilmi too could see the value of taking turns in the vantage points to assess the problem of dyadic motor control in a social dance; that is, through multidimensionally and dynamically switching the frames of reference.

Vilmi also had the computational skills to develop the model I had in mind using the new tools we acquired with the Phase Space. In fact, she had already had a major accomplishment in her native Greece. There she helped Stella, a 17-year-old girl suffering from quadriplegia gain her voice through an interactive computer interface, a story that made headlines in many newspapers in Greece (Figure 6.4**B**).

FIGURE 6.4   A multifaceted student: (A) Vilelmini performing ballet: As a soloist, as Carmen in dyadic partnering and as part of an ensemble in modern dance. (B) Computational skills came in handy to help Stella communicate with others. A friend led Vilelmini to Stella and upon their meetings she developed interactive interfaces for computer-driven communication. Their success got much press coverage in Greece.

I was happy to finally have all the elements to build a co-adaptive dynamic interface and address the social issues in autism using the theoretical geometric model I had proposed so many years ago. My problem now was of a different nature: We were *bankrupt* beyond repair! I said to the lab, *nobody prints*! Until we replenish our funds we have enough to keep us afloat with very basic expenses to cover subject's compensation and those sorts of things; but we do not even have funds to publish the slew of papers we already had accepted by various peer-reviewed journals and others ready to go under review. With my tenure consideration around the corner, I was afraid the timing was a bit off. In a miraculous moment, Rutgers came to the rescue with internal funds to cover our publication expenses and give me some time to seek external funds.

By April 2013, our autism paper came out in Frontiers in Integrative Neuroscience[1] as part of a broader Research Topic entitled *Autism: The Movement Perspective*[3] that we launched a year earlier. The online interest spiked with such a speed, with an interest particularly driven by parents and self-advocates, that I can safely say they essentially changed the faith of our lab. This combination of serendipitous interest and planned events in turn caused the Editorial office to temporarily embargo the paper and do a formal Press Release of it later in partnership with the Nature Publishing Group. By the end of July 2013, the word on our work went out through many media outlets. A first wave of interest came in through more media interviews and promotion of the work. It was all very overwhelming and very exciting for us, but also very poignant. I received many personal emails from parents and self-advocates from all over the world. They sent me pictures of their children and desperate questions about possible solutions and possible avenues for treatments. I remained utterly in shock during those days and months thereafter. My main concern was that autism was conceived by the scientific and clinical communities as a deficit model. There was no room for positive hope in their diagnostics or for meaningful accommodations in their treatments. The affected individual was spoken about in third person and treated as a lessened human being. It was awful to see this; but it was also part of *the dark matter of science*. People were so busy trying to define these phenomena and solve the problems they posed under certain a priori preconceived notions and assumptions, that they could not see that their propositions were robbing these children of free will. They were being robbed of their humanity. I could see it clearly because I am one of *them*. I felt it in my own skin and it deeply hurt somehow.

The *media storm* waned after 2 or 3 months, and a second wave of interest came in, but this time from the private sector. Notably, for a variety of different reasons we became acquainted with researchers from Eli-Lilly, Johnson & Johnson and SRI-International. I went to lecture about our autism research program at their headquarters in Bloomington-Indiana, Titusville-New Jersey and Princeton-New Jersey, respectively. Yet, the most incredible outcome of this wave of interest came from the *Nancy Lurie Marks Family Foundation*. The Foundation Director and Chief Scientific Officer, Clarence Schutt (Emeritus Chemistry Professor at Princeton) and their Special Advisor for Research Programs, Kenneth Farber (JD), contacted the lab and actually paid us a visit. Cathy Lurie had read our 2013 paper on Micro-movements and taken an interest in our work. She saved my lab.

The NLMF Foundation is one of the largest autism foundations in the country. They have funded autism research for many years, but as in every field I have worked on thus far, I tend to remain unfamiliar with the core research community. As such, I did not know about the foundations related to autism or any of the social networks underlying their decision-making for funding of people working in the autism field. This foundation seemed different as they took the time to actually come to the lab and experience first-hand what we were doing. I am certain now that in the past, my lack of social skills and networking abilities have hurt the advancement of my scientific career, but this time I had the unique opportunity to talk one on one with them and that was all I needed.

When I received their phone call, I was thrilled. I assembled the lab and we organized everything to show the visitors what we had been "cooking" for the last 3 years of our autism research. They spent all day at the lab and upon their departure back to Boston—where they have their headquarters; they encouraged me to submit an application for a Development Career Award. I did so and the Foundation granted the award in 2014. Their funding of our research program saved the lab, literally saved my scientific career and enabled us to continue the work we had initiated under the NSF grant. This **chapter** is about the implementation of our promise to this incredibly generous Family Foundation.

BUT FIRST, MIND THE GAP!

We have seen in the previous chapters that the field of neural control of movements has studied the sensory-motor phenomena under a rather restrictive *assumed* theoretical lens. Under this lens, theoretical assumptions follow linear models; enforce normal distributions and impose stationary processes to make the modeling and analyses easier. Nonetheless, this approach is not only inadequate to assess natural behaviors at the individualized level; it poses a burden to advance progress in the context of mental illnesses as well. The approach is ill-suited to study social exchange during social interactions. Being the issues with social interactions one of the emergent problems that most neurodevelopmental disorders pose to the scientific community, it is a pity we do not have a research program that directly defines the types of problems we should be studying from a biological stand point. The paucity of methods available to address even the most basic issues of social dynamics from the vantage point of nervous systems' anatomy and physiology, prevents the field of neurological and neuropsychiatric disorders from moving forward into the realm of Precision Medicine (PM) depicted in schematic form in Figure 6.3 (reprinted with permission from AAAS). Even the nascent interdisciplinary field of Computational Psychiatry or Precision Psychiatry remains close to pseudoscience because observation scaffolds all their data analyses. Lack of truly objective science or mathematical models to address problems of the nervous systems in biologically plausible ways will continue to be a stumbling block to allow bridging the gap between the bottom layers of the knowledge network for PM containing clinical records and the top layers related to genomic data.

The broad gap in the knowledge network will continue to prevent progress toward target therapies. Such therapies ought to start with objective measurements of the dyadic behavior during the initial diagnostics stages, before any treatment is prescribed. In this sense, a personalized approach that balances out the two vantage points that we mentioned in Chapter 4 to study and model behaviors is critical to succeed. Recall that they are the *external* vantage point of the observer (the clinician or the experimenter in the lab) and the internal vantage point of the patients' nervous systems that we harness activity from (Figure 6.6).

Our research program is aimed at providing tools that enable carving out a new path to study naturalistic behaviors interchangeably from these two vantage points, while using the very taxonomy of the nervous systems that nature has provided us with (depicted in Figures 6.11–6.13 Chapter 3).

We have seen in previous chapters that there is taxonomy in the nervous systems with phylogenetic order of formation and maturation that we can use to define and classify atypical developmental paths. Clinicians defining disorders of the nervous systems by observation call issues related to these internal layers of the taxonomy *"comorbidities"*. This is likely because they have no way to connect them at present with their eyeballing of behavior. Yet, the critical relevance of considering these issues in the development of treatments that accommodate the child's needs cannot be underscored enough. The so-called comorbid conditions are vital components of the disorder. We cannot possibly design effective treatments without considering them. Effective treatments are not those which make the person look *"socially acceptable"*. They are those which contribute to the overall wellbeing of the affected person and ultimately make autonomy, agency, self-sufficiency and self-control a true reality. Effective treatments are those which help the person have a happy co-existence with the rest of society and urge society to adopt for a moment the vantage point of the affected person to understand what it would be like to have these so-called *comorbidities*, function in daily life and on top of that having to deal with being treated as a lessened human (due to ignorance and lack of proper interdisciplinary training in biomedical professions.)

I lose my temper and tend to go into a rant when I see the tremendous success certain groups have had at misleading the public into believing there is scientific evidence behind clinical treatments like ABA that completely ignore the nervous systems of the person being treated; or for that matter, cognitive theories from basic science claiming the autistic person has mind-blindness[4,5] or God only knows how many other nonsense, no-good-for-treatment ideas lacking any sort of objective foundation. How people can get away with such pseudo-science and be so successful at promoting those ideas remains the real mystery to me. It is also very difficult to see this in the US, because such pseudo-science does influence legislation that directly affects the lives of the affected people and their families by, for example, denying insurance coverage for treatments that address basic sensory motor needs in the autistic nervous systems and are potentially conducive of the development of functional autonomy and self-independence.

Taking a multilayered biologically sound approach to address all issues of autism is critical. Yet, no literature on this exists, given the skewed ways the problem has been posed and addressed as a mental illness or a cognitive-social phenomenon. We designed our own comprehensive taxonomy and dynamic classification scheme to acquire a comprehensive profile of all levels of the nervous systems. The example we provided involving cross-talk among the deliberate, spontaneous and autonomic processes by the various nervous systems components is one aspect of such comprehensive and biologically plausible approach to the assessment of nervous systems disorders.

Another important component of the problem that has not been addressed by the scientific or clinical communities comes in the form of age-dependent rates of change in the statistical signatures of biorhythms' micro-movements. We saw the characterization of such signatures in Chapter 3. Figure 6.5 is used here as a reminder of what we are up against

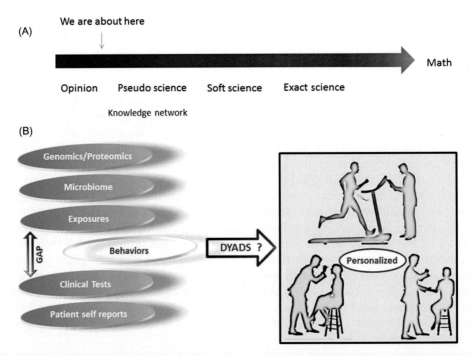

FIGURE 6.5   The need to improve our science. (A) While mathematics is the backbone of science, we are very far from its exactness in the clinical arena. Indeed we are far from leading a true scientific quest in autism and other neurological and neuropsychiatric disorders. Based primarily on opinion, the diagnostics system of clinical disorders has stalled the scientific quest. Forced to rely on such biased and incomplete accounts, basic and translational science cannot apply the scientific method which depends on objective quantification of physical phenomena, proper statistical inferences and blind reproducibility of results worldwide across labs. Instruments like the ADOS, DSM and ICD drive basic research, but because of their lacking of standard scales and faulty statistics, they create a rather weak foundation for any scientific attempt of true discovery. (B) The layer of behavior in the knowledge network is missing the objective angle required to track dyadic interactions between patient and clinician but also to track social exchange in the lab between experimenter and participant.

when examining human biorhythms under the *"one size fits all"* statistical model prevalent across the field. Essentially, the stochastic signatures of micro-movements in biorhythms change at different rates for different ages in general. In particular they change differently and with irregular rates in disorders of the nervous systems. And they are affected by medication in ways we need to systematically characterize, individually and as a group across different disorders. It is simply impossible to reach any conclusion from this *"one size fits all model"* prevalent across the health and mental sciences.

As our research program took off supported by the NLMF award, we became immersed in an ocean of behavioral data with new discoveries every day. By disregarding the old theoretical assumptions and adopting instead an empirical and age-sensitive approach to behavioral analyses, we discovered that the methods in use could not possibly address the inherent nature of the actual physical data.

While the discrete clinical data primarily grounded on subjective opinion was examined through the linear, normal and stationary lenses; the actual physical data that we continuously harnessed from the nervous systems' biorhythms were nonlinear, variable-skewed and nonstationary. More important yet to the issue of neurodevelopment was the fact that the statistical shifts and their rates of change were dependent upon the individual's coping biology and very different across developmental stages.

A fetus heart chamber begins developing by 25 days upon conception (**Notes** [2]). This sets a fundamental biorhythm in motion while supporting the accelerated development of the nervous systems and the rest of the body. It is estimated that these weeks of growth, when the embryo's heart beats twice as fast as the mother's, are the most rapid in the fetus development. This is to the extent that if that rate of change and growth were to remain steady for all 9 months, the final outcome at birth would be 1.5 tons.[6] Clearly, the rates of change and growth slow down as in full term; healthy babies weigh a few pounds. Yet, relative to a 4–5 years old child, the newborn baby is in an accelerated rate of growth that goes in tandem with the neurodevelopment of motor control.[7] By 3–4 years of age, there is a phase transition that shifts the stochastic signatures of bodily biorhythms from exponential like to skewed, narrowing the dispersion of the probability distribution functions (PDFs) as well. There is an acquired stability in these signatures after 5 years old; and a definite marked change again around puberty Figure 6.6.

All inflection points of stochastic change over time suggest that the organism of the infant is poised for nonuniform accelerated changes at different stages of life. Yet, these evolution and maturation processes are currently obviated by the theoretical assumptions of linearity, normality and stationarity that the field persistently imposes *a priori* on the data. As such, we changed the old lenses altogether and put on new lenses for a personalized, developmental-time-dependent approach. Under a new perspective we could then accommodate the variability inherently present in the data, the shifts in its stochastic signatures and the age-dependent rates of change in those shifts that we examined back in Chapter 3. As the lab moved toward a personalized agenda for PM, we realized that our methodology had the potential to fill in a nonobvious (but rather wide gap) in the PM schema of Figure 6.3. We then initiated the path toward a new discipline altogether, the nascent field of *Precision Psychiatry*.[2]

In PM the idea is to integrate information across the various layers of the knowledge network to ultimately tailor treatments to the patient's individualized needs. These needs are partly shaped by the genomic information unique to the patient and partly contributed by the environmental exposures of the person throughout life (e.g., food intake, geographical and industrial regions the person developed and lived in) In the context of cancer-target therapies the scheme of PM in Figure 6.3 seems adequate to assess various options for personalized treatment and the compilation of information for target therapies and individualized treatment design. However, translating this schema to the realm of disorders of the nervous systems seemed to pose a bit of a problem because of the gap between the bottom and the top layers of the knowledge network. As we have stated this broad gap is given by the lack of proper objective means to continuously assess behavior in the individual as explained in Figure 6.5.

More important yet, this gap in methodology to bridge across the bottom and top layers of the PM knowledge network extends as well to the assessment of dynamically

[2]TED-talk https://www.ted.com/talks/alexander_tsiaras_conception_to_birth_visualized#t-205910

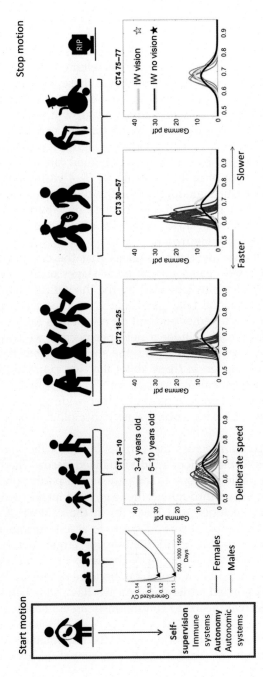

FIGURE 6.6 The age-dependent evolution of stochastic signatures empirically derived from our nervous systems' biorhythms. Probability distribution families capturing the inherent variability of our motions are skewed; their skewness changes with age and within a given age group, the rates of change in the skewness are nonuniform. The newborn growing at an accelerated rate manifests patterns of neuro-motor control that change by the day and very early on separate their signatures by sex (by 230 days in females and 254 in males). Between 3 and 5 years of age a major stochastic transition occurs in NT development that is absent from neuro-motor development in autism spectrum disorders (ASD). ASD remains at the level of deafferented participant Ian Waterman with similar signatures to those of his distributions' shape and noise (yellow during visual guidance and black in the dark). These signatures reappear in the elderly above 75 years of age; but during college age the patterns of motions in young humans have the lowest dispersion and highest symmetry. By 30 year of age the dispersion begins to widen indicating higher levels of noise. It has been our proposition that these age-dependent patterns from peripheral somato-motor signals are a form of kinesthetic reafference contributing to our intelligence since the conception of life, when the immune systems endow the embryo with self-supervision and the autonomic systems confer autonomy to the developing body. Motion in its broadest sense, across all nervous systems' biorhythms is present in every behavior of the living organism from conception until death, when motion stops.

coupled dyadic behavior. Such behavior is inherently present in diagnoses through clinician-patient exchanges, in therapies through the therapist—patient interactions and in basic research, through the interactions between the researcher (experimenter) and the participating subject (Figure 6.5). As it turns out, all clinical practices, treatments and basic scientific research provide a one-sided account of phenomena because they leave out some *dark-matter*, namely (1) the internal vantage point of the person's nervous systems and (2) the hidden coupled dynamics behavior and the influences the clinician, therapist or researcher inevitably inflict on the responses of the patient (or participant) when gathering the self-reports or the clinicians/experimenter accounts that scaffold the bottom layers of the PM knowledge network. *How could we fix this problem and bring more balance and completeness to our enquiry?*

Disorders of the nervous systems include neuropsychiatric and neurological conditions. Both give rise to deficits in social interactions with bidirectional properties: the individual does not assimilate into the surrounding society and society fails to embrace the affected individual. At the core of this problem is the basic unit of social exchange: the social dyad.

The social dyad initiates interactions from the moment the affected person visits the clinic and is evaluated by a clinician through questionnaires that rely on both self-reports and explicit questions to the patient by the clinician. At that very moment, the interaction becomes two-sided but the evaluation and its outcome are currently one-sided and skewed by various agendas extraneous to the nervous systems under evaluation. One never knows how the clinician arrived at a diagnosis and why /or how a given treatment was prescribed. Furthermore, the clinician may have a conflict of interest with the insurance coverage system or with the pharmacological company pushing for the use of a given drug; or the clinician may simply dislike the patient from the start (for no particular reason), perhaps even have a bad day and take it out on the patient. What we all started to wonder in my lab was: How come the clinician-patient dyad exchange is not evaluated in the clinical settings? And better yet, how is that lacking even factored into basic clinically-related research?

The sad reality we found was that while the methods in basic science have not been properly developed to analyze continuous physiological data underlying natural behaviors, the discrete clinical inventories continue to lead the scientific enquiry. Dr. Caroline Whyatt joined my lab in 2014 to spearhead a project we will describe in the second part of this chapter. In preparation for that project addressing the very question of the interplay between Psychology/Psychiatry and Physiology in the broader context of diagnoses and autism research, she performed a meta-analysis of the peer-reviewed papers in the field of autism spanning 21 years, that is, published in that field since 1994 until the year 2015 (Figure 6.7). She found that up to the year 2004, the papers published in the 10 top journals, that is, those leading the field, were primarily from Psychology. However, interdisciplinary research started to kick in and after 2004 dominated the upward trend. This is automatically shown by two big coincidental inflexion points in the year by year change: one with a maximum on the Interdisciplinary-research papers curve and the other contrasting with a minimum in the Psychology-driven papers curve. Interestingly amidst this change were papers with a focus on Physiology focusing on research related to Genetics, Neuroscience and brain imaging fields. When she computed the trending slopes of these curves, she found a positive slope in the Interdisciplinary-research case, a negative slope in the case of pure Psychology and a near flat slope in pure Physiology. The data

Rank	Publication: Author (Year), brief description	Classification
1 (2122)	American Psychiatric Association (2000). *Publication of the DSM-IV-TR*	Psychology/Psychiatry
2 (1814)	Lord, Rutter, and Le Couteur (1994). *Publication of the Autism Diagnostic Observation Schedule.*	Psychology/Psychiatry
3 (1565)	(American Psychiatric Association,1994) *Publication of the DSM-IV*	Psychology/Psychiatry
4 (1321)	Lord et al. (2000). *Publication of the Autism Diagnostic Observation Schedule –2ND Ed.*	Psychology/Psychiatry
5 (730)	Kanner (1943). *Seminal text describing behavioral observations of Autism*	Psychology/Psychiatry
6 (555)	American Psychiatric Association (2013). *Publication of the DSM-5*	Psychology/Psychiatry
7 (506)	Bailey et al. (1995). *British Twin study illustrating potential genetic etiology of ASD.*	Physiology
8 (447)	Sparrow, Balla, Cicchetti, Harrison, and Doll (1984). *Publication of the Vineland adaptive behavior scales.*	Psychology/Psychiatry
9 (443)	Mullen (1995). *Publication of the Mullen Scales of early learning.*	Psychology/Psychiatry
10 (389)	Lord et al. (1994). *Publication of the Autism Diagnostic Interview-Revised.*	Psychology/Psychiatry
11 (362)	Lord et al. (1989). *Precursor publication of the Autism Diagnostic Observation Schedule.*	Psychology/Psychiatry

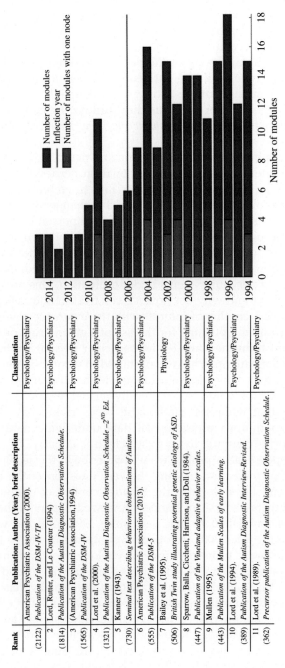

FIGURE 6.7 Twenty years of autism research and its trends. Text analyses of the corpus of peer-reviewed journals and the classification of top journals publishing just Psychological work (based on observation and opinion), just Physiological work based on measurements (e.g., electrophysiology, fMRI, and genetics) and Interdisciplinary work combining areas of psychological and physiological interest with new technologies and analyses. Network analyses of links connecting keywords and self-emerging modules indicate the prevalence of research topics driven by Psychological or Physiological research. Table shows classification of top cited publications across the corpus 1994–2015. Graph shows an overview of self-emerging interconnected modules (based on connectivity patterns of co-occurring keywords) across the yearly assessment of the corpus. Note the yearly modules that include a single keyword—these are clearly indicated.

generated by a corpus of research spanning over 20 years of peer-reviewed papers indicated a clear trend shifting toward interdisciplinary collaborative efforts in the field of autism spectrum disorders with unambiguously emerging hubs connecting the research (see also Chapter 7 Figure 5) (Figure 6.7). This was very encouraging and provided a glimpse of hope to the scientists that try to advance new quantitative ideas and face the resistance of the opinion-driven establishment. In particular, we received confirmation about the deep concern we had that discrete clinical inventories were driving the scientific enquiry and likely posing serious obstacles to scientific progress through legislation and distribution of available funds at the federal and state levels, but also from private foundations unaware of this pervasive use of subjective, opinion-based data in science.

The very notion of a *neutral objective observer* automatically detecting trends in the physiological/psychological data does not exactly exist in these fields. As such, most research involving any sort of dyadic exchange relies on subjective observational accounts of overt behavior of one person being evaluated or judged by another. These accounts occur at a macro-level of descriptive discourse. Such methods use a *top–down* approach to the definition and study of phenomena whereby overt, unambiguous aspects of the dyadic exchange are consciously recollected and registered by the researcher. Then, rather pedestrian means such as the hand-coding of videos and verbal description of observable aspects of complex behaviors incur in gross data loss. With a sole focus on the affected person, these methods ignore the important role of the clinician who is posing the question, scoring the person and charging money for it.

Information contained in these observational descriptions are often computerized and adaptive methods from, for example, machine learning used in an attempt to capture and quantify obvious aspects of the social exchange. This is at the expense of leaving out important covert dynamic information contained in minute fluctuations inherent to natural actions. While they are treated as *noise* by current *"one size fits all"* approaches, to us they are precisely *the signal*.

The gap that we identified at the level of *Behaviors* in the PM knowledge network will help us begin the path of defining new methods specifically tailored to assess dyadic interactions within the realm of Precision Psychiatry.[2] Such methods have the chance to not only assist clinicians and patients with the process of giving and receiving a diagnosis or treatment; they will also assist the community of mental health in providing longitudinal outcome measures of disorder progression with and/or without interventions. With these notions in mind, we proceed next to examine the somatic-motor physiology underlying various types of dyadic interactions that occur in actual partnering dancers. We then translate such methods to the realm of diagnoses and treatments. This is in the hopes to help us assess the more subtle social dance that takes place beneath awareness, as we engage in the type of social exchange a diagnosis test gives rise to.

BALLET PARTNERING: THE DANCE OF COUPLED BIORHYTHMS

As we set up the stage for motion caption with the Phase Space and started to collect pilot data using their acquisition software; the company sent us data to test our biometrics on. The data was collected from complex dancing routines performed by well trained professional dancers in partnering. Figure 6.8 shows the setup with dancers in T-Pose for

FIGURE 6.8   Dancers in T-pose for calibration of the Phase space cameras and acquisition system. The suits comprise 78 light emitting diodes (LEDs) distributed across the body and head (1−39 on the female and 39−78 on the male dancer).

calibration. They are wearing suits with 78 light-emitting diodes (LEDs) distributed evenly across the two dancers (1:38 in the female and 39:78 in the male). The cameras sampled at 480 Hz, so we had enough data across a variety of highly complex ballet routines and pauses, that is, moments when the dancers rested or planned their next moves. Figure 6.9 shows different outputs of their software (Notes 1) helpful to understand the context in which the data was collected. A sample movie (**Notes** [3]) helps visualize the actual dance and the avatar representation.

We used the skeleton tool the Phase Space software provides to visualize each dancer and map the LED numbers to the corresponding body parts (Figure 6.10**A**). Borrowing my previous model of the arm explained in Chapter 4, Vilmi extended the map to the rest of the body and we built a kinematics map and a graphic representation of it in an avatar (Figure 6.10**B**). We could then later use this avatar in a real-time co-adaptive interface to help children in the spectrum interact with the image of their own movements, to causally trigger sounds and visuals using their own bodily postures and movement speeds. In this way, we could later help them establish cause and effect between the motions self-generated by their bodies, the motions they could visualize in the avatar, and those generated by the dynamic stimuli they triggered (sounds and movies). We will discuss these data at the end of the chapter, but first we need to see how we designed new ways to visualize the unfolding complex dynamic coupling by adapting the types of network connectivity analyses that are commonly used in brain imaging research to the peripheral networks innervating the body.

[3]Movie link to dance in partnering.

FIGURE 6.9 Various outputs from the system aid visualization of complex dances and precise behavioral modeling using robotics driven animation. *courtesy of, Kan Anat of PhaseSpace and Tarik Abdel-Gawad of Bot & Dolly.*

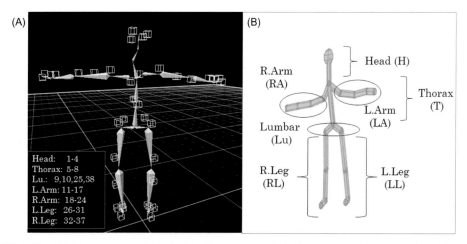

FIGURE 6.10 Building our avatar for visualization purposes. (A) Phase space skeleton to help calibration and tracking of LED's. Inset on lower left corner shows a sample grouping definition for bodily regions of interest and investigation of coupling dynamics. (B) Avatar with similar skeletal structure (sensors 1–38) and labeling of body regions of interest defined for explanatory purposes.

The data from the dancers served as a springboard to develop analytics amenable to dynamically track dyadic exchange between the child and the avatar, as well as between the child and the clinician. In one case, the game-like environment helped the child develop self-confidence while having fun. In the other case, we designed a new type of personalized dynamic diagnostic tool. This tool could track the coupled interactions between the child and the clinician as the interaction unfolded minute by minute.

The position data from the LEDs provided complex kinematic trajectories. For example, Figure 6.11 shows the trajectories from a routine involving complex acrobatics (featured in the movie of Notes 3) captured by the 78 LEDs and shown in two rotated views. Such positional trajectories provide very rich information (curvature, torsion, velocity fields, speed profiles, acceleration, etc.) However, in some instances occlusion of markers (e.g., by clothing) would incur in data loss for the paths generated by other body parts. This required the use of interpolation and spline-based techniques that we developed in the lab to recover the missing frames. Figure 6.12**A** provides an example of one LED trajectory segment with some data loss. This LED is located at the hip of the female dancer and captures the positional trajectory in three dimensions during a complex acrobatic movement. The missing frames can then be recovered and shown in Figure 6.12B where we resampled the original data to have a very fine resolution. Since the motions are so rich in curvature, we used a bending metric whereby we locally measured, from frame to frame, the deviation from the straight line joining the frame at time t to the frame at time $t + 1$. We used local curvature metrics from differential geometry of curves[8] to assess these complex motions. Figure 6.12**C** shows an example of the bending profile for a segment trajectory. This is a small segment, but when we glued all segments continuously performed by the dancers, we can appreciate a full curvature profile for each dancer with all the fluctuations in a complex choreographic routine (Figure 6.13) from one of the curves displayed in Figure 6.11 generated by Marker 23 on the female right arm. These fluctuations in local

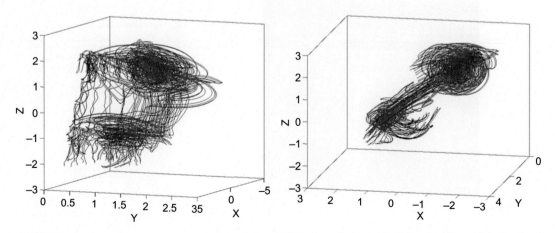

FIGURE 6.11   Complex positional trajectories from 38 LEDs on each dancer (red female, blue male) of a short segment from a ballet partnering routine. Two rotated views show the highly complex and rich kinematics of the dyadic interaction.

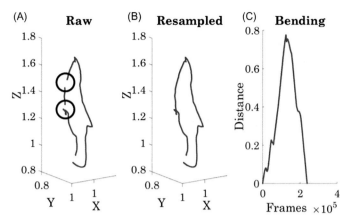

FIGURE 6.12   A new data type to quantify complex trajectories such as those in Figure 6.11. (A) Raw data from the Phase Space positional trajectory of one LED shows discontinuities due to occlusion. (B) Interpolation with splines and resampling techniques developed in the lab helped us reconstruct the positional trajectories, yet the resampling at equally spaced intervals resulted in the loss of the temporal dynamics from the speed. (C) Geometric measure of local curvature profiled over all frames of panel B constructed the data type we used to study the complex kinematics of ballet routines from the dyadic partnering.

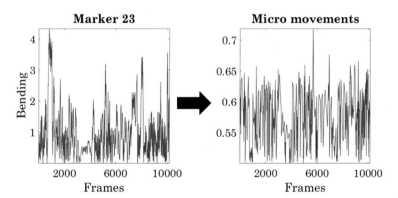

FIGURE 6.13   The local curvature (local bending of the straight line tangent to the local curve) data-type defined in Figure 6.12 taken across thousands of frames provides a waveform for analyses (sampled from one LED marker). The micro-movements waveform extracted from the bending profile normalizes the fluctuations in amplitude from 0 to 1 thus providing a standardized data type of use across the 78 sensors from both dancers interacting in tandem. Note that these fluctuations in amplitude are time-independent. They preserve the order in which the changes in the peaks occurred but say nothing about when they occurred.

bending of the curve (i.e., the curve's moment-by-moment's departure from the local straight line) were then transformed to the micro-movements waveform by normalization between 0 and 1 using formula 3 in Table 1 of Chapter 3. The result is shown for one LED of the female in the right panel of Figure 6.13.

The micro-movement trajectories were obtained for all markers. Then we performed power spectrum analyses and pairwise coherence analyses for all 78 LEDs. The pipeline of data analyses to obtain the coherence profile and the cross-spectrum phase portray of the data across all frequencies and lead-lag shifts are shown in Figure 6.14A−C for 2 sample markers from the female (red) and male blue). Once the cross-spectrum coherence analyses were completed, we built adjacency matrices to convert the data to a weighted directed graph and perform various connectivity analyses. The entries of the coherence-related matrices on the top panels of Figure 6.15A reflect the pairwise coherence, phase lead-lag values and frequency values for entry [i, j] in the matrix. In the case of the coherence they have a value ranging between 0 and 1, with 1 representing completed synchronized wave-forms and 0 representing asynchronous data (color coded by the color bar). Lead-lag phase information is reflected in the Phase matrix ranging between −180 and 180 degrees (see also color bar for values). In this case, we retained the positive range with the convention that an entry in [i, j] contains the leading phase for the given coherence level of the [i,j]-pair. As such, the arrow of the directed link in the graph will point from node i to node j. Lastly, the entry in [i, j] for the Frequency matrix (see color bar) represents the frequency value for which the coherence and lead phase information were registered. The bottom matrices are those corresponding to the data for the positive phase lead information used to build an adjacency matrix of a weighted directed graph. Using our convention, the weight is the lead-phase angle. Node i leads node j according to the positive phase value, so the arrow points from i to j. The node value is the coherence value for the given frequency. Each frequency in this case generates one such an adjacency matrix such as those in Figure 6.15A. Then Figure 6.15B provides a snapshot of the female and male dancers in T-Pose whereby each node size is the coherence; each arrow is the leading direction and the thickness of the arrow is given by the weight (i.e., the value of the lead phase angle for [i,j]). This configuration can then be tracked across the dancing routines for the various frequency bands.

This weighted directed graph representation is expanded in Figure 6.16 where we show the three types of networks we track. First, we focus on each individual dancer, as the inner links provide information about which body parts lead which body parts, as well as the degree of intensity with which one dancer is leading the other. The number of incoming and outgoing links in a given node provides a measure of the connectivity strength of each node and its potential role as a hub of the network, so we can use this number in each node to visualize the leading information of a given body part over another. Likewise, we can track which body parts in one dancer leads which body parts in another dancer. The modules emerging in the inner network of one dancer provide information about internal bodily synergies of that dancer. Modules are sub-networks within the body which have maximal inner connectivity and minimal outer connectivity with the rest of the network. Their information is a read-out of that dancer's motions broadcasting it to us (external observers) from the internal vantage point of the dancer's nervous systems. We will soon see how we use modularity as a metric that helps us track synergistic relations in each dancer's body but also in the coupled network of the two dancers as this network's activity dynamically unfolds.

We can track connectivity in two ways: Top down (externally defined and directed by us) and bottom up (letting the internal self-emerging patterns lead us). Ultimately the choreographer and the dancers have to figure out the goals of their performance, so they can

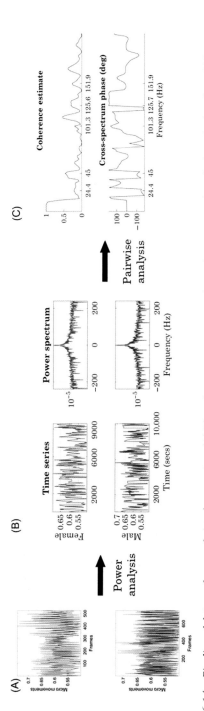

FIGURE 6.14 Pipeline of data analyses and signal processing. (A) The micro-movements waveforms from two markers (red on the female and blue on the male) are used for a pairwise estimation of coherence upon power spectrum analyses (B) and coherence estimation (C) Further cross-spectrum phase analyses are used to identify peaks at relevant frequencies and lag-lead information.

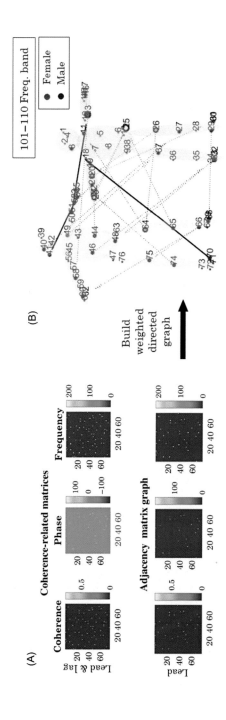

FIGURE 6.15 Representation of the peripheral network to track individual patterns of synergies and coupled synergistic dynamics between the male and female dancers. (A) Coherence, phase lead-lag and frequency matrices whereby each [i,j] entry represents a value from the pairwise coherence analyses between the micro-movements waveforms between the activities of two LED markers. Coherence values, lead-lag phase values and frequency values are represented in the corresponding color bars of the top rows. The bottom rows represent the adjacency matrix for the representation and dynamic tracking of the activities of a weighted directed graph. These nodes represent all nodes of the network and links from the coherence matrix values. Node i connects to node j with a leading arrow whereby i leads j when the phase angle (positive value) i leads j. Further the thickness of the arrow reflects the amount of lead phase angle with increasing thickness representing higher positive values. Nodes' size reflects coherence value and marker face color represents modularity information. Other connectivity metrics can be used to represent each state of the network for a given frequency band.

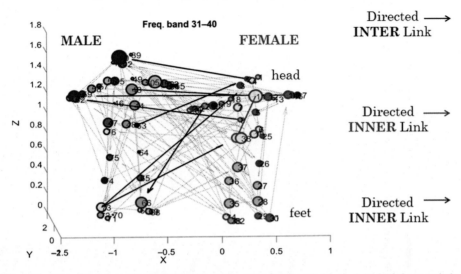

FIGURE 6.16 Visualization tool to track the dynamics of each network (dancer) and of the coupled network as well. Here a frame is shown for a given frequency band comprising values between 31−40 Hz. Colors of the marker represent modules while links are entry [i,j] connected by a coherence value reflected as well on the size of the marker. Directed arrows represent lead information of node i over node j. Blue links are within the male dancer network, red are within the female dancer network and black links are the coupled network.

adjust their performance iteratively toward that goal, while balancing the top−down *vs.* bottom up types of feedback that these patterns offer as they unfold during the performance. For example, using the *top−down* approach, we can define body blocks informed, for example, by the choreography, to learn how the dancer's rhythms entrain those body blocks, so as to produce harmonious coordination with very smooth transitions from one part of the piece to the next. In any given moment, the choreographer and the dancers may want to know how the upper body, the lumbar region and the lower body connect during a routine (for instance). Network connectivity metrics can provide this information. Leading information can be extracted from the incoming and/or outgoing strength of a node, defined by the number of incoming or outgoing links respectively, as obtained in the nodes comprising the coupled network (marked by the black arrows inter link in Figure 6.16).

Figure 6.17 provides an example of tracking the information on *"who leads who"* for these selected body parts for the female and male dancer, during the dance and also during the resting periods while they plan ahead the next routine. In this case, the upper body includes the arms, the head and the trunk area, while the lower body includes the hips, legs and feet. The lumbar area is tracked separately to examine upper−lower body leading patterns in relation to the lumbar area.

This is an arbitrary selection just to show an example of network activity as it unfolds across different frequency bands to identify which bands are informative of a particular feature in the partnering routine. In this case, the lower extremities of the male dancer are

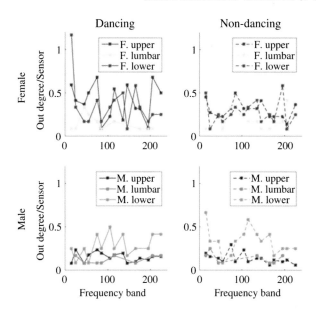

FIGURE 6.17 Top–down tracking of which body parts lead which body parts in each dancer during dancing routines or during breaks, while planning ahead other routines. Bodily regions of interest to track leading activity are as defined in Figure 6.10. Activities are summarized from the network states corresponding to each 10-based frequency band.

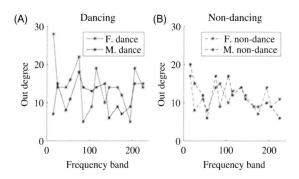

FIGURE 6.18 Who leads who? Activity profiled from the out degree metric (measuring the net number of outgoing links from the female or male nodes of the coupled network) during the dance for each of the 10-base frequency band under examination. (A) Dancing routine reveals female leading male dancer for some frequency bands and lagging behind the male in others. (B) Planning ahead other rehearsals reveals male's generally leading activity over female's activity.

leading over the lumbar and upper extremities, while the upper body in the females leads over the lower body and lumbar areas.

The patterns can also be obtained for the breaks when the dancers are not dancing but planning ahead. Likewise, Figure 6.18 shows an overall lead-follow profile of the female and male dancing patterns taken on average across the full body. In this case, the patterns are profiled across some frequency bands where the female motions lead the male motions, so the user can "zoom in" and examine the information transmission for selected frequency bands depending upon the patterns or body parts one may query about.

While selectively examining body parts and coupled patterns is very useful to deliberately plan ahead and finesse the performance practice, other connectivity metrics may be more helpful in providing information from a *bottom up* perspective. That is, using a perspective that emerges independent of our deliberate selection or overt decisions. This type

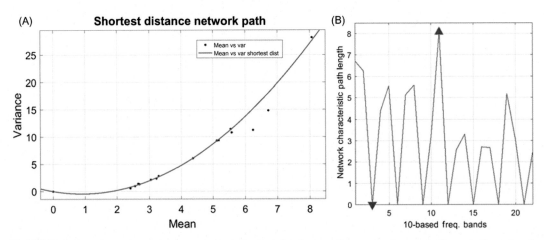

FIGURE 6.19   Network connectivity metric reflecting the shortest distance path averaged across the network reflecting the characteristic path length to communicate one node to another. As the mean increases over the nodes included, so does the variance with a quadratic trend. (B) Difference characteristic path lengths for different 10-based frequencies reveal extrema useful to further examine network's states and automatically detect self-emerging patterns in bottom—up analyses.

of information that is contained in spontaneously self-emerging patterns of the network (as they unfold over time) is also very important feedback to help the choreographer and dancers adjust performance toward desirable goals. For example, it is possible to use the connectivity weights and traverse the network node by node along the links with maximal weight leading to the shortest distance connectivity path from one body part to another.

   This would be useful to identify synergies in each dancer's body. It would also be useful to identify synergies between body parts of one dancer and the other dancer. Figure 6.19 provides information along these lines for the network's characteristic path length. This is the average shortest distance path connecting the network's node. In this case, the nodes are linked body locations (nodes) with specific node-to-node coherence values that we can track explicitly through phase leading values (directed weighted links represented by arrows in Figure 6.16) to reveal which nodes are maximally entrained, as well as which nodes lead performance. Recall here that the weights represent the phase leading amount. Thus, navigating the coupled network along the maximally synchronized nodes with maximal lead gives us a sense of the leading dancer in each frame. It also tells us which body parts in that dancer are responsible for the lead. An interesting pattern in Figure 6.19**A** is that the shortest characteristic path length has the lowest variance. As the mean characteristic path length increases, so does the variance along a quadratic trend. In Figure 6.19B, we can identify the frequency band with the maximal characteristic path length and the frequency band with the minimal characteristic path length. Then, we can examine in more detail the patterns of connectivity corresponding to those frequency bands. Figure 6.20 unfolds the corresponding patterns in the adjacency matrix of coherence values (Figure 6.20A maximal and 20D minimal characteristic path length, respectively) and in the matrices providing information about the shortest distance paths, that is, the paths that connect body parts that are highly synchronized (along the nodes with higher coherence values) (**20B** maximal, **20E** minimal). Recall here that the

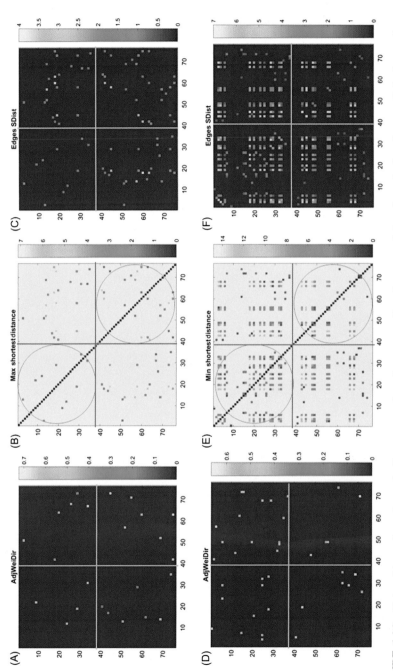

FIGURE 6.20 Network's configurations of adjacency matrix and corresponding shortest path lengths and edge distance. Color bars reflect range of values in each case. (A-C) Dancing routines; (D-F) planning rehearsal while standing around or walking but no dancing. (A and D) Adjacency matrices corresponding to one network state (one frame) reflects the entry [i,j] pairwise coherence value for dancing routine and rehearsal plans. Upper left quadrant (markers 1–38) reflects pairwise coherence in the female nodes, while lower right quadrant (markers 39–78) represents pairwise coherence in the male nodes. The upper right quadrant (cross talk among 1–38 and 39–78) and lower left quadrant (39–78 and 1–38) represent pairwise coherence of the coupled networks female-male and male-female, reflecting synergistic activity of the coupled dynamics of the two dancers as one leads the other. (B and E) The pairwise shortest distance paths. (C and F) The pairwise shortest edge's distances.

weights are given by the phase lead angles which inform about synchronized body parts that lead other body parts by a certain amount. The larger the amount the *"faster"* to entrain ahead a body part with another. This can be quantified between dancers or within each individual dancer. Panels **20 C** and **20 F** provide the patterns in terms of the edge-weight information. As with panels B-E, the circles highlight the information of the quadrant representing the inner networks of the female dancer (upper left quadrant) and the male dancer (lower right quadrant), whereas the other two quadrants of the matrices are the coupled-networks of the female leading the male (upper right) and the male leading the female (lower left).

The frame in Figure 6.16 represents the female network's directed links in red, while blue links represent the male's directed links. Then, the coupled network is represented through the black directed links between the two dancers (i.e., the third neutral observer that self-emerges from the entrainment of their biorhythms). In this configuration, the feet of the male lead the lumbar and the head areas of the female, while the trunk areas of the female lead the right arm of the male. Further the head of the female leads the head of the male in this configuration of the network. Clearly this is still a very complex way to examine the coupled motion data. As such, other tools are required to analyze the various configurations of the network in rather dynamic ways. The movie online (Notes 3) provides ways to visualize the unfolding dynamics of who leads who and what body parts lead which body parts in each dancer individually and as a coupled dyad.

As we have seen, one of the advantages of visualizing the peripheral network in motion as an unfolding interconnected weighted directed graph is that we can utilize various connectivity metrics to summarize and track the patterns of synchronicity and coordination that emerge and dynamically shift throughout the dance. For example, frame by frame,

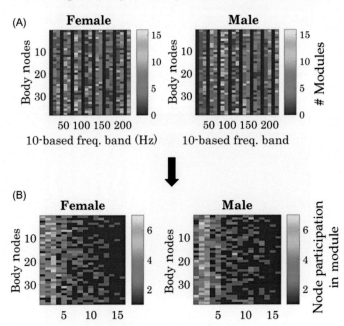

FIGURE 6.21 Bottom—up approaches using network connectivity metrics such as modularity to quantify synergies as they unfold in type. Modularity metric defines subnetworks for each network state whereby nodes are interconnected maximally within the subnetwork and minimally with other external nodes. (A) For each dancer the modularity metric is obtained across every 10-based frequency band (color bar reflects the number of modules found in each band for each participating node). (B) Counting the participation of each node in each module yields another matrix of use to track various types of coupled dynamics.

we can obtain a measure of modularity for each dancer's inner network and for the inter-dancer's dynamically coupled network in Figure 6.21A–B. Modules in this case are semi-independent sub-networks that are maximally internally connected and minimally externally connected with the rest of the network. In Figure 6.16, we showed a representation of the modules using different colors. Some modules emerge as synergies within the individual's network, while others represent synchronous activities that cross over and extend to the partner, thus revealing coupled synergistic patterns.

An example of the latter was given in the cross-over pink module of the coupled dyad in Figure 6.16. The size of the node in this case corresponds to the number of links entering and leaving the node. This representation provides a visual sense for hubs in the network (i.e., important nodes that if eliminated impede overall connectivity and information transfer throughout the network). Using the dynamic unfolding of the network for 10-based frequency bands, for instance, we saw that we could profile the shortest-distance paths connecting one node to another (Figures 6.19–6.20). This connectivity metric reveals the most efficient path to carry information from, for example, the feet to the head, or the arms to the trunk.

Another metric of modularity can reveal systematic evolution of modules in the network as they change across frequency bands. An example of such profiling is provided in Figure 6.21A for each dancer. Once we obtain the modules, we can count the participation of each node in each module, that is, the number of times a node (a body part) participates in each given network module. The outcome of this example is displayed in Figure 6.21B for each dancer. This information can then be used to examine for each body region that we define involving different nodes (e.g., the head, trunk, lumbar region, left arm, right arm, left leg, and right leg) which nodes participating within a given region, do so simultaneously in the male's and the female's body. To that end, we can choose (for example) as a threshold the 50% value of the maximal number of node participation (Figure 6.22A) and count the simultaneous participation of body parts of the female dancer and the male dancer (marked with stars in the figure) to build a metric of coupled activity. Figure 6.22B provides a representation for the first Module found by such a metric. Since we came up with it, we coined it a metric of "togetherness" of the dancer's body parts and Figure 6.22C provides the visualization of the coupled body parts in the avatars of the dancers. We highlight the simultaneous synergies in yellow for both dancers. These are the body parts of the dancers that are synchronized in the first Module (out of the 16 detected across the network comprised by all 78 nodes).

We came up with this representation to provide a new language to describe synergies in complex dyadic interactions such as those of the partnering routine. As the movie unfolds frame by frame we compute the modules' synergies and display them on the dancing avatars (see movie in Notes 3). Underlying these modularity-based coding of synergies are also stochastic signatures of the biorhythms we obtain from the Gamma process we use to empirically model the fluctuations in the peaks (i.e. the amplitude) of curvature. We profile the stochastic signatures including estimates of the shape, dispersion (scale), mean, variance, skewness, and kurtosis of each body part empirically estimated using a Gamma process. These are plotted on the Gamma parameter space in Figure 6.23 along with the estimated PDFs for the dancing routine (panel **23 A**) and for the breaks, when the dancers plan ahead their next motions (panel **23B**). The color bar using a gradient of green shades represents the level of noise to signal ratio (NSR) in the fluctuations of local

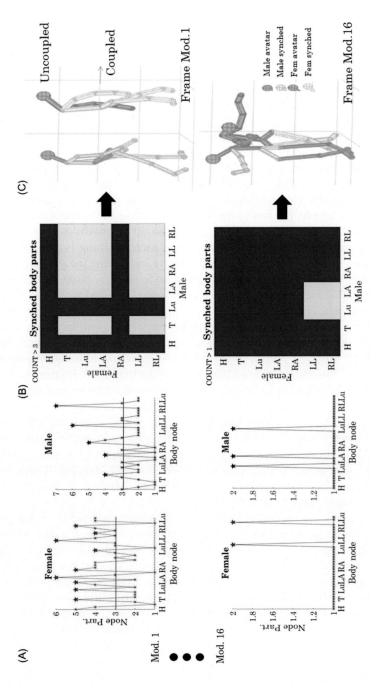

FIGURE 6.22 Our definition of a top–down metric of "togetherness" across the dancer's bodies. (A) For each of the regions of interest defined in Figure 6.10 we obtain the number of times that a dancer's body region is coupled with the other dancer's body region. If the body part exceeds over 50% of the count of node participation a star marks that region. Then regions that are participating together above half the time are marked as "together" in yellow (1) or as (0) in blue to mark they are not synergistically coupled. (B) Binary matrix used to represent togetherness. For example, in this case, the thorax/trunk region of the male is synergistically coupled with the thorax/trunk region of the female and also with the lumbar and left arm, and with the left and right leg, but not with the head and right arm. (C) The avatar representation of this frame color coded with the togetherness information derived from the node participation extracted from the modularity metric. This analysis is performed for other modules as well (e.g., shown here is module 16 information and dancer's code.)

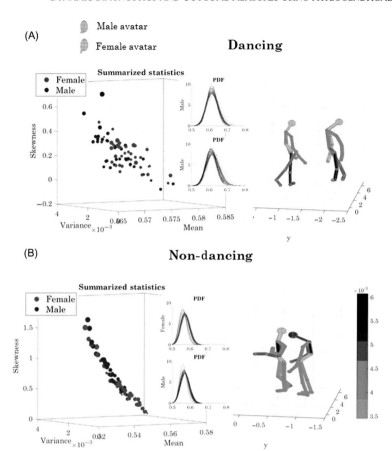

FIGURE 6.23 The stochastic signatures underlying the micro-movements fluctuations in amplitude of the local curvature (bending metric) are profiled for the dancing and nondancing examples discussed throughout the chapter. (A) Gamma process yields the summary statistics space encompassing all Gamma moments with insets showing the empirically estimated probability distribution functions (PDFs) and corresponding avatars color-coded according to noise to signal ratio (Gamma scale parameter) in (A) dancing and (B) nondancing. Color code shows the noise range from lower to higher coloring the dancers by togetherness metric in Figure 6.22. Movies showing all frames in movies available through link in Notes 3.

curvature peaks. According to the synergies we uncovered in Figure 6.22 we can also reveal the unfolding NSR of those coupled body parts and color code the avatar accordingly. This provides another visualization tool that objectively and automatically profiles the complex coupled dynamic motions of the partnering dyad as these motions unfold. The color-coded avatars are shown in Figure 6.23.

DYNAMIC DIAGNOSTICS AND OUTCOME MEASURES USING PHYLOGENETICALLY ORDERLY TAXONOMY

The computation of the unfolding stochastic signatures harnessed from the nervous systems along with the network connectivity patterns revealing synergistic interactions in the dyad allow us to use real-time feedback whereby the biorhythms read out of the nervous systems could be statistically parameterized in a personalized manner. In turn, this parametric signal could be used as guiding feedback input to the system to steer the performance of the person toward neurotypical (NT) regimes. Since we were tapping into different levels of

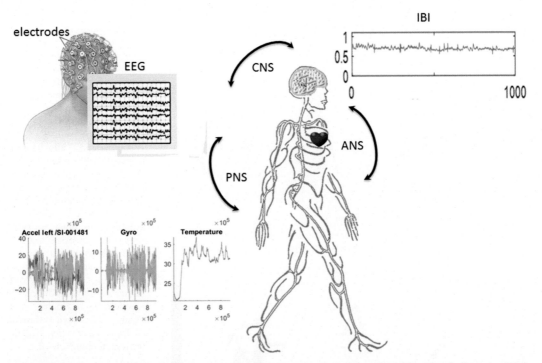

FIGURE 6.24   Acquiring data from the nervous systems in tandem. Different biorhythms can be harnessed from the central (cerebro-cortical EEG), peripheral (somatic-motor IMU activity) and autonomic nervous systems (heart rate interbeat interval timing ECG) using wireless instrumentation and streaming live synchronously with LSL.

the nervous systems (voluntary, automatic, involuntary and autonomic) we were able to simultaneously, in real time, probe for the most effective channel, among the central, somatic-motor and autonomic channels to measure the changes we were evoking in the person's systems (Figure 6.24). Most important yet, we can measure if the change is toward a good stochastic regime or a bad stochastic regime (Figure 6.25). In the case of a good regime, that is, consistently trending toward low noise and high predictability in the right lower quadrant (RLQ) of our Gamma map, we would encourage the process and at the same time identify the stimuli that caused the system such systematically positive change. In the case of bad regime, that is, consistently trending toward the left upper quadrant (LUQ) of our Gamma map, we would switch to different stimuli and conditions. Since the stochastic process is also dynamically changing, it is important to have changing criteria as the system co-adapts with the stimuli and conditions. These include as well interacting with the experimenter or therapist, which in our setup is being measured as well and this information considered as an important (social) source of sensory input.

In Chapter 3, we introduced our proposed taxonomy to dynamically classify nervous systems disorders and track their progression in response to treatments. Here we use this dynamic classification scheme to probe the person's systems and determine the capabilities and possible coping strategies the system has already developed, so that the interventions

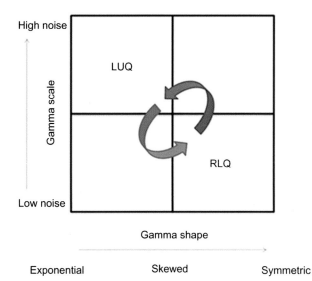

FIGURE 6.25 The dynamic Gamma parameter plane to track the evolution of the stochastic signatures of the biorhythms' micro-movements as they unfold in time. High noise-to-signal ratio (Gamma scale axis) and exponential signatures in the LUQ indicate elevated levels of spontaneous random noise thus warning of a problem if the biorhythms express a tendency to remain stuck in that quadrant. Symmetric shapes and low noise in the RLQ are expressing positive outcomes if the biorhythms' signature shifts to that quadrant often. As such the amplitude and frequency of these shifts over time can be used to assess the neuroplasticity of the system or the level of readiness or the effectiveness of a treatment.

we design are tailored accordingly to what the person's nervous systems feel comfortable with and codes best. For example, we first learn about the cross-talk among the fundamental processes of the nervous systems that we defined in Chapter 3 based on the phylogenetically orderly taxonomy involving the central nervous system (CNS), the somatic-motor peripheral networks and the autonomic and enteric systems of the peripheral nervous system (PNS). To that end, we use a simple experimental paradigm that probes each system noninvasively. The task involves naturally walking under three different conditions: baseline condition; metronome condition; paced-breathing condition set to a metronome (Figure 6.26) hoping to elicit activities of various kinds across the central and peripheral networks of the nervous systems.

These three conditions were designed to explore deliberate processes likely driven by the CNS via precise goal-directed instructions; spontaneous processes involving the peripheral somatic-motor networks operating largely beneath awareness but nonetheless subject to the central influences and having as well *"a mind of their own"*. Along these lines, peripheral activity is seldom taken seriously in its own right; but we believe that this network of efferent and afferent nerves is doing more than just passively transmitting feedback from movement, pressure, touch, temperature and pain. In our conceptualization of the nervous systems functioning, we see these somatic-motor components as a form of distributed intelligence with a degree of autonomy and self-supervision necessary to be not just a pipeline of information to feed central controllers in the brain and spinal cord, but actually to gait and modulate the information with some degree of autonomy and independence. There is some evidence of that in mouse models of nociception[9] but the idea as of today may seem too far-fetched to be taken seriously in other research domains. The pain network is rather old and runs parallel to the kinesthetic network, so a question we have is whether they develop similar gating and modulatory mechanisms to autonomously control the flow of information through self-supervision.

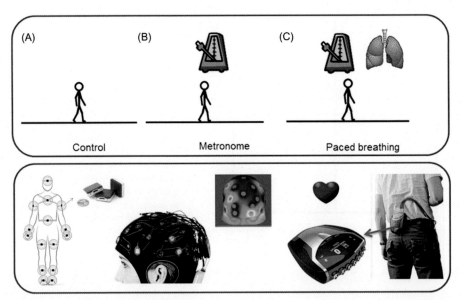

FIGURE 6.26 Experimental paradigm and instrumentation to assess deliberate, spontaneous and inevitable processes of the nervous systems as cross-talk among the CNS, PNS, and ANS unfold. (A) Walking automatically without a care in the world provides the control (baseline) task to evaluate the interactions across these systems. (B) Setting a metronome in the background provide means to spontaneously (without instructions) entrain the biorhythms of these systems and processes. (C) Instructing to deliberately breathe at the pace of the metronome give a way to assess voluntary control or action at will (volition) in the person. Bottom panels show wearable sensors: (APDM IMUs positioned across the body and synchronously sampling at 128 Hz); wireless EEG from Enobio sampling at 500 Hz and wireless Biotrace Nexus sampling at 256 Hz the heart rate variability. All wireless sensors are listed in LSL with their APIs and code to synchronize their output and record activities in tandem with full knowledge of millisecond delays during live data streaming.

In my mind, self-supervision is acquired by the immune systems of the embryo during gestation, when the systems cross-talk with the mother's immune system and learn somehow to get along well. A nervous system with poor, underdeveloped or absent self-supervision may not go on to acquire the ability to distinguish self-generated information from externally generated information. This in turn would impede fully autonomous and intelligent self-control, and predictive control of the sensory consequences from actions generated by the self and by others. I discussed some of these ideas within the context of the mirror neuron systems theory in Chapter 2, but it is good to revisit them here since autism may be an instance where at some point in neurodevelopment these mechanisms take a different route.

We do explore our hypothesis of distributed intelligence in the peripheral networks. Along these lines, we also believe that a coping mechanism of neurodevelopmental disorders could help one aspect of the problem but in so doing "disconnect" the peripheral networks from the central controllers; or obstruct the cross-talk with centrally driven processes. We thought it was worthwhile exploring that idea in autism and besides the CNS and somatic-motor peripheral networks, assess the autonomic heart signal in tandem with the rest of the signals. More specifically we defined *inevitable processes* of the autonomic nervous system (ANS) using the waveforms form interbeat interval (IBI) timings and their signatures of variability while participants performed the three conditions of the walking task in Figure 6.26**A**−**C**.

It was challenging to create a synchronous setup whereby all instrumentation could be aligned to simultaneously stream the data. But during the spring semester of 2016 I had a team of biomedical engineering students completing a senior design project in the lab and one aspect of the project was precisely to simultaneously connect various instruments and stream life data in tandem. They contributed to the Lab Stream Layer community (LSL) (**Notes** [4]) and incorporated various aspects from instrumentations of our lab to the community website. Joe Vero, one of the team members remained in the lab while completing his master's degree in Biomedical Engineering and taught the other students and postdocs how to accomplish such feat, so they could synchronously acquire data from multiple instruments. This helped us integrate the data in novel ways to learn co-dependencies and cross-talks among the three fundamental processes we defined. Joe wrote the Python code to parallel the Matlab code in the companion website of this book, so our biometrics can also be used by those who do not have access to Matlab.

My own PhD student Jihye Ryu spearheaded the electroencephalographic (EEG) project and was very happy to have her work accepted at the Qualcomm Innovation Fellowship Competition in 2016, where she and Vilelmini Kalampratsiduo were finalists. We all flew to beautiful San Diego and they presented their work to a very interesting crowd. Upon their return to the lab, each one followed a slightly different project of their own that stemmed from the original idea of exploiting all these signals in tandem for real-time stochastic feedback-based selective control designed in a personalized manner. The word *selective* is very important here because it refers to the ability our methods offer to allow probing multiple processes and automatically (without our intervention) discover the best process endowed with the capacity for change—that is, identifying the channel(s) with the highest plasticity.

In the case of Ji's project, she pursued the classification scheme proposed by the walking experiment so as to learn where the performance of different people fell along the deliberate, spontaneous and inevitable nervous systems processes. The criteria were based on the evolution of noise and randomness detected in each bio signal along with levels of cross-talk among these processes illustrated in Figure 6.24. For example, if the noise and randomness levels were highest for the heart's IBI this would place the subject as a good candidate for us to monitor such autonomic processes and measure the somatic-motor and EEG signals relative to the heart's signals. If the heart signals, at the core of autonomy in the system, are not properly working and the communication of adequate pacing of other systems is weak, then the problem is far more serious than not having representational thoughts or not feeling the body properly because of excess noise and randomness in the other bodily biorhythms. In fact, when we examined the outcome of the signatures of the heart's IBI in the group of autistic patients we studied, and found excess randomness and noise in their heart biorhythms, we suspected something far more difficult was preventing proper entrainment with biological motions present in the social scene. It then made sense

---

[4]Movie link to EEG-body networks.

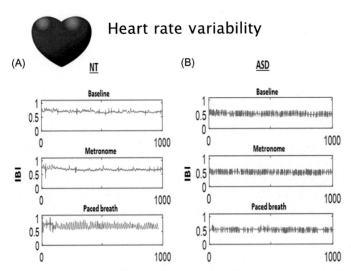

Heart rate variability

FIGURE 6.27   Autonomic signals that do not entrain in the autistic system. (A) Neurotypical (NT) representative IBI raw signals for each of the three conditions changes as the person walks automatically or spontaneously entrains with the metronome, or deliberately attempts to breathe at the metronome's pace. The biorhythms then acquire periodicity and visibly change signatures. (B) The same task does not evoke such changes in ASD representative (a pattern later found more systematically in others) whereby the biorhythms remain random and noisy throughout all three conditions. No visible cross-talk among the deliberate, spontaneous and inevitable processes manifest in the ASD case. This is very worrisome as the heart autonomy is at the core of all other biorhythms.

to have found all along the excess noise and randomness at the voluntary and involuntary levels of deliberate and spontaneous processes in thousands of autistic participants.[2,10] If the heart was not timely functioning and entraining with other biorhythms, designing a proper noise-cancellation therapy would require more thought on our part (though influencing the heart rhythms through other channels would not be impossible.)

A sample case is shown in Figure 6.27, where we contrast the IBI signals of a typical representative participant to those of a participant with ASD. We thought it was interesting to see that controls such as this representative participant modulated the IBI signal across conditions during the walking task, but the ASD participants' signals did not change in the same way as those of NT. For example, the metronome brought spontaneous periodicity to the signal of the NT and then the periodicity was even more evident when deliberately breathing at the pace of the metronome upon being instructed to do so. The ASD representative did not have such modulation. No matter what the condition was, the IBI signal remained random and noisy when we plotted it on the Gamma parameter plane. Figure 6.28 shows this for the representative NT participant and for two representative ASD participants of different sex. This lack of handshake or communication between the deliberate voluntary control from the CNS, the peripheral somatic-motor signals from the kinesthetic flow and the temporal dynamics of the heart rates was very troublesome for us. If the heart is not properly talking to the other components of the nervous systems, then that child's biorhythms are bound to be in disarray. *How could we help such cases regain autonomy and auto-regulation?*

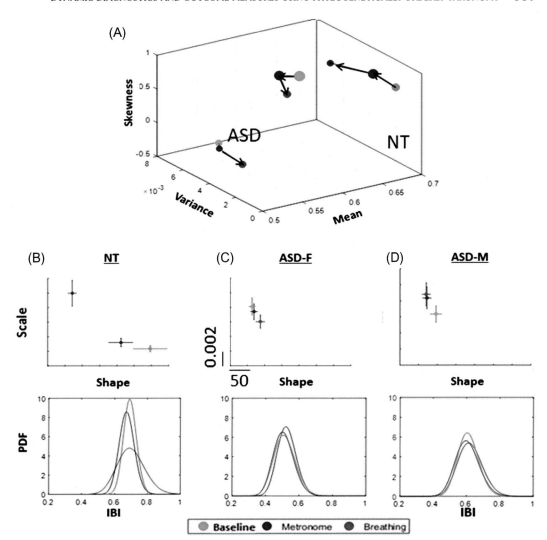

FIGURE 6.28 Gamma process representation of the empirically estimated stochastic signatures from the IBI traces (e.g., such as those in Figure 6.27). (A) The Gamma summary statistics space spanned by the estimated mean (X-axis), variance (Y-axis), skewness (Z-axis) and kurtosis (marker size). NT control spans different signatures for different conditions while ASD representatives have very narrow bandwidth of values across all axes. (B) Gamma parameter plane reveals high noise and randomness in the IBI micro-movements while controls have a broad range of features with automatic walking revealing the lowest dispersion and highest symmetry in the empirically estimated signatures. ASD participants show very similar PDs across conditions suggesting poor corss-talk between inevitable processes of the autonomic system and spontaneous process from the periphery and/or deliberate processes from the central nervous systems.

Wireless technology has advanced to the point that we can easily use wearables and stream live the signal from the nervous systems across multiple layers of autonomic, automatic and voluntary control, including as well spontaneous involuntary motions embedded in these biorhythms. Figure 26-botom panel shows some sample instrumentation and sensory positioning that allows for naturalistic tasks and harnessing of nervous systems' biorhythms in noninvasive and mobile ways. We overcame the challenge of synchronizing such signals to bring them to a similar scale. Then, we were able to integrate them in ways that optimized detecting possible cross-talk, randomness and noise levels across the proposed layers and processes' taxonomy.

We realized during the signal processing stage that somehow, we were facing a similar problem to that of the sensory-motor systems; namely sampling at different frequencies and transmitting transduced signals throughout the nervous systems at disparate time scales. As the nervous systems do, our instruments too sample at different frequencies and different times. Thus, a set of accelerometers embedded in IMUs may all sample at the same frequency but start-stop independently thus rendering very challenging the pairwise comparison of, for example, comparing what the right hand does with respect to the left foot. Their output signals may be shifted in time despite similar sampling rates. This was part of the challenge we faced but were able to alleviate using LSL. By synchronizing the output of multiple sensors streaming live simultaneously we could estimate with millisecond time precision the disparity in instrumentation speed and transmission times from one instrument to another synched to the same computer CPU clock cycles. In this way we could perform cross-coherence and phase locking analyses and use different lead-lag combinations to find adequate frequencies best broadcasting the changes in the signals across the brain, body and heart.

How the nervous systems manage to do something similar to what we did with our instrumentation, that is, synchronize all flows of information, process them in tandem and then somehow time the signals transduction and transmission to perceive the physical world at once must be one of the most difficult problems to resolve. In this sense, I think proper cross-talk among all these systems is very necessary. If they function in a disjointed way (like the way the heart IBI signal was suggesting for the ASD participants) then the whole perception of the physical world (emerging from the processing of sensory motor signals) is bound to become compromised. To a person with such disconnected signals perceptually experiencing the simultaneity of physical phenomena must be challenging. Social interactions including physical bodies in motion and micro-gestures experienced through multimodal sensory motor channels at disparate time scales and frequencies must therefore suffer or be somewhat impeded under such conditions—as we suggest in Chapter 2 while rethinking the mirror neuron systems theory for autism spectrum disorders.

There were many products in the market to choose from to harness biorhythms from the nervous systems in noninvasive ways. However, we wanted to be economical in our budget without sacrificing quality. It took some searching and enquiring, but we settled on a subset of instruments we could afford while maintaining high research grade quality. For this study, the acceleration of the participant's bodily motion was captured using Opal IMUs (APDM Inc., Portland, OR) at 128 Hz sampling rate, and acceleration signals streaming in real time were acquired with Motion Studio (APDM Inc., Portland, OR).

Participants wore ten-twelve opal IMUs with Velcro belts on the wrists, ankles, foot, upper arm on both right and left sides, and on the posterior trunk and anterior chest (shown in Figure 6.26 bottom panel left). We captured the cortical potentials using the Enobio wireless EEG device (Barcelona, Spain) at 500 Hz sampling rate with 32 sensors positioned across the scalp. The EEG recording device was positioned on the back of the participant's head (yellow device in Figure 6.26 bottom panel center), which contains an inertial measurement unit (IMU) that records head acceleration at 500 Hz. Data from this device were recorded by NIC Neuroelectrics (Barcelona, Spain). The heart signals were obtained from a wireless Nexus-10 device (Mind Media BV, The Netherlands) and Nexus 10 software Biotrace (Version 2015B) at a sampling rate of 256 Hz. Three electrodes were placed on the chest according to the standardized lead II method, and were attached with adhesive tape (shown as well in Figure 6.26 bottom−right). As mentioned, we synchronize the signals from all three devices—EEG, electrocardiography (ECG) sensors, and Opal IMUs, using open source package LSL. We chose LabRecorder and Mouse to event-mark the EEG by mouse clicks on the display screen of the computer, from where the software interfacing the sensor devices (e.g., Motion Studio, NIC, and Biotrace) were running.

Here we underscore that one of the challenges of mixing natural movements such as walking and EEG is the types of motion- and mechanical-artifacts introduced in the EEG signals. Such artifacts may emerge from possible movements of the head which in turn may move the head cap and alter the positions of the leads; from eye blinks and jaw motions (simply swallowing could generate artifacts), among others. To help remove motor and mechanical artifacts we not only used conventional methods in the literature[11] but also added our own methods to automatically detect various problems with the signal. First, we band passed the signal to remove 60 Hz AC current. Then, in order to remove motion-artifacts in the 60Hz-band-passed cortical data (aside from the usual ICA and automated eye-blinking removal[12]), the EEG data were further band passed at 16−31 Hz (i.e., beta frequency band) and that band signal examined over time. Specifically, we measured the magnitude of the rate of change of the cortical signal against the magnitude of the rate of change in the head IMU's acceleration values (head jerk). Coherence analyses of their amplitude-fluctuations' micro-movements informed of their coincidences. More precisely, acceleration peaks of head jerks were followed by peaks of change in cortical signals at the beta band frequency after approximately 40Ms, implying that these cortical signals reflected head motion artifacts. When 0.1% highest peaks of head acceleration (i.e., head jerks) occurred, the acceleration signals were compared against the beta band cortical signals via cross correlation, and if the cortical signals lagged the head acceleration peaks, these instances were excluded. Overall, this resulted in eliminating approximately 0.001% of the entire data. Subsequently, the cortical data from each channel were referenced by the channel that had the least noise according to the Gamma process outcome. Specifically, for each channel, the peaks extracted from the fluctuations in the amplitude of the cortical waveform were studied as a Gamma process and the channel with the lowest scale value (NSR) was chosen to have the lowest noise and set as the reference. This automated rereferencing helped us isolate the highest signal-to-noise content in the EEG to effectively examine it in relation to the other instrumentations' outputs.

To bring all signals to a common scale we up sampled to 500 Hz the acceleration data obtained from the Opal IMUs and the ECG heart data using cubic spline interpolation.

In this way, they could later be analyzed at the same sampling rate as the EEG brain potential data. Once we had all the data synchronized, sampled at similar rate and denoised from the artifacts, we proceeded to create the micro-movements of each signal and examined the pairwise cross-coherence of all the sensors involved in the EEG and body. We examined the heart signals separately for the time being to illustrate the methods in a simpler way. Yet everything in the methods applies to the heart as well. We would just have to add another channel to the analyses, that is, examine the pairwise comparison of the heart rate data with all other leads so they would be another arrow in the resulting matrices (see below).

The micro-movements provide the normalized amplitude spike train representing the fluctuations in the order in which they occurred. That is, in a series of micro-movements, the order of the original signal's peak values is preserved, but the actual frame/timing values are lost. To recover all the frame values and preserve the sampling resolution of the original signal, we set the nonpeak values to 0 and superimpose the micro-movements (i.e., amplitude (peak) fluctuation values) on the original frames. In this way, the cross-coherence analyses are done on signals with similar number of frames. An example of a spike train in the original frame order is shown in Figure 6.29**C**, which is based on the micro-movements signal (Figure 6.29**B**) extracted from the raw data signal (Figure 6.29**A**). This transformation of the data to create the appropriately normalized data type is done for all sensors' signals, thus allowing us to integrate them from different levels of the nervous systems, for example, the CNS (EEG-spikes) and the PNS (somatic-motor-spikes). Figure 6.29D shows an example of two channels (left parietal and right wrist) where we perform a power analysis and obtain in Figure 6.29E the coherence estimates and the cross-coherence phase plots. As with the dyadic data from the partnering dancers above, here we use the weighted directed graphs that we can build using the coherence matrix, the phase lead-lag matrix and the frequency matrix. This is a parameterization of the micro-movements data that we perform on the continuous signal (i.e., using the micro-movements embedded in all the original frames). As before, we use a triangular filter and a sliding window of $\frac{1}{2}$ a block for blocks of 1 minute's worth of data to preserve the original times of the peaks in the raw data. The size of the block depends on the sampling resolution of the sensors. With 500 Hz, we can afford shorter blocks to sweep through the data and still get enough statistical power for our Gamma process estimation. Yet 1-minute windows are amenable to our tasks taking place within 15–20 min sessions. These are aspects of the analyses and visualization tools that we can further test and explore according to different sensors and combination of bio sensing instrumentation.

In this case, we examine the coupled dynamic network between the brain and the body. Figure 6.30**A** (top panels) shows the original matrices while bottom panel shows the matrices restricted to the positive (lead) phase angles adopting as before the convention that in [i,j] the node i leads the node j. In Figure 6.30B, we show the network representation in three forms: (1) the coupled network of connections across the brain and body (43 nodes minus the reference lead in the EEG); (2) the bodily network of connections across the 11 nodes (including the head which has the Enobio accelerometer) and (3) the brain network (32 nodes). We are interested in tracking the synergies across the body, the brain, but most importantly we are very curious about the *coupled brain—body synergies*.

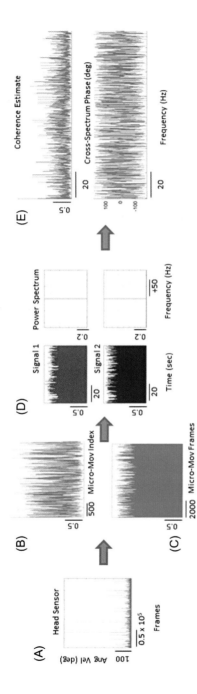

FIGURE 6.29 Pipeline of analyses for the coupled brain–body dynamics. (A) Sample raw signal from IMU's. (B) Micro-movements parameterization with peaks obtained in the order in which they occurred. (C) Micro-movements conserving the order and the original number of frames from up sampled data at 500 Hz. (D) Two such signals from the brain and body are used to perform power spectrum analyses. (E) Pairwise coherence estimation of these signal and cross-spectrum phase analyses to determine frequency and phase lead-lag information for adjacency matrices and weighted directed graphs used in connectivity network analyses.

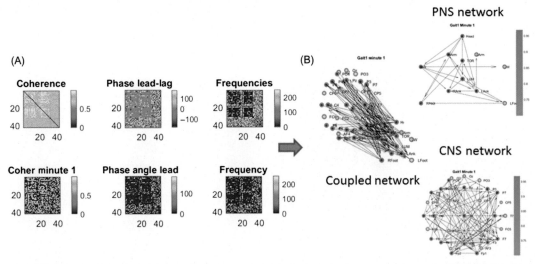

FIGURE 6.30   Weighted directed graphs for connectivity network analyses. (A) Coherence, phase lead-lag and frequency matrices to parameterize the micro-movements data derived from 32 nodes from the EEG and 12 nodes from the accelerometers in the IMU. As before, top rows are the full matrix and bottom rows are the matrices restricted to the positive phase angles (lead information) to adopt the convention in matrix entry [i,j] that node i leads node j. (B) Adjacency matrices in A are converted to weighted directed graphs and connectivity network analyses performed on the coupled brain—body network (black directed links) and on the individual networks of the body (red directed links) and the brain (blue directed links). Color bars reflect the coherence values of each node (matrix i,j-entry) used to color marker edge and marker face's color reflects node participation in a module. Activity corresponds to a module from a frame.

As before, we have two ways to examine the network behavior: top down *vs.* bottom up (Figure 6.31). In the top—down case, we define the regions of interest (RoI) we want to examine. For example, perhaps we want to study the left side *vs.* the right side of the brain; or the front *vs.* the back of the brain. Perhaps we want to examine upper *vs.* lower body regions. Accordingly, we can define such RoI and examine the activity as the three networks (brain, body and coupled) unfold minute by minute during each of the 15minute-walking tasks. We can also, as in the previous case with the dancers, examine bottom—up processes that automatically emerge from the evolution of the connectivity patterns of the weighted-directed graph. As before, we can use, for example, the modularity metric and determine the modules of the coupled network along with the participation of each node on the various modules. In this case, we are interested as well in self-emerging reciprocal connections between the body and the brain.

Combining RoI and nodes' participation in modular sub-networks can provide us with a metric of *"togetherness"* as we did with the dancers, useful in this instance to track which body parts entrain best with which brain nodes. Lastly, as we did before with the dancers, we can examine the stochastic patterns that emerge across all nodes and empirically map the activity in a parameter space to track the shifts in the amplitude fluctuations. Similar visualization tools involving avatars and brain representations can be used

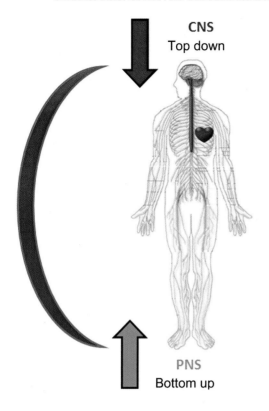

FIGURE 6.31 Top–down vs. bottom–up approaches to query the network connectivity across the CNS network (cerebro-cortical patterns from EEG), the PNS network (somatic-motor patterns from IMU's activity) and the ANS (heart activity from ECG activity). Monitoring efferent biorhythms dynamically unfolding and serving as afferent-reafferent sensory feedback activity to the brain. Importantly this model aims at uncovering self-emergent bottom–up synergistic activities as well as monitoring regions of interest the user defines to query the processes and the systems' cross-talk activity.

here to see the coupled dynamic activity, for example, as in the movies in (**Notes** [5]) and a new vocabulary of synergies developed to describe the complex evolving dynamics of the networks. In what follows, we describe the various tools we invented to track these biorhythms of the brain–body interactions in the hopes of changing the *disembodied brain* or the *brainless body* approaches that have plagued the study of the nervous systems and their pathologies.

Part of our goal in this book is to offer new ways to draw a comprehensive profile of all aspects of the nervous systems of the person. In this way we can push for a personalized approach to dynamically diagnose the person with a coping nervous system, changing and evolving by the day. We will expand on this new concept of a dynamic diagnosis to contrast it with the type of static diagnosis we have in place today. It will be really important to acquire the notion of rapid change and age-dependent, nonuniform rates of change in the developing nervous systems so we learn to appreciate the power of dynamic and stochastic tools that measure such changes in a *continuous* way. As of today (2017) they do not exist in the *discrete* observational clinical inventories that drive scientific research in autism, so progress is really stalled.

[5]LSL labstreaminglayer https://github.com/sccn/labstreaminglayer

Reciprocally Connected Network

The network of coherence among paired nodes can be visualized by looking for self-emerging patterns from sub-network (module) synergies within the coupled network (Figure 6.32**A**−B). During each session of recordings, for each node, we count how many minutes that node participated in a particular module and compute the proportion of times across a condition that a given node stayed in each module (e.g., bar plot in Figure 6.32B bottom panel). If a pair of nodes had the same proportion of time staying in each module during the entire session (i.e., node *i* led node *j* and node *j* led node *i* equal number of times within the module), then the two nodes were considered reciprocally connected. Thus the network graph can be represented with double arrows pointing in both directions for those reciprocally connected nodes (Figure 6.32**C**). Essentially, reciprocally connected pairs of nodes exhibit the same pattern of modularity during the recording session. This simultaneity implies coupled synergies between these pairs of nodes, which we can track from condition to condition (see movie in Notes 5) and Figure 6.33 comparing NT performance and performance from a representative participant with ASD. Each pattern is personalized and as such unique to the person, yet one thing is clear, the automatic condition requiring no mental effort, just walking around without a care in the world, typically engages a few coupled areas. In marked contrast, in autism this is an effortful computation when many brain areas are recruited to perform a simple task that should be otherwise automated. When a deliberate effort is required between brain and body to pace the breathing on command at the given rate of the metronome, a nearly disconnected scenario emerges in various ASD participants we tested on this.

By assessing these brain-body biorhythms' outcomes, one would think the two systems; the CNS and the PNS are not cross-talking enough to modulate each other through stochastic closed feedback loops. When we examined the heart IBI activity this was indeed apparent at the core of the autonomic system (see previous **Figures** 6.27−6.28**)**. The work alerted us once again on the need to provide a more comprehensive profiling of the CNS, PNS, and ANS in order to be able to build proper accommodations to help the children in the spectrum gain autonomy and volitional control. In this sense, the spontaneous condition, where the metronome spontaneously entrained the brain−body networks, seemed to be our entry point (Figure 6.33B) to more effectively interface and communicate with the nervous systems of a person with the autism diagnosis. The spontaneous-process of this condition showed in both the NT and the ASD participants with somewhat similar patterns of self-emerging bidirectional coupling. The sound of the metronome seemed to have automatically entrained some body nodes with some brain nodes. Because of this, we chose sound as one of our sensory inputs to target for therapy. Besides the empirical results from our studies, we also understand from the literature on auditory processing that sound is processed rather fast by brainstem structures. For example, complex auditory brainstem responses are measurable a few milliseconds upon stimulus onset[13] with detectable peaks at about 7−8 Ms of corresponding landmarks in the stimulus.

The field that studies auditory brainstem responses has well-established normative ranges, so we may be able to benefit from bringing sound as another form of sensory stimuli for closed-loop interaction, so as to guide us in profiling the responses of the nervous systems within a neurodevelopmental disorder like ASD. As such, we moved toward the

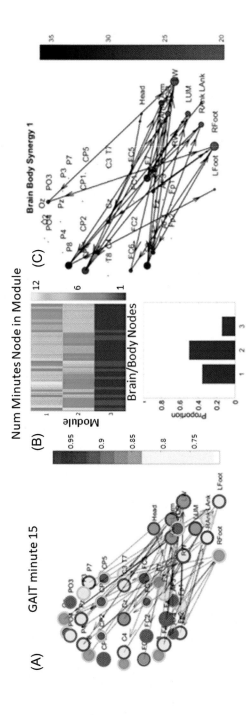

FIGURE 6.32 Connectivity metrics to quantify top—down and bottom—up activities across the coupled brain—body network. (A) Brain network and body network weighted directed connectivity. Color bar expresses the coherence values used to color the edge of the markers while modularity metric is used to color the marker face and provide visualization of the coupled network modules. (B) Modularity metric across all nodes (horizontal axis) reflects 3 modules (vertical axis) with node participation (number of times a node participate in a module) reflected in the color bar. Bottom panel shows the proportion of nodes participating in each module is well distributed across the coupled network. (C) Reciprocally connected network reflecting bottom—up brain—body synergies whereby when a pair of nodes had the same proportion of time staying in each module during the entire session (i.e., node i led node j and node j led node i equal number of times within the module), then the two nodes were considered reciprocally connected. Reciprocally connected pairs of nodes exhibit the same pattern of modularity during the recording session and serve to automatically quantify self-emerging brain—body patterns.

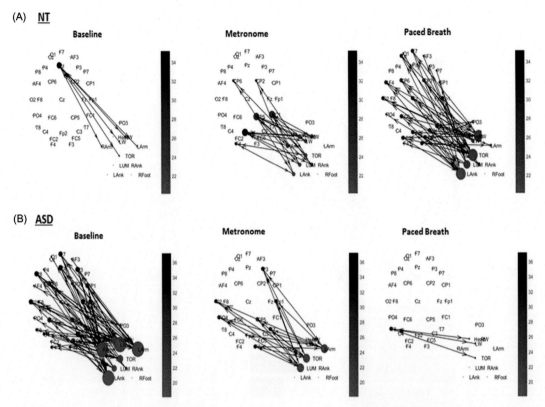

FIGURE 6.33   Dissimilar reciprocal brain—body connections in NT controls and ASD representatives across all three tasks. Whereas automatic walking is effortless in the NT control, it takes a lot of effort in the ASD partici-pant. In marked contrast, the deliberate breathing typically entraining the brain and body and suggesting high levels of cross-talk between the CNS and PNS in the NT control appears absent in the ASD participant, confirm-ing as well the ANS disconnect previously discussed in **Figures** 6.27—6.28. This metric enables the dynamic track-ing of the unfolding activities of the three processes of interest (deliberate, spontaneous and inevitable as driven by the CNS, the PNS, and the ANS, respectively).

sonification of various self-generated biorhythms in order to incorporate them as another form of sensory feedback. Unlike setups where the auditory stimuli were externally pro-duced by some sound collection (e.g., pre-recorded consonants), in our setup the auditory stimuli would be internally self-generated by the person's nervous systems: by the CNS (EEG waves), by the PNS networks (kinesthetically derived from kinematics output) and by the ANS (from the heart waves). In the next section, we show the type of co-adaptive, sound-driven feedback we intend to use in ASD to habilitate volition with the goal of evoking vocalization in nonverbal children.

The visualization tools we have designed allow us to understand the connectivity of the dynamically coupled brain—body—heart networks as patterns self-emerge from their unfolding activities, stochastically and dynamically changing from session to session.

Additional Tools to Summarize Coupled PNS–CNS Modules as a Measure of Sub-Networks' Togetherness

The modularity metric of network connectivity was further used to visualize the patterns of *brain–body togetherness* for each self-emerging module and according to RoI that we could flexibly define. For example, Figure 6.34**A** shows modularity matrices for each of the three networks of interest (brain, body and coupled). Here (as we did with the dancers' dyadic network), we count the number of times the node participated in the same module (Figure 6.34B) and use this information to create representations of togetherness across the networks and according to our definition of RoI. Then, we categorize these nodes by regions (e.g., parietal region is comprised of nodes in the right/left parietal lobes; Arms region contains the right/left upper arms and wrists). This regional sub-division is arbitrary (i.e., it can be sub-divided in other ways, depending on what we want to see) but allows us to flexibly query the networks and ask how two known regions may relate to each other as their self-emerging coupled dynamics evolve.

For each region, we examine whether those regional nodes participate in a certain module for more than half of the maximum time counts (in Figure 6.35**A** stars mark peaks if they exceed the 50% threshold). Then, we examine one region from the body and another region from the brain in pairs, to see if the two regions exceeded the threshold (i.e., participated together). If both regions do not exceed the threshold, they would be considered 'disjointed'. This can be represented in a binary matrix shown in Figure 6.35B for all three modules, where yellow indicates the brain–body region togetherness (1) and blue indicates disjoint-ness (0). Figure 6.36 is a graphical network representation of the first Module (out of 3 in this case), where the double sided arrows indicate the togetherness between the two given regions from the brain and the body. This type of detection of coupled activity is automatic once the user queries about a given RoI.

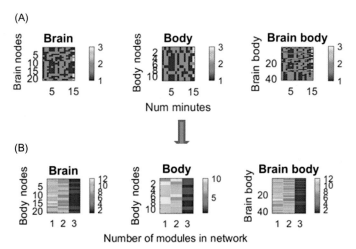

FIGURE 6.34 Modularity patterns of the brain, the body and the brain–body networks (A) quantified by node participation in each module (B) reflected in the color bar. These serve as an intermediate step to compute other metrics to quantify top–down activity patterns.

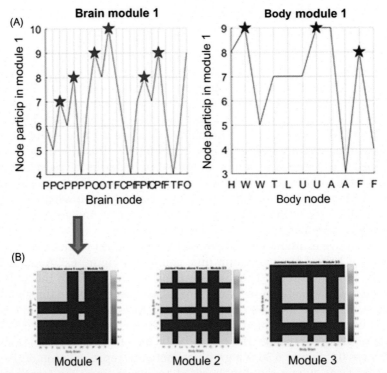

FIGURE 6.35   Quantification of top−down activity patterns across the brain−body networks and user-defined regions of interest to query about patterns of "togetherness" in the coupled network. (A) Sample profile from one module provides information about the number of minutes (in a minute-by-minute profiling of the network) that each node participated in a region of the brain (red traces) and in a region of the body (blue traces). Those nodes with participation about 50% of the time are marked with a star at the RoI. The RoI with stars in both the body and brain are marked in yellow (1) in the binary matrix representing the patterns in B (e.g., the arrow points to the binary matrix corresponding to patterns in A) and if not co-occurrence took place the matrix entry in marked blue (0). Togetherness binary matrices corresponding to other modules are also shown. Figure 6.36 describes the axes containing the body and brain regions (head (H), thorax (T), lumbar (Lu), legs (L) and feet (Fe) for the body). Then the brain regions are frontal (F), prefrontal (Pf), central (C), parietal (P), occipital (O), temporal (T).

Summary Statistics Profile

Underlying each node is a stochastic signature of spike trains in amplitude fluctuations. We define such spike train of fluctuations in amplitude for each node and use them as input to a Gamma process, where the Gamma parameters are empirically estimated. One way to visualize the statistics of each node's spike trains is a four-dimensional graph, as is shown in Figure 6.37 with the estimated Gamma PDF in the top insets of Figures 6.37A−B and the corresponding estimated Gamma parameters plotted on the Gamma parameter plane in the bottom insets. In the four-dimensional graphs, the empirically estimated mean, variance, and skewness of the fitted Gamma PDFs for each node during each condition are plotted along the x, y, and z axes, respectively. The size of the marker reflects the level of kurtosis, where larger size indicates high kurtosis level of the fitted PDF. The

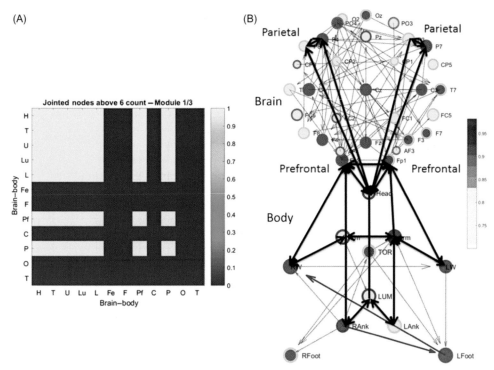

FIGURE 6.36 Visualization tools (A) Binary matrix representing brain–body togetherness. For example, the prefrontal region (Pf) and the parietal region (P) entrain synergistically with the head, thorax, upper body (right and left arms and wrists), lumbar and lower body (right and left ankles), a pattern that is captured by the double arrows connecting the brain and body as well as the RoI within the brain and the body in (B).

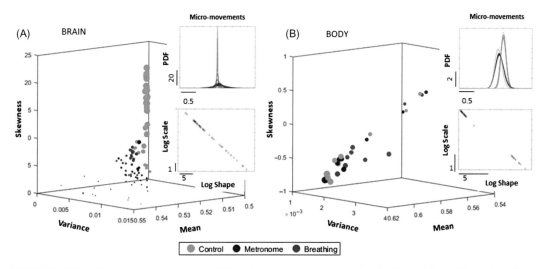

FIGURE 6.37 Underlying patterns of variability in the micro-movements of the biorhythms of the brain and the body are captured by the empirically estimated Gamma signatures for each condition (color coded). Insets show the PDFs and the Gamma plane maps where variability patterns of the brain and body separate well for each task. Notably the automatic walking stands out as the condition with highest symmetry in the PDFs and lowest dispersion.

colour of the marker is differentiated across conditions. This graph allows us to visualize the statistical features of each node and understand how the stochasticity changes across different conditions for the brain nodes and for the body nodes.

The profound differences we found between ASD and NT in this work alerted us of a possible new way to classify autism according to levels of cross-talk among the three types of processes we proposed in Chapter 3 and the level of randomness and noisiness we find in each of the cerebro-cortical (EEG), somatic-motor (movements' kinematics) and autonomic (heart variability) biorhythms. With this in mind, we set to profile the person by building a more comprehensive phenotypic portray and evaluating the degree of impairment and the capabilities of the coping system. Those processes which appeared strikingly different from NT were initially avoided in our experimental therapies. In contrast, those processes showing promise as they resembled typical ranges were prioritized as the entry point to interface with the person's nervous systems.

CLOSING THE FEEDBACK LOOPS IN PARAMETRIC FORM

The research on dancers and the development of analytics and visualization tools to track coupled dynamics opened a new line of research in the lab. Once we had the ability to stream live and simultaneously harness the outputs of different processes and levels of control, we were in a position to estimate in real time the stochastic signatures of the biorhythms and parameterize their inherent variability. This real-time data streaming ability also gave us the possibility to integrate various types of outputs with various forms of external sensory guidance, particularly using different media. We proceeded to build closed-loop co-adaptive interfaces that helped alter the biorhythms in the end-user in a well-informed manner, whereby we were able to estimate the output and measure the stochastic changes in these biorhythms while tracking their rates of change from moment to moment. We created a new form of objective behavioral analyses in real time.

As we explained in Chapter 2, we first characterized the normative data to learn the typical ranges of the micro-movements across different parameters from different instrumentations measuring biorhythms at different levels (e.g., voluntary, automatic, involuntary, and autonomic). Then we tested individuals with ASD and learned the disparities in parameters' range in relation to NT controls matched by age and sex. Given the elevated levels of noise and randomness in their biorhythms across multiple levels of the nervous systems, we proceeded to use noise-dampening techniques that we developed, by augmenting our setup with feedback that included (1) veridical signatures empirically derived from their own motions under a variety of sensory feedback conditions; (2) signatures derived from the motions of others; (3) combination of media whereby the child would be in control and the child's motions would trigger the various media types (video, sounds, self-videos, etc.)

We visited some of these studies in Chapter 2. In the first set of experiments adapted to a school setting, we carried on the assessment of ASD sensory-motor plasticity using a closed-loop co-adaptive computer-body interface that we built to that end[14]. We used media that the parents identified as the child's favorite at home and media that the teachers identified as the child's favorite in the school context. The child's task was to spontaneously (without instructions) self-discover a magic spot in peripersonal space whereby if the hand passed through it, the media would play. This is analogous to those sensor-driven soap dispensers in public restrooms, or to the sensor-driven faucets that open up and provide water when we move the hand by the right spot and trigger the soap or the water. In this case, we had a very precise and continuous positional location of the moving hand of the child. As such, we could obtain at any given moment the distance from the hand to a spot in space that we would define (unknowingly to the user). We could code the position of the spot in space and the moving-hand position relative to a global "world" coordinate frame of reference, and also relative to the child's position. We were continuously measuring the child's upper body position as the child naturally behaved. If by chance, the moving-hand of the child entered the vRoI, the computer program monitoring all interactions and automatically measuring the distance from the child's moving-hand to the vRoI would trigger media. Figure 6.38**A** shows in schematic form the setup we built to that end and a representative child performing the task. Figure 6.38B shows the evolution of the frequency histograms and the points of the Gamma parameter plane representing the estimated dynamically evolving PDFs, whereby red are derived from hand trajectories in the vRoI and blue are derived from hand trajectories exiting the vRoI or outside it while the hand is in search mode. Figure 6.38**C** gives the evolution of the shape and scale (dispersion) of the final PDFs for typically developing children (TD) and their ASD peers. Each point represents a child's signature with the corresponding PDFs in the bottom panels.

The performance of all children with ASD, despite nonverbal status, and as such our inability to instruct them was remarkable. Unanimously, one by one, they all engaged in the task and *spontaneously* self-discovered the goal without instructions. We set up the computer and suited them up, and then we hid behind the room door to watch what happened from the distance. We had cameras recording the session the whole time and could later on, in the lab, watch the entire performance in detail; but while at the school, when they were engaged in this self-discovery, we did not want to interfere. In most cases, we merely hid and watched.

I recall that it was very emotional to see some of the school aids actually watch in disbelieve what was unfolding in front of us. As they gained autonomous control of their motions, some children spontaneously vocalized some words (*"what's happening?"*), an utterance they would not normally articulate in the proper context. Rob (Dr. Isenhower) and I caught ourselves with teary eyes trying to hold back our emotions real hard, because we witnessed these children really excel at the task on their own. They had gained a type of volition and autonomy foreign to them.

The Developmental Douglass Disability Center where we ran the experiments is an ABA school. The children receive intensive ABA training every school day. As such, they are exposed to mostly top—down instructions to do tasks such as discrete trials, pivoting and other heavily structured, goal-directed actions with very little room for *spontaneous exploration*. They receive external reward in the form of food or tokens when they comply with the instructed goals; yet, there is very little internally triggered satisfaction from spontaneous self-discovery, self-regulation and self-control. These are the types of abilities babies come to discover on their own, while their nervous systems are still in a nascent stage. In our setup, much as babies would do, the children with ASD essentially self-discovered cause and effect by watching and experiencing the consequences of their self-generated actions. Eventually they too learned to predict such consequences before even carrying on the action. And that was evident in the dampening of the noise-to-signal ratio in the variability of their exploratory and eventually (voluntary) goal-directed motions.

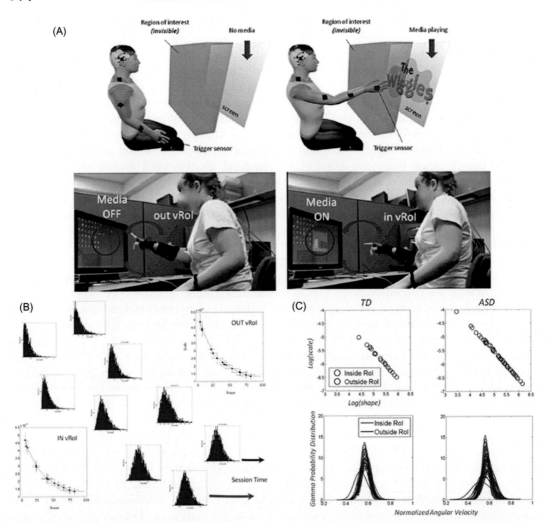

FIGURE 6.38 Co-adaptive body-computer interface to spontaneously (without instructions) evoke goal-directed actions through self-exploration and self-discovery in children of ages 4–14 years old. This interface enhanced their volitional control and resulted in the dampening of somatic-motor noise, the development of anticipatory statistical patterns and the retention of these features 5 weeks later without training. (A) Avatar representation of the task drafted by PhD student Polina Yanovich. Four elecro-magnetic sensors (Polhemus Liberty, 240 Hz) were attached to the child's arm and hand on the marked locations. The hand sensor was used to continuously track the positional trajectories and an invisible (virtual) vRoI was created (pink volume) so as to trigger media once the hand passed by or remain within its boundaries. This volume could be programmed to shrink and move within the child's peripersonal space, but it was invisible to the child. Actual set up involving the child and the media are shown in panel A-bottom area for the case when the child is exploring and the case when the media plays because the hand enters the vRoI. (B) Evolution of the frequency histograms of the hand speed micro-movements during exploration and trial and error until the child systematically places the hand inside the vRoI to continuously play the media. (C) All the children shifted the Gamma signatures from the LUQ to the RLQ as shown in the Gamma plane and corresponding PDFs.

The fundamental difference with ABA was that in our setup, motions spontaneously (without top—down instructions) transitioned into goal-directed mode. This was not "rat- or pigeon-conditioning" within a Skinner-box approach that behaviorists so hard push for with ABA. This was spontaneously self-emergent, self-discovered volitional control—as it should be in human beings!

We granted these children in the brief interaction they had with our interface a sort of freedom, a sort of free-will they could rarely experience in their daily school lives because of the pervasive excess of noise and randomness in their sensory-motor systems and the lack of knowledge their therapists have about this basic scientific discovery. Unknowingly perhaps, the therapies they receive leave very little room for such type of freedom so fundamental to nurture the person's autonomy and self-drive. I could not imagine my own existence subject to what they put these children through every day. And since I felt from my first encounter that I am one of them, I vowed to come up with a fundamentally different way to help improve therapeutic interventions using our new objective behavioral analyses (OBA) to continuously measure the treatment outcome. This would be a way in which we would interface with their nervous systems and communicate with them at a synergistic level that we could quantify. To that end, we used our dyadic interaction biometrics and tracked multisensory inputs that were both externally generated by us, but

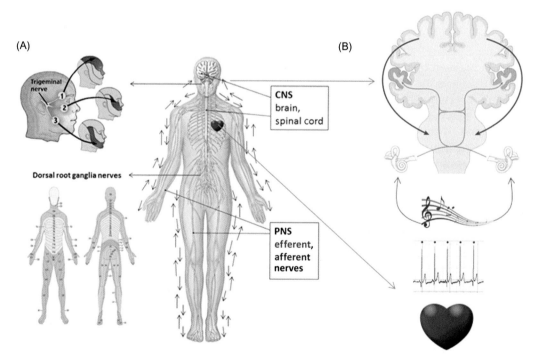

FIGURE 6.39   Playing the biorhythms back to the participants by sonification of motion patterns during the use of co-adaptive interactive interface with an avatar. (A) Taxonomy of the somatic-motor peripheral networks carrying efferent and afferent information through different nerves and involving different ganglia from the neck up and from the neck down. (B) Proposed schema to convert autonomic heart activity into music patterns (streaming in near real time) and feed it back to the participant while evaluating the brain activity during this neurofeedback.

most importantly internally generated by the children's own nervous systems as well (Figure 6.39). We proceeded to map out the unfolding stochastic signatures across the brain—body—heart efferent motions (**39 A**) and made sounds and visuals to provide these inputs as reafferent visual, auditory and kinesthetic feedback (**39B**). Our hope was to tap into the very primitive centers of the brainstem guided by fundamental biorhythms of the heart, so as to reach high-level neocortical areas of the brain. Then, in turn, we would read out in real time, the efferent output of the person's nervous systems through the micro-movements streamed live from the various biorhythms of the brain (EEG), body (kinematics) and heart (IBI).

Notably, all children (TD and ASD alike) followed the following order of events:

1. Random triggering of media that made them very curious;
2. Trial and error to search for the magic spot that triggered the media;
3. Systematic search honing into the vRoI and eventually holding the hand inside the vRoI to continuously watch the media (a video or a cartoon of their preference) or a video of themselves from a webcam feed facing the child.
4. Transition from spontaneous-random noise to well-structured predictive moving-hand's speed micro-movements signals.

When we used a camera in front of the child and had the computer program toggle it ON upon the hand's entrance inside the vRoI, the child could see his/her upper body and face on the screen. This made several children very curious and playful (making funny faces, for instance.) Owing to these reactions and the stochastic shifts in the biorhythms' micro-movement data, we felt very confident that we had tapped into bottom—up spontaneous processes the autistic system could perform rather well.

In all 25 (nonverbal) children that we tested we saw the progression of the self-discovery process, the curiosity and the *joy* this co-adaptive interface brought to them in a brief period of time. We came back 5 weeks later and retested the children to see if the effects were transient. We had seen that the ABA training (top—down driven by well-structured settings, with unambiguous, discrete instructions and externally rewarded with food or tokens) did not generalize from task to task. We also witnessed how after the children returned from the Christmas break there was a rather effortful comeback to the level of ABA performance they had when they left. We wondered if the bottom—up, spontaneous self-discovery process with initially rather ambiguous goals and then evoking -through the process of self-discovery- the internally rewarding experience, would also fail to transfer 5 weeks later. We wondered about this because we had a new room assigned by the school to do our experiments during the breaks between classes, and knew that novelty takes some time to get used to for all of us, much more so for the child with ASD with high levels of uncertainty in the sensory inputs their nervous systems must process.

We witnessed each child in the group pick up the pace immediately and have as much fun as they did the first time (ever) they did this task. More important yet, while ABA was drilling these children with hundreds of rote repetitions of the task, ours lasted 15 min at most and within the first 5 min they had already figured out the magic spot and were playing with the video or sound playback. They had not practiced doing this *at all* during those 5 weeks. Yet, they were able to do it immediately and when we measured the somatic-motor physiological signatures, we learned about the outcome of the task in rather objective ways (not by pencil and paper opinion). The changes our sensors recorded were indeed positive. The stochastic signatures of their motions immediately shifted from random and noisy to well-structured, predictive and with high signal to noise ratio[14]. This was the case for NT controls and ASD children alike.

Endowed with high confidence from our proof of concept, we proceeded to build a more elaborate setup to exploit all DoF of the body, multiple sensory inputs and media to really explore this discovery further.

TOWARD DYADIC INTERACTIONS BETWEEN CHILDREN AND AVATARS

The Nancy Lurie Marks Family Foundation awarded me a Career Development Award in 2014 in response to a research proposal I submitted to their agency outlining a program to develop co-adaptive interfaces. These interfaces would be aimed at enhancing several axes required for social interactions, including the possibility of evoking vocalization through the use of movements, music and video feeds. As we grew more knowledgeable of ASD and the many issues we encountered with the PNS of individuals affected by this condition, we began to better understand the need to empirically derive a comprehensive profile of NT controls and determine normative ranges. Then in turn, we could begin to better understand the autistic PNS so underexplored in basic research, in clinical practices and in treatments that at best had the goal of making people in the spectrum behave in a *"socially acceptable"* manner. The behavioral therapies were designed to *extinguish bad behaviors*. These could be odd postures or repetitive movements that—although awkward to the external social world—seemed comforting to the affected child. There was no effort to understand why the child repeated a ritual or moved in an awkward way. The notion of compensatory coping strategies self-discovered by the child's nervous systems was absent from such treatments. Without a single Neuroscience-based subject matter in their curricular activities for certification to become a Board Certified Applied Behavioral Analyst (BCBA) (see **Notes** [6]), there was no way these army of behavioral psychologists could possibly imagine or understand that there was *neurophysiology* and *neuroanatomy* behind all of that they were so diligently trying to *reshape* at all cost (including punishment schedules that no Institutional Review Board would ever approve for a researcher.)

Other types of therapies like occupational therapy (OT) or speech therapy offered at public schools within early intervention programs (EIP) or Individualized Educational Programs (IEP) were also aimed at making the child look socially acceptable: for example, carrying a tray in the school cafeteria in a certain way, and climbing the stairs of the school in a certain way or pronouncing certain socially acceptable phrases. The type of OT that focuses on sensory-motor integration issues were not part of those offered at schools or covered by insurance. As such, we were seeing over and over in the lab, many local families that came to our studies and told us how they were desperately trying to take

[6]BCBA site https://bacb.com/
No *neurophysiology* or *neuroanatomy* courses offered or required in the curricula. No mathematics, statistical analyses, signal processing courses, instrumentation courses, etc. What does their word analysis mean in the context of human behavior?

their children out of ABA to try other therapies, but could not gain insurance coverage for them. These were middle class and upper middle class families who could not afford the hourly rates of those therapies at an average rate of $150–350 per hour. Those alternative therapies targeting sensory-motor issues seemed to work better than ABA at integrating the functioning of the child's nervous systems and the activities of daily life. The frustrations of these families transferred to our lab. We grew impatient too at the narrow-minded methods of basic research and the lack of basic neuroscience knowledge clinicians and practitioners in the field displayed. While they were trying to change the child's nervous systems, they had no knowledge of the structure and function of basic components underlying autonomous behavior. None of that neuroscience training entered in their curricular activities toward their accreditation. *How would they acquire this basic knowledge now?* We decided to work really hard, day and night to come up with possible solutions that solved the most fundamental problem of not having insurance coverage and diversification in therapies that treated the nervous systems in the first place.

The lab had developed all sorts of dynamic outcome measures to track change and provide a new type of continuous and longitudinal assessment of the nervous systems. This comprehensive profiling was now ready for real-time biofeedback in the context of a co-adaptive interface where we could uncover the plasticity and adaptive capacity of the nervous systems of each individual child. Combining the dyadic biometrics we designed for complex dance partnering in ballet and the methods to track and integrate the coupled dynamics of brain–body and heart, Vilmi and Ji deployed their platforms and we started to profile the NT controls first, and then the children in the spectrum. There was nothing in the literature close to what we were doing, so we had to build everything from scratch. It was arduous and slow, but fun.

The first thing was to develop the platform for real-time display, interaction and modification of the biorhythms. Our program had three initial parts: (1) profile the child under a plethora of sensory inputs and manipulations to determine baseline ranges and their rates of change; (2) Identify the effective source of sensory guidance under the conditions we were testing according to the different types of biorhythms we were exploring (i.e., the most plastic process); (3) Use the noise signatures extracted from the biorhythms to modify them and feed them back in modified form.

We endowed the person's avatar with the veridical motions of the person's biorhythms in the movement domain (as we did in Chapter 2). Then, we could do several things to alter the patterns of the avatar as a function of the end user's motions and their consequences. For example, we defined a RoI (as in the simpler case of the upper body in Figure 6.38) to automatically trigger other stimuli (e.g., sounds and videos) (Figure 6.39B) and built a game-like environment to engage the end user to interactively search for that *magic spot* that played music (Figure 6.40). The person learned how to establish cause (the action) and effect (the action sensory consequence consisting on the triggering of media). As we had seen before with the children, the biorhythms changed and the changes and their rates were quantifiable in real time.

This closed loop interaction became a platform for testing and training volitional control in the children with autism (see **Notes** [7]). Indeed, as they gained control over their bodies and

---

[7]Movie link to interactive co-adaptive interfaces.

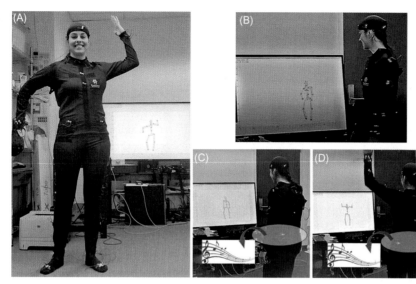

FIGURE 6.40 Co-adaptive closed loop interface for stochastic-based feedback control. (A) Dr. Carla Caballero testing the interface that Vilmi built. Avatar in the background moves in real time with the veridical motions of the end-user as captured by the cameras across the LEDs distributed throughout the body and head. (B) Searching for a "magic spot" that plays music once the hips enter within a volume position that Vilmi programs across the workspace where Carla moves. (C) She found the spot and triggered music. (D) Happy to have found the music and controlle its playing speed!.

grew certain of the sensory consequences of their actions in response to environmental dynamically changing stimuli, their patterns shifted signatures toward desirable statistical ranges (i.e., found as well in NT controls). Yet, one subset of the children we saw in the lab and for whom we recorded the heart rates as they walked around and performed different tasks, provided a hint that noise and randomness in their biorhythms likely had deep origins in their fundamentally atypical heartbeat timing patterns. The autonomic systems of this group of children were certainly in disarray. Their parents informed us of their systematic struggles with the GI system since birth; a hint that the types of bodily autonomy the autonomic and enteric systems typically scaffold the nervous systems with since infancy had been compromised from the start of their lives. In such cases we thought the taxonomy for classification we proposed in Chapter 3 invited to cluster these participants in the group of altered inevitable processes, that is, those processes identified with the *autonomic and enteric nervous systems*. These cases were more severe than we had seen before. At the core of the excess noise and randomness in their bodily biorhythms were these two fundamental systems without which a nervous system cannot properly function with autonomy and self-regulation.

We concluded from these studies that the full spectrum of autism required different categorization according to the degree of cross-talk among these processes: deliberate (driven by the CNS); spontaneous (driven by somatic-motor peripheral networks) and inevitable (driven by the autonomic and enteric nervous systems.) As such, we adopted the path of transforming the diagnosis from a static to a dynamic concept, one that considered as well the unfolding of the dynamic coupling between the child and the clinician.

THE ADOS DYADIC EXCHANGE: ANOTHER FORM OF SOCIAL DANCE

As the lab moved toward the development of new data types, analytical tools and experimental paradigms to understand the complex coupled dynamics of dyadic exchange, we came across the limitations of the ADOS test. At every conference we attended to, we found an increasing uneasiness among geneticists and people who try to develop animal models of target drugs. The high heterogeneity of the autism spectrum was beginning to call for new classification schemes that allowed these scientists better test their theories using the neurophysiological and neuroanatomical principles of the nervous systems. All tests available for clinical use were based on opinion without a single physiological measure. This was very disconcerting for most scientists forced nonetheless to use the a priori psychiatric or psychological classification the person came with from the clinic: that is, a label attained from a one size fits all model that left out the physiological phenotypic features of the individual's nervous systems so necessary for tailoring treatments to fit the needs and capabilities of the person.

At the basic science arena, research grade versions of these tests would presumably take care of other issues concerning reliability and reproducibility of scores and cut-off criteria. In reality though, the research grade ADOS (for example) was just as bad. Science remained stuck with a test with many inconsistencies, culturally dependent, extremely expensive and difficult to administer. Most important of all, the test made the children feel very uncomfortable when the emotional component of it was administered. This component consists of a series of questions concerning the person's plans for the future, issues concerning loneliness, or how others may think of the person, bullying issues at the school and other very unsettling matters that would only serve to emphasize how difficult fitting in the social world must be for the affected person. More than once I had to leave the room during such interviews while observing clinicians scramble over the questionnaires and even not finishing the questions at times because the child became visibly so upset. *Wouldn't you be upset if someone tried to peak into your most private fears and life uncertainties without any warning?*

I decided to apply for a grant at the state level to build a physiological version of the ADOS inclusive of both the clinician and the child within the interacting dyadic rhythms emerging during this test. I had developed already the biometrics to track the dynamic coupling of two nervous systems using graph theory and connectivity metrics. I had already developed as well the methods to track natural behaviors continuously and profile noise-to-signal bodily maps from various biorhythms. All I needed then was to deconstruct the ADOS test and set up the lab to administer it systematically in NT children and children with a neurodevelopmental disorder.

The idea of profiling NT children while performing the ADOS test came from a problem I identified with their score system. They call this test *"the gold standard"* of autism diagnosis, so I wanted to see how they had built their *"standard"* scale. To my dismay, when reading their seminal paper on the ADOS-G test[15] (G stands for generic), I realized they lacked a normative data-set to refer their scores to. In their own words, *"…Replication of psychometric data with additional samples including more homogeneous nonautistic populations. . .will add to our understanding of its most appropriate use"*. Their scores are absolute values someone gives to a child while observing the reactions to prompts and questions posed and asked by the same person scoring the child (and by the way, charging quite a bit of money to that end). Those absolute scores are not invariant to things like cultural nuances or clinician's biases in posing the social overtures to try and get a spontaneous response from the child. The expectations of the test to score the responses of the child had been a priori predetermined by elements of the Western culture, including as well a profound conflict of interest between the examiner, the test provider and the scoring system in general. They were based on the perception of what is supposed to be a "normal" child in our side of the world.

It was not surprising then that when they took the test to North Korea to screen over 20,000 children for autism, they tagged three-quarters of the children as being affected by the condition. Although the screening tools these researchers used were flawed,[16] such tools continue to drive the criteria that defines research in autism at large. It is really mind boggling.

A characterization of typical ranges for ADOS score scale does not exist. As such, within the research realm, despite maximizing the homogenization of disparate groups to ensure statistical significance in reported differences, there is still large heterogeneity that in many cases prevents unveiling statistical differences. The tests driving the scientific quest are plagued with biases from the administrating clinician.[17] When we examined demographics data from the Autism Brain Imaging Data Exchange (ABIDE) repositories including well over 2,000 individuals, we found conflicting results between the ADOS-2 and ADOS-G versions, thus rendering the data useless for any further attempt to include the clinical scores in the development of biomarkers of ASD subtypes.[18] Behavioral biomarkers are needed to help close the broad gap between behavioral and genomic data, so as to assist in the development of personalized target drug according to the tenets of PM.[19] Unless we close the behavioral gap we cannot progress on this endeavor.

TWO WINGS OF THE SAME BIRD

While immersed in the task of raising funds to address the ADOS issues, I received an email from Dr. Caroline Whyatt from Queen's University at Belfast. Caroline (as we now call her dearly) had read our work on the micro-movements in the 2013 paper.[1] She was very interested in coming to the lab to learn more about the new techniques. She had completed a PhD thesis on autism and was simultaneously doing research involving patients with Parkinson's disease and autism—as were we in the lab at the time. When we talked over the phone and later on in person, we immediately found many things in common with our ways of connecting the two seemingly disparate disorders. Autism manifests at the start of life but PD does so toward the end of life; yet there was a common denominator manifested in their sensation of movements. By then I had examined both the children's patterns and those of PD patients in relation to the signatures of the deafferented participant Ian Waterman. In both cases the signatures of noise and randomness were comparable to those a person who could not feel movement, touch and pressure; thus suggesting an underlying issue with the motor efferent signal that did not preclude lack of kinesthetic sensation of continuously self-generated motions in both disorders. Caroline and I could see this very clearly. Both had experienced the reaction by others on the out-of-the-box nature of this notion. Her colleagues back home were willing to listen to her; while here in the States I had spent nearly 2 years before we could even get our work read by an editor of an autism journal. Somehow we were both wings of a same bird.

I decided to apply for the funds of the New Jersey Governor's Council for the Research and Treatments of Autism so I could bring Caroline to work in the lab with us. She understood very well the clinical tests because she had already deconstructed the movement assessment battery for children (MABC)[20,21] and derived objective metrics of this battery based on kinematic analyses of continuously recorded motions during its complex tasks. I was stricken by her diligence and self-drive. She visited the lab while visiting NYC in transit to Washington DC and gave a wonderful presentation. I wanted to ask her to join the lab immediately but at that point we had just learned that we could not apply for a NCE on our NSF-ARRA grant, so my plans had to wait.

By April 2014, I learned that I was awarded the Governor's Council funds and contacted Caroline. She came from the UK and joined the lab immediately. Caroline spearheaded the ADOS project. With 20 children in the spectrum longitudinally tracked for four visits and 10 NT controls we had a good set of questions to address while characterizing the somatic-motor physiology underlying this test.

We both became ADOS certified with the purpose of understanding this test inside and out. Then we deconstructed it using our biometrics to automatically isolate the coupled dyad physiology underlying the segments corresponding to each task. We called it *the ADOS dyad*. As with the previous dyadic data from the dancers and the data from the brain—body—heart setup, we were able to develop a data analyses pipeline to track each member of the dyad and more importantly to track the coupled network dynamics minute-by-minute. Figure 6.41 shows the pipeline for data collection in the ADOS experimental setup, while Figure 6.42 shows signal processing steps toward building the dyadic network (Figure 6.43) best captured by movies unfolding the network activity from task to task (**Notes** [8]). We could identify synergies across the body of each member of the dyad denoting the cooperation or lack thereof in their social exchange. And we could determine who would lead who for each segment of the ADOS across multiple segments over time. Figure 6.44 shows an example taken from four representative dyads, two for NT controls and another for two participants who went on to receive an ASD diagnosis. Clearly, in the case of ASD the clinician leads most of the time, unlike the cases when the child is NT. Then the child is in control and commands the interaction most of the time despite being questioned by the clinician.

The results from tracking the minute by minute interactions of the dyad were very encouraging in the precise sense that we could really summarize the interaction and possibly shrink the test. A test that lasts between 45 min to a full hour was not really necessary to find out there were issues with social exchange. In fact, upon examination of these data sets we realized that even two components of the test were sufficient to set the children apart when they had social and communication issues: (1) the enacting of a story by the

---

[8]Movie link to ADOS-dyads movies.

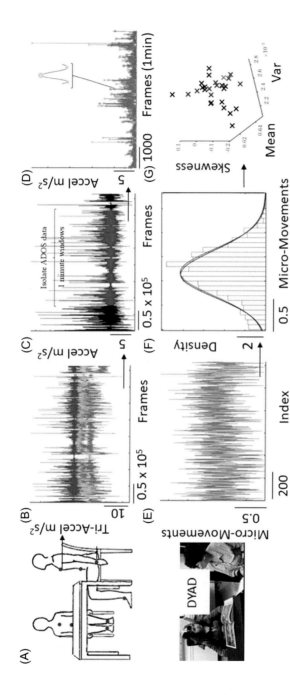

FIGURE 6.41 The use of network connectivity metrics and stochastic analyses in the physiological characterization of the ADOS test. (A) Clinician-child dyad with 6 IMUs each outputting activity in real time synchronously across all 12 sensors. (B) Raw tri-axial acceleration traces from one sensor (APDM Opal, 128 Hz) with 1 hour recording. (C) Acceleration signal corresponding to the ADOS test time. (D) Normalization of raw signal to derive the fluctuations in peak amplitude. (E) Micro-movements extracted from the acceleration signal. (F) Estimation of PDF for one frame (G) Estimation of Gamma moments and tracking on Gamma parameter space.

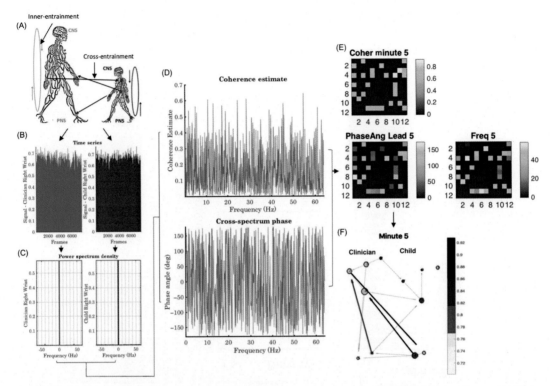

FIGURE 6.42 Pipeline of data analysis, signal processing and network connectivity analyses. (A) Inner entrainment analysis of each individual of the dyad and cross-entrainment analyses of the coupled network. (B) Micro-movements across all frames for two sensors to do pairwise analyses of synchronous activity. (C) Power spectrum density distribution across frequencies. (D) Coherence estimation and cross-spectrum phase analyses. (E) Adjacency matrices for weighted directed graph. (F) Weighted-directed graph involving all 12 nodes for minute 5 of the hour with clinician (red links), child (blue links) and black links denoting the coupled network. Color bar denotes the coherence values. As before, modules are denoted by the face color of the marker while the edge of the marker denotes the coherence value.

child to the clinician and (2) the emotional interview. These two tasks could already reveal excess noise and randomness in the somatic-motor physiology underlying the overt (deliberate) and covert (spontaneous, involuntary) motion patterns of the children who went on to receive the ASD diagnosis. They differed dramatically from NT controls.

Interestingly, given that we would pioneer the normative data sets with physiological somatic-motor data, it was important to attain this notion of a relative biomarker invariant to all those nuances of cultural biases, clinician's conflicts of interests and overall lacking of a truly standardized scale. In that sense our study provided not only the very first characterization of NT children who were subject to this test but also the range of physiological activities standardized in a way that remained invariant to the disparate sampling bias that different testers could have. Figure 6.45 reveals the stochastic signatures of two representative dyads whereby one is the NT child and the other is a child who ended up

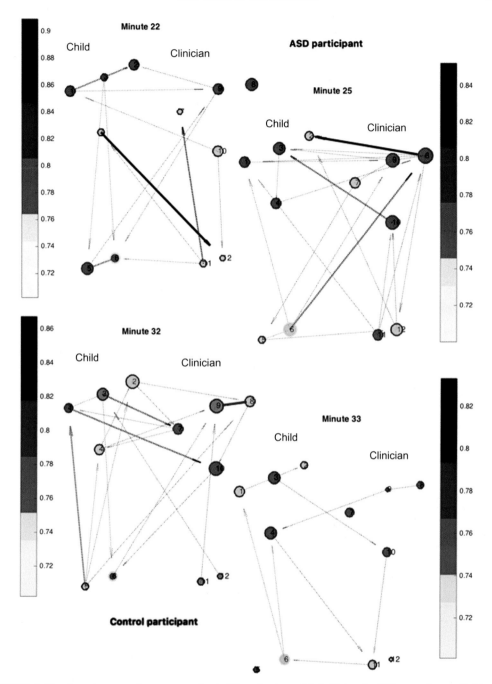

FIGURE 6.43 Sample network states at minute 22 and minute 32 (left) for the NT control and clinician in contrast to ASD child and clinician on the right (for minute 25 and 33 of their ADOS session)

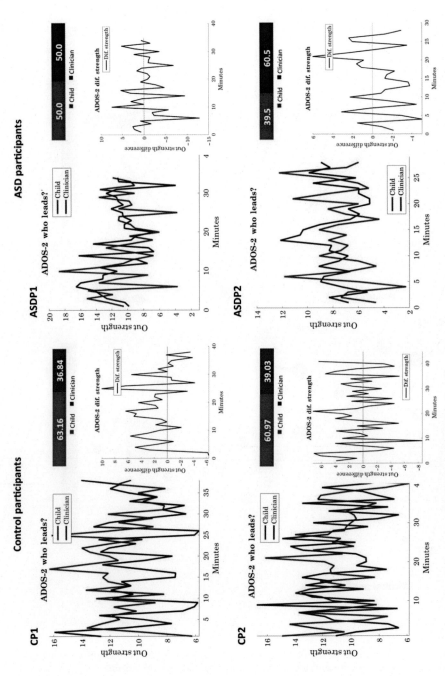

FIGURE 6.44 Who leads who and when? During ADOS performance it is possible to determine based on the dynamically unfolding network activity who leads at each task of the ADOS and also overall who leads or lags. Two NT controls are used here to illustrate the leading patterns overall with over 60% of the time leading the session. In contrast the ASD representatives are either led most of the time by the clinician or sharing equal leadership. In all cases Module 3 of the ADOS-2 was used and the same clinician conducted the test.

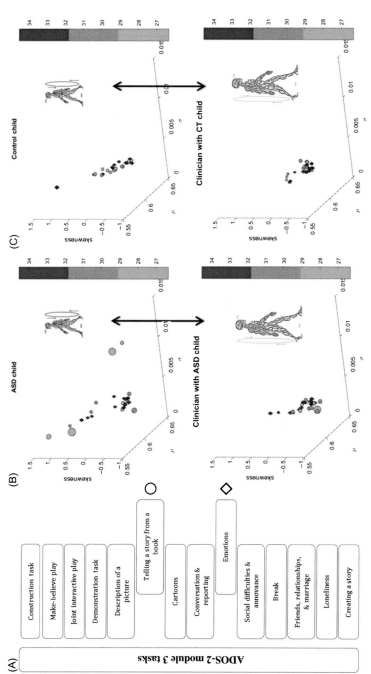

FIGURE 6.45 Automatically tracking physiological somatic-motor signatures of an ADOS session and their shifts in stochastic patterns. (A) Two tasks in the sequence emerge as the most informative ones. Representatives are two female participants (ASD: 13 yrs. and control 10 yrs.). Points of maximal differences in out-strength connectivity automatically isolated the tasks of interest cross-referenced as corresponding to the 'emotion' and 'telling-a-story' tasks (B) Summative estimated Gamma parameters of the underlying physiological signatures for the corresponding timestamps were profiled (ASD in panel B, NT control in panel C). Maximum temperature values were integrated with the motion (color coded markers), higher temperatures mark active, overt actions while lower values are spontaneous minute motions. ASD participant shows higher variability and more so during the emotional task when very little movement was required. Underlying the seemingly passive task was a higher accumulation of involuntary (spontaneous) motions with higher uncertainty during the emotional task than during the tell-a-story task when the child actually had to move quite overtly.

having the ASD diagnosis. These are the stochastic ranges of a child and clinician for the two tasks we highlight from the Module 3 of the ADOS-2 test. Basically the ASD child is in a separable statistical landscape altogether. To assume a theoretical normal distribution to analyze these behavioral data is a gross mistake.

We completed all 100 evaluations under the criteria of two clinicians. This was to address issues of reliability and influences of the clinician on the administration and scoring of the tasks. Reporting the results of the full study in detail are out of the scope of the book since our objective here is rather to describe the philosophy behind the new biometrics and methods to alert researchers of the new platform to track neurological conditions.

We nevertheless present in the next chapter a summary of the results from the comparisons between the ADOS-2 and ADOS-G for over 2500 males and females with ASD and AS diagnoses[18]. *How could you have such a major test called "standardized test" or "gold standard" and have never characterized the normative data or provided a normalized age-dependent scale?*

This ADOS test has driven the research and treatments in autism since its inception. Besides the cost of certification and materials, and the cost of clinicians administering and scoring it (see **Notes** [9]), this test is utterly difficult for the children who receive it. My heart shrinks each time I witness them taking this test. I feel their pain. I can see how this child with profound sensory-motor issues stresses over all those questions and task requirements. The intentions to diagnose the child and have the family gain access to EIPs or IEPs (in the US) may be good, but the outcome of the entire operation is awful.

My hope is that at least the objective and condensed version of it that we provide serves some day to do basic science so that in the end, geneticists can have access to precision phenotyping of the child's neurophysiology as many times as needed to construct a personalized dynamically changing profile, rather than a static one-size-fits-all assumed profile. This chapter closes with an invitation to examine the code and simulations we provide in the companion website of the book so we can improve the system that diagnoses ASD for clinical and for research purposes.

Perhaps the superb infra-structure that ADOS- and ABA- providers have created to deliver knowledge and certify people to do what they do will serve someday to educate those willing to learn about the nervous systems neuroanatomy and neurophysiology. The human nervous systems are after all what they are trying to diagnose and treat. *Why not measure it while they are at it?*

---

[9]The cost of ADOS is prohibitive for any study seeking to establish a result with statistical power. Consider a conservative estimate at $200.00 per hour paid to the clinician administering it and $2,095.00 per ADOS kit per study, ADOS-2 Software upgrade package per study $630.00 and $53.50 per booklet per child (http://www.wpspublish.com/store/p/2648/autism-diagnostic-observation-schedule-second-edition-ados-2). This practice results in a very costly operation to science. So, for example, a study involving 50 individuals with ASD and 50 controls would with this conservative estimate add up to ~$28,075.00, not including transportation costs and participants' compensation —not to mention "double billing" in many cases where the same child undergoes the test by the same clinician at the clinic and also in the research lab.

References

1. Torres EB, Brincker M, Isenhower RW, Yanovich P, Stigler KA, Nurnberger JI, et al. Autism: the micro-movement perspective. *Front Integr Neurosci.* 2013;7:32.
2. Torres EB, Isenhower RW, Nguyen J, Whyatt C, Nurnberger JI, Jose JV, et al. Toward precision psychiatry: statistical platform for the personalized characterization of natural behaviors. *Front Neurol.* 2016;7:8.
3. Torres E.B., Donnellan A.M., Torres E.B., Donnellan A.M., eds. Frontiers in Integrative Neuroscience, 2012, vol. Research Topic, chap. Neuroscience, pp. 1–374.
4. Frith U. Mind blindness and the brain in autism. *Neuron.* 2001;32:969–979.
5. Baron-Cohen S, Leslie AM, Frith U. Does the autistic child have a "theory of mind"? *Cognition.* 1985;21:37–46.
6. Werth B, Tsiaras A. *From Conception to Birth : A Life Unfolds.* 1st ed. New York: Doubleday; 2002:283. p.
7. Torres EB, Smith B, Mistry S, Brincker M, Whyatt C. Neonatal diagnostics: toward dynamic growth charts of neuromotor control. *Front Pediatrics.* 2016;4:1–15.
8. Carmo MPD. *Differential Geometry of Curves and Surfaces.* Englewood Cliffs, N.J: Prentice-Hall; 1976. pp. viii, 503 p.
9. Du X, Hao H, Yang Y, Huang S, Wang C, Gigout S, et al. Local GABAergic signaling within sensory ganglia controls peripheral nociceptive transmission. *J Clin Invest.* 2017;127(5):1741–1756.
10. Torres EB, Denisova K. Motor noise is rich signal in autism research and pharmacological treatments. *Sci Rep.* 2016;6:37422.
11. Nathan K, Contreras-Vidal JL. Negligible motion artifacts in scalp electroencephalography (EEG) during treadmill walking. *Front Hum Neurosci.* 2015;9:708.
12. Kilicarslan A, Grossman RG, Contreras-Vidal JL. A robust adaptive denoising framework for real-time artifact removal in scalp EEG measurements. *J Neural Eng.* 2016;13:026013.
13. Skoe E, Kraus N. Auditory brain stem response to complex sounds: a tutorial. *Ear Hear.* 2010;31:302–324.
14. Torres EB, Yanovich P, Metaxas DN. Give spontaneity and self-discovery a chance in ASD: spontaneous peripheral limb variability as a proxy to evoke centrally driven intentional acts. *Front Integr Neurosci.* 2013;7:46.
15. Lord C, Risi S, Lambrecht L, Cook Jr. EH, Leventhal BL, DiLavore PC, et al. The autism diagnostic observation schedule-generic: a standard measure of social and communication deficits associated with the spectrum of autism. *J Autism Dev Disord.* 2000;30:205–223.
16. Pantelis PC, Kennedy DP. Estimation of the prevalence of autism spectrum disorder in South Korea, revisited. *Autism.* 2016;20:517–527.
17. Whyatt C, Torres EB. *Fourth International Symposium on Movement and Computing, MOCO'17.* vol. 4. London, UK: ACM; 2017:1–8.
18. Torres EB, Mistry S, Caballero-Sanchez C, Whyatt CP. Stochastic signatures of involuntary head micro-movements can be used to classify females of ABIDE into different subtypes of 3 neurodevelopmental disorders. *Front Integr Neurosci.* 2017;11:1–17.
19. Hawgood S, Hook-Barnard IG, O'Brien TC, Yamamoto KR. Precision medicine: beyond the inflection point. *Sci Transl Med.* 2015;7. 300ps317.
20. Whyatt C, Craig C. Sensory-motor problems in Autism. *Front Integr Neurosci.* 2013;7:51.
21. Whyatt CP, Craig CM. Motor skills in children aged 7-10 years, diagnosed with autism spectrum disorder. *J Autism Dev Disord.* 2012;42:1799–1809.

C H A P T E R

7

Different Biometrics for Clinical Trials That Measure Volitional Control

"It is difficult to get a man to understand something when his salary depends upon his not understanding it." *Upton Sinclair, Jr.*

PART I OPENING PANDORA'S BOX

Since ancient Greek times, the virtues of hard work have been exalted and contrasted with the acquisition and squandering of others' wealth through bribery and corrupted decisions. In Hesiod's poem *The Works and Days* the ancient Greek poet writes instructions to his brother Perses on how to acquire and preserve wealth through the virtues of *hard work*.[1] According to the poem, Hesiod does so in response to bribed lords who favored Perses' squandering of their joint patrimony. In the story, it may appear at first that Perses got away with the squandering of their bequeathed farm, a source of wealth he did not create. But somehow his brother Hesiod brings balance to the impunity of Perses' actions by exalting the value of hard work and condemning the action of stealing the fruits of someone else's efforts.

Hesiod's poem underscores that somehow bad deeds are never exempt of castigation. His poem illustrates this through the story of Zeus punishing Prometheus for the theft of fire by offering Pandora to his brother and unleashing Pandora's action, actions that would irreversibly change the course of humanity.

The story serves to further illustrate the uncertain consequences of harmful actions, like stealing *knowledge* and *technology* from others. In the story, fearing further punishments from Zeus, Prometheus warned his brother Epimetheus not to accept any gifts from the gods. Epimetheus however, did not listen. He did accept Pandora (the first woman) and

Objective Biometric Methods for the Diagnosis and Treatment of Nervous System Disorders
DOI: https://doi.org/10.1016/B978-0-12-804082-9.00007-1

she dispersed the contents of her *jar*, leaving only hope inside. In Hesiod's writings (emphasis added):

> *"For ere this the tribes of men lived on earth remote and free from ills and hard toil and heavy sicknesses which bring the Fates upon men … **Only Hope remained there in an unbreakable home within under the rim of the great jar**, and did not fly out at the door; for ere that, the lid of the jar stopped her, by the will of Aegis-holding Zeus who gathers the clouds. But the rest, countless plagues, wander amongst men; for earth is full of evils and the sea is full."*

In Hesiod's story, by stealing the fire, and giving it to the men, Prometheus ended the Golden age of an all-male society of immortals when Zeus brought Pandora to Epimetheus. Indeed, the story asserts that the opening of Pandora's jar marks the beginning of the Silver Age of mortal men and women, who can give birth to men, thus initiating an era of death and rebirth cycles. Pandora's opening of the jar unleashes the origins of a new era, through a radical departure from the past. Not only bad deeds do not go unpunished, they may also have a silver lining leading to new improved beginnings.

My lab is all about opening Pandora's box, unleashing transformative change to improve people's quality of life, and then closing it tight to keep *hope* safely secure.

THE HOPE KEEPERS OF SMIL

The women from my lab these past few years have repeatedly confirmed to me that science is most rewarding when it ultimately has the chance to help others in need. These young ladies (Fig. 7.1) converged to the Sensory Motor Integration Lab (we call it SMIL) through different paths. Some came from Europe, while others came from the New York and New Jersey areas. They knocked at the lab's door once, eager to learn and work on the problems we were interested in at the time. Some went on to industry, others started medical school and one went on to pursue worldwide humanitarian causes. Others remained in academia, doing research and teaching new generations of students whom, like them once, became interested in the work we do. All of them have left an indelible mark in my life and made what I consider will be profound contributions to our understanding of the human condition. Their enthusiasm, brilliance, and kindness have been the forces driving the lab since its beginnings, nearly a decade ago now, in 2018.

Together, my lab members and I have made fun discoveries. And, as in any exploratory journey, as part of that discovery process, we also opened Pandora's box more than once. Sometimes we did so following very deliberate detective work along well-defined, logical steps. Other times, the discoveries were rather serendipitous. We realized along the way, that fundamental assumptions had been enforced on complex phenomena under investigation to simplify their study, but such assumptions, in many cases, had never been empirically verified. Over time, these oversimplifications had become an obstacle to progress and in some cases, we could even foresee bad consequences for policy making and other more tangible aspects of our communal lives within society at large.

As a team, we embarked into some serious inquiry and found unexpected facts, hidden in the Methods section of many peer-reviewed papers. But in this process of research and verification, we also had crazy fun (Figs. 7.2 and 7.3).

FIGURE 7.1 **The HOPE keepers of the Sensory Motor Integration Lab (SMIL) at Rutgers University.** (A) From left to right: Fulbright Scholar Sejal Mistry (from Biomathematics-PreMed), Future Dr. Ushma Majmudar (Biomedical Engineer and Wiz Hacker at the Albert Einstein Medical School in NYC); Dr. Caroline Whyatt from the UK our Research Assistant Professor (we were twins separated at birth and have identical sense of direction—she is my right hand and my wisdom); Dr. Jill Nguyen (my very first doctorate student, so proud of her! A genius analyst at the NYC Finance world today somewhere in Wall Street); Jihye Ryu (doctoral student who reversed the path from finance analyst top executive to PhD student—a mummy today who joggles career and family with amazing discipline, we look up to her for advice on how to organize our lives); Vilelmini Kalampratsidou (the joy of our lab who brings sunshine and warmth from Greece, Vilmi is a classical ballet dancer featured on the cover of the book. She is my doctorate student from CS who helps people with her magical interactive human machine interfaces); (B) Four detectives who solved big problems in SMIL when we opened Pandora's box more than once. Jill discovered children in the spectrum of autism do not have insurance coverage for occupational therapies that focus on sensory motor issues and our biometrics in this book could get their families the coverage they need; (C) Sejal and Caroline independently discovered with me that the WHO-CDC growth charts are not based on incremental (velocity) data but rather on absolute length or size-based and weight-based values without a proper metric. The Normal distribution imposed to build the percentiles pediatricians measure are not appropriate. We read hundreds of pages of methods until we uncovered the mystery. Newborn babies are being measured with the wrong ruler! Sejal, Caroline, and Carla helped me verify an odd finding by independently checking the ADOS-2 vs. ADOS-G scores of ABIDE. They found discrepancies and lack of a proper metric scale. Together we discovered that these "gold standard tests" were never tested in neurotypical children. They have no baseline scale.

FROM DELIBERATE AUTONOMY TO A MEASURE OF QUALITY OF LIFE

One of the most poignant aspects of doing research that encompasses diverse neurological disorders is to see the person that you are interacting with lose some of the most basic control of autonomy over his/her actions and thoughts. Seemingly simple acts that we

FIGURE 7.2   **Solving the puzzles of one of those escape rooms in NYC.** Laughing hard when I suddenly (in total shock) realized that the cues to escape the "Scape room" were being provided to us all along on a computer screen (here reflected on the glass) that had been following us (like a robot) and that the top of the screen with a message stating "The Alien Scape" was the name of our escape room and not a message that we were done and out of there. I thought we had finished but we were trapped. What a riot!

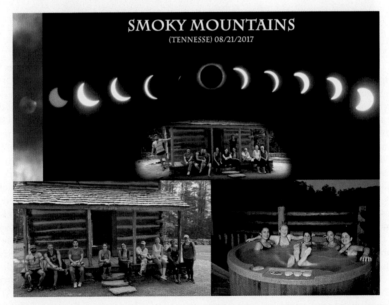

FIGURE 7.3   **The SMIL and friends at the total solar eclipse at the Great Smokey Mountain National Park of Tennessee** (posing by the John Oliver house where we watched the eclipse). In a total spirit of the moment and after watching a TED talk in the middle of the early hours of the morning (about 2:00 a.m.) of August 19, 2017, I got inspired by the message of the TED talk. This message was that we owe it to ourselves to experience a total solar eclipse once in our lifetime. I sent a link of the talk to the lab suggesting we go see the solar eclipse somewhere along the path of totality. The next day, they had rented a cabin at the Smokey Mountains, a car rental was ready, packed with all the food and gas; and they called me up to take off. What a memorable trip! We are all going to Argentina to chase the next one! Relaxing at the Jacuzzi after a long day of hiking, biking, and swimming (from left to right, Richa Rai our lab manager, Caroline Whyatt, myself, Vilelmini Kalampratsidou, and Carla Caballero).

take for granted deteriorate to the point that the person loses control over the physical execution of intentional thoughts that once were so automatic, that very likely, the person did not even know they took place. But the problem was even more disconcerting in the young infants and children we saw in the lab; because their nervous systems had not yet developed the type of volitional control that we, as adults, so much enjoy. When examining their nervous systems' functioning, we could detect elevated levels of noise and involuntary motions overcoming the brain's attempts to deliberately and autonomously control their actions and decisions. We thought in the lab that one of the ways to help the children and adults cope with such dysregulated nervous systems was to quantify the amount of motor control they could exert over their bodies in motion vs. the amount of spontaneous involuntary random noise they inherently had. If we could build indexes expressing the balance between these states of controlled and dysregulated nervous systems, we could then track the changes in that balance over time, progressing with or without treatment.

The relevance of building such indexes is that knowing this information would enable us to guide the person along a path of motor control to better cope with the loss of deliberate-autonomous control. As we saw in Chapters 2 and 6, using this approach we could habilitate and rehabilitate volition. Acquiring volitional control and agency helps the person better cope with activities of daily living. Further, using these indexes of volition could help us researchers and clinicians design and acquire biomarkers that are more precisely aimed at improving the quality of life of the person and of the care givers and supporting family members. As such, it became our goal in the lab to find ways to develop biometrics for closed loop performance. These biometrics are aimed at quantifying ways in which the nervous system could bypass the neuromotor control issues and steer the person's system to find the path to self-discovery and self-healing. After all, the one property all nervous systems that have some degree of autonomous mobility retain is that of self-supervision and self-correction. We know that kinesthetic reafference is critical to maintain self-supervision and self-correction; but we need to provide the proper form of feedback and dampen the elevated levels of noise and randomness in their nervous systems. To do so appropriately we need to understand the physical ranges of various forms of sensory input the person receives (including those of spontaneous involuntary motions beneath the person's awareness) and build standardized scales that consider age-dependent statistical variations in the human nervous systems across the human lifespan.

To that end, we combined the analyses of in-lab and in-home data with the use of data from large repositories publicly available. We initiated the path to examine biophysical rhythms of the nervous systems across multiple layers and levels of control. Together, these data could inform us of ways to derive indexes evaluating the amount of controllable autonomy each person's nervous systems has and use those ranges in closed-loop methods such as those explained in Chapters 2 and 6.

The combination of empirically determined parametric multilayered approach and closed loop methods has helped us in the lab optimize ways to help the system regain volition on its own, by augmenting and/or compensating the missing or corrupted feedback information with other controllable means. The essence of the success of our approach has been the design of a new way to parameterize the output activities of the nervous systems and influence it (or altogether change it) *in real time*.

At the upfront of our data acquisition pipeline, we have managed to synchronize the instrumentation to gather signals in tandem from various layers (autonomic, automatic, deliberate) and for various states (voluntary and involuntary). We used the lab-streaming-layer (LSL) platform, openly maintained by the research community in collaboration with industry. We applied signal processing techniques to rid the data of instrumentation noise and mechanical artifacts. Then, further layers of analyses and biophysical data integration helped us create a standardized parameterization of the nervous systems output to address in real time the changes across multilayered dynamics and the stochastic signatures of these time series signals.

The conceptualization of new forms of sensory—motor augmentation and sensory—motor substitution combined with indexes of adaptiveness operating in real time during naturalistic behaviors, helped us help the person in more effective ways. Whether a drug intervention or a form of behavioral/physical/occupational therapy, we designed objective biometrics—derived from the nervous systems—that ultimately led us to profile the amount of deliberate autonomy of the brain over the body. As such, we learned new ways to help the person maximize volition and with it develop or restore the senses of self-agency and self-sufficiency. We learned new ways to help increase the person's well-being and with it, augment the person's quality of life.

FROM INVOLUNTARY HEAD MOTION IN FMRI TO FAULTY DIAGNOSIS OF ASD

The advent of open access to fMRI data has initiated new avenues of autism-related inquiry in my lab. An example has been our use of the Autism Brain Imaging Data Exchange (ABIDE) databases: ABIDE I (http://fcon_1000.projects.nitrc.org/indi/abide/abide_I.html) and ABIDE II (http://fcon_1000.projects.nitrc.org/indi/abide/abide_II.html) to address motor sensing issues in autism spectrum disorders (ASD).

These databases provide many clinical scores drawn from observation. These scores are currently used to diagnose autism and ASD. The demographic records also include information about medication intake by the patients. The thousands of patients in these data repositories range between 5 and 60 years of age. The imaging data of ABIDE is rich in information. Yet, the important aspect of these data sets is that they contain the original time series of the images, rather than the *scrubbed* version of them. The term scrubbing refers to the cleaning procedure that fMRI researchers perform on the images to remove frames with motion artifacts. Since the head moves, the images become contaminated with noise, but an open access computer software package (the Analysis of Functional NeuroImages (AFNI) software packages[2]) allows extraction of the head's linear displacements and angular rotations in three dimensions (Fig. 7.4). It is then possible to set a motion threshold (e.g., using linear or/and angular speed) to eliminate all frames above that threshold. This is a common practice in fMRI research, which ends up eliminating a large percentage of the original images in any given study. This *garbage* of Cognitive Neuroscience turned out to be a goldmine for our lab.

Those frames contaminated by artifacts from the spontaneous involuntary head motions are not thrown away in ABIDE. As such, it occurred to me that a parameterization of these

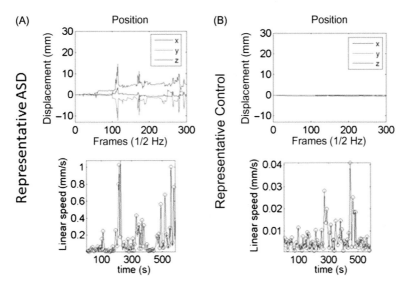

FIGURE 7.4 **The garbage of Cognitive Neuroscience is our goldmine.** Sample raw head motions extracted from resting state fMRI data and corresponding speed profiles with kinematics landmarks of interest (peaks marked by open circles). (A) Displacement and rotation kinematics extracted using SPM8 from raw resting-state image files (NifTI format) provided in ABIDE (yielding 3 positional and 3 orientation parameters). Representative ASD participant's linear displacements and angular rotations of the head registered with respect to the first frame. Speed profiles obtained by computing the Euclidean norm of each 3 dimensional velocity vector $(\Delta x, \Delta y, \Delta z)$ displacement at each point of application (x, y, z) from frame to frame, for 300 frames $speed_{frame} = \sqrt{(\Delta x)^2 + (\Delta y)^2 + (\Delta z)^2}$. To obtain velocity vector fields with corresponding speed scalar temporal profiles, the position data was analyzed using different methods and the results compared. One method filtered position data using a triangular filter to preserve the original temporal dynamics of the first rate of change data (i.e., the original timing of the peaks) while smoothing the sharp transitions from frame to frame[115] (using triangular window $v'(i) = \dfrac{\sum\limits_{k=-d}^{d}(v(k+i)\cdot(d+1-|k|))}{\sum\limits_{k=-d}^{d}(d+1-|k|)}$ for velocity v of frame i, k summation index from $-d$ to d and testing various values of d, e.g., up to 6, to build a symmetrically weighted sum around the center point, frame by frame). We also used regular derivative functions in the Matlab spline toolbox to transition from position to velocity. We obtained similar results as those with the triangular filter. (B) Representative control data. Notice the differences in magnitude between these two representative participants. For clarity the speed data of the control is not plotted at the same scale of the ASD panel so as to be able to see the patterns (notice that there is a large difference in the y-axis scale from A to B).

raw data using the micro-movements was possible, in combination with the SPIBA analytical framework. In this way, using the demographic data and the outcome of SPIBA from thousands of individuals in the spectrum of autism, we could gain several insights on these involuntary motions for a cross-section of the general population. By first building a normative data set, we could see the departure of the autistic SPIBA signatures from the neurotypical cases. In the process of analyzing these precious data (useless to other fields), we could begin to answer several questions that to this day had remained a mystery in ASD research and clinical practices (Fig. 7.5).

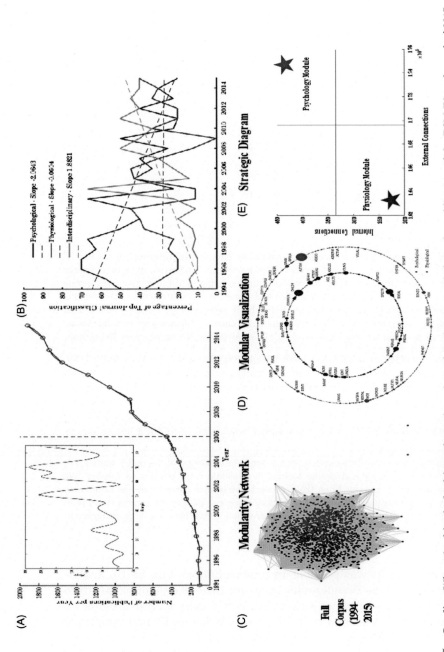

FIGURE 7.5 **Caroline Whyatt's objectively driven thematic review of Autism research** from the publication of the DSM-IV to the end of 2015 with over 17,620 peer-reviewed papers. Results (17,620 articles) illustrate that while there has been a growth of autism research, there has been a simultaneous consolidation of research focus on psychology and physiology approaches with emerging interdisciplinary trends. (A) Objective identification of a critical inflexion points in year of publication rates (2006 is maximal change). Time series data showing the derivative of number of publications per each year, used to calculate the slope of the trend line of publications (slope = $\Delta y / \Delta x$). The year 2006 automatically identified turned out to be the year of The Combating Autism Act, Pub. Law No. 109-416, of the United States Congress (Senate Bill 843) that was signed into law by President George W. Bush on December 19, 2006. (B) Most frequently occurring journals for publication per year were categorized into broad thematic clusters: Psychological, Physiological and Interdisciplinary to profile and examine the transition of research focus. The slope of the curves demonstrates a nascent growth in interdisciplinary publications. (C) Summary of co-keyword analysis completed across the corpus of texts (1994–2015). Keyword co-occurrence matrices are visualized using graph theory methods, including connectivity and modularity analyses applied to identify self-clustering networks. Here the cluster on the left is physiology-driven papers and the cluster on the right is psychology-driven networks. (D) Visualization of the modules of these two subnetworks with node size reflecting normalized Eigenvector centrality (identifying hubs in the networks). (E) Parameter plane displaying the prominence of the psychological module across the corpus, with notably higher levels of internal and external connectivity relative to both the corpus normalized median, and the associated metrics of the neighboring physiological network. This implies a well-established field of psychological research that acts as a "driving" force within the broader autism basic research field including work on physiology.

Some of the questions were: *Why is there a disparate ratio of five boys to one girl diagnosed with ASD? Why are we forced to use a clinical test called the ADOS (the Autism Diagnostic Observational Schedule*[3,4]*) in basic research and What is the metric scale of that test? What are some effects of psychotropic medications on the developing, nascent nervous systems of these children and adolescents with ASD and ADHD? Can we use our biometrics to distinguish Asperger's syndrome from ASD? And can we track outcomes in clinical trials using such biometrics?*

Before addressing these questions, we need to understand how autism is diagnosed and how the diagnostics system is tied to the treatments offered in the United States. These treatments consist of pharmacological interventions (recommended by the psychiatrists) and behavioral interventions (recommended by the psychologists). They are done on a trial-and-error basis, in *open loop* mode (without measuring or considering feedback from the nervous systems of the person). As emphasized throughout the book, these observational inventories follow a one size fits all model. And in the case of clinical trials, these traditional methods do not quantify the internal dynamics of the child's nervous systems. Indeed, they use primarily a static approach.

When examining traditional ways to study the nervous systems, we felt that at best they could tell us about possible differences between handpicked homogeneous groups, whereby one of them may be neurotypical. This type of "open loop" concept seemed insufficient to intervene in real time and reassess the reactions of a person's nervous systems to a given intervention. To that end, we considered other ingredients to enable *close loop* interactions.

In these new co-adaptive settings (as we saw in Chapter 6), we considered the internal dynamics of the nervous systems and the feedback they produce (e.g., through self-generated biorhythms that contain predictive codes within motion patterns echoed back to the brain as kinesthetic reafference). We had to create proper metric scales and standardized methods that worked across different ages and anatomical features. This meant that in addition to scales comprising absolute values denoting the size of the phenomena, we needed to have scales based on incremental data. Such increment data (also known as derivative or velocity-dependent data) give us a chance to assess the rates of change of neurodevelopmental aspects of the nervous systems. As such they bear dynamics and enable us to better quantify change. Without complying with these requirements, progress in our understanding of nervous systems disorders would be stalled, particularly in the nascent and rapidly changing system of an infant. We set to acquire a bird's view of the societal infrastructure for diagnosis's and treatments in general, so we could better understand these inherent problems and design proper parameterizations of the real-time changing signals.

DUAL DIAGNOSIS AND RECOMMENDED TREATMENTS IN NEURODEVELOPMENTAL DISORDERS UNDER PRE-IMPOSED ASSUMPTIONS

There are currently two main official avenues to *diagnose and treat* ASD. One is driven by the American Psychiatric Association (APA). It involves the Diagnostics Statistical Manual (DSM-5).[5] This manual has a compilation of criteria to define, diagnose, and provide pharmacological interventions for mental illnesses via medication prescription. ASD is among those so-called mental disorders. As such, many of the individuals receive medication upon a diagnosis of ASD (e.g., see Table 7.1 of some medication classes[6]).

The other avenue for diagnosis in ASD is the Autism Diagnosis Observational Schedule (ADOS).[3] In this case, trained psychologists under an accreditation system are certified to use this method in clinical and research settings to diagnose ASD. For example, Caroline and I got certified in April of 2016 and the cost would have been 12,000, so we could have both research and clinical certificates and could ourselves acquire accreditation to certify others in the lab. The New Jersey Governor's Council for the Research and Treatments of Autism waived the 12,000. We were lucky! At present, all research in ASD requires the use of the ADOS (e.g., the more than 2500 demographic records of ABIDE I and II, and ADHD-200 combined contains ADOS scores). This system is also used to recommend behavioral intervention.

In the United States, once a child receives the diagnosis, the school district system can prescribe early intervention program (EIP) or an individualized educational program (IEP)

TABLE 7.1 Some Psychotropic Medications Taken by Participants with ASD Reported in the ABIDE Repository. Here They Are Listed by Medication Class Along with Reported Motor and Bodily Related Side Effects

Class (Psychotropic Medications, N)	Specific Medications (across the seven Sites)	Motor and Bodily Related Side Effects
Antidepressants	Fluoxetine, Sertraline hydrochloride, Trazodone, Escitalopram, Citalopram, Bupropion, Mirtazapine, Duloxetine hydrochloride, Venlafaxine, Paroxetine	Tremors; paraesthesia; dizziness, drowsiness (Paraesthesia is the sensation of itching, burning, numbness, prickly feeling on the skin, or the feeling of "pins and needles")
Stimulants	Amphetamine and Dextroamphetamine, Lisdexamfetamine, Methylphenidate Extended release, Dexmethylphenidate, Dextroamphetamine sulfate	Dizziness, drowsiness; twitching; convulsions
Anticonvulsants	Oxcarbazepine, Valproic acid, Lamotrigine	Tremors; drowsiness
Atypical antipsychotics	Risperidone, Ziprasidone hydrochloride, Asenapine, Quetiapine, Aripiprazole	Tremors, twitching; restlessness
Benzodiazepine anticonvulsant	Lorazepam	Drowsiness; muscle trembling
Alpha agonists	Guanfacine, Clonidine	Restlessness; shakiness; dizziness
Atypical ADHD medication (NRI)	Atomoxetine	Tremors; dizziness, drowsiness
Nonbenzodiazepine sedative-hypnotic	Eszopiclone	Clumsiness; difficulty with coordination
Nonbenzodiazepine anxiolytic	Buspirone	Nervousness

Additional information is provided in Refs. 7–15. Sources of reported side effects for each medication class: https://www.nlm.nih.gov/ medlineplus/druginformation.html, http://www.drugs.com, and http://www.medicinenet.com.

if the child is already attending mainstream school. The child will receive behavioral interventions at the school (e.g., Applied Behavioral Analysis (ABA) though their poor outcomes have questioned their pervasive use in the past and in recent years[16–19]) and some form of occupational therapy and speech therapy. The aim of these behavioral interventions is to make the child *look good* in social situations. As such, they may teach the child how to carry a tray in the school cafeteria, or how to climb upstairs *in a socially appropriate manner*, or how to hold a pen, etc. They do not address the numerous sensory–motor issues these children have; issues that by now we know impede the development of autonomous motor control. In fact, consistent with the views of the psychological and psychiatric constructs, the American Pediatrics Academy discourages any form of occupational therapy that addresses sensory or sensory–motor issues.[20] Such sensory–motor oriented therapies have no insurance coverage in the United States, despite the claims by parents, practitioners, and self-advocates that they are effective in helping the affected person better cope with activities of daily living.

Behavioral methods that attempt to teach children in the spectrum how to be *socially appropriate* include punishment-reward reinforcement schedules within the framework of classical/operant animal conditioning that behaviorists employ in their research paradigms. But this research has never been properly validated using adequate statistics derived from human biophysical rhythms. They have never been reproduced outside the inner circles of behavioral practices, e.g., reproduced in other fields like movement neuroscience, or in developmental studies of motor control. They resonate with the diagnostics system emphasizing a deficit model of the child; a model that points at social deficiencies owing to the failure of the child to comply with some (culturally dependent) social expectation. And they leave out any hope to assume that the child has competence and a very plastic nervous system with highly adaptive capabilities. Indeed, the very fact that the autistic system is a coping nervous system surviving an early insult or glitch in neurodevelopment, denotes how much more flexible the autistic system is than previously assumed by these observational tests and behavioral interventions.

The term behavioral analysis within the context of animal research models really means that you stand there and observe the animal as it behaves in isolation, or within a group. Then, you write down by hand what you think the animal is doing. You may say "it is happy," "it is depressed," "it is antisocial," etc. This is a subjective opinion. You may come up with some ordinal scale, e.g., spanning a range of values from 1 to 10 and assign 1 to sad, 5 to somewhat happy, and 10 to happy. You may take 20 (or an arbitrary number of) measurements on the same animal and then take a grand average and claim that this average is the average behavior of the animal—and call that an objective metric! Then you may go on and do this for other animals and even pass those scores through some machine learning algorithm and go on to publish your "behavioral analysis" on a prestigious journal. Of course, when others try to reproduce your results, it is very likely they come up with different numbers based on their subjective appreciation of the behavior of the animals that they trained to do something along the lines of what you did. As such, it becomes extremely hard to verify any of the results you publish and as a field, true progress towards illuminating possible mechanisms underlying disease escapes us all. Understand here that what I have just described is the common method used to create *animal models of a disorder*.

Researchers build a genetically modified animal and then perform behavioral manipulations under some pharmacological agent. The outcome of that pharmacological treatment is quantified via observation and subjective description such as those above. In rare cases, a *fancy* camera-based method may be used to "quantify" behavior. But ultimately there is a human behind the camera hand coding the videos to help isolate epochs of the behavior that the researcher renders relevant to the phenomena under investigation. The outcome of the behavioral analysis is once again hard to verify by others, because it is rooted in subjective observation. And yet, based on that description and qualitative data, this animal model and *treatment* may be translated to a clinical trial that *a human child* will undergo.

Despite the lack of verification by other fields outside behaviorism and the total lack of objective definition of behavior in animal models, these experimental methods, derived from animal studies, have been translated to actual behavioral treatments for developing humans in the case of ABA. And further research for drug development is based on such opinion-based pseudoscience. The behavioral interventions like ABA are grounded on the works by Pavlov and Skinner using methods that largely replaced Freudian Psychoanalysis in the United States. The profits of such practices are in the billion-dollar range and the lobbying systems are impossible to defeat for obvious reasons (https://www.beaconhealthoptions.com/beacons-autism-spectrum-disorder-services/).

BEACON: Over 350 clients, including 200-employer clients, 41 Fortune 500 companies and partnership with 100 health plans; 5000 employees nationally, serving more than 50 million people). When we found this out through over 100 interviews conducted in the ecosystem of autism in the US, we were shocked and saddened by what these children are put through based on opinion-driven *science*.

Because of the basic stimulus-response association principles these behavioral interventions use, and the top-down instructions they pair with external rewards, there is no proper way to *spontaneously* elicit in the developing nervous systems, internally, from the bottom up, the type of prospective code necessary to function in a dynamically changing social medium. Such predictive code, particularly anticipating the sensory consequences of impending social overtures via reafferent feedback, is a fundamental ingredient for the development of deliberate brain control over autonomous bodily processes that take place in any social exchange. These processes are present in both the young nervous system[21,22] and the nervous systems undergoing neurodegeneration.[23] As such, there is a critical need to disrupt the paradigm altogether. Objective biometrics can play a fundamental role in the upcoming years. They are the *hope* we keep tight in the Pandora's box we opened.

The effects of these behavioral and pharmacological interventions on the developing nervous systems of the children with ASD are entirely unknown today (including the psychological and physiological effects of behavioral treatments that include aversive therapies using, e.g., cattle prods to discourage self-injurious behaviors (see for example, http://www.saukvalley.com/articles/2007/03/15/news/state/300595337906853.txt;

http://www.autismweb.com/forum/viewtopic.php?f = 6&t = 10143&sid = ea2b06f7922 e539707153edb2f7186cb).

Both the methods of diagnosis from psychiatry and those from psychology are purely observational and descriptive. The persons performing the diagnosis have not received training in neuromotor control and neuromotor development because these bodies of knowledge from neuroscience fields are not an integral part of the psychiatry or clinical psychology curricula. The certified personnel performing the diagnosis and recommending treatments have not received mathematical or computational training either, not even training on the types of statistical methods used in engineering, physics, and other applied math fields. As such, physiological measurements and objective analytics are not part of their paradigms.

Indeed, to understand the balance between the psychological/psychiatric vs. the physiological constructs of the current autism research models, Caroline Whyatt performed a metaanalysis of the autism-research literature published in top peer-reviewed journals since 1994 to 2015. She designed text processing methods to quantify trends and self-emerging patterns evolving over the years. She found that even the most recent trends point at a prevalence of psychological methods based on observation over the physiological methods based on objective quantification of phenomena. Part of the reason is that although the trend is pointing at interdisciplinary research as a new emerging force to eventually drive the science; the psychological and psychiatric tests continue to drive. Every single study of this literature required at least one such observational test. Further, every study is required to report correlations of continuous physiological data with discrete ordinal observational scores—a practice that is mathematically invalid.

As mentioned above, the methods these clinicians or researchers use to diagnose and treat ASD employ ordinal data (data obtained counting by hand or tracking isolated behavioral events with a clicker, using discrete scales they have made up without any physiological or statistical foundation as pertaining to human development). Consequently, the methods exclude motions that are occurring largely beneath the observer's awareness, i.e., occurring at time scales and frequencies outside the range of conscious vision.[21] They lack proper scientific criteria to detect and objectively quantify the evolution of sensory impairments and/or impairments in the development of somatic-motor physiological milestones. The methods simply impose the fundamental assumption that there are no sensory or somatic—motor issues in ASD. As it turned out, a bit of detective work led us to the outrageous discovery that "**we have been taken for a ride all along**". We will next discuss this in more detail.

THE DSM

The diagnosis of ASD following DSM criteria excludes motor or movement problems because many individuals in the spectrum, including infants and young children, are on psychotropic meds. These meds were developed for adults. In fact, drug development and FDA-approved pediatric clinical trials that involve psychotropic meds and/or combinations of meds are rare. This is perhaps because arguably, even if a parent consented to the trial on behalf of the child, the affected child could in principle sue the pharmaceutical company later in life, when aversive side effects unambiguously manifest, e.g., early symptoms of Parkinsonism.

The side effects of these drugs on the adult nervous systems are well known (e.g., Table 1, https://doi.org/10.1176/appi.books.9780890425596.MedicationInduced). They are written on the labels of the drug bottles and commonly explained on the websites of drug companies and government health agencies. However, their side effects on the rapidly changing nervous systems of a young developing child are, given the lack of clinical trials, unknown. Their effects on the development of somatic-motor and sensory systems of young children and adolescents are unknown. It is also unknown what dosage or combination may benefit a child, or for how long the child should take the medication before it turns harmful to the development of his/her nervous systems.

The makers of the DSM argue that owing to the above well-known confounds, the motor issues may not be core symptoms of the disorder, but rather a byproduct of the meds. This ambiguity has up to now kept somatic-motor issues out of the DSM-defined core symptoms of ASD. Indeed, under the DSM-5 section entitled "Medication-Induced Movement Disorders and Other Adverse Effects of Medication", several disorders are listed as byproducts of adverse effects from psychotropic meds intake, but none of these include developmental disorders like ASD or ADHD that under DSM-5 (but not under DSM-IV, According to the World Wide Web, in answer to the question, how many mental disorders are in theDSM 5? "The DSM-I, from 1952, listed 106; the DSM-III, from 1980, listed 265, and the current DSM-IV has297. (Complaints about this ever-increasing total led the chair of the DSM-5 task force, David Kupfer, toannounce that the total number of disorders in DSM-5 will not increase.)") can be comorbid.[5,24] Both developmental disorders are heavily medicated since infancy in the United States[25–27] and also abroad[28] (with uncertain sensory–motor consequences). I quote from the DSM-5 (emphasis added):

> "Sections: Neuroleptic-Induced Parkinsonism Other Medication-Induced Parkinsonism | Neuroleptic Malignant Syndrome | Medication-Induced Acute Dystonia | Medication-Induced Acute Akathisia | Tardive Dyskinesia | Tardive Dystonia Tardive Akathisia | Medication-Induced Postural Tremor | Other Medication-Induced Movement Disorder | Antidepressant Discontinuation Syndrome | Other Adverse Effect of Medication
>
> Excerpt
>
> Medication-induced movement disorders are included in Section II because of their frequent importance in 1) the management by medication of mental disorders or other medical conditions and 2) the differential diagnosis of mental disorders (e.g., anxiety disorder versus neuroleptic-induced akathisia; malignant catatonia versus neuroleptic malignant syndrome). *Although these movement disorders are labeled 'medication induced', it is often difficult to establish the causal relationship between medication exposure and the development of the movement disorder*, especially because some of these movement disorders also occur in the absence of medication exposure. The conditions and problems listed in this chapter are not mental disorders".

Our recent analyzes of involuntary head motion patterns present in ABIDE participants with ASD vs. age- and sex-matched controls, helped us uncover a simple way of evaluating such effects in cross-sectional data. We examined first ABIDE I and found excess *involuntary head excursions* (Fig. 7.6). These motions, when taken as the net amount of physical movements the person self-generated at rest, are indicative of excess energy consumption by the system with ASD. Indeed, it must be exhausting to try to remain still on command, when the nervous system spontaneously generates so much involuntary motion. I fail to understand how an entire field of research in Cognitive Neuroscience eliminates these artifactual frames and tells us a story on brain function upon interpretation of the inferences drawn from the remaining frames. This is done without acknowledging that physical motion did take place; and that it may affect the BOLD signal in some way we need to know as the excess motions are different for different people. Wouldn't you be curious to know these things?

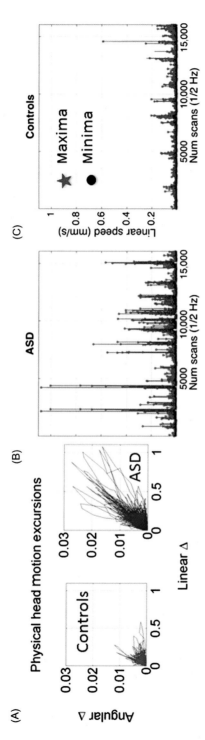

FIGURE 7.6 **Spontaneous involuntary head motions in human participants of fMRI studies while trying to remain still.** (A) Normative data from controls and departure of ASD participants from normal range of head excursions obtained by accumulating the rates of change of head's positional displacements and rotations while laying down at rest with the head padded (motion extracted from nonscrubbed resting state fMRI data in ABIDE I). (B) Time series of linear speed with peaks marked for ASD participants and (C) age- and sex-matched neurotypical controls.

We detected excess in head excursions that in ASD was accompanied by a higher rate of noise accumulation in the involuntary rotations and translations of the head. Then, a group of ASD participants were outliers with significant departure from normative data and from others in the spectrum of autism (Fig. 7.7).

These irregularities in the data prompted me to parameterize the ABIDE data by the micro-movements. This standardization was critical because multiple sites from the United States and other countries contributed data to this repository. Such data were collected using different sampling resolutions. Some scanners sampled above 1 Hz while others sampled below 1 Hz. I wondered how or if the differences in sampling resolution would affect the inherent variability of the speed data we were examining.

Carla Caballero (third in Fig. 7.1 C) came from Spain that year to do postdoctoral work in the lab and to train on the new methods we had developed. She was quite familiar with a broadly used method coined detrended fluctuation analyses (DFA),[29] a popular method for examining stochastic processes and gaining insights on the self-affinity/stationarity (or lack thereof) of biophysical time series data.[30] The DFA method had been used to analyze for example, motion biorhythms extracted from signals harnessed from heart rate[29] including local-scale of shorter time series[31] than the original application published by Peng et al.[29]; gait[32,33] including short time series too[34,35], finger tapping[36], among others. The parameter of interest is the alpha exponent (α), describing the quality of physiological noise that has been characterized in the literature (Fig. 7.8) giving us a sense for the color of the noise derived from the sampling resolution of a scanner.

The steps of DFA are listed and illustrated in Fig. 7.9. These graphs are implemented in MATLAB (version R2014a, The MathWorks, Inc., Natick, MA). In addition to MATLAB, we reproduced the outcomes using Python. Python code to compute DFA is provided in the following link: https://gist.github.com/JVero/9bb4921eeaefba8f0edff41cb584b460. Joe Vero, a Master student from Biomedical Engineering doing a rotation in the lab at the time, implemented the Python version of DFA. In this way, we double checked Carla's results independently—a practice I have in the lab because it ensures we are all on the right track and can blindly/independently reproduce our own results.

We faced another issue with the ABIDE sets, namely that each group type (e.g., ASD, AS, PDNNOS) had different sizes and when pooling across different sites with the same sampling resolution to perform our comparisons, we had cases where the disparity in sample size was too large. As such, statistical comparison was not possible. To avoid such problems, we used bootstrapping methods with the additional constrain of building our groups for comparison while keeping them age matched.

Sejal Mistry (the first in Fig. 7.1 C, who worked with us on the Growth charts in Chapter 1) came up with the code to do the bootstrapping with appropriate age matching. She also figured out how to work out the wrinkles in the publicly available AFNI code and modified it to automate the whole process of head motion extraction over thousands of records; organize it by various inclusion/exclusion criteria and taught others in the lab how to do all of that before she went to India to fight malnutrition as a Fulbright scholar. This was all happening while she shadowed all those pediatricians in the low-income districts of New Jersey, volunteered her time at various hospitals helping children with disabilities and aced all the classes of her pre-med program. We are still in awe with her amazing kindness, altruism, and overall sense of work ethics. Indeed, Sejal is our angel.

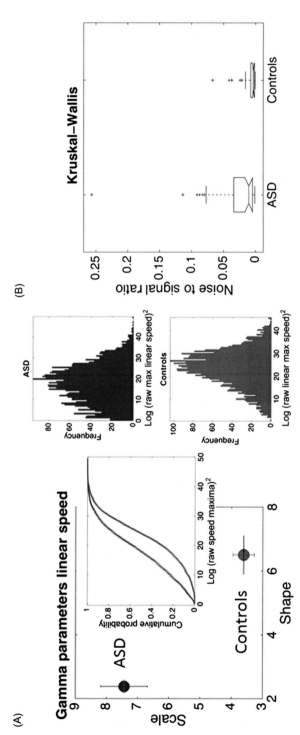

FIGURE 7.7 **Excess noise accumulation underlying higher rates of head excursions in data from (Figure 7.6).** (A) Empirically estimated Gamma parameters from the fluctuations in linear speed peak amplitudes normalized to account for disparate anatomical features across ages. Data pooled over sites of similar sampling rate. Points represent the shape and scale (dispersion of the estimated Gamma PDFs best fitting the data in the maximum likelihood sense). Points are from summative data across the ASD and TD groups. Inset shows the cumulative distribution functions (CDF) and frequency histograms contrast their accumulations with ASD showing a prevalence of small peaks (tremor like motion). (B) Significant statistical differences in the scale parameter (dispersion of the distribution) capturing the NSR.

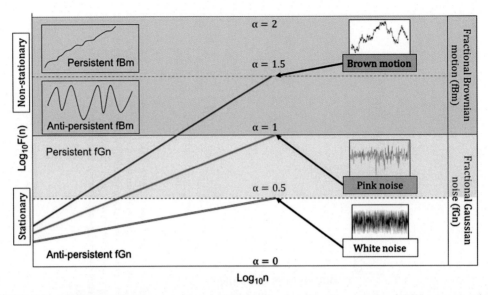

FIGURE 7.8 **Relationship between α values and types of noise.** Graphic example of different types of noise and their corresponding α value. Fractional Brownian motions (fBm) are a family of processes defined by Mandelbrot and van Ness[123] in which the successive increments in position are correlated, that is, a positive correlation means that an increasing trend in the past is likely to be followed by and increasing trend in the future (persistent series). Conversely, a negative correlation signifies that an increasing trend in the past is likely to be followed by a decreasing trend in the future (antipersistent series). These families correspond to α exponents ranging from 1 to 2. When α exponent is equal to 1.5, the process is named Brownian motion. It is a stochastic process defined using the integral of white Gaussian noise (α = 0.5). Contrary to fBm, the increments in position of this process are uncorrelated (each displacement is independent to the former, in direction as well as in amplitude). Fractional Gaussian noise is a family of fractal processes, defined as the series of successive increments in fBm. This family corresponds to α exponents ranging from 0 to 1. These two processes are interconvertible but they have different properties: fBm is nonstationary with time-dependent variance, while fGn is stationary and has constant expected mean value and variance over time[124]. Finally, we have a special case that is simply a statistically reliable departure from white noise in the direction of persistence (positive correlated). That case is when α exponent is equal to 1, corresponding to 1/f or pink noise. This noise is characterized by a form of temporal fluctuation that as a power density is inversely proportional to the frequency of the signal. This means that fluctuations at one-time scale are only loosely correlated with those of another time scale, showing a relative independence of the underlying processes acting at different time scales.

She is now debating between joining the Utah or the Yale medical school programs. But is good to have options. We look up to her for inspiration. The steps for the bootstrapping method are explained in Fig. 7.10 and the MATLAB code is in the website companion. It is easily adaptable to do this in other settings where the size disparity is a problem. In this case, we did the bootstrapping with and without replacement from the original sets we sampled from.

The main results remained regardless of whether we did the bootstrapping with or without replacement: the answer to our question was that indeed, different sampling rates affected the variability of the raw speed profiles (linear and angular) as shown in Fig. 7.11.

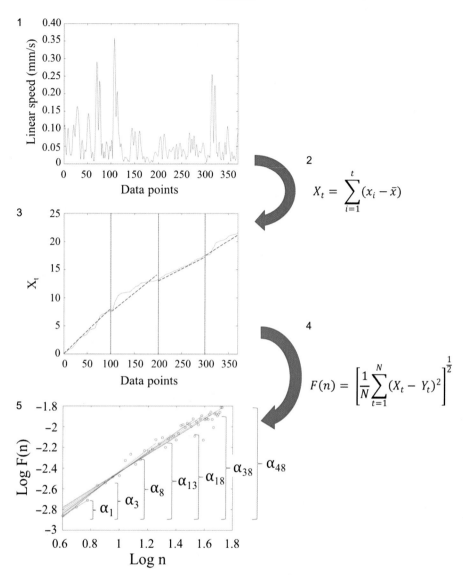

$$X_t = \sum_{i=1}^{t} (x_i - \bar{x})$$

$$F(n) = \left[\frac{1}{N} \sum_{t=1}^{N} (X_t - Y_t)^2 \right]^{\frac{1}{2}}$$

FIGURE 7.9 **Schematic steps to perform the detrended fluctuation analysis (DFA) using the linear and angular speed profiles of the involuntary hear excursions.** (Step 1) Sample raw linear speed data extracted from linear positional displacements of the head along the x-, y- and z-axis. The magnitude of the rate of displacements frame by frame (the linear speed temporal profile) is obtained and the profile resampled at 2 Hz. The data is truncated to 370 points for all participants to ensure equal number of points; (Step 2) Given the time series of length $N = 370$ (the minimum number of points across the data set), we obtain the integration or summation within 100-point window for each $X_t = \sum_{i=1}^{t} (x_i - \langle x \rangle)$, where t denotes the size of the window, x_i is each point in the series within the window, and $\langle x \rangle$ is the overall mean across the entire time series with linear speed (empirical range bounded between 4.28e-05 and 10.14 mm/s over the entire data set). The X_t is the cumulative sum or profile and the summation converts from a bounded time series to unbounded process. (Step 3) The cumulative profile X_t is divided into nonoverlapping time windows of equal length n (range $4 \leq n \leq N/10$), where N is the total

Whether ASD or TD control groups, the alpha values denoting the quality of the noise in Fig. 7.8 was very different for sites with scanners that sampled above 1 Hz in relation to sites with scanners that sampled below 1 Hz. This is shown in Fig. 7.11A for ASD participants diagnosed with DSM criteria and for age-matched TD controls in Fig. 7.11B in the case of linear speed. The four panels in each case show different metrics. Panel 1 is the frequency histogram of alpha values obtained from the DFA method applied to the time series of the linear speed of the involuntary head motions (500 values generated with replacement form the larger group of participants sampled below 1 Hz using the bootstrapping method) compared to the alpha index obtained from DFA applied to the linear speed time series of the group of participants in the smaller set above 1 Hz. Panel 2 is another way of visualizing the differences in the variability of the alpha using the space of empirically estimated moments of the PDF that best fitted the data in a maximum likelihood sense. We tried different distributions and the continuous Gamma family once again was the best fit, so we estimated the Gamma parameters (shape and scale) and plotted them on the Gamma parameter plane with 95% confidence intervals (panel 3). Then we also obtained the PDFs and plotted them in panel 4. The differences are not just statistically significant. The problem is that these are different probability distributions altogether. And they denote different quality of the noise derived from the signal variability.

Carla went on to test other features of ABIDE for the different demographics, taking into consideration that the raw signal had to be separated according to sampling resolution to ask if within a given noise color there were differences between groups with different demographics. These included ASD, AS, participants in meds (MEDS) or not on meds (NoMEDS), males and females. The overall results are shown in Fig. 7.12 at a glance, for both linear (LS) and angular (AS) speeds of the involuntary head motions during resting state fMRI. The sampling resolution of these scanners do have an impact on the variability of the raw data.

Given this issue, we decided to parameterize the raw data of ABIDE by the standardized micro-movements. They capture fluctuations in the amplitude of the original raw data, which describes the inherent speed variability independent of the temporal structure that the sampling resolution of the instrumentation imposes a priori. Fig. 7.13 shows the micro-movements waveform for the ABIDE I repository including data from three sites

◀

number of points in the signal, which in our case is $N = 370$[125]. In each interval, a local least-squares straight-line fit (which is the local trend) is obtained using minimization of the least squares errors in each window. The resulting piecewise sequence of straight line fits is denoted Y_t, then we calculate the root-mean-square deviation

from the trend, i.e., the fluctuation: $F(n) = \sqrt{\frac{1}{N} \sum_{t=1}^{N} (X_t - Y_t)^2}$. (Step 4) The above process of detrending and obtaining

the fluctuation metric is repeated over a range of different window sizes and a log–log map of n vs. $F(n)$ obtained. This map provides a relationship between $F(n)$, the average fluctuation as a function of box size, and the box size n. As explained in Ref. 126, the straight line of this log–log relation indicates statistical self-affinity expressed by the scaling exponent $alpha$, $F(n) \propto n^{\alpha}$. The exponent alpha (a generalization of the Hurst exponent,[127] is a measure of long time memory in a time series) is the slope of the straight line fit to the $log(n)$ vs. $log(F(n))$ relation using least squares. (Step 5) To obtain a series of alpha values for each participant, we windowed the data starting with 3 points, then 4 points, then 5 points, etc. to the maximum number of points (370) we had.

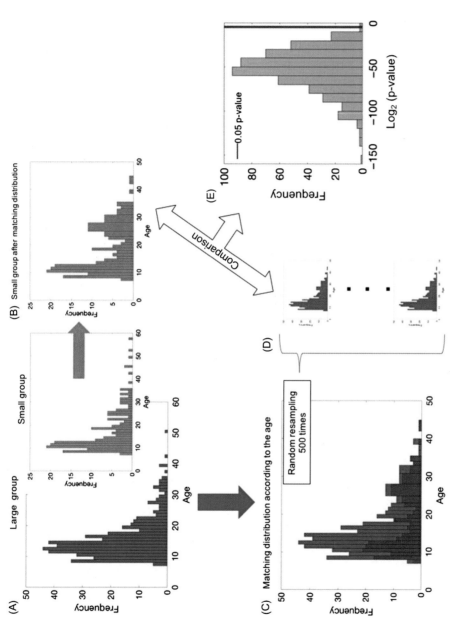

FIGURE 7.10 **Explanation of the Bootstrapping method.** (A) Inconsistent group sizes commonly found in the ABIDE datasets. The group with fewer samples is labeled "small group" (colored in red) in contrast to the group comprised of more samples, coined "large group" (colored in blue). For each group, we built frequency histograms binning the age such that each small-size subgroup drawn at random with replacement from the large group would have the same age composition as the small group we use for reference. (B) Age distribution of the small group after the bin sizes of the small and large groups are overlaid to maximize overlapping in age composition; (C) Example of resulting age-matched large and small groups upon overlapping analyses ensuring age-bin composition matching. (D) Random subsamples drawn from the large group while preserving age-bin composition, $N = 500$ age-matching subsamples with number of participants matching the small sample size. (E) Frequency histogram of $\log_2(p$ values) from the 500 comparisons. The red line represents the cutoff value for significance at 95% confidence.

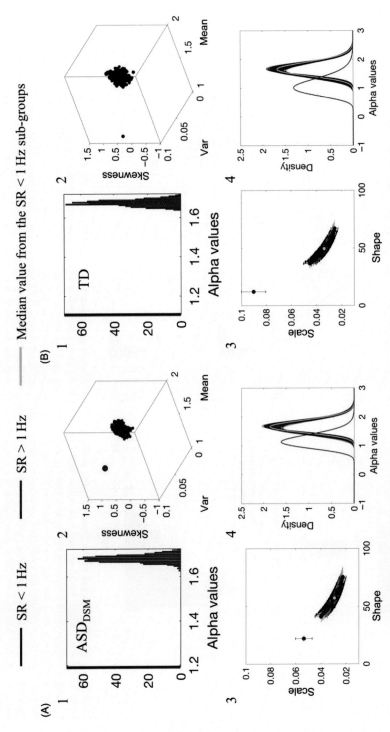

FIGURE 7.11 **Comparison between the distributions of α values of the linear speed for the groups with different original SR.** (A) 1— Frequency histogram of the mean α values from the 500-subgroups extracted from ASD$_{DSM}$ participants from SR0 group (large group) using boot- strapping, compared with the mean of 44 ASD$_{DSM}$ participant from SR1 group (small group, represented by the red vertical line). 2—Empirically esti- mated Gamma moments (marker size is the kurtosis). Red dot is estimated from the ASD$_{DSM}$-SR1smaller group. Blue cluster is the ASD$_{DSM}$-SR0 large group with 500 sub-groups built using bootstrapping method described in Fig. 7.3 while preserving the age binning composition of each sub- group to match that of the small group. 3—Estimated points on the Gamma parameter plane with 95% confidence intervals. Red point is the median value. Cyan is the median value. 4— PDFs ASD$_{DSM}$-SR1 small group while blue dots are from the ASD$_{DSM}$-SR0 500 subgroups from the large group. Cyan is the median value. 4— PDFs obtained from the estimated shape and scale Gamma parameters. (B) TD group participants with similar format as (A).

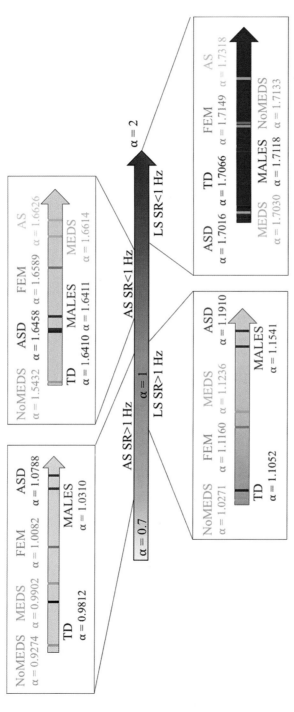

FIGURE 7.12 Summary of the mean of alpha values from each group from all sites from ABIDE I and ABIDE II and their relationship with the different types of noise in Fig. 7.8.

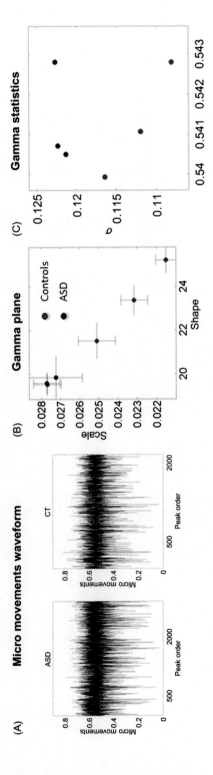

FIGURE 7.13 Automatic classification of ASD vs. TD participants using the standardized micro-movements and Gamma process under iid assumption. (A) Micro-movements of peak speed amplitude for ASD and TD controls. (B) Gamma parameter plane with points representing the data from three different sites of ABIDE using similar magnets (similar sampling resolutions). Points are plotted with 95% confidence intervals. ASD groups manifest excess noise and higher skewness in the distributions. Data from controls is systematically less noisy and characterized by more symmetric PDFs. (C) Empirically estimated Gamma first and second moments underscore the separation with higher variance in the ASD group (contributing in the numerator, the Gamma estimated variance, to higher NSR within comparable estimated Gamma mean means in the denominator of the ratio).

with scanners of similar sampling rates. The speed-dependent data (linear speed in this case) clearly separated between the ASD and TD groups.

When using the micro-movements data denoting the fluctuations in the signal's amplitude for each participant in these sites, I noted some outliers (Fig. 7.14) with much higher levels of noise-to-signal ratio (NSR) in the ASD cohort than in the TD cohort. The empirically estimated Gamma parameters for all individuals showed a great deal of overlapping (Fig. 7.14 A) and the log–log plane gave me a scatter that was well fit by the line (Fig. 7.14 A inset) with marked differences nonetheless between the family of PDFs that TD revealed and the family of PDFs that ASD revealed. (Fig. 7.14B).

I obtained a power relation and found different fitting errors between ASD and TD, so I decided to measure the departure from the TD fit and plot the NSR as a function of this error. The histograms in Fig. 7.14C revealed large differences in relation to the NSR. The latter were also plotted according to the empirically estimated Gamma's first and second moments in Fig. 7.15. The consistency of the differences led me to query the data some more in search for answers about the outliers in the ASD group.

Closer inspection of these participants in the demographics records revealed that they were on multiple psychotropic meds, though some were taking only one medication. The preliminary results were published[6] and motivated us in the lab to examine other aspects of the data. When pooling their data according to medication intake (no meds, one med, and two or more meds) we found a systematic trend whereby, as the number of meds increased, so did the NSR (Fig. 7.16). Further, the shape of the distribution of the peak speed values became more skewed and had a larger departure from that of TD controls. The normalized micro-movements took care of possible anatomical differences across the cohort spanning 5–60 years of age. The trends were clearly depicted on the Gamma parameter plane where we could also see the 95% confidence intervals providing nonoverlap between the TD and ASD with no meds; and the TD and ASD with meds. Fig. 7.16 shows these patterns, which were consistent for both the involuntary linear displacements and angular rotations of the head at rest. The insets depict the empirically estimated Gamma PDFs using the normalized micro-movements data. Clearly, these biophysical rhythms could distinguish, across the population, differences in meds and no meds intake in ASD. *But how about a breakdown of the data into different age groups? And different age groups where the participants only took one med? Would we still find differences?*

PANDORA'S BOX LET OUT SOMETHING AWFUL

A research faculty at Columbia University back in 2015, an alumni from the Rutgers Psychology Department came to my lab looking for help with her career. Let's call her Dr. X. During her time at Rutgers we did not cross paths because my research on neuromotor control is very different from her research in Perceptual Science (Fig. 7.17A vs. B). We met at a Simon's Foundation lecture in New York City, a lecture concerning issues in autism, an area then common to both our research interests. The 2013 paper on the micro-movements in ASD had gained some momentum and large press coverage. Because of this, several people had asked my lab to collaborate on various areas of autism research, spanning from clinical trials to therapeutic interventions. Dr. X was part of that crowd.

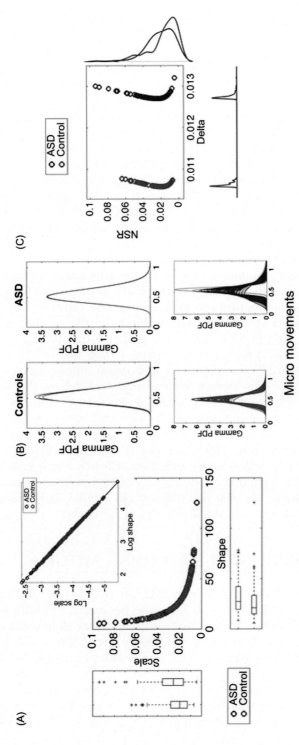

Micro movements

FIGURE 7.14 **ASD outliers on psychotropic meds.** (A) Individually estimated Gamma parameters for each participant in the summative data groups of Figs. 7.12–7.13 manifest a power relation (inset) whereby as the shape of the distribution is more symmetric, the NSR (the Gamma scale parameter) decreases. The linear polynomial fit to the scatter was obtained for TD and departure of ASD from normative ranges of the power relation was obtained. In the ASD participants, the scale values are significantly higher than controls and the shape of the distributions significantly more skewed than found in controls. ASD has several outliers with excessive noise and very skewed distributions of the speed peak amplitudes. (B) Estimated PDFs for ASD and TD also reflect the outliers. (C) Departure from the power relation by the ASD participants can be appreciated using the delta error (each point's deviation from the line).

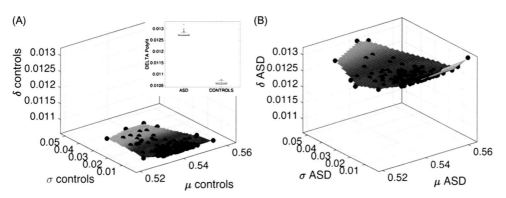

FIGURE 7.15 **Quantifying outlier's patterns.** (A) Three-dimensional surface fitting the parameter points of the controls and (B) of the ASD. Inset shows result of the Kruskall—Wallis (one-way nonparametric ANOVA) test with statistically significant differences at the 0.01 level for comparison of differences between the two groups on the delta residual from the polynomial fit to the scatters in A—B (ranksum Wilcoxon test, $P < 10^{-4}$ for shape and noise, and $p < 10^{-41}$ inset).

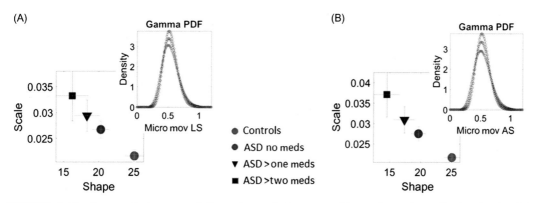

FIGURE 7.16 **Systematic increase of dispersion and skewness with psychotropic medication intake.** (A) Estimated Gamma shape and scale parameters with corresponding PDFs for the linear speed peaks in groups of TDs, medication naïve ASD participants and participants with ASD taking two or more medications. (B) Similar trends were found for the angular speed peaks.

She came over to my group during the break following the Simons' Foundation talk and asked me if I would be interested in presenting my new methods at Columbia sometime. She then requested an abstract and some ideas for a workshop. I sent both the abstract and ideas to her, but somehow, the workshop never took place. I never heard back from her until many months later, when she showed up at my lab.

The platform introduced in the autism paper[37] describe a general approach to assess and track disorders of the nervous systems. They provide a standard biometric scale to work with any biophysical rhythm from the nervous system. For that reason, any parameter that intrinsically has motion (change of the parameter position over time) can be

(A)

(B)   ACKNOWLEDGMENTS

We thank children, adults, and their families who generously gave their time to this study. We thank Yunsuo Duan, Feng Liu, and Satie Shova for their technical support. We also thank Molly Algermissen for conducting ASD assessments. We thank Liz Torres for helpful advice and discussions on the stochastic analyses, Jack Grinband for suggesting time-on-task modeling and insightful discussions during preparation of the manuscript, and

FIGURE 7.17   **Interdisciplinary collaboration that led to appropriation of intellectual property.** (A) Word cloud of Dr. X's papers before our 2016 collaboration with my lab did not contain the word movement in it. Her work was focused on perceptual science and vision. In stark contrast work from my lab was focused on movement and sensory—motor issue, particularly on motor variability and sensory—motor differences in neurological disorders. (B) My name was used in the acknowledgement section of a paper without my consent or knowledge. Without my knowledge, Dr. X used the analyses I did for their paper and in addition to that, without my consent and under an NDA, she used methods I developed to address the anonymous review process of our paper. She did not cite our paper even though it was being produced in parallel to hers at Columbia. Then, another paper using our methods and data type omitted the reference to the original micro-movements paper of 2013. Her intentional omissions let to a blatant attempt to appropriate intellectual property invented at my lab.

statistically modeled and studied using our approach. The lab had considerably grown and it was now commonplace to have these kinds of visitors. As such, I was not surprised that she came all the way from NYC and took the time to ask me about our current projects. My lab had many ongoing collaborations and we were now very well funded by the Nancy Lurie Marks Family foundation, the Henry Wallace Foundation for the study and treatment of Sensory Processing Disorders (SPD) and by the New Jersey Governor's Council for the research and treatment of autism. This NJ State fund for autism comes from a dollar-donation from every traffic violation. Pilot projects compete for state funds to advance research and treatments in autism within the state of New Jersey: A good use of those traffic violation tickets!

She asked if I could help her with ideas for a project where we could use the new metrics from my lab. It had become the norm by then, that people from other institutions would come to us with questions about autism data and the movement sensing angle my lab had taken. She had worked on Perceptual Science problems and fMRI research for the past 5 years but had no publications yet and had no funding left. Her situation was dire. She asked me to help her with a Sackler Foundation application—a prestigious Foundation at Columbia. I gladly sponsored her application and helped write the proposal using data and paradigms from my lab. Times are hard when junior researchers have difficulties publishing, so I thought of her as I do of any other junior person who stops by my lab for advice. I helped.

There was no concrete problem we could collaborate on because our fields of work were so different; but perhaps I could find something of interest and offer my lab's instrumentation and expertise on human behavioral analyses.

To discover a problem that we could work in common while applying both skill sets from vision and neuromovement science, I decided to follow the advice and infinite wisdom of my former student Jillian Nguyen. Jill (Fig. 7.1B) had an amazing insight when we conducted the NSF Innovation Corps (I-Corps) 120 interviews in search for a market-fit for our technology. Namely, she asked open-ended questions and let the information flow spontaneously lead us to something of interest. I reasoned that if I did this with Dr. X, the critical need for my skill set or more specifically, for my lab's technology to address a problem in fMRI research, would *spontaneously* emerge.

Jill and I had the most incredible experience with the NSF I-Corps and did discover the market fit for our technology using open-ended questions. Jill led the effort and her strategy kept us away from the type of confirmation biases that tend to emerge when the questions are too constraining or prompting the person to tell us what we already know.

I applied the open-ended question strategy and I asked her to walk me through the fMRI settings (as if I was a 4-year-old child who was going to do an experiment in the fMRI). I asked her to tell me, step by step, what she had been doing for the past 5 years of her postdoctoral training, i.e., what she had done every time that she brought someone in for an experiment in the fMRI setting.

She explained: "I ask the person to lay down and keep still." (Then I asked why? Why is the person to keep still?) She explained that artifacts from head motions contaminate the fMRI image if the person moves freely. This brought back memories of my Professor Marty Sereno (see the story in Chapter 7) when he designed a bite bar to restrain head motion inside the magnet and let me use his original design to modify it and use it to hold targets for my PhD thesis orientation matching task.

I asked, how do you know the image is contaminated? and she said because it is blurred; but motion thresholds indicate which frames to eliminate—a process we call scrubbing. At that point, I asked, *how do you know the motion threshold*? She said, it is obtainable from the image frames using open access methods. Then I asked? *Do you actually keep track of the motion frame by frame?* She said *yes*. What parameters do you record? I asked. Head displacement and rotation per frame—she said. *So, you do have a time series, don't you?* Yes, of course. (OK, we have motion data and *that*, I know how to analyze using my micro-movements approach—I replied.)

There was a long pause and she then realized that there was a possibility for a joint project and jumped of joy. She could bring her expertise related to the fMRI data, whereby she could extract the head motion parameters, and I could analyze the data using my micro-movements and Gamma process approach.

Within days Dr. X came back from NYC to my lab with a data set comprising a little over 100 participants, half with autism and half age-matched controls. The data was part of a paper that had been rejected from *The Journal of Neuroscience* in 2014. She brought the paper along too. The rejection seemed to be related at least in part, to the behavioral analyses they performed at Columbia using the head motion data. They took grand averages under assumptions of normality and homogeneity of variance—commonly done, and claimed the participants had comparable head motions as one of the conditions to go on and compare other parameters while excluding other possible confounding effects. Such approach was common and often would not yield any statistical significance. They could not find any substantial distinction between the participants with ASD and the controls; *how could that be, when it was common knowledge that the participants with autism tended to considerably move their heads in the fMRI much more than controls?*

Lack of statistical significance in autism research was not uncommon and rather puzzling. But I could see why: The main reason for this was the above treatment (discussed at length in Chapter 3, e.g., Figs. 4 and 9): the data were traditionally subject to a grand average under the a priori assumption of a theoretical normal distribution. As explained in the previous chapters, this approach smooths out as noise the minute fluctuations in the data, which our research has shown over and over, are the actual signal you want to look at in autism. Averaging the motion data under enforced theoretical assumptions of symmetric distributions is also problematic because the empirical distributions are skewed. In fact, the range test is violated by increment (derivative) data such as the speed-dependent data one examines in the case of head motion: see Fig. 4 in Chapter 3, where I discuss these issues plaguing the autism literature attempting to analyze bodily biorhythms of various kinds.

I analyzed the data set from her rejected paper and wrote a detailed report explaining to Dr. X all the steps I took along with my interpretation of the preliminary results. Plotting data and explaining preliminary results is common in motor control research because it is the first of many steps before modeling the phenomena. One must discuss intrinsic dynamics and statistical features in the data first, then one decides what platform to use for modeling and further analyses. I found in this and other collaborations with people in autism that this step does not exist in their research and in more than one occasion now they have run away with the explanations and methods thinking *this is it, it's mine to keep!*. Very odd! There is both a lack of ethics in the whole exchange; but more importantly there is a lack of interest in understanding in depth the phenomena under study. These biometrics and platforms all have caveats that we—the researcher who model them and adapt them/invent them—understand very well. They are not meant to be blindly applied to everything. It is a pity Dr. X did this, but the story is worthwhile telling for it will not go entirely unpunished. As Prometheus upon stealing the fire, or Epimetheus taking Pandora, or Pandora opening the box, bad deeds unleash uncertain consequences. Nobody escapes *Zeus eyes*.

I found a set of participants showing elevated levels of NSR. I read the Methods section of the paper in detail and noticed that nearly half of the autistic participants were on psychotropic meds. I then asked her to bring me the data of their demographics (age, sex, clinical scores, and most important of all, medication intake) to see if I could find something different or special about them. I did find something interesting: these outliers were those on *psychotropic meds*!

I had just finished working with Jill on a cohort of patients with schizophrenia. Her PhD thesis involved this patient population. Because ours was a study focused on their motor control, their medication intake was important to avoid confounds between medication intake and core sensory—motor issues. As such, in the spirit of the DSM, Jill and I tried to separate patients who were on meds from those who were not. We had worked with the team of Dr. Steve Silverstein at Rutgers/Robert Wood Johnson to isolate medication intake and shed light on possible motor symptoms from such meds. In the end, only two patients in our cohort were on meds. The rest were outpatients who had long ago stopped the meds. However, the question the DSM-criteria raised, prompted us to compile the meds information. Thus, when the fMRI data arrived in the lab, we had already compiled a table with possible meds from mental illness in adult patients. As it turned out, several of these psychotropic meds were being used in autism and ADHD patients across all ages.

The ABIDE demographics contained information about medication-intake and medication-naïve participants with a diagnosis of ASD that we could use to understand the patterns of involuntary head motions—present despite the padding of the head.

The resting state fMRI head motion data that I analyzed from Dr. X, the same data that her group could not publish in *The Journal of Neuroscience* because the parametric methods they used yielded no statistical differences, suddenly "came to live" with the micro-movements approach. Very clear separation emerged between the participants with ASD and the TD controls.

To my dismay, after all that work, Dr. X had omitted an important fact: this was not her data to share. Her title of Research Assistant Professor at Columbia did not carry the same type of independence that a title of tenure-track Assistant Professor at Rutgers would. A tenure-track Assistant Professor has an independent lab. The data generated in her lab could be shared in collaborative projects upon following data-sharing agreements and Institutional Review Board (IRB) protocols. This turns out not to be the case when there are other Principal Investigators (PIs) who direct a given project—an omission on her part that costed me dearly.

She informed me then (only after I revealed the results and shared the detailed report on the analyses) that these data could not be published because she was not the PI of the study. The PI of the study had in fact left Columbia University and was now in California, at the University of Southern California—according to her. Dr. X informed me that it was complicated to reach out to him and explain all of it to her group in Columbia. After all my effort, and the time I put into this project, the interesting results were not publishable. I was disappointed but hopeful we could use the same methods on other data sets. I had effectively obtained proof of concept for a possible biometric distinguishing meds intake in ASD and their departure from medication-naïve ASD and medication-naïve TD controls.

Using my SPIBA technology and the micro-movements parameterization of the ABIDE data, I had discovered a set of motion parameters sensitive enough to minute involuntary fluctuations in the autistic nervous system, that the biometrics could broadcast fundamental differences between medication naïve and medication-treated autistic patients. The implications in my mind were broad, given the important motor ambiguity raised by the DSM criteria.

I was deeply disappointed, but with many other projects in my mind, did not think twice about it. I moved on to the other ongoing projects in the lab and simply suggested to Dr. X that perhaps the National Database of Autism Research (NDAR), maintained by the NIH, was feasible to perform my tests. I explained I was busy to work further on this but she insisted on NDAR. All that was needed was the head motion data. However, since I had no idea how to extract it from the images, Dr. X offered to do so. She went back to NYC and worked on getting head motion data from public repositories. I thought this was a good collaboration and moved on with the suggestion. I came to deeply regret this decision later.

The NDAR was not as organized as we thought and some data sets had been scrubbed. This prevented the use of that data because it is in the excess motion and its noise accumulation over the session's time period that the signal separating ASD from TD resides. Dr. X came up with the idea of using the ABIDE repositories, which were not scrubbed and were much more organized than the NDAR. She communicated to me that she processed all the images—though months later she said a graduate student (let's call him student Y) helped her. As soon as the data-sharing agreements were in place, I analyzed it and found similar trends as with the previous data set containing fewer cases. The numbers were now in the range of thousands of participants. This was a real statistical phenomenon.

Given the strength of the finding and the fact that I was sharing with Columbia technology that Rutgers was in the process of patenting, I suggested we both signed a two-way Non-Disclosure Agreement (NDA). In this way both parties were protected (or so I thought). With the NDA in place, and all Institutional Review Board documents for data sharing ready, I proceeded to analyze the ABIDE I data set in full. I also shared all my designs and results with Dr. X. Of course, this was a collaboration. *How was I to imagine what came later?*

Under my guidance, and based on my results from the previous set involving the data from the paper that had been rejected from *The Journal of Neuroscience*, I asked Dr. X to select from the demographics information of ABIDE I, the participants who were medication naïve, those who took only one med and those who took more than one med. I asked her to group them by age, so we could see these patterns in cross-sections of the population at large. As with the unpublishable data that Dr. X brought to my lab, the results using SPIBA on these data parameterized by the micro-movements were very interesting. With larger numbers of participants and multiple ages, they revealed systematic trends worthwhile exploring further.

Unfortunately, the collaboration ended up rather bad. My new graphs to visualize the data and all the analyses I did alone, mysteriously appeared in a publication she and others at her Columbia group authored.[38] Most of them were her co-authors in the paper rejected from *The Journal of Neuroscience*, but other new names appeared on the list of co-authors (e.g., student Y mentioned above). Using Rutgers intellectual property behind my back and without my consent, they published that very data set that I had originally analyzed and used to teach Dr. X some of the aspects of SPIBA. They repeated verbatim some text and even cited references only I knew about as I had used them to build the motivation behind my normalizations and so forth. She took everything from my explanations and published my work under her name.

Google scholar alerted me of their publication, on December 2016. The paper had come live on line ahead of our own! In other words, I saw my own analyses, visualization graphs, and theoretical ideas in someone else's paper;[38] but that *someone else* was collaborating with me on this other paper.[6] What was more perplexing yet, was that these were the very head motion data I had analyzed. My heart shrunk in disbelief and anguish. *How could someone do this?*

Only then I understood why the review process inexplicably stalled that summer. Despite my efforts working on the revisions, and performing additional analyses to convince the reviewers and the editor of Springer Nature Scientific Reports that the methods were adequate—just unfamiliar to them; I could not get the revisions out because many words I had fixed kept changing and with that, the meaning of the text also changed. She erased the term micro-movements from the paper. It became a nightmare somehow to work out the revisions. I backed out of the collaboration and refused to go on; but a professor from the Columbia group (let's call him Dr. Z) contacted me to encourage finishing the process. He kindly offered to pay for the publication fees in full—an offer that I did not accept because my lab is very well funded and we can pay for the papers we publish. I did accept that he mediated the exchange, so the interesting results saw the light in that journal.

Dr. Z never mentioned though, the other work taking place in that Columbia group in parallel to mine, i.e., using the very analyses I had developed. To this day, my hope is no one else in Columbia Psychiatry really knew my work was being brought over without my consent. I must believe this! Science is much too sacred for me to believe otherwise.

That summer was very sad. My mother had become terminally ill. I went back home to San Diego and worked there throughout the summer while I took care of her. I could not explain with words how magical this time with my mother was. She passed very peacefully and rather happy the 21st of July that year of 2016. A large part of me went with her and remains somewhere out there wherever she went. I have never recovered from that loss.

I returned to New Jersey from San Diego in August and as soon as I had enough energy to get back to work, I pursuit the revision process. This was to no avail—even while being mediated by Dr. Z. Later, I learned the time line of these events. It was not until the Wiley journal of *Brain and Behavior* had the Dr. X et al. paper accepted (September 11, 2016), that I was able to submit my last revisions to the Springer Nature Scientific Reports.

During the review process, I had designed new analyses and visualization tools that my presumed collaborator took and ran away with. My name appeared on the acknowledgements of a paper I never saw (Fig. 7.17B), as if I had consented to the use of my intellectual property by those authors on the paper, all of whom were members of Columbia University, a prestigious institution with tremendous power in the field of mental illness.

My chances of initiating and winning a dispute about this incident remain slim. As the editor of the journal succinctly put it, *this was the case of the giant swatting the fly* and I was the fly in this case. My ideas and hard work went their way without any consideration or hesitation by any of the members of this group. No one asked, how come the very data they got rejected from *The Journal of Neuroscience* made total sense all the sudden. Not even the genuine curiosity of a scientist came to the rescue. The use of my name without my consent, the appropriation of my methods and visualization graphs without giving proper credit to my work and effort were very upsetting to me. By the time that my paper with Dr. X was out, I thought anyone could see the striking similarities between the graphs I used for visualization of the results in the Scientific Reports paper and the graphs in their own paper. These graphs were up to that point unique to my group and very distinct indeed from the rest of the field doing autism research.

An important issue came later when a second paper came out of that group by Dr. X and student Y.[39] That paper left out entirely any reference to the work from SMIL. The two people whom I had suspected were complicit on the first Wiley *Brain and Behavior* paper revealed their true faces. Their acknowledgement section in that paper reads "...contributed to analytic streategy...". They appropriated the analytics I invented and even shared it on this occasion.

Pity these two did not realize one fundamental mistake. The methods they described and so easily trivialized when the Wiley Editor reached out to Columbia for an explanation, do not appear in a text book or specific paper (other than previous papers from SMIL and a book edited by Caroline and I[40]) The partition and interpretation of the Gamma parameter plane that appears in those two papers described as generic, is something that we at SMIL built empirically from years of research profiling the human motor control systems. The only justification for the type of inference they claimed (copied straight out of our papers) is empirical. The only source for that interpretation is unique to SMIL. In this respect, "Zeus eyes" were mysteriously looking upon these two. It is a matter of time before they are caught for what they did and judged by the scientific community.

The outcome of all of it was very disheartening for me to see, particularly because my lab had welcomed Dr. X as one of our group. Far from it though. She was not one of us. None of my students and postdocs would have ever done something like that—much less lure a graduate student into such collusion.

At SMIL we work hard together and share ideas and respect each other's creative thinking. To this day, we remain in shock that someone could do such a thing. Sadly, it seems to be common in the world of interdisciplinary collaboration. In this case, it was unambiguous that some oddity had occurred: *out of magic*, the Wiley paper contained all the analyses I had created during many years of research in neuromotor control. Fig. 7.17A left panel shows a word cloud predominantly from Perceptual Sciences with a focus on vision extracted from the PubMed peer-reviewed papers of Dr.X up to 2016. The word cloud of Fig. 7.17A right panel is derived from my PubMed peer-reviewed papers up to 2016. These are from movement science—the theme of the new behavioral analyses of the Wiley paper in question.

It takes many years of work and research to come up with an idea and the methodology to execute it. It is so sad to see it all disappear into the wrong hands because those hands do not understand where these ideas came from. They do not understand the caveats and the alternative ways in which one can attain these results. Using the methods in the blind, merely regurgitating SMIL's previous findings is a mistake.

This unfortunate event certainly made me aware of these issues in interdisciplinary collaboration. But research must go on despite such awful incidents. In the end, the important thing is to get the results out, so those affected by autism and their care givers know that psychotropic medication intake have quantifiable effects on the developing nervous systems. I somehow have the feeling though, that the awful stuff Dr. X let out of Pandora's box, the stuff that Columbia group is brewing, will get back to them in due time. It's just the way things mysteriously work out in science. In the end, the truth always prevails. One way or another, this will be known and self-corrected.

EXCESS NOISE IN AUTISM

When I examined the effects of medication across the population, they appeared to be complex and markedly different within each age group. Fig. 7.18 shows different patterns for the translational (T) and rotational (R) speed of the involuntary head motions for the case when the participant was taking multiple medications. Here we could gain a sense of the prevalence of the medication in the ABIDE data set. For example, Group 1 composed of children between 6 and 11 years of age, had a prevalence of alpha agonists, but Group 2 composed of children older than 11 years of age, but younger than 13 years of age, had a prevalence of anticonvulsants. Both of those cases had elevated NSR, well above TD controls. The figure caption explains the details but it suffices to say that for the cases of combination treatments with multiple psychotropic drugs, the older the group, the farther away they were from the age-matched medication-naïve ASD participants and even farther away from the TD controls.

The patterns were different for the case when one psychotropic drug was taken in isolation. This is depicted in Fig. 7.19. The prevalence of antidepressants and stimulants in group 5 of participants 17 years old and above, along with the systematic increase in the NSR—relative to the age-matched medication-naïve ASD was interesting. Yet, in this case, the patterns for the younger groups were more thought-provoking in that they showed to be closer to the TD controls than to the medication naïve ASD. After puberty, the stochastic signatures and their patterns seemed quite different than they were in childhood. The effects of psychotropic medication intake appeared to be very complex. The important message was that these biometrics could capture shifts within a group and cross-sectional changes in the population over time. As such, they could be used as outcome measures to assess treatment effects on involuntary motions.

One of the messages from the preliminary work with the ABIDE I records was that medication-naïve ASD individuals manifested noisy and random involuntary motor signatures across all ages, sex, and degree of severity (severity in this case, as described using observational means). This result strongly suggests that the excess noise and randomness in the involuntary motions is a biological core feature of the condition that can be isolated from effects of psychotropic meds. In this sense, DSM makers could now reconsider motor issues as part of the diagnosis. The important point their manual raised had been addressed by this research.

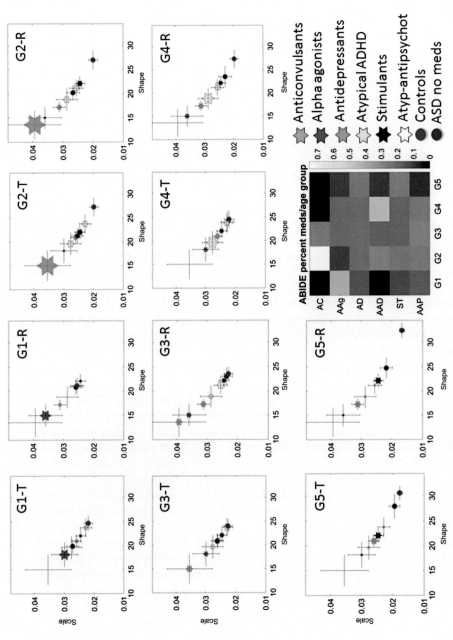

FIGURE 7.18 **Speed-dependent stochastic signatures of head micro-movements as a function of medication status and age, for medication classes when taken as part of a combination treatment.** Each age group number and letter (e.g., G1-T and G1-R) corresponds to the linear/translational (T) and angular/rotational (R) speed-dependent signatures across medication classes shown with 95% confidence intervals (CIs). Groups by age are G1 (6–10.99), G2 (11–12.99), G3 (13–14.99), G5 (15–16.99), G5 (above 17) years old. The empirically estimated Gamma parameters are obtained from the pooled group data comprising ASD individuals "on" medication and each point is cast against the members of age-matched reference groups (medication naïve ASD: blue and CT: red). Note that controls have the lowest NSR and the highest shape (most symmetric) value across all age groups. The size of the marker on the Gamma parameter plane represents the percentage of that medication type within the group based on the reported information in the ABIDE. The color-coded matrix presents these proportions for each age group (columns) and medication class (rows). The most commonly prescribed medication class for each group is as follows: G1: alpha agonist, G2: anticonvulsant, G3: anticonvulsant, G4: atypical ADHD, and G5: stimulant. No marker means that the medication intake is not reported in the group (darkest brown entry in the matrix).

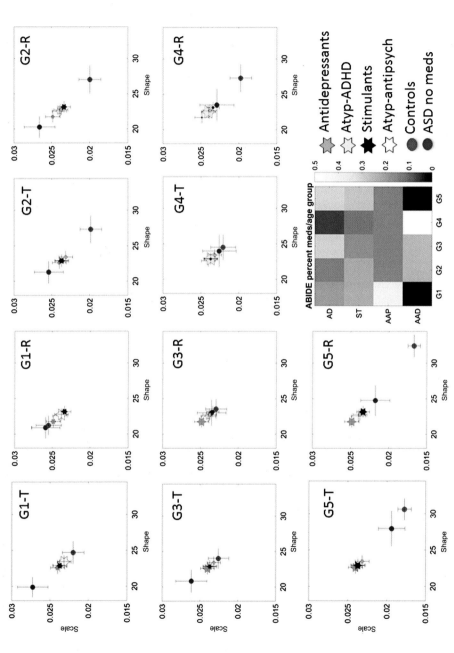

FIGURE 7.19 **Speed-dependent stochastic signatures of head micro-movements as a function of medication status and age, for medication classes when taken as part of a combination treatment.** Each age group number and letter (e.g., G1-T and G1-R) corresponds to the linear/translational (T) and angular/rotational (R) speed-dependent signatures across medication classes shown with 95% confidence intervals (CIs). Groups by age are G1 (6–10.99), G2 (11–12.99), G3 (13–14.99), G3 (15–16.99), G5 (above 17) years old. The empirically estimated Gamma parameters are obtained from the pooled group data comprising ASD individuals "on" medication and each point is cast against the members of age-matched reference groups (medication naïve ASD: blue and CT: red). Note that controls have the lowest NSR and the highest shape (most symmetric) value across all age groups. The size of the marker on the Gamma parameter plane represents the percentage of that medication type within the group based on the reported information in the ABIDE. The color-coded matrix presents these proportions for each age group (columns) and medication class (rows). The most commonly prescribed medication class for each age group is as follows: G1: alpha agonist, G2: anticonvulsant, G3: anticonvulsant, G4: atypical ADHD, and G5: stimulant. No marker means that the medication intake is not reported in the group (darkest brown entry in the matrix).

SIGNIFICANCE AND POTENTIAL CONSEQUENCES OF THESE RESULTS

From the stand point of basic research, these involuntary fluctuations can be considered as vibrations echoed back to the CNS as a form of returning kinesthetic afference. In the case of ASD, with or without meds, such noisy random signatures provide persistently corrupted feedback. This raises the possibility of deleterious effects of kinesthetic reafference on the autistic nervous systems—even when the person had not been exposed to psychotropic meds. But if those who were on meds manifested systematically increasing levels of noisiness and randomness that shifted the probability distributions with one med, two meds, three meds, etc. in relation to those individuals with ASD who were not on meds (as in Fig. 7.16); one wonders about the critical need to measure these effects during early childhood. It should be interesting to know when symptoms of movement disorders begin to manifest on a child that is exposed to such psychotropic drugs since early infancy. My words in fact convey a sense of urgency. Something doesn't feel right after this discovery.

Atypical motor patterns are symptoms of ASD. Contemporary basic research has demonstrated that they manifest across many levels of the nervous systems[18,41−46] and early symptoms in neonates foretell future problems with neurodevelopment.[47,48] One confounding factor has been that movement has been essentially conceived as a form of efferent motor output, with a unidirectional flow from the central nervous system (CNS) to the periphery[49−61] neglecting in more than one way the internal dynamics of the individual's nervous systems and unintentionally ignoring that this self-generated activity flows back to the CNS along afferent channels. This approach has also overlooked spontaneous behavioral variability patterns inherently present in our motions[62−64] and their nonstationary statistics—all issues that were much earlier pointed out by Bernstein[65] and Thelen in her seminal work on neurodevelopment.[66]

The new biometrics we describe in this book are by no means the only way to track these symptoms. However, they do provide a new standardized data type to more properly combine discrete, ordinal scores from clinical observational inventories with continuous real-valued data from physiological biorhythms. Such biorhythms can be more generally harnessed noninvasively with biosensors or using other noninvasive means that are nowadays commercially available.

By providing a standardized scale (ranging from 0 to 1), the micro-movements parameterization summarizes the continuous analogue data into a continuous point process, while maintaining the real value range and normalizing for anatomical discrepancies. Such a parameterization enables integration of signals from multiple instrumentation and analyses of such signals under a similar scale. Consequently, the various layers of the person's nervous systems can be continuously sampled in tandem using the personalized approach SPIBA affords us. This method considers the intrinsic dynamics of the person's nervous systems and the physical ranges of amplitude of such signals. As such, the comparison across individual signatures and their shifts across the general population are now possible under a common statistical umbrella.

Along those lines, the ABIDE demographic records contained absolute IQ scores for many individuals of different ages. When plotting the frequency histograms of these absolute scores (Fig. 7.20A left panel) they spanned a symmetric distribution. But the physical

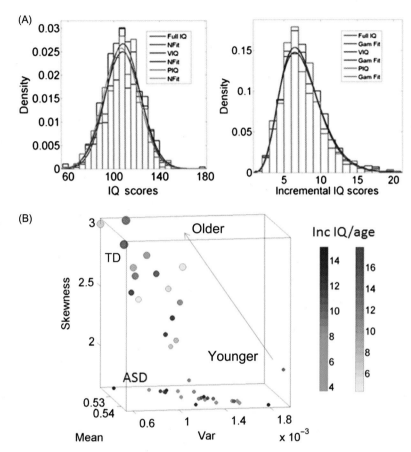

FIGURE 7.20 **The need to examine increment (velocity) data in neurodevelopment.** (A) The absolute IQ scores distribute normally, but the incremental (velocity-dependent) data that we obtain by dividing the person's score by the person's age produces *skewed* distributions with very different stochastic features than the assumed symmetric Gaussian distribution. The nonlinear stochastic dynamic nature of human neurodevelopment can be better captured by nonlinear processes with families of probability distributions that fit the nonlinear dynamics of developmental trajectories better. (B) Instead of attempting to directly correlate ordinal discrete IQ scores with continuous biophysical data, here median-ranking the data provides a way to separate head-motion patterns corresponding to participants with IQ scores above the median and compare them to the movement patterns from participants whose IQ scores fall below the median. The results show stunting in the evolution of the probability distributions of ASD participants, a remarkable departure from the evolution of controls. As TD controls age, they acquire PDFs with more symmetry (symmetric distribution has skewness value of 3 along the z-axis). In contrast ASD participants remain with skewed PDFs with long right tails indicating an accumulation of small values of the micro-movements extracted from the linear speed. Another parameter that remains unchanged in ASD is the kurtosis, in marked contrast with the TD groups, which tend to have peakier distributions as they age. As with our previous research involving voluntary motions, here we witness the stunting of ASD development (5−60 years of age).

age of these participants, which incurs in gradual physical anatomical and neurological changes of the brain and peripheral networks, is bound to impact IQ. The incremental changes in anatomical sizes are not considered by such absolute scores that report just the size of the phenomena—not the rate of change of the phenomena. In the case of IQ scores, this static approach to neurodevelopment considers instead "mental age" to design their questionnaires. This is a psychological construct based on an a priori assumed expectation of what the person's performance ought to be at a given age. It is not a physiological approach considering the dynamics of nervous systems' evolution.

To capture the incremental (derivative) nature of physical age, whereby the neuroanatomy and neurophysiology of the aging population dynamically changes, I converted these absolute scores into increment (derivative) data. This is attained by simply normalizing by the age of the person at the time of the test—another entry available in the demographics data. As we had seen in Chapter 1, Fig. 21, such transformation of the physical growth data in the neonates revealed fundamentally different PDFs between the pre-term and the full-term babies that were masked by the (static) absolute scores merely reflecting physical size and neglecting its rate of change over time.

As in the physical growth chart data of the baby study, the resulting frequency histogram was rather skewed, whereas that of the absolute values was symmetric. This contrast is shown in Fig. 7.20A (right panel) for all IQ score types superimposed (full IQ, Verbal IQ VIQ, and Performance IQ (PIQ). These empirical frequency histograms were well fit by the Gamma family (fits are also shown in Fig. 7.20A) These results motivated me to examine the changes in incremental IQ per age groups. To that end, the ABIDE sets were split by age groups as in Figs. 7.18 and 7.19 (G1 (6–10.99), G2 (11–12.99), G3 (13–14.99), G3 (15–16.99), G5 (above 17) years old and the scores were median ranked according to the normalized IQ scores. This way of clustering the data gave 10 groups in the ASD and 10 groups in the TD cohorts—two per group denoting participants with IQ above the median and participants of another group below the median. Then, the micro-movements parameterization of the involuntary head motions for each of these groups were pooled to build a group time series of normalized speed amplitude fluctuations. These 20 sets of micro-movements served as input to a Gamma process under the IID assumption. The empirically estimated Gamma parameters (shape and scale) were used to estimate the Gamma moments and plotted on a parameter space using five dimensions. The empirical Gamma mean was plotted along the x-axis, the empirical Gamma variance was plotted along the y-axis, and the skewness along the z-axis. The kurtosis was plotted as the size of the marker and the color of the marker's face was the change in the incremental IQ value normalized by age (see color bars in Fig. 7.20B).

The points in Fig. 7.20B distributed differently for ASD than for TD. The normative data showed an evolution by age whereby the skewness of the distribution approached the symmetric value increasing as the participants' age increased (this metric is called "excess skewness" so a value of 3 is perfectly symmetric in this case). Likewise, the kurtosis value increased with the increase of the age of participants in each group. These higher values in kurtosis are indicative of peakier distributions. Less dispersion was also evident in the decrease in variance and the increase in the mean values. For this reason, a systematic decrease in the NSR (variance/mean) was revealed for the TD cohort as they increased in age. Further, the range of incremental changes in IQ (i.e., the incremental differences in IQ with aging) was broader in TD than in ASD.

The ASD cohort revealed very different patterns. All points remained on the bottom of the graph with lower values of skewness and fixed lower kurtosis (flatter PDFs), denoting highly skewed distributions with long right-hand side tails and broad dispersion. This implies an abundance of high frequency values of speed peaks (i.e., tremor like motion). Further the low kurtosis values remain unchanged, indicating a lack of evolution on this parameter in ASD as well. With much higher variance and lower means their NSRs were much higher and remained so across all ages.

Consistent with other motor control research we discussed in Fig. 13 of Chapter 1 and Fig. 30 of Chapter 3, the maturation in stochastic signatures stalled in ASD. Previous research in the lab had probed goal-directed voluntary pointing motions toward visual targets and contrast their signatures with those emerging from rather automatic motions of retracting movements toward rest, upon communicating the decision in a match-to-sample task. Here the patterns of involuntary head motions also marked the lack of evolution in the movement signatures. The range of probability distributions measured across the population was much narrower for ASD and changed at a different rate than TD. This was a new way to examine discrete, ordinal clinical scores of IQ in relation to continuous biophysical data.

Since the results of the full IQ (Fig. 7.20) were so informative, and revealed age-dependent changes, I decided to examine the ABIDE sets with two additional questions in mind: (1) are there differences between verbal VIQ and performance PIQ between ASD and TD? And (2) Are there general age-dependent differences in ABIDE denoting an evolution of the stochastic signatures of the head excursions?

The first question could yield a different scenario from the full IQ scores. Perhaps the full IQ hid the age evolution of the incremental scores with age; but the summary statistics of the subscores provided by the VIQ and the PIQ of the ASD group revealed other differences.

The second question seemed obvious from the signatures of the G1−G5 split, but that split concerned only ABIDE I participants and medication intake. A year later, ABIDE II had been populated with new data and we had the possibility of doubling the number of participants to address the question of age-dependent rates of change of the stochastic signatures that we had already seen in a smaller cohort of 180 participants using goal-directed voluntary movements. With more than 2500 participants and utilizing individual time series of speed (one per participant) rather than pooled data from various groups, we could tackle several issues, including whether different families of probability distributions characterized the spontaneous involuntary head motions of TD and ASD. Further, we could examine the rates of change of these signatures from age group to age group.

The VIQ and PIQ analyses included pooled data from all participants with ASD and TD controls for which VIQ and PIQ scores were available in ABIDE I. As before, I median ranked the scores properly normalized by age. I pooled the head motion data from all participants above the median and all participants below the median for each participant type. This resulted into 8 subgroups, 4 ASD and 4 TD controls. I did this for both the linear and the angular speed time series, then normalized the fluctuations in speed peak amplitude and built the micro-movements data. The Gamma process analyses yielded estimated parameters with 95% confidence intervals (plotted on Fig. 7.21). They revealed a very different picture for ASD and TD.

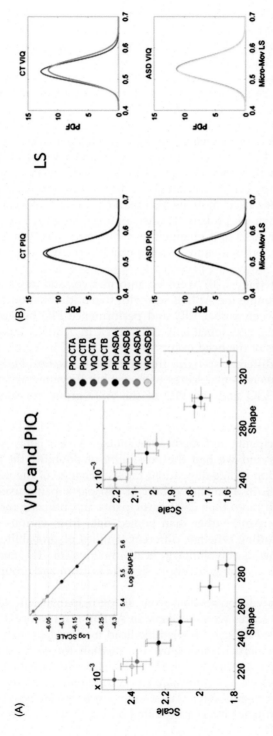

FIGURE 7.21 **Incremental IQ scores performance (PIQ) and verbal (VIQ) automatically separate the ASD and TD groups according to head motion speed stochastic signatures.** (A) Linear speed (left) and angular speed (right) related distributions reveal a trend to have higher NSR and skewness for ASD than TD (see legend). This trend is captured by a power relation in the inset whereby the more symmetric the distribution and the lower the dispersion, the higher the incremental IQ score in general. (B) Broad range of PDFs for VIQ than PIQ in TD are captured for the linear speed case (LS) using the same color code as with the Gamma parameter plane in (A). The opposite trend shows in ASD whereby PIQ has broader range of PDFs than TD.

The PIQ of the TD controls had a narrower range of PDFs than the VIQ (both in the linear and in the angular domain). But in ASD the pattern was opposite. The PIQ range of PDFs was broader than the VIQ. As before, the NSR, measured in this case through the Gamma scale parameter, was much higher for the ASD cohort. The departure of the median-ranked ASD PDFs from typical ranges was quite significant. Indeed, the spontaneous fluctuations in the amplitude of the involuntary head motions spanned a different statistical landscape altogether. This was not just a shift in the mean or the variance of PDF. This was an altogether different family of PDFs. As such, the results underscored the inappropriateness of correlating these discrete ordinal scores with continuous data under the assumption of normality for the general population. ASD and TD controls had different statistical regimes with different types of random processes characterizing each group. For the first time, a clear relationship between involuntary motions and IQ scores could be established, with a trend that the higher the dispersion and skewness of the PDFs, the lower the increment IQ score. The flip side was that the lower the dispersion and the higher the symmetry of the PDFs, the higher the IQ. This relationship between cognition and involuntary motions was characterized by a power law and I fitted the polynomial of one degree to the points and found the exponent (slope) of this relation close to (negative) 1. The emergent power-law relation shown in the inset graph of Fig. 7.21A between the median-ranked IQ scores and the log−log of the empirically estimated stochastic signatures for the rate of change in linear displacements and angular displacements respectively was: $f(x) = p_1 x + p_2$, where the coefficients (with 95% confidence intervals) are $p_1 = -1.06(-1.07, -1.04)$, $p_2 = -0.29(-0.37, -0.22)$ with goodness of fit: SSE: 1.304e-05, R-square: 0.9998, adjusted R-square: 0.9998 and RMSE: 0.001474 and $f(x) = p_1 x + p_2$, where the coefficients (with 95% confidence intervals) are $p_1 = -1.02(-1.03, -1.02)$, $p_2 = -0.47(-0.52, -0.42)$ with goodness of fit: SSE: 6.42-06, R-square: 0.9999, adjusted R-square: 0.9999 and RMSE: 0.001034. Panel B highlights the broader PDF ranges for the groups below and above the median for the PIQ in ASD and VIQ in TD.

THE AGE-DEPENDENT SHIFTS IN PROBABILITY DISTRIBUTION FUNCTIONS

The results on the IQ analyses alerted us of the importance to consider the age of the groups under study when examining any parameters of movement, imaging or demographics in ABIDE. After the work with ABIDE I was published[6] and I realized what Dr. X. had done, I opted for extracting the head motion data from imaging repositories ourselves in the lab.

I wanted to continue exploring ABIDE (and other repositories), but I had no idea how to extract the head motion data. Around that time, I had an fMRI-related grant that I had obtained as seed funding for a pilot project within the Rutgers campus at Newark. The grant was to do a study involving recognizing one's own motions in avatars (the NSF-IGERT project described in Chapter 2). However, some technical issues were keeping us from completing the project and the recruitment had turned out to be rather difficult. We requested a change in project and contacted a psychology faculty member of the Newark campus to collaborate in this project. Professor William Graves was happy to join us and use the fMRI grant to further an idea he had for a pilot study. In return, we could have the data and learn the ropes to analyze it and extract the head motion parameters from the images.

Sejal came to the meeting and offered to learn the steps and test them on our new data from Newark-Rutgers University Brain Imaging Center (RUBIC). She did so in one afternoon between good jokes and café latte. We loved Dr. Graves lab and personnel. In a single visit, he kindly explained the methods and pointed us to the open access sites fMRI researchers use to extract the head motions and determine the thresholds for artifact removal. Dr. X had refused to teach members of my lab how to do this—a pipeline of analyses that was widely accessible to everyone.

Sejal taught other members of the lab how to do all of it before she departed to India for her Fulbright project. She downloaded and processed all ABIDE I and II unscrubbed images to extract head motion and other parameters of interest. She also processed the demographics and organized all the repository data in the lab. We then had all the head motion data ready to further probe other important questions and in the meantime, Dr. Graves finished his experiments at the RUBIC and secured a nice pilot grant to continue his quest about autism and language. It all worked out perfect! This was the silver lining of this awful incident with the group at Columbia.

In due time, Carla analyzed all micro-movements of the fluctuations in head speed amplitude for each participant of ABIDE I and II. She found clear shifts in stochastic patterns with changes in age groups. Fig. 7.22 shows an example of such a map using the linear speed and tracking the evolution of these cross-sectional data from involuntary head motions across age groups. These plots also contrast TD and ASD participants. **The color-coded circles represent the ranges of physical head excursions (along the gradient in the colorbar), while the micro-movements parameterization enables us to examine these involuntary motions and their empirically estimated probability distributions across ages, i.e., despite anatomical differences. The results confirmed that there is a shift in probability landscape across ages and that the rates of change of these shifts are different for each age-group under examination. Further, they are different between TD controls and ASD participants. Once again, we cannot assume a one size fits all theoretical models for the human population. These results have further implications for imaging analyses because criteria for scrubbing the images depend on the head motion data and its stochastic signatures.**

Further, we can ask a person to measure the level of spontaneous involuntary motions during sleep, when the brain is not trying to deliberately control the body Fig. 7.23A. And then, we can contrast those patterns with spontaneous involuntary motions that are present when the person is awake and deliberately trying to control the body (as in the resting state fMRI case). This is the example in Fig. 7.23B, where we plot a map of the person's spontaneous involuntary bodily responses present upon asking the participant to remain still (as they did in the resting state fMRI scenario). In this case, we contrast different maps found in different participants.

These maps of NSRs empirically derived from a grid of accelerometers—such as those embedded in smart phones—and obtained by asking the person to lay down and remain still for the span of 3 min can be easily obtained within the time frame of a clinical visit to the doctor. At a sampling rate of 128 Hz (commonly found in android and mac OS

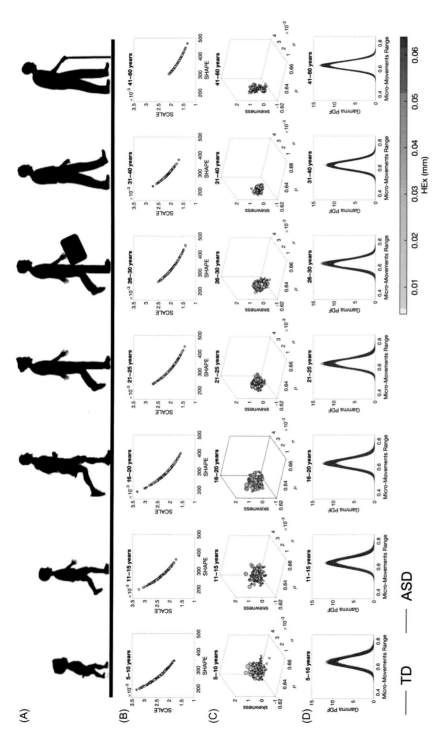

FIGURE 7.22 **Age-dependent shifts in stochastic signatures of involuntary head micro-movements underscore the need to consider the different dynamics of the nervous systems as the person undergoes neurodevelopment and ages.** The evolution in probability distribution functions underscores the need to not use a one size fits all statistical model. Here the micro-movements parameterization of angular speed extracted from the involuntary head motions permits to build a standardized statistic scale that shifts PDFs as a function of age and with different rates from age group to age group across the general population. Consider the traditional model with a single assumed theoretical PDF, whereby only shifts in theoretical moments (means and variance) are studied under the significant hypothesis testing model. Clearly, a family of PDFs exists and needs to be empirically determined for each of the phenomena under consideration.

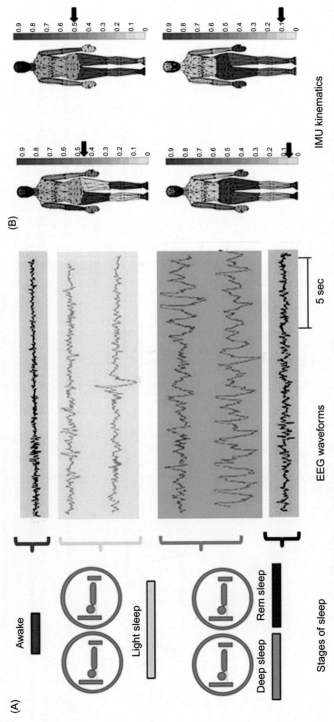

FIGURE 7.23 **Fundamentally different dynamics of spontaneous involuntary activity generated by different states of the nervous systems.** As a simple experimental paradigm is to ask the participant to remain still while laying down and then measure the levels of spontaneous involuntary motions that interfere with the person's volition thus impeding autonomy of the brain over the body. (A) The unconscious brain at rest can be recorded across the well-known sleep cycles to examine their different waveforms in relation to those of awake-active states. Further, activity from a grid of bodily nodes can inform us of spontaneous involuntary motions when the brain is unconscious *vs.* (B) when the brain is conscious and trying to remain still, the spontaneous involuntary motions across the head and body can be used to build a map indicating (for example) levels of NSR as they change over time.

phones), there is enough data from the acceleration peaks and the inter-peak-interval times to fit distributions and estimate the parameters of interest. Such maps can reveal fundamental departures from the person's baseline data, e.g., obtained from involuntary motions across the body during sleep.

We became interested in these comparisons and data types when we studied the sleep cycles of an athlete and a classical ballet dancer during their training for a competition and a performance, respectively.[67] The results from that work pointed at sleep as a good state of the nervous systems to use as a basis of comparison to understand patterns of activity in other awake states. Such information helped us infer other activities taking place during the previous day. For example, we could learn that excess involuntary motions that manifested in the trunk of the dancer during sleep were indicative of excess self-imposed stiffness on the trunk motions in the 4-h rehearsals that occurred during the day. The self-imposed control of trunk motions demanded by the choreography the dancer was rehearsing inevitably led to excess involuntary motions at night, during sleep. Likewise, the dancer's ankles and feet revealed excess heat, as shown by temperature maps we derived from the biosensor's signals. Research and textbook material from exercise science informed us that such increase in temperature is normal and commonly quantified as part of the metabolic processes of muscle repair at rest, upon intense exercise regimes such as those the dancer underwent before her performance.[68]

The information and paradigms that we developed for athletes and performing artists served as a test bed to help design the paradigms for ASD and other neurodevelopmental disorders. We mapped, for each age group, the expected levels of bodily involuntary motions when the person was at rest and had been instructed to remain still (see Fig. 7.23). This led us to initiate the path to build age-dependent dynamic charts of neuromotor control indicative of the level of involuntary motions inherently present in the person at a given age with and without conscious control. That is, we mapped the person's spontaneous involuntary responses while at rest, in the awake state and under the instruction to remain still. We also mapped the person's spontaneous involuntary responses at rest during sleep, when the person is not deliberately trying to control the body. Such charts are still *under construction* and beyond the scope of this book. However, I mention them here to underscore the need to halt the static treatment of the nervous systems' functioning in general, but more so in the rapidly growing and developing nervous systems of an infant or a young child.

This dynamic chart can be conceived as a matrix whereby each entry gives us a quantity built from a quotient. This quotient can be obtained using the magnitudes of spontaneous involuntary motions while the person is awake at rest and trying to remain still (denoted V) vs. when the person is resting during sleep time (denoted S). We can build two quantities:

$$q_1 = \frac{V}{V + \hat{S}} \text{ and } q_2 = \frac{S}{S + \hat{V}}$$

The scale goes from 0 to 1. V represents the peak of involuntary displacements and/or rotations per unit time (the unit time is to be determined by the clinician, e.g., 5–15 min in a visit); $\hat{V}$ is the average amount of displacements and/or rotations during that time. S is like V but recorded during sleep hours at night, in the home environment. To collect S, we need to consider all well-known cycles of sleep (Fig. 7.23A), namely light sleep, deep sleep, and REM (rapid eye movement) sleep. In the case of q_1, we can learn how much involuntary motions during awake volitional states are affected by involuntary motions during sleep cycles. Higher values of V (toward 1) indicate lower values of involuntary motions S during the given sleep cycle (on average). In the case of q_2, we will have to examine each cycle separately and estimate how much involuntary motions during the awake period tend to affect the involuntary motions of each sleep cycle.

By forming a personalized matrix with such entries and color-coding it according to these values, we can have a normalized scale of involuntary motions tailored to the person's nervous system (e.g., Fig. 7.23B). This scale can help us make further inferences about the person's neuromotor control in the deliberate and spontaneous modes of behavior. We can also identify the person's regular states across different activities of daily living and detect departures from regular states within a given time frame (e.g., a month prior to meds intake in a clinical trial; then during meds intake in a clinical trial to monitor departures from baseline and lastly during the washout phase upon completion of the trial).

Likewise, such a personalized dynamic chart would then be used in a more general sense, to evaluate the effects of psychotropic medication intake on the person's nervous systems. Given the systematic effects we found across more than a thousand participants,[6] it may be possible to quantify interaction effects from multiple drugs, effects according to dosages and overall time of exposure to a drug, etc. In the part II of this chapter we examine some dynamic biometrics to capture change in the context of a clinical trial involving a rare neurodevelopmental disorder called Phelan—McDermid syndrome a.k.a. SHANK3 deletion syndrome.

Charting such changes dynamically is important at all stages of the human lifespan, but critical during neurodevelopment, when the nervous systems are in a nascent state and subject to accelerated rates of change. We can better understand the notion of plasticity and adaptiveness of this nascent system if we measure it during awake and sleep stages. We would then be better informed about the child's neuromotor control progression, the ability to develop volition, agency and deliberate autonomy over his/her body in action. The responses of his/her nervous systems to medication intake will help us intervene in a better-informed manner. Clearly, this schema would be ideal for evaluation of clinical trials. But it would also help design a new dynamic diagnosis systems scale.

If motor symptoms could be part of the DSM criteria, new objective methods to measure motor physiology imported from the fields of movement neuroscience and adapted to measure change, could be used to monitor rates of change in the rapidly changing, developing nervous systems of an infant. More generally, under a proper standardized scale, such methods could also be used to monitor progressions/regressions of disorder stages given various therapeutic and drug interventions.

Because DSM makers have well-known financial conflicts of interest with the Big Pharma system[69–73]; because disclosure of such conflicts is now enforced by the healthcare overhaul legislation[74]; and because the DSM guarantees prescription of psychotropic meds upon diagnosis; a neutral third party with technical (mathematical and computational) knowledge and training in neurophysiology would be needed to monitor the making of the DSM. This would be particularly so for the cases of heavily medicated neurodevelopmental disorders such as ASD and ADHD.

Industry could help in this regard because most activity trackers already detect sleep cycles automatically. A problem in this regard is that large data throughput are not used yet to their maximal potential. The lack of standardized biometrics prevents us from exchanging results and discussing them using a common standard scale appropriate for all ages, conditions and instrumentation-waveform. Our methods add a layer of statistical analyses built from empirical results with standardized criteria derived from large data repositories and neurological case studies. These biometrics are not the only ones one could use to track change and their rates; but they already permit further comparison across members of the population at large by providing a standardized scale amenable to build a proper population statistical metric.

The first part of the recent studies from my lab established motor noise and randomness in ASD as core biological symptoms that in some form ought to enter the DSM criteria. The science behind these effects strongly suggest that the APA and pharmaceutical companies behind the DSM should be under a moral obligation to make these findings and drug interactions known to parents and pediatricians even before prescribing them. But medication effects were not the only issue we found when datamining ABIDE. As it turned out, the other (psychological) branch of the mental illness model has a test that was found to be highly problematic.

THE ADOS

The second part of our studies of ABIDE[75] uncovered fundamental flaws in the Autism Diagnostic Observational Schedule (the ADOS). This is the psychological test providing criteria for autism diagnoses. It is called *the gold standard to diagnose autism*. The makers of the ADOS are connected to the psychological testing companies (e.g., the Western Psychological Services (WPS) Publishers) which sells the product. Somehow scientific practices are expected to (1) buy this product and (2) use it to correlate results from experiments that harness physiological data registered with high grade instruments.

There is a system in place that enforces the administration of this very lengthy and arguably harmful test to the children in every peer-reviewed study we examine. Indeed, the ADOS (and other observational inventories) drive the science behind autism. But there are many problems with such tests. Some of these were self-evident, but others were hidden in nonobvious ways. It took some detective work to find out the many flaws of this particularly popular test that is used to diagnose autism.

To start the detective work, Caroline and I got certified (The cost of certification as of April 2015 was of $6000.00 for clinical and $6000.00 for research grades. Once certified, the person can certify others in the same University. They do not need to have a degree in clinical psychology). We learned how to administer the test, how to score it and how to train others in the lab to do so. Our purpose was not to do any of the above, but rather to learn every single aspect of it, deconstruct it, convert it to an objective physiological test and provide neutral-observer criteria from self-emerging patterns in the dyadic social exchange this test evokes. We presented some of these methods in Chapter 6. Here we examine the thousands scores we studied in ABIDE within the context of contemporary (unresolved) problems we identified in the autism literature.

THE 5:1 FEMALES TO MALES (STATISTICALLY IMPOSSIBLE) RATIO OF ASD

One of the most puzzling outcomes from observational inventories used to diagnose autism is the disparate ratio of 5 males for every female diagnosed with the disorder.[76,77] Because the diagnosis precedes any group selection for comparison in the design of basic science studies, with such a low number of females being diagnosed, we know very little about the autistic female phenotype. If a lab carries on a study with 100 participants (a very ambitious and costly number due to clinical assessment—see below) with the ongoing ratio, only 20 females may be part of the study (unless one biases the inclusion criteria to boost than number instead of having a random pick from the general population).

To include more females with autism in our study of decision making under variable cognitive loads,[78] we biased the inclusion criteria and were able to recruit more females than we would have done at random. The task was simple. It entailed to touch a screen and match the stimulus that popped up on the screen, with one of two choices, by pointing to the choice and touching the screen location of the image. This is called a match-to-sample task paradigm and we already saw it in other chapters (e.g., 2, 6, and 7) when we discussed the results of studies concerning neurological disorders of various types, including autism, Parkinson's disease and deafferentation. Fig. 7.24 shows the trajectories of the child's hand while making the decision, along with the layout of the stimuli the child had to use to decide. In this case a banana on the center bottom of the screen oriented as that on the upper left-hand corner provided the matching samples. As such, the correct answer was for the child to point and touch the stimulus on the left-hand upper corner and ignore the stimulus on the upper right-hand corner. A lot of hesitation can be appreciated, but embedded in all those trajectories are the trajectories relevant to the decision making. These can be automatically extracted from behavioral landmarks that our instruments trivially capture. The paths are shown in isolation on the bottom panel while Fig. 7.24B shows the corresponding speed profiles of the trajectories indicating the decisions. We varied the complexity of the stimuli and provided different types of cognitive loads to probe the decision making in terms of accuracy and speed, but also as the neuromotor performance evolved from highly uncertain (with many peaks from accelerations and decelerations to ballistic with a single peak, bell-shaped speed profile).

Using this cognitive-motor task, we discovered fundamental differences between the males and the females in the spectrum, quantifiable through the linear speed patterns of the hand, as the hand approached the targeted choice of the comparison they made (Fig. 7.25). However, the number of females with autism was still relatively low (60 participants total with 20 females in the group). We wanted to learn more about the autistic female phenotype and characterize the volitional control of their nervous system.

To attain our aim, we used the ABIDE I and II data bases combined and could sample over 300 females from a cross section of the population spanning from 5 to 60 years of age. We had at our disposal females with a diagnosis of ASD, and because the records included DSM IV criteria whereby angular speed (AS) was still considered part of the autistic spectrum; we also had females with AS.

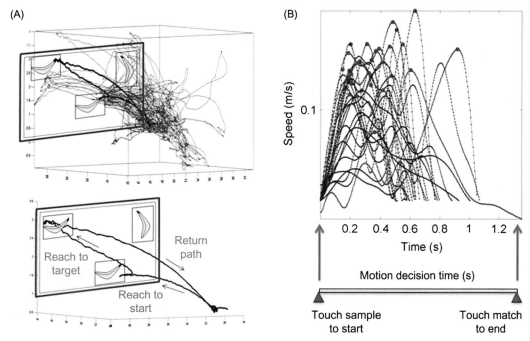

FIGURE 7.24 **Simple pointing behavior can be converted to an experimental paradigm to study embodied decision making under different levels of cognitive loads.** (A) Hand trajectories of a TD child while making decisions in a naturalistic environment at the school settings. Bottom panel shows automatic extraction of hand trajectories to initiate the task (reach to start, which triggers the sample to match); decide and point to the chosen stimuli matching the sample (reach to target) and return path to rest. Automated extraction of these experimental epochs from naturalistic motions is possible because of landmarks such as distance to target, target location, 0 velocity, among others (see text). (B) Sample speed profiles from decision making trajectories and land marks delineating the temporal information concerning the decision.

The use of open access brain imaging data banks had already revealed differences in brain signal variability[79] and patterns of connectivity between the typically developing (TD) brain and the ASD brain.[80,81] And more recently, the data served to highlight specific sex-based differences[82], including differentiations in structural organization of the motor systems. Some of these differences had been discussed considering repetitive behaviors,[83] cortical volume, and gyrification,[84] among other morphological parameters. Such growing body of evidence supports fundamental, physiological differences in ASD expression between the sexes hinted at the utility of objective biophysical data to provide better characterizations of the female phenotype with ASD than those provided by observational scores. The latter are severely biased by cultural expectations that perhaps *girls should be quieter and less rambunctious than boys*. Indeed, one of those questionnaires was administered to 55,000 school children in South Korea and although it was initially reported that 1 in 38 were affected by ASD, later on it became evident that these results and the prevalence estimates derived from them were flawed.[85]

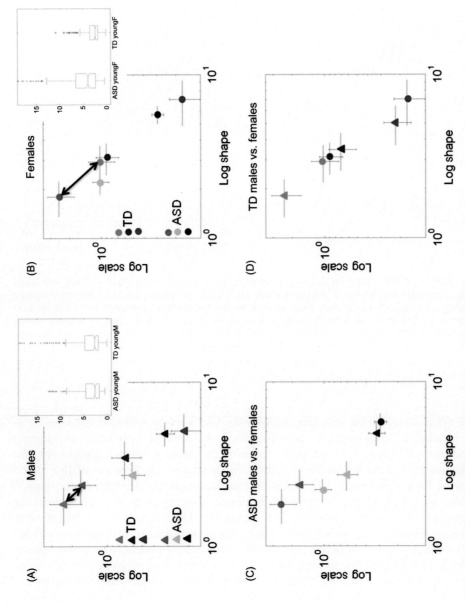

FIGURE 7.25 **Movement decisions to communicate the target under cognitive loads serve to separate sex in ASD and TD.** (A) ASD males of different age groups (3–4, 5–10, 17–25 years old) can be separated from TD males within those age groups. (B) Likewise, ASD females can be separated from TD females in each of these different age groups. (C) ASD males vs. ASD females separate according to their signatures of speed-dependent fluctuations in peak amplitude. (D) Similarly, females separate, particularly within the 3–4-year-old group. Recall that each point represents a different Gamma PDF obtained for the speed-dependent data normalized to avoid allometric effects from anatomical differences.

Using the ABIDE data, we conducted analyses of involuntary head motions and asked if within the females in the spectrum there were any obvious differences that could help us differentiate them from age-matched neurotypical controls. We reasoned that since the questionnaires and ADOS-based tests could not detect the females within a random draw of the population, perhaps within a large pool of females we could identify normative levels and see departures from it. We had access to over 300 females total (63 neurotypical controls and a mixed group of ASD, AS, other pervasive neurodevelopmental disorders and some of them on meds or no meds). These numbers provided enough female participants to probe some questions about spontaneous involuntary head motions and their signatures across several parameters.

One of the parameters we examined was the net head excursion across the resting state fMRI session. To that end, we first obtained the linear displacements and angular rotations frame by frame. Then, we accumulated these values by adding them across the session. This net motion describing the total pathlength of the head translation (mm) or head rotation (deg) could inform us of possible differences across these participants (as we saw in Fig. 7.6).

To try and understand these potential differences, we built clusters based on the clinical labels we had from the ABIDE demographics: i.e., the information on ASD, AS, and typical development (TD). We found that the ASD group moved much more than the TD group. But to our dismay, we also found that the AS group was the one with the most involuntary head motions. Fig. 7.26 contrasts the females with ASD and AS with the controls. This graph clearly shows that females in the spectrum move involuntarily much more than controls. Indeed, the rates of accumulation and the levels of NSR empirically determined using the Gamma process approach that we described in the introductory chapter, revealed a systematic trend that we show in Fig. 7.27. Namely, the empirically estimated probability distribution functions have higher dispersion as we move from TD to ASD to AS, with significant differences between the females in the spectrum and the TD females (Fig. 7.27A). Further, the noise and skewness of the PDF systematically increase in that direction too: TD, ASD, and AS (Fig. 7.27B).

The group trends we uncovered while using the clusters labeled with the ABIDE information confirmed differences that the DSM diagnoses of the females provided a priori. Yet, we wanted to know if it would be also possible to blindly determine clusters using a data-driven method. Our question was whether clinical criteria provided by the DSM had a correspondence with the somatic-motor measures. If the answer was positive, we also wanted to learn if the trends would correspond to clinical criteria provided by the ADOS. Both criteria actively avoid somatic-motor issues, yet the field of Psychiatry had a history of using (in its early inception) motor-based criteria[86] and the Psychological test (the ADOS) has repetitive behaviors and joint attention tasks (both of which implicitly require movements and physical entrainment). We reasoned that somehow clinicians may be implicitly using motor disturbances in their observations as they compile criteria for diagnosis. The question then was, which test (DSM or ADOS) best reflected the departure from typical levels of involuntary motions that we found using the somatic-motor criteria.

To address this question, we profiled each female participant separately and empirically estimated the Gamma parameters from her spontaneous involuntary head motions. This procedure produced a scatter of points that we plotted on the log–log Gamma parameter

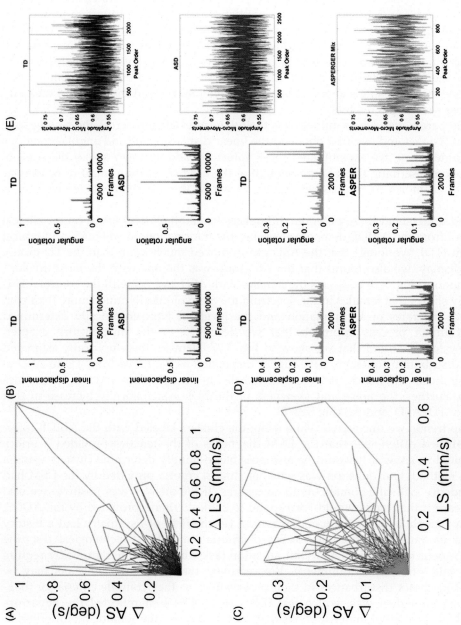

FIGURE 7.26 **Involuntary head excursions were detected during the resting state fMRI sessions.** (A) Linear displacements vs. angular rotations of the head contrasting ASD females vs. age-matched TD controls. (B) Raw data consisting of linear speed and angular speed extracted from the position and orientation tri-axial trajectories contrasting TD vs ASD participants. (C,D) Same as in (A,B) for the females of ABIDE I and II with a DSM-IV diagnosis of AS and PDDNOS vs. the age-matched TD controls in (A). (E) Micro-movements extracted from rs-fMRI head linear displacements in TD, ASD and AS combined with PDDNOS participants.

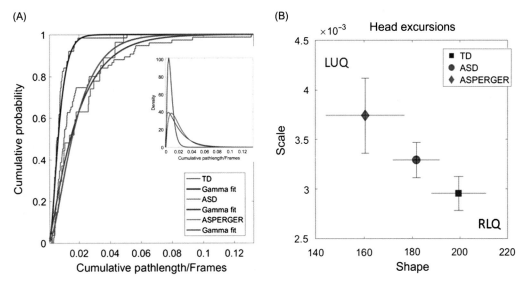

FIGURE 7.27 **Differences in relative physical head excursions manifest in female participants during rs-fMRI session upon instruction to remain still.** (A) The empirical cumulative distribution functions (eCDFs) estimated using Gamma fits to the empirical data pooled across all subjects per pre-labeled group separate age-matched TD participants from ASD and AS. Note that eCDFs from ASD and AS participants also separate, but the separation has non-statistical significance (see text). The inset shows the estimated Gamma PDFs. (B) Using the SPIBA Gamma process the involuntary head micro-movements of each labeled cohort localize AS and ASD on the LUQ of the Gamma plane with more elevated NSR relative to TD controls localized on the RLQ of the Gamma plane with lower dispersion and more symmetric shapes of the estimated distributions.

plane (Fig. 7.28A). We then fitted a line through that log–log of the scatter using the MATLAB curve fitting functions *polyfit* and *polyval*. These functions determine the coefficients of the first-degree polynomial (slope (p_1) and intercept (p_2)) describing the best linear fit ($f(x) = p_1 \cdot x + p_2$ with parameters and confidence intervals, $p_1 = 1.02$ (−1.024, 1.05) and $p_2 = −0.423$ (−0.43, −0.38)). Then, using that information we could compute the residuals (denoted delta) to obtain the departure of each point from the line best fitting the TD controls. Here remember that each dot represents the parameters of a PDF that we have empirically estimated (shape and dispersion). As the points on the log–log plane follow a power relation, the departure from the line informs us of the level of dispersion in the variability of the scatter. As it turned out, we first plotted on the parameter plane spanned by Delta vs. the log Scale the scatter of points in the blind (i.e., randomly chosen without label). We noted (Fig. 7.28B) that three clusters self-emerged from these error patterns, one of which departed maximally from the other two. When we colored the points of the scatter by the clinical labels, they coincided with AS being the one with the highest dispersion—denoting the highest departure from the line spanned by the scatter derived from the TD controls. Further, the ASD females departed from the TD females too (Fig. 7.28C). The three-dimensional plot of Delta (*x*-axis), log Scale (*y*-axis) and the cumulative pathlength of the head excursion (*z*-axis) we could see these clusters automatically self-emerge from the random distributions of scatter points in Fig. 7.28A.

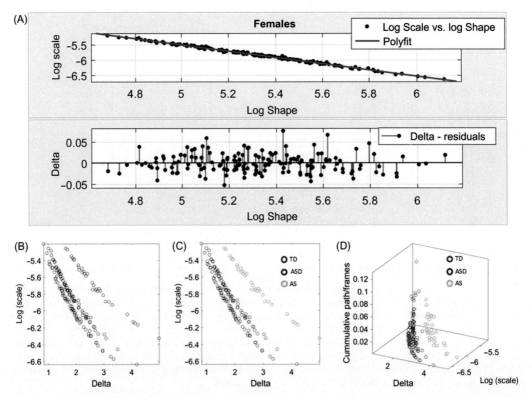

FIGURE 7.28 **Data-driven approach for cluster detection based on stochastic properties of the head micro-movements data of the females in ABIDE.** (A) Individually estimated Gamma probability distributions of the females and power relation fit using polynomial of degree 1 on the log-shape vs. log-scale parameter plane (top panel). Bottom panel shows the residuals (delta) obtained from the error between the polynomial fit and the actual scatter points. (B) Parameter plane distinguishes three clusters along the Delta vs. log (scale) or NSR. (C) Scatter colored by DSM labels reveal clusters congruent with the diagnosis. (D) Further separation of the groups emerges when using the relative head excursion (cumulative path length per frames), with the AS group singled out as the farthest apart from the age-matched TD controls.

This result was very surprising, since the noise-departure from the line blindly classified females into TD, ASD and AS. The data-driven division agreed with the DSM criteria; but we were expecting the AS group to be closer to TD than the ASD group. This was not the case. As it turned out, the AS females had far more involuntary motions of the head and higher levels of dispersion and randomness according to the estimated PDFs than the females with ASD. The pattern then prompted us to ask how the same analysis would pan out in the males of ABIDE. Recall that the ratio is 5 males for each female. As such, we had five times more males than females in ABIDE (over 1500) to perform this test. Fig. 7.29 revealed the answer: the pattern in the males was different than the pattern in the females.

The AS group was closer to the TD group than the ASD group. This was intuitively what we had expected, given our experience in autism research. The ASD group had a

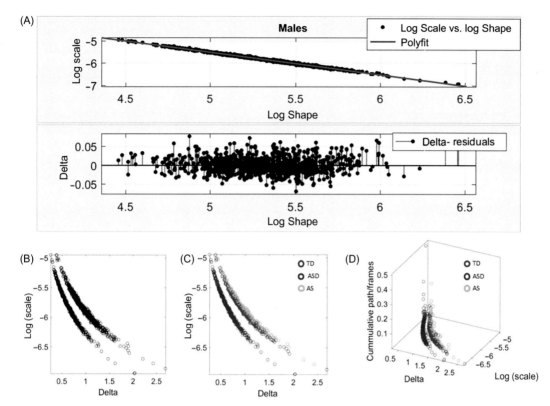

FIGURE 7.29   Data-driven approach for cluster detection based on stochastic properties of the head micro-movements data of the males in ABIDE. (A) Individually estimated Gamma probability distributions of the females and power relation fit using polynomial of degree 1 on the log-shape vs. log-scale parameter plane (top panel). Bottom panel shows the residuals (delta) obtained from the error between the polynomial fit and the actual scatter points. (B) Parameter plane distinguishes three clusters along the Delta vs. log (Scale) or NSR. (C) Scatter colored by DSM labels reveal clusters congruent with the diagnosis. (D) Further separation of the groups emerges when using the relative head excursion (cumulative path length per frames), with the ASD group singled out as the farthest apart from the age-matched TD controls and AS subgroup overlapping with the TD controls.

higher departure from TD than the AS group, which overlapped a great deal with the TD group. This result was in line with the DSM argument that AS and ASD should not be classified as the same disorder. Clearly, the criteria driven by males confirmed this opinion. Yet, at the same time, these results informed us that females are being measured with a *male ruler*. That male ruler is a wrong metric according to these analyses. The outcomes of our analyses are in line with the fact that male and females have different genetics and different physiology altogether. AS females depart maximally from TD females. It appears to be the case that in the best interest of a diagnosis system capable of detecting females, it will be best to keep the diagnosis of females separate from that of males. In other words, it is best to establish separate autism criteria for each sex. Intuitively this makes sense. After all, autism is defined by social and cultural expectations and socio-cultural expectations are different for males than for females. Why use a *one size fits all model* to drive the sex quest in autism, when even at the underlying physiological level they are fundamentally different?

The DSM-based criteria agreed with the somatic-motor features that we empirically derived from these data sets. To further understand observational criteria employed by psychologists, we needed to examine the ADOS scores provided by ABIDE. There we had two types of scores, the ADOS-G (generic)[4] and the ADOS-2 (a more modern version of it with multiple updates)[87−91]. To use the scores in an age-appropriate manner, we converted them to incremental (derivative) scores and normalized them to be on a scale from 0 to 1. Then, we studied the cumulative distributions and the probability density functions that empirically fitted the score-data best in a maximum likelihood estimation sense. We also used the Kolmogorov−Smirnov test to compare two empirically estimated distributions. Fig. 7.30 shows the differences between ASD and AS for each sex using two sub-scores of the ADOS-2: the severity subscore and the total score.

Using the ADOS-based psychological criteria, we learned that the demographics information provided by this test also separated the clinical labels within each sex. However, when we mixed sexes these differences waned. Once again, these results strongly suggested the need to consider females and males as different groups and derive separate diagnostic criteria for each sex phenotype. In other words, we needed to build metrics to measure the departure from typicality in males and females separately.

The ADOS is considered the gold standard for ASD criteria, but it does not apply separate criteria for males and females. It currently misses the females. Unfortunately, the test is pervasively present in all research, ranging from behavior to genetics. As such, one cannot conduct a basic science research study about ASD without conducting the ADOS. The cost of this operation is prohibitive for any study seeking to establish a result with statistical power. Consider a conservative estimate at $200.00 per hour paid to the clinician administering it and $2095.00 per ADOS kit per study, ADOS-2 Software upgrade package per study $630.00 and $53.50 per booklet per child. This practice results in a very costly operation to science. So, for example, a study involving 50 individuals with ASD and 50 controls would (with this conservative estimate) add up to ∼$28,075.00, not including transportation costs and participants' compensation—not to mention "double billing" in many cases where the same child undergoes the test by the same clinician at the clinic and in the research lab (Note that in both cases one may incur on the use of tax-payer's dollars through the Medicare system and the federal/state grants systems that fund research, so in the United States these are societal and policy making issues concerning the general population at large). Yet, like the DSM, the ADOS explicitly excludes somatic-motor issues (despite the complaints and recommendations of self-advocates, i.e., affected people and their caregivers, and despite the evidence presented by researchers who, like those in my lab, investigate sensory and somatic-motor issues in ASD, see for example[37,49−61,92−99], among many others.

To be precise, the ADOS-2 manual under the section *Guidelines for Selecting a Module* proposes the following caveat:

> "Note that the ADOS-2 was developed for and standardized using populations of children and adults without significant sensory and motor impairments. Standardized use of any ADOS-2 module presumes that the individual can walk independently and is free of visual or hearing impairments that could potentially interfere with use of the materials or participation in specific tasks"[100]

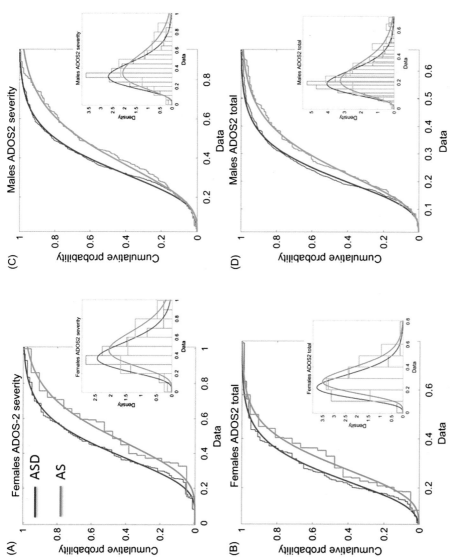

FIGURE 7.30 Age corrected (incremental) ADOS-2 scores mark statistically significant differences between ASD and AS observational pheno-types in the cross-sectional data form ABIDE I and II of sites that reported medication status. (A) Females with an AS diagnosis have higher age-corrected ADOS-2 severity scores than ASD females (Table 1 in Ref. 75 reports the Gamma fit first (mean) and second (sigma) moments from highly skewed distribution of incremental scores considering physical age of the person at the time of the test (i.e, this is different corrective criterion than adjusting for mental age, already factored into the module selection process). (B) Same trend as in (A) for the age-corrected ADOS-2-total reveals worse scores for AS females. (C,D) The analyses of (A,B) for females were performed on the males. Similar statistical features were detected for the incremental age-corrected ADOS-2 scores: skewed distributions with higher mean values for AS in relation to ASD.

The above statement implicitly assumes that the person administering the test and a priori selecting the module to use with a given child knows the extent to which the child has significant sensory and somatic-motor issues that could impede performance of the test's tasks. But how could this be the case. Conscious vision is limited. As such, the observer dictating the scores has limited capacity to make that determination with any degree of certainty. Even instrumentation will miss the problems if no proper analytics are used. *On what scientific basis can the ADOS designers make such a claim?*

As a field, it is possible to overcome the limitations of relying exclusively on human observation-based methods. Previous chapters of the book show that we can complement observation with high-grade biosensors and appropriate statistical methods to analyze the biophysical signals continuously output by the nervous systems of the person under examination: e.g., in the case of the resting state fMRI setup, we had access to the time series of fluctuations in amplitude and timing of the involuntary head micro-motions as the person laid down in the scanner. This gave a sense of the lack of voluntary control that the brain had over the peripheral nervous system innervating the body of the person. Yet, any other instrumentation reading out the bodily biorhythms could do so noninvasively. For example, Fitbits or smart watches could be used to that end and then a simple study of the involuntary motion patterns during sleep could help us refine our characterization of each sex phenotype.

Our largest surprise in the ABIDE studies we carried on came when we compared the outcome of the scores from ADOS-G with those from the ADOS-2 (Fig. 7.31). When examining the profile of significant differences between individuals with ASD and AS across the ADOS-2 and ADOS-G scores for both the female and male cohorts, we found other differences. Specifically, we found fewer axes of both the ADOS-2 and ADOS-G that significantly differentiated between female ASD and AS. This was in contrast to the patterns in males, which revealed more axes differentiating the ASD and AS groups (see Tables 1 and 2 of the Supplementary Material in Ref. 75). We interpreted this pattern as a measure of the lack of sensitivity of the clinical assessment tools that are currently used to quantify and classify symptomatology of ASD in females. We can see at a glance that the patterns these age-corrected normalized scores make are different between males and females (Fig. 7.31). The tails of these distributions are very different between AS and ASD female participants; but also between females and males overall.

These results hint at a different statistical landscape altogether for the ASD female phenotype in relation to the ASD male phenotype. Combined, the results concerning somatic-motor patterns and age-normalized observational scores caution that it may be inappropriate to continue the use of a social-behavior male ruler imposed by such social- and culture-dependent questionnaires to diagnose females.

Further analyses pertaining to the male cohort revealed that according to the theoretical population assumptions (i.e., a normal distribution) underlying the ADOS-based scoring systems, if we were to have a random draw of one male individual with a diagnosis of ASD and one male individual with an AS diagnosis from the ABIDE population that reports medication, we would find that the ASD individual is likely to be worse off than the AS under ASD-G criteria but better off than the AS under the ADOS-2 criteria. In contrast, if we were to do this with females, both ADOS-based criteria would yield a better outcome score for the ASD than the AS DSM-driven phenotype. These results are

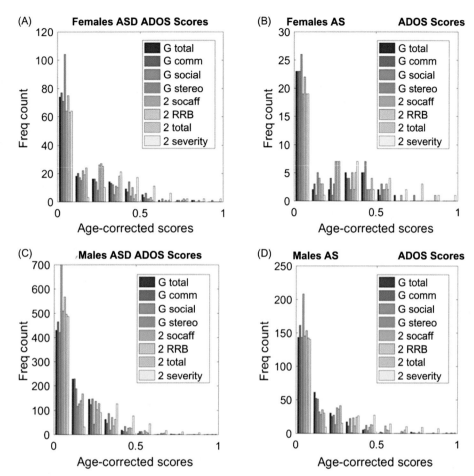

FIGURE 7.31 **ADOS-2 and ADOS-G scores at a glance taken from all subjects we included in the analyses.**
(A) Females with a non-DSM-IV diagnosis of ASD. (B) Females with a DSM-IV diagnosis of AS. (C) ASD Males as
in (A). (D) AS males as in (B). Note the skewness of these distributions and the differences in their tails separating
the two sexes. As detected by the nonparametric Rank-Sum test, the ADOS-2 severity and the ADOS-2 total do
not tapper off as the individual ages, physically grow and develop at irregular rates.

summarized in Fig. 7.32. They were rather puzzling to me when I found them, since these
questionnaires are used interchangeably in the field. When I summoned the lab members
and explained that I had a finding that seemed contradictory; we decided to independently analyze all the demographic ABIDE ADOS-related data again and see if any contradictory results emerged from their analyses. Caroline Whyatt, Carla Caballero and Sejal
Mistry (in Fig. 7.1C) led the effort. They constructed graphs and did their analyses independently. Then they came back to a lab meeting and presented their conclusions: these
distributions are far from normal; in the male cohort, there are no consistent results when
we examine AS vs. ASD using ADOS-2 vs. ADOS-G; in the female cohort, we have

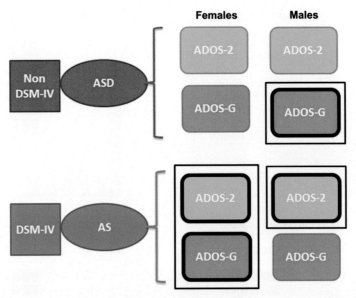

FIGURE 7.32   **Contradictory results between the ADOS-2 and ADOS-G reported demographics in ABIDE I and II in relation to the non-DSM-IV ASD and the DSM-IV AS phenotypes.** According to the theoretical population assumptions underlying the ADOS-based scoring systems, if we were to have a random draw of one male individual with a diagnosis of ASD and one male individual with an AS diagnosis from the ABIDE population that reports medication, we would find that the ASD individual is likely to be worse off than the AS under ASD-G criteria (enclosed rectangle) but better off than the AS under the ADOS-2 criteria (enclosed rectangle). In contrast, if we were to do this with females, both ADOS-based criteria would yield a better outcome score for the ASD than the AS DSM-driven phenotype (enclosed rectangle comprising both ASD-2 and ASD-G). Note here that these are the outcomes of statistical tests. We are not interpreting here the data beyond that outcome from the ABIDE data (see *p*-values and empirically estimated summary statistics in Appendix Tables of Ref. 75).

entirely different distributions of scores in AS and ASD subgroups; the shape, dispersions and tails of the subscores' distributions for males and females hint at different probability landscapes altogether. Then we retrieved the original ADOS-G paper and studied it in detail. It wasn't until we reached the very last sentence of the final paragraph in the paper that we understood how fundamentally flawed using this test has been. I quote from the paper (emphasis added):

> "Replication of psychometric data with additional samples including more **homogeneous non-autistic** populations and more individuals with pervasive developmental disorders who do not meet autism criteria, establishing concurrent validity with other instruments, evaluation of whether treatment effects can be measured adequately, and determining its usefulness for clinicians are all pieces of information that **will add to our understanding of its most appropriate use**."

Basically, this is a disclaimer indicating that the test has never been validated in neurotypicals. This implies that the ordinal scale the test uses to construct the scores (which has different ranges for each subscore) has no baseline value (perhaps assuming 0 as neurotypical value). These ordinal scores without a proper metric space are used in the clinical domain to diagnose the children and recommend treatments. More puzzling yet is that those ordinal scores without proper metric space to derive point-to-point distances in the scoring space, or similarity indexes (i.e., to measure departures from typicality) drive all basic research in autism. They are ubiquitous in all peer-reviewed papers that try to explain physiological measures of other aspects of the nervous systems functioning, from brain imaging studies to genetic studies to clinical trials. *What a disaster!* (we all thought when we learned this fact.) Our hearts shrunk even more when Sejal computed a conservative estimate of the cost of the ABIDE cases that we examined along with an estimate of what the lab would have paid had we had administered the tests ourselves (see Appendix for detailed calculations):

Costs if Our Labs Were to Run These Experiments

Item	Cost (US dollars)
Baseline Cost for Equipment + Training	15,144.00
ADI-R cost	285,418.00
ADOS cost	184,330.00
ADOS_Gotham cost	112,585.00
ADOS_G cost	114,742.50
ADOS-2 cost	92,698.00
Total	**804,917.50**

(This is the cost for running basic diagnostics. Includes cost of booklets, clinician hours, + 1 Baseline equipment's cost)

Total Cost of Running These Experiments

Item	Cost (US dollars)
ADI-R cost	321,610.00
ADOS cost	441,778.00
ADOS_Gotham cost	279,169.00
ADOS_G cost	326,758.50
ADOS-2 cost	244,138.00
Total	**1,613,453.50**

(This is the cost overall among all sites. This includes all costs sites may have had in purchasing equipment and obtaining certification)

Caroline raised the point that these costs to US researchers and more broadly to US taxpayers are merely a drop in the ocean. These tests that do not have a proper metric scale, are based on the wrong theoretical statistical assumptions, lack a standardized proper scale for males and females and drive our science in autism, have been translated in one version or another to Arabic, Croatian, Czech, Danish, Dutch, Finnish, French, German, Hebrew, Hungarian, Icelandic, Italian, Japanese, Korean, Mandarin (traditional characters), Norwegian, Polish, Romanian, Russian, Spanish, Swedish, and Ukrainian, a table that can be found in the WPS website where one buys it: https://www.wpspublish.com/app/OtherServices/PublishedTranslations.aspx.

We are shocked in my lab and remain skeptical about the true utility of such a fundamentally flawed test. What a waste of resources in science and society! In fact, when we turned the test into an objective instrument by using inertial measurement units and measured the social dyad as the child and clinician interacted in real time[101]; we found that (1) the clinician tends to lead the child and as such biases the child's responses—that bias enters the score the clinician gives the child while s/he is paid for it; (2) we could lower the time of administration to merely 15 min and (3) summarize the large number of subtasks to just a couple (at most three) to sufficiently identify areas of strength the child had to interact socially with others.

Instead of using this traditional deficit model of diagnosis, we could point out the potential of the child; we could do so while building a proper metric scale standardized for physical age, anatomical differences, sex differences and considering a dyadic interaction with scoring systems based on physical entrainment abilities, independent of spoken language. The continuous physiological signals we obtained from the clinician and child in tandem helped us redefine the diagnosis and transform it from a static to a dynamic tool. Indeed, we could use this tool in a personalized manner, to track the progression of the disorder with or without interventions. Our proposition was a totally different lens to examine autism. Armed with a statistical tool that did not assume normality, linearity, and static features, we proceeded to use it in a clinical trial that involved children in the spectrum of autism of genetic origins: these were children with SHANK3 deletion syndrome who had also received a diagnosis of ASD and were about to initiate a clinical trial involving insulin like growth hormone factor I (IGF-I) at the Mount Sinai Medical School of New York City: https://clinicaltrials.gov/ct2/show/NCT01525901?cond = shank3&rank = 1.

PART II BIOMARKERS FOR CLINICAL TRIALS

Alexander Kolevzon, a researcher and physician from the Seaver Autism Center for Research and Treatment at the Icahn School of Medicine, at Mount Sinai, New York City, call me at some point, came to the lab and introduced himself. He asked me if I would be interested in measuring children and adolescents with 22q13 deletion syndrome while they underwent a clinical trial. The trial involved IGF-I and had a double-blind placebo-controlled crossover design. I said yes, but to be honest, his words sounded like gibberish to my ears. I had no idea what he was talking about. As it is common in my lab, we prepared for this *clinical trial adventure* from scratch. My group and I proceeded to educate ourselves on this syndrome and read everything we could find that had been peer-reviewed about it.

This is a very important step of any experimental design we do in ASD. Since it is a spectrum and the diagnosis is so broad, many different features can be found within any given group of children with similar scores. The reality is that each child is unique and each group of children within a given genetic disorder that also receives such a diagnosis is very different from any other group. The one thing in common for us is to discover how to make the children that come to the lab feel at home in the lab so the experiments are more like a game than like a boring and perhaps tiring study.

We learned that there were delays in all aspects of their development, from walking to pointing behaviors, and in many cases language would not develop at all. This posed a tremendous challenge and required brainstorming a bit to come up with a design of data gathering and analyses that allowed us to detect change along the longitudinal study we had just accepted to be part of. Further, we could care less for the terms *double blind cross over* and such, because we were going to do this in total blindness anyways and see what we got. Then much later, Dr. Kolevzon could lift the blind and tell us when the placebo (saline solution) vs. the drug (IGF-1) was administered.

The traditional design that people seemed to use stroke me as nonsense because implicit in their assumptions was some linear correspondence between the nervous systems' reaction to the drug and the time when the drug or placebo were given. Knowing that a typical nervous system is rather complex and has extremely complex and nonlinear intrinsic dynamics, making such assumptions already hinted at a problem with the design of the paradigm. But this was *the state of the art* of clinical trial design out there.

The intrinsic dynamics of a nascent nervous system that underwent a neurodevelopmental glitch are even more unpredictable than one undergoing neurotypical development. A coping biological system with unknown adaptive capacity must be monitored relative to itself to form a self-defined metric scale. Then, normative data and data from other systems with pathologies that give rise to somewhat similar phenotypes can be used as anchors to measure the departures of the system we are interested in (in this case the system with 22q13 deletion syndrome) from neurotypical development. As such, the best bet was to profile each child individually relative to the baseline measurement we would take of their behavior before the drug trial started. Examining the rate of change over time and the shifts on the stochastic signatures of their biorhythms' fluctuations as they underwent the treatment was something we could do with our biometrics and standardized scales. The challenge was how to engage them in a task that was not boring or too taxing for them.

We surveyed the literature and learned that although delayed, walking was something that by 5 or 6 years of age was generally attained in SHANK3 deletion syndrome. We decided to measure their walking patterns as they would naturally pace back and forth on a platform. This task evoked turning at each end of the platform, before returning to the other end, where one of the parents and/or one of us would be waiting. To make it more interesting and amenable to the child, we had iPads running their favorite movies at each end of the platform. One of us would hold the iPad and encourage the child to return to that end. The experiment was more like a game where parents actively participated too. Each visit brought us so much joy! The lab became a family with the families we interacted throughout this trial. Although this was not our study per se, to this day we have kept in touch with several of the participating families. They left a beautiful mark on each one of us across the two years and a half we participated in this endeavor.

SHANK3 DELETION SYNDROME

The 22q13 syndrome goes by several names. Some refer to it as Phelan–McDermid syndrome (PMS) while others call it SHANK3 deletion syndrome. It is a complex neurodevelopmental disorder believed to emerge from underlying impairments in synaptic transmission and synaptic plasticity.[102,103] The origins of such problems can be traced back to heterozygous deletions of chromosome 22q13.3,[104–106] which encodes for the *SHANK3* gene. SHANK3 codes for a scaffolding protein located at the post-synaptic density (PSD) of glutamatergic synapses. The protein plays a role in the formation and stabilization of synapses, as they assemble glutamate receptors with their intracellular signaling apparatus and cytoskeleton at the PSD.[107] Other *SHANK* genes in different locations of the genome also play a role in neuronal development. The literature reports that in neurons, *SHANK2* and *SHANK3* have a positive effect on the induction and maturation of dendritic spines, whereas *SHANK1* induces the enlargement of spine heads.[107]

When we reviewed this literature in the lab, we realized that the corpus of research findings had a focus on the CNS, but said very little about the PNS. Synapses go beyond the brain and spinal cord into the peripheral nerves. As such, a natural activity such as walking could reveal several layers of information. From gait patterns to minute fluctuations in sensory–motor activity, we could learn and infer much more about brain control over the body than just examining the synaptic activities of the brain in isolation. This disembodied approach to brain research was commonplace in every neurodevelopmental disorder we had come across thus far—autism being one of the main ones we examined. But what good is a clinical trial if it says nothing about the improvements in behavior that can lead to improvements in activities of daily living. After all, the ultimate goal of an intervention ought to be to improve the quality of life of the person being treated. But quality of life depends intrinsically on the person's mental control over the body as it performs regular activities with agency: i.e., as the brain exerts volitional control over its self-generated behaviors and learns to adapt them on demand, with *deliberate autonomy*.

In principle, any of the *SHANK*-related disruptions reported in this literature review could alter synaptic flow, increase synaptic noise throughout the periphery, and consequently compromise the re-afferent flow of peripheral sensory feedback that emerges from self-produced movements.[108,109] Some evidence for peripheral disruption appeared in *SHANK3* mouse and rat models.[110] Paired with cerebellar deficits in 22q13 deletion syndrome,[111] such peripheral dysfunction could interfere with the formation and maturation of internal dynamics. Furthermore, the hypotonic issues that the children present with at birth,[112] could further delay maturation and negatively impact motor variability. Along these lines, excess randomness and noise in the stream of reafference from self-generated motions could compromise the child's corporeal self-awareness, and delay achievement of sensory–motor developmental milestones. The notion that social interactions were disrupted seemed obvious to us, because they inherently depend on movement dynamics and proper neuromotor control. While we could do very little to improve the child's quality of life with a diagnosis of autism by the DSM-V or ADOS-based criteria, we could certainly gain insights on how much neuromotor control they had over their bodies in motion, as they performed a seemingly simple task: walking back and forth. I say here "seemingly" because a task that we take for granted as adults—walking up straight in a controlled manner without intermittently falling—is one of the major accomplishments of the human species. In typical development, full gait maturation typically manifests later, after 6 years of age.[113,114]

Self-produced movements under CNS control, such as those embedded in walking, contain minute fluctuations that shift signatures moment by moment. This form of kinesthetic input is coded through a plethora of receptors. Among them are touch receptors on the skin, joint receptors, muscle receptors, etc., gating and flowing information across the peripheral and central nerves. They serve as an additional source of peripheral afferent feedback that can be harnessed with noninvasive means, such as wearable biosensors of various kinds.

As in other work in the lab, here we could measure such activities with millisecond time precision at the motor output level.[37,115] We had already profiled disruptions in the flow of micro-movements in children with a diagnosis of idiopathic autism spectrum disorder (iASD).[37] Here we could characterize such patterns of micro-movements in children with PMS who also receive an ASD diagnosis. This was important to try and build a general model of autism treatments. Indeed, PMS is estimated to accounts for up to 2% of ASD cases.[116] As such, PMS may serve as a valid model of such phenotypical manifestations in subgroups of iASD.[117] We reasoned that the sensory and motor issues associated with PMS[112,118] likely interfere with social interactions and contribute to the ASD diagnosis. The phenotype of PMS had been described using primarily parent-report measures and subjective observational assessments. Here we had an opportunity to expand those profiles and complement them with objective physical biometrics under a standardized scale. In this way, besides building the personalized profiles of intrinsic dynamics for each child, we could also examine this small cohort as a group whereby self-emerging patterns could tell us more about the patterns of motor noise and their shifts with the treatment they were about to undergo.

Detailed sensory–motor phenotyping in a single gene form of ASD could increase the likelihood of linking deficits in social behavior to physical sensory–motor disruptions caused by specific synaptic problems due to underlying specific genetic factors. There are many parameters and may different aspects of this task that we could study while the children visited the lab. With these new notions in mind, we set to study gait patterns and develop new biometrics and new visualization tools for longitudinal tracking.

TRACKING A CLINICAL TRIAL USING GAIT BIOMETRICS

The Critical Baseline Measurement

We gained access to 16 children with PMS. The families came from afar. Some came from Pittsburg, others from Virginia and the Washington DC areas. Other came from the up-State New York regions and some from New Jersey. Each single family connected with the members of our lab in amazing ways. The visits took place primarily on weekends to accommodate the traveling schedules and the battery of tests they had to undergo at Mount Sinai, in NYC the previous day. My lab members raised to the occasion and were on call for the weekends we needed to conduct the gait studies. The day prior to the visit we prepared the lab and calibrated the equipment to be ready for the study the next day (Fig. 7.33A). Setting up the sensors and making sure all interfaces were properly functioning for real-time recordings and calibration procedures became routine for undergraduate students who rotated through the lab on their way to medical school. They served as subjects to gather pilot data and these data provided the test bench to further design various biometrics.

In the first phase of the study, we received 10 children in the lab. Jill took care of all technical issues concerning the recording equipment and proper functioning of the computer interfaces. She also taught Sejal how to calibrate, record, and upkeep the data, to prepare her to take over these duties after Jill graduated and moved on to industry. One of the features of the lab is the comradery and pedagogical value the research we do raises. Because what we do is so unique, the only way to keep it going and improve it, is to maintain a harmonious and collaborative environment where we all help each other and learn from each other. I am the PI of my lab, but I am also a student of my students. I learn from them and encourage them to lead too. Somehow, this philosophy keeps us moving autonomously toward better and more ambitious goals. It is always "our" research program, "our" project, "our" aims. The students create and maintain the manuals to conduct the research and there are many steps to follow in each project manual; so, upkeeping all of it and adding to it is very important.

By the second phase of the study, we received an additional group of six children, and added inertial measurement units to the positional (electromagnetic) sensors we were using in the first phase (Fig. 7.34). These new sensors served to initiate the steps towards degrading the higher resolution data into data commonly harnessed by commercially available sensors. The purpose of this exercise was to promote methods that could be used in clinics and home environments, away from the constraints of a laboratory environment. In neurodevelopmental research, it seems important to initiate such a transformative path because children are easily bored in the lab. Further, the environment of the lab is somewhat artificial. To improve their quality of life we need to be able to take these measurements unobtrusively across different naturalistic contexts, as they move around and behave in the home, the school, and the clinics they go to receive therapies and various forms of treatments. In previous chapters of the book we have previewed various ways to record data in the home or school environment with wearable biosensors. The clinical trial gave us the opportunity to do so in such a way that we could provide physicians with new ways to complement their observational means with new visualization tools that helped us easily see departures from normative data.

In the case of PMS, we had to measure their patterns without the drug intervention to characterize their baseline gait and compare it to the levels of age- and sex-matched controls. Further we needed to understand these patterns in relation to idiopathic ASD, since the phenotype of PMS had been already associated with the ASD diagnosis. To that end, we recruited controls in the neurotypical population, but also examined individuals with idiopathic ASD. The numbers in our study were modest (30 participants in total) because the idea was to use the trial as a test bed for new biometrics while gaining some insights on longitudinal trends.

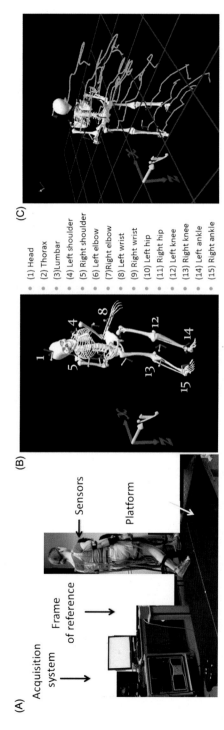

(A)

Acquisition
system

Frame
of reference

Sensors

Platform

(B)

1
4
5
8
12
13
14
15

(1) Head
(2) Thorax
(3) Lumbar
(4) Left shoulder
(5) Right shoulder
(6) Left elbow
(7) Right elbow
(8) Left wrist
(9) Right wrist
(10) Left hip
(11) Right hip
(12) Left knee
(13) Right knee
(14) Left ankle
(15) Right ankle

(C)

FIGURE 7.33   **Experimental set-up and sample parameters.** (A) Gait platform where subjects paced back and forth and performed other natural-istic interactions during walking. The acquisition system, the wearable sensors affixed across the body, and the frame of reference used to obtain the kinematics from the movement trajectories are also shown. (B) Digitization of the full body using 15 electromagnetic sensors attached to body parts enumerated on the right and the orientation of the frame of reference. (C) Sample kinematic trajectories in green showing the positional changes over time as the person walks in one direction.

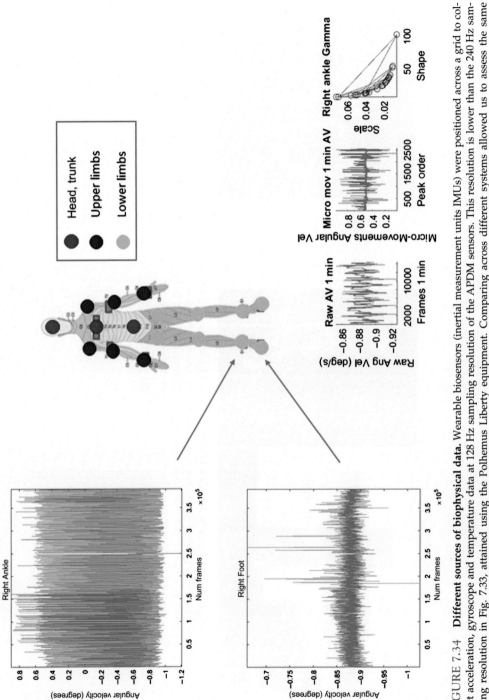

FIGURE 7.34 **Different sources of biophysical data.** Wearable biosensors (inertial measurement units IMUs) were positioned across a grid to collect acceleration, gyroscope and temperature data at 128 Hz sampling resolution of the APDM sensors. This resolution is lower than the 240 Hz sampling resolution in Fig. 7.33, attained using the Polhemus Liberty equipment. Comparing across different systems allowed us to assess the same analytics and standardized data type of micro-movements for different biosensors in the market. (A) Raw data from the right ankle and right foot shown as representative of the grid we collected in (B) schematics. (C) Data from a 1 min block used to examine the fluctuations in peak amplitude according to the Gamma process under iid assumption and compare it to the data obtained with overlapping sliding windows of various sizes. Such comparisons allowed us to examine co-dependencies between past and future events bound to exist in the nervous systems (kinesthetic reafferent) activity. The Gamma parameter plane shows the stochastic trajectory of the estimated points under non-iid assumption (i.e., with overlapping sliding windows of overlapping activities varying up to 0.5 min).

The first thing we noted was how difficult it was for the children to sustain a regular pace in their walking patterns. They seemed to nearly float and turning was very tricky for them. This gave us a first clue as to what to start looking for. Kinematics trajectories are very rich in parameters. To discover the manifold of relevant parameters outputting information with patterns that maximally separate participants into self-emerging groups is nearly an art. It requires some exploration of tasks and contexts to evoke appropriate signals with a chance of raising information from the inherent variability of the many parameters one could consider. Walking was no exception to this "more and art than a science" phenomena. We had to have a good eye for possible movement features inherently informative of change.

At the same time, we needed to find ways to avoid fatigue when using movement driven experiments. A common pointing experiment would require over 100 trials to gain statistical power for analyses if the parameter of interest were the linear speed peak (for example), because each ballistic pointing experiment has a single peak. Since we had an interest in the speed-dependent stochastic signatures of the fluctuations in the peaks' amplitude (and their evolution with the IGF-1 drug over time), we wanted to examine speed. To increase the number of peaks per unit time, we examined instead the angular speed and the angular acceleration. Both parameters relate to joint rotations and with a grid of 15 joints across the body (see Fig. 7.33B), we could map out the levels of NSR for several body regions using only a few minutes. For example, Fig. 7.35A shows 30 s worth of gait data, whereby we identify two different regimes of rotational amplitude: small to the left and large to the right panel. We noted that across PMS subjects this was a feature of the gait patterns. And since we had a mixture of young children 5–6 years old, older children 8–10 years of age, and adolescents 12–15 years of age, their anatomical measurements also varied. Since anatomy has an impact on the range of motions (e.g., longer limbs can travel longer distances per unit time and exert more force with higher mass), we normalized the peak speed values (as explained in the other chapters) to avoid such allometric effects on the speed data. This is illustrated in Fig. 7.35B, where we mark the local maximum surrounded by the local minima that we use in the normalization. The equation we used was introduced in the Table 1 third row of Chapter 3.

This data, are continuous real values scaled between 0 and 1. Such parameterization of the kinematics data was then used to build several biometrics that allowed us to place all children, PMS, iASD and age- and sex-matched controls on the same parameter plane to search for self-emerging patterns of the baseline visit. Given the enormity of the parameter spaces and the many degrees of freedom we had to choose from, we decided to examine several locations of the body in search for informative patterns differentiating PMS from the other participants. Clearly the legs are critical to walk in a controlled manner, so that was a good starting point.

The first thing that we noticed when we plotted the normalized angular velocity denoting the micro-movements parameterization of the amplitude fluctuations from the sensors at the feet, was the fundamental differences between representative female and male PMS participants, and the non-PMS participants in the cohort. These differences can be appreciated in Fig. 7.36, where I plot the patterns corresponding to the right and left foot. All controls and one of the iASD participants had multimodal frequency histograms; but the representative PMS and one of the iASD did not. There was a unimodal frequency

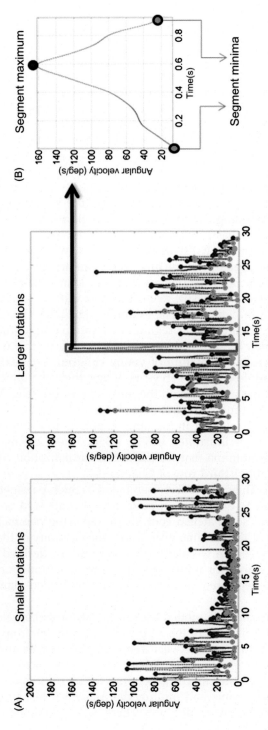

FIGURE 7.35 Raw data from angular speed capturing joint rotations were normalized and micro-movements of different frequencies used to characterize the fluctuations in amplitude and time of the peaks and valleys of the time series data. (A) Contrast smaller and larger rotations; (B) uses the local peak to divide it by the sum of the value of the local peak and the surrounding minima. This normalization produces the micro-movement data in the angular speed domain of interest.

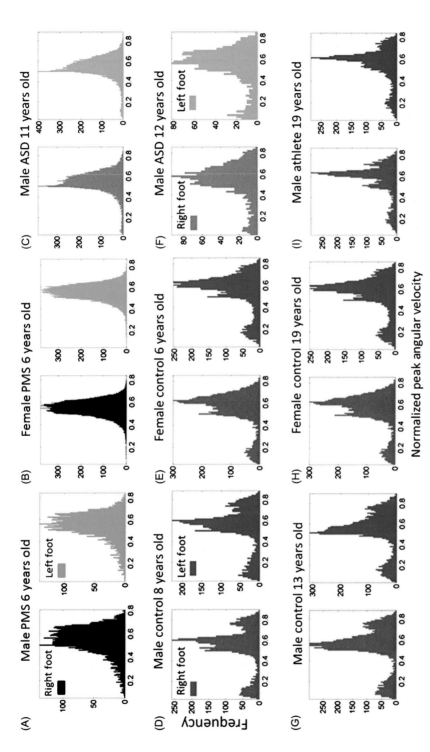

FIGURE 7.36 **Frequency histograms of the normalized peak angular velocity for the footsteps.** (A, B) Representative patterns from PMS participants (6 years old) show a unimodal distribution of the normalized joint angular velocity index. (D, E, G, H, I) Representative controls of a broad range of ages including ages and sex of the PMS participants show multimodal distributions with a systematic bump of smaller values of this index indicating a density of faster rotations on average (the denominator of the index containing the average angular velocity between the two local minima is higher). (C, F) Participants with idiopathic autism spectrum show both a lack of faster rotations in the turns (C) quantified in the PMS and the presence of them (F) quantified in controls. (I) This panel shows the participant with ideal patterns, a martial arts expert with fine control, timing, and joint coordination in his movements.

distribution of their normalized amplitudes, missing micro-movements of lower value centered in the 0.1–0.2 range between the full range of the normalized data, from 0 to 1. These values of lower magnitude missing from the PMS data are those peaks with higher rotational speed on average. This is because of the normalization (Table 1 third row of Chapter 3) which divides by the sum of the local maximum and the average rotational speed between the two local minima. Lower values indicate higher rotational speed on average for the local segment. This feature belongs in the turns, which were visibly very different in the PMS children and adolescents. Whereas the neurotypical children could rotate at the turns in a controlled manner, the PMS children did it in a very uncontrolled way, seemingly floating. We were afraid of them falling, but somehow, they did not. However, their turning patterns were extremely different from those of controls.

We used the Hartigan dip test of unimodality and confirmed that all the neurotypical children failed the Hartigan's dip unimodality test ($p < 0.01$) as they manifested multimodal distributions (Fig. 7.36). In contrast, all the PMS children had $p > 0.05$ for both or one feet, as did the iASD children we tested. This can be appreciated in Fig. 7.37 where I plot the p-values for the ensemble of children and adolescents along with the representative frequency histograms color coded as in Fig. 7.36. These graphs in Fig. 7.36 are frequency histograms of the normalized values but we also plotted the proportions of frequency count taken with respect to each person's performance in Fig. 7.38 for PMS participants and in Fig. 7.39 for neurotypical participants.

This was important to see if the standardized scale revealed similar patterns for the group when we considered the individual nuances of the participant's gait. The answer was positive, the standardized scale did not mask any of the individual nuances. We could use the feet's data to distinguish unique baseline features present across the PMS group. As such, this parameter was identified as a potential variable to track longitudinally during the trial. Would the IGF-1 treatment change this parameter? And more importantly, given that we sampled normative data too, would it shift the signatures towards the normative neurotypical ranges?

The angular speed data from the feet were very informative of fundamental differences in gait patterns between PMS and the rest of the group; but the nonunimodal nature of the frequency histograms prevented us from examining the micro-movements of the fluctuations in angular speed peak amplitudes as a Gamma process. We wanted to do so to map the empirically estimated Gamma parameters for each participant on the Gamma parameter plane. In this way, we could identify the locations of each estimated probability distribution function and see if any patterns spontaneously emerged.

We divided the participants by age into PMS-Y (younger children in the group, between 5 and 9 years of age) and PMS-O (older children in the group above 12 years of age). The Gamma parameter plane derived from the angular speed (rad/s) revealed 2 quadrants of interest, according to the median values of the shape and scale parameters of the full group (including neurotypical and iASD participants) (Fig. 7.40A). The left upper quadrant (LUQ) and the right lower quadrant (RLQ) are interesting because they contain PDFs with very different features.

The PDFs of the LUQ have higher NSR (higher dispersion) than those of the RLQ. Further, the shapes of their distributions are more skewed. All PMS-Y grouped below the median shape value and most localized above the median scale value. This indicated

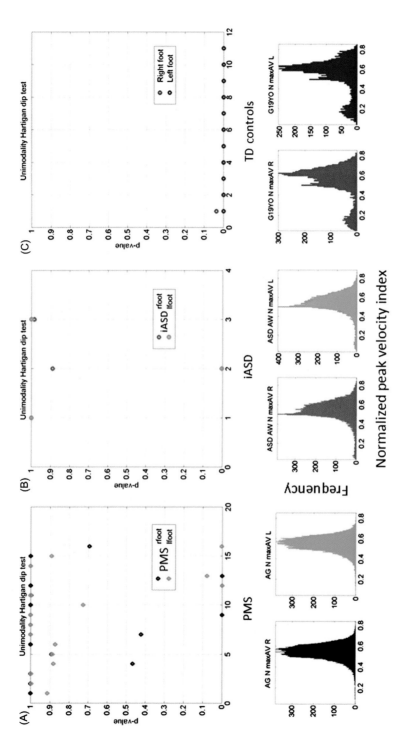

FIGURE 7.37 **Outcome of the Hartigan's unimodality dip test.** (A) The p values > 0.05 from the Hartigan's dip test of unimodality from the distributions of the normalized peak angular velocity index in 16 PMS children ranging between 5 and 16 years old show systematic lack of multimodal distributions in one or both feet. (B) Patterns of iASD children show consistent $p > 0.05$ patterns for one or both feet. Representative histograms for right and left foot are also shown. (C) Patterns from control participants (5–19 years old) had $p < 0.01$ thus failing the unimodality dip test for both feet. Representative histograms for both feet of one participant are shown.

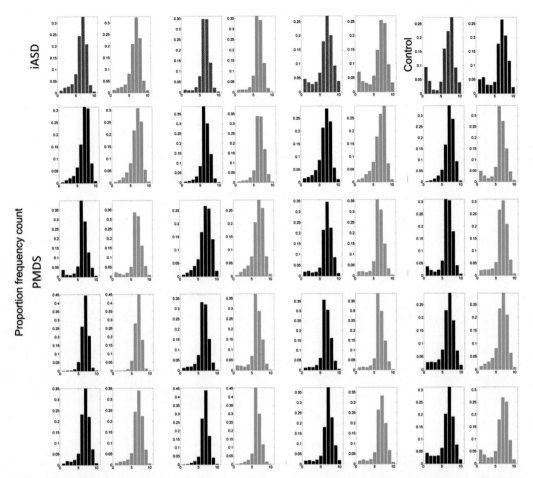

FIGURE 7.38   Bar plots of the proportion of count of actual angular velocity peaks for representative children in each group under study complementing Figure 7.36 (containing the histograms of the normalized peak velocity index, normalized to account for possible allometry effects).

higher noise and more skewed shapes for the younger children in the trial (at baseline value). The older PMS fell in different locations, away from the younger group. All but one fell above the median shape value denoting less skewed distributions and all fell below the median scale value, indicating lower NSR. Neurotypical participants were divided into males and females and color-coded accordingly. All neurotypical males fell in the RLQ, while the females split into younger (<7 years old) and older (>7 years old). The older females were in the RLQ. The others fell above the median scale value, with elevated levels of noise, but more symmetric distributions that separated them from the PMS group. The three iASD we tested were closer to the PMS-Y than to neurotypical controls. However, they were 10–12 years of age, much older than the PMS-Y group. In this regard, their levels of controllability of the legs' angular rotations were not as mature as those of their age- and sex-matched neurotypical controls. They were certainly below the age- and

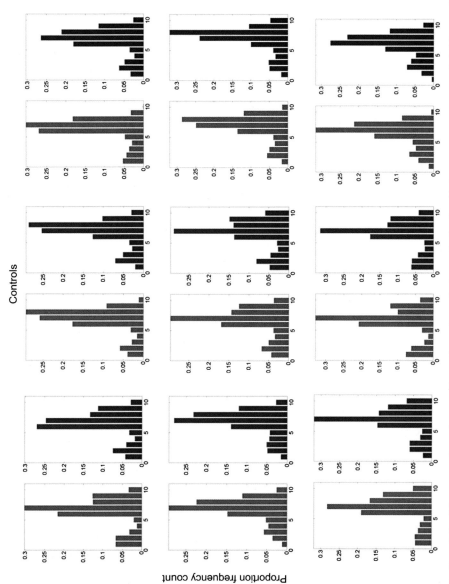

FIGURE 7.39 Bar plots of the proportion of count of actual angular velocity peaks (in deg/s) for representative typical children and adolescents complementing those in Fig. 7.36 (containing the histograms of the normalized peak velocity index, normalized to account for possible allometry effects). Notice the multiple bumps in the distributions absent in most PMS children (who have either a total absence of multimodality in both feet or a modest additional bump in one foot). No PMS has significant evidence of multimodality in both feet.

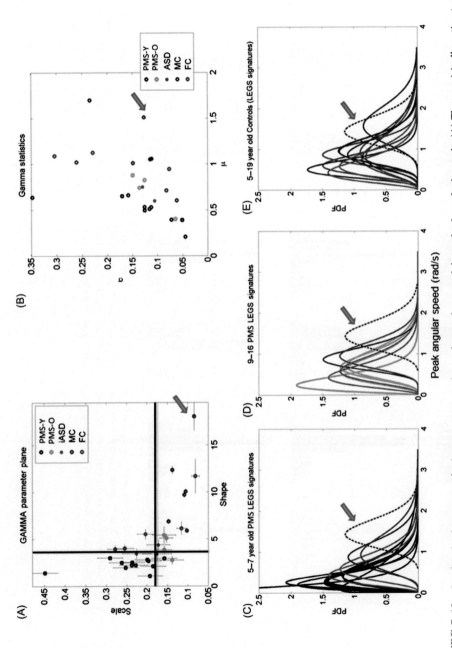

FIGURE 7.40 **Stochastic signatures of micro-movements in peak angular speed from the legs during gait.** (A) The empirically estimated shape and scale Gamma parameters from the normalized peak angular velocity index of the legs are plotted for each child with 95% confidence intervals. The lines denote the median shape and scale values automatically dividing the Gamma parameter plane into four quadrants. All but one younger PMS children (PMS-Y) 5–7 years old fall in the left-upward quadrant with highest noise and shape values close to 1 (especial case of the Exponential distribution). The farthest to the right (symmetric distribution) and lowest scale value (lowest NSR) is a professional male athlete (19 years old) marked by the arrow. Green dots are older 9, 12, 14, and 16 year old PMS (PMS-O). (B) Empirically estimated Gamma mean and variance with similar color scheme as (A). Higher values of the mean normalized peak velocity index indicate slower angular velocity on average. (C–E) Empirically estimated Gamma PDFs for each participant using the same color code scheme as in (A, B) according to age for the PMS and sex for the controls. ASD participants are plotted as reference. In each panel (C, D) with the PDFs from PMS children, we plot the PDFs of age- and sex-matched controls. Red (FC) and blue (MC) are typical female and male controls respectively. (E) All controls and ASD participants are plotted and the PDF of the professional athlete stands out as the Gaussian shape with lowest dispersion in the cohort (see his shape and scale parameter range in (A)).

sex-matched PMS. Fig. 7.40B shows the empirically estimated Gamma statistics of the actual angular speed range (rad/s) whereby we can see higher mean rotational values for the neurotypical participants (with more variable patterns for the youngest females and lower mean values for the PMS-Y). I included an athlete in the group (a professional martial artist) to compare his parameter values to the rest. The PDF is marked with an arrow in all panels of Fig. 7.40. His estimated Gamma PDF was the most symmetric and his scale value was in the lowest range of the RLQ, denoting the lowest noise to signal ratio. This parameter range provides the empirical range of the best-case scenario, one that we can use as anchor to compare changes towards or away from it, as the values of the PMS change with the drug or placebo stages during the longitudinal clinical trial. Fig. 7.40C shows the PDFs of the participants according to age groups available. In each panel, we plot the sample PDFs from the iASD as well as the sample PDF from the age- and sex-matched controls. This helps us locate the PDF features relative to the controls and evaluate their departure from the neurotypical data. Further, the PDF of the athlete is also noted as an ideal optimal one that in the last panel is contrasted with all controls in the group as the most symmetric with lowest dispersion. Notice that unlike in the Figs. 7.36 and 7.37, where we used the micro-movements of angular speed parameterization, here we used the values of the peaks (rad/s). This is because we wanted to examine the physical ranges of the angular rotations across the diverse range of ages and anatomical features.

We were also curious about the motions of the upper body as the participants walked and paced back and forth on the platform. To examine the upper body patterns, we used the angular acceleration and empirically estimated the best probability distribution function that fit the frequency histograms of the inter-peak-interval times (ms) and of the peaks' amplitude (deg/s^2). The scatters of the estimated Gamma shape and scale can be appreciated in Fig. 7.41A while Fig. 7.41B shows the estimated Gamma statistics. From the plots, it is evident that the neurotypical controls span a broader range of shape values with low scale. Indeed, the histogram of the shape parameter distribution is rather broad with high dispersion (top of Panel A). The departure from this feature in ASD is evident as the distribution is narrow, suggesting a very narrow range of PDF shapes across the upper body of the representatives iASD. The PMS participants are somewhat in between, with broader ranges in PDF shapes than those found in the iASD, but lower than controls. These graphs are used to illustrate different biometrics and parameters that allow us to visualize self-emerging patterns from empirically estimated statistical ranges. Although we merely had 30 participants, we included 10 body nodes for each (i.e., the upper body nodes) so that with 300 data points we could better understand their underlying stochastic patterns. Figs. 7.41C,D show the empirically estimated cumulative distribution functions (CDF) of the inter-peak-interval times (in Ms) of the angular acceleration for all participants split by sex. They differed significantly according to the Kolmogorov−Smirnov test for empirically estimated CDFs. In Fig. 7.41D, the iASD males stood apart from the PMDS males when comparing the CDFs of the peak angular acceleration (deg/s^2).

Visualization Tools

One of the challenges of studying the nervous systems outcomes is to find ways to represent its changing dynamics. In any one given visit, the activities captured by the grid of

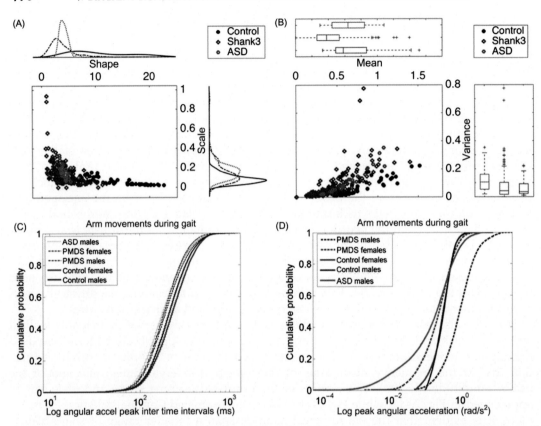

FIGURE 7.41   **Additional examples of atypical stochastic signatures in PMS and iASD.** (A) Distributions of the empirically estimated shape and scale parameters of the continuous Gamma family of PDFs across the 14 joints of the body using the micro-movements embedded in the angular accelerations. (B) Scatters of estimated first (mean) and second (variance) moments of the Gamma PDFs using the estimated shape and scale values in (A) and statistical comparisons using nonparametric ANOVA reveal statistically significant differences at the alpha 0.01 level. (C) Differences in the timing of the peak angular acceleration captured in the cumulative distribution function across subject types. (D) Differences in patterns of angular acceleration micro-movements captured by the cumulative distribution functions across subject types (see legends).

biosensors that we used to measure motion could be represented as a peripheral network with different states reflecting synergistic patterns across the upper and lower body. It occurred to me that network connectivity analyses commonly used in fMRI data could be adapted to analyze the body in motion. The initial steps to that end are featured in Fig. 7.42 whereby we use analyses of phase locking value (PLV) to identify patterns of synchronicity across the bodily nodes.

PLV is a statistic used to quantify the phase coupling between two biological nonlinear signals in time-series, e.g., of electroencephalographic signals.[119,120] In the present study, PLV was adapted to quantify the level of coupling (phase synchrony) in the time series of angular velocity values between each one of the 14 joints and all the others. Specifically,

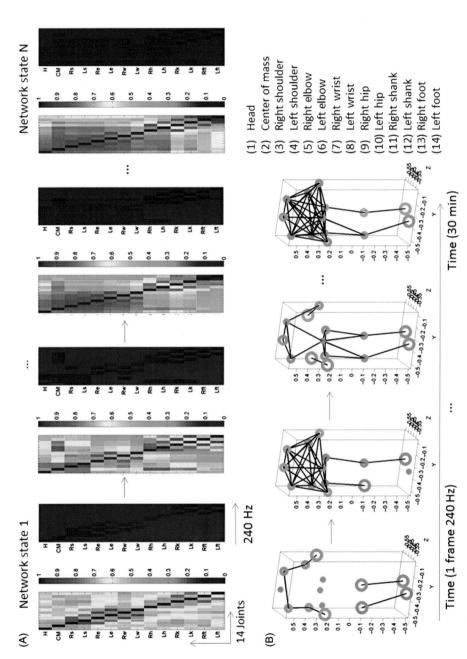

FIGURE 7.42 Visualization of dynamic states of the peripheral network connectivity across 14 joints of the body. (A) PLV and binary matrices (at 0.85 threshold) showing individual frames from a representative PMS participant. (B) Network representation of peripheral joints (nodes) and links (edges) between highly synchronized joints. Circles represent joints and lines are active links within a frame corresponding to highly synchronized joints within a ½ hour period taken in frames of 240 Hz each. The size of the circle represents a measure of node neighbor-clustering (see text for details). The color of the circle is based on the modularity metric (see text for details). Circles with the same color represent modules of nodes that maximize the number of within-group edges, and minimize the number of between-group edges, i.e., a modular community structure within the network.

the PLV nears 1 in cases where the instantaneous phases of the two joints' angular velocities time series are synchronized. Conversely, if they are unsynchronized the PLV tends to 0. Greater detail regarding the procedure for computing PLV can be found elsewhere.[9,28] Here we obtained the *14 joints x 14 joints* PLV matrix every 240 frames (240 Hz sampling resolution of the sensors). We used a high threshold of synchronization value (0.85) to create a binary matrix. Entries in the original PLV frame that were above or equal to the threshold were set to 1 and those below the threshold were set to 0 (see Fig. 7.42A). This instantaneous binary representation of the synchronicity can then be converted to an adjacency matrix representing an unweighted, undirected graph. The graph can be used to represent the peripheral network bodily nodes and a number of connectivity metrics commonly used in brain research, can be imported to represent patterns of dynamically changing activities across the network.[121,122]

To visualize the temporal profiles of emerging modules and connectivity patterns across this peripheral network of rotational joints we adapted the following indexes and used them to represent several metrics in these plots:

The *connectivity index*, which is given by the degree of each node (joint), that is, the number of links connecting the node to other nodes in the network. In this case (an undirected, unweighted binary graph), we provide a simplified representation to illustrate densely vs. sparsely interconnected nodes across the network, to identify critical differences between controls and PMS–ASD cases.

The *modularity index* provides a sense of the instantaneous subdivision of the network into nonoverlapping groups of nodes that work together within a "community." This metric assesses self-emerging structuring of the network by maximizing the number of within-group edges, and minimizing the number of between-group edges. Thus, modularity is a statistic that quantifies the degree to which the network may be subdivided into such clearly delineated groups. For example, Fig. 7.42B shows the contrast in modularity (green vs. cyan node colors) as the connectivity differs between the upper and lower body and changes frame by frame. This is better appreciated in Fig. 7.43 unfolding the walking session over time, frame by frame within a 30 min walk period, as well as helping us identify synergies across the network as its dynamics unfold.

The clustering coefficient: Another metric that we can use is the clustering coefficient, i.e., the fraction of triangles around a node, or equivalently, the fraction of node's neighbors that are neighbors of each other. For example, in Figs. 7.42B the size of the circles at each node is given by the clustering coefficient value. Large circles indicate nodes whose neighbors are neighbors to each other. Fig. 7.42B shows snapshots of the evolution of the network for different frames. In summary, lines represent links between nodes. Circles at the node track the modularity (color coded) and the clustering coefficient values (size of the circle).

LONGITUDINAL TRACKING OF SOMATIC-MOTOR CHANGE ACROSS THE GROUP

The PLV parameterization of the grid of bodily activity to study peripheral network's connectivity patterns helped us study the evolution of the changing dynamics during the

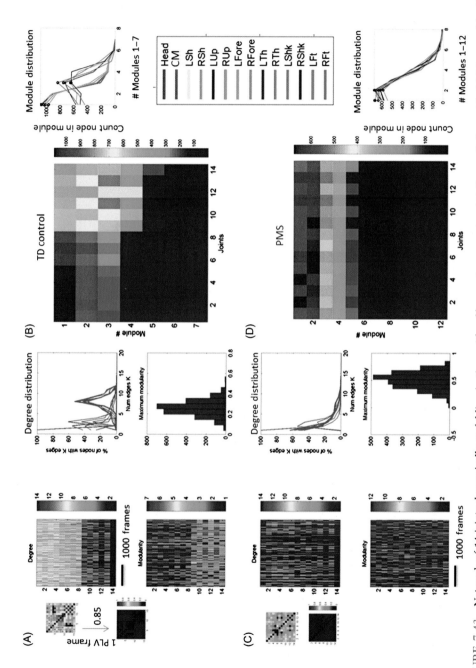

FIGURE 7.43 Network of 14 joints dynamically unfolding in time with self-emerging synergies. (A) Connectivity matrix 14 joints × 16 s (4000 frames at 240 Hz) from representative control. One frame of the full and the corresponding binary PLV (determined at a 0.85 threshold) are shown as insets. Color bar represents the connectivity. The network has 14 nodes (the 14 joints in the legend of Fig. 7.42 where the center of mass is estimated from thorax and lumbar sensors located as 2 and 3 in the trunk of the avatar in Fig. 7.33) for two representative subjects, one from the typical control group (Fig. 7.43 A, B) and one from the PMS group (Fig. 7.43 C, D). The degree distributions of the network and the modularity distributions are obtained from a 30 min walk registering fluctuations at 240 Hz. Notice the contrast between the two representative participants. Panel A also shows

on the left the full and binary PLV matrices of one frame at a 0.85 threshold (high synchronicity). In the 14 joints × time (frames) matrix each entry provides the degree for each node (joint) as reflected in the color bar. There is higher connectivity in the upper body (numbers 1–14 are as in legend of Fig. 7.33) than in the lower body for this typical participant. The degree distribution of the network for a segment of the 30 min walk is also shown on the right-top panel. Notice that the curves represent the degree distribution of the joints and that a pattern emerges corresponding to the upper and lower body as well. Specifically, the upper body has distributions centered farther to the right, with higher number of links than the lower body. The lower body has more nodes with fewer links and fewer "hub" nodes (in the tail of the distribution). Notice that the upper body has higher connectivity than the lower body in this participant (joints are as in the legend of Fig. 7.33). Degree distribution identifies two distinct groups of nodes. Corresponding modularity matrix identifies up to 7 modules (color bar). Frequency histogram of maximal modularity is also shown. (B) Matrix quantifying the joint participation per module (color bar shows the counts). Inset shows the module distribution with identification of modules with maximal count per joint participation. Two modules self-emerge (1 and 3) as the ones with maximal joint participation denoting two main synergies summarizing the complex 14 joint patterns of the gait in this participant. Colors in the legend identify the joints participating in the two synergies. Notice that one contains the upper body joints and the other the lower-body joints, consistent with the degree distribution in (A). (C, D) The same information as in (A, B) is shown for a PMS participant. Notice the striking differences with the control and the lack of synergies in the PMS case.

baseline visit, prior to any drug intake. Once the trial started, I designed a different parameterization using frequency domain coherence analyses. Examples of these analyses were introduced in Chapter 6 within the context of brain-body coupled dynamics (Figs. 29, 30, 32–36 in Chapter 6) and dyadic interactions taking place during the administration of the ADOS test (Figs. 41–45 in Chapter 6). Here, these weighted directed graphs provided dynamic ways of examining synergistic patterns across the body and several connectivity metrics helped us track immediate and delayed effects of the drug on such patterns. We quantified the underlying activity of each node and built a portray of the estimated Gamma moments from visit to visit. This graph helped identify effects of the drug on the overall network and showed how the patterns of variability across the grid of nodes changed with the drug. For example, the top row of Fig. 7.44 shows the PDFs of each of the 14 nodes in the head (red), the upper body (blue), and the lower body (green). The star denotes when the drug was received. The rectangle highlights the pattern closer to the control. The top row shows a participant that had an immediate effect of the drug. In contrast, the participant in the bottom row had a delayed effect of the drug. Both participants approached at some point the patterns of the neurotypical participant matching their age and sex. Yet these effects were transient in the first one and remained in the second one (we measured as the participant returned 12 weeks later). It would have been very interesting to have tested the participants a year after the trial ended. Across the body, we also quantified network synergies and their evolution with the drug. In this case, the two participants showed an immediate change in the distribution of nodes participating in three modules detected within the network. These parameters suggest that while some effects can be immediately quantified, others take time to manifest. Further, they manifest through different delays across different subjects.

Fig. 7.45 shows the state of the network at the 7th minute, i.e., after 7 min of walking (for a full movie on 30 min walking see link) for the two participants in Fig. 7.44. The top panel plots the first participant's patterns for each visit. As shown in Fig. 7.44A, this participant received the drug in the second visit, and had an immediate response throughout the body moving the signatures of angular speed towards those of the representative control. Here the patterns of connectivity across the network also changed, connecting the lower to the upper body nodes, in contrast to the baseline, when they were disconnected.

The second participant in the bottom row does not change the patterns as much during the drug intake. As in Fig. 7.44B, here the patterns change more during the week 12 after the placebo phase of the trial. Notice the dramatic contrast between the patterns of connectivity of the PMS participants and those of the age- and sex-matched control. To further illustrate the different delays in these PMS individuals from Figs. 7.44–7.45, Fig. 7.46 shows the link (edges) distributions of their networks. As with other metrics, participant on the top row has an immediate change with the drug; whereas participant in the second row did not manifest a change until the post visit (once the trial had ended). In both cases, as in all other cases, these patterns were very different from those of the age- and sex-matched control.

Consistent with the network's analyses, the analyses of the motion kinematics that we identified as good predictors of change revealed shifts towards the patterns of neurotypical subjects. Notwithstanding the significant changes, these changes did not bring their gait within normative ranges of the age- and sex-matched controls. The drug did affect the

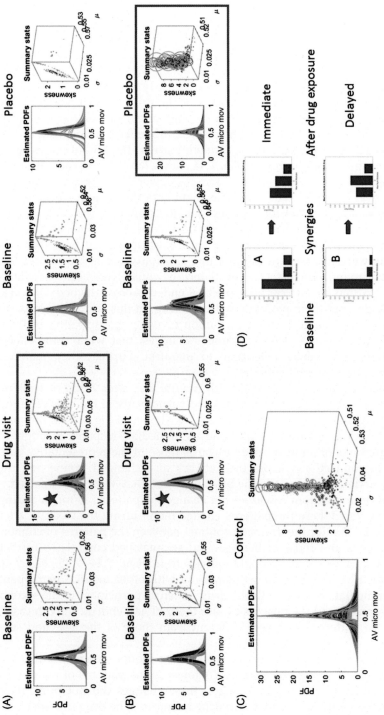

FIGURE 7.44 **Tracking different pharmacodynamics in the IGF-1 trial of SHANK3 deletion syndrome (PMS).** (A) Estimated PDFs for body parts in Figs. 7.33 and 7.34 during 4 visits spaced by 12 weeks in a 9-year-old male child undergoing the clinical trial (top row) and a 14-year-old female (bottom row). First visit without any drug intake allows us to take a baseline measurement to consider departures from this state later, when the drug or placebo stages take place. The child was administered the drug right after the baseline visit (the next day). In the next visit, 12 weeks later, the change in the stochastic patterns across the body was noted. There was an increase in the variability ranges of the angular speed of the lower body that coincided with a change in the patterns of the ankles' rotations and in the bodily synergies. In the next two visits, the patterns return close to the original baseline, suggesting a transient effect of the drug in this participant. (B) Another participant with the same drug schedule shows a very different, delayed reaction that does not show up until the fourth visit in the placebo stage of the trial. (C) Age-matched control patterns across the body. (D) Change in synergistic activities across the network manifest sooner for the first participant than for the second but a similar pattern of reorganization takes place in both.

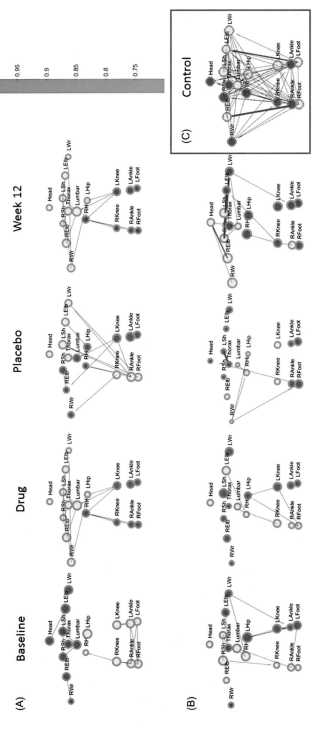

FIGURE 7.45 Network activity patterns for the 7th minute block of activity (out of 30 min walk) for the participants featured in Figure 7.44. Color of the circles represents self-emerging modules maximally connected internally and minimally connected externally with other nodes of the network. The edges of the circles represent the level of coherence with the other nodes the node is connected to. Arrows indicate who leads who and the thickness of the link represents the leading phase value. (A) 9-year-old male. (B) 14-year-old male. (C) The control participant patterns within the 7 min block.

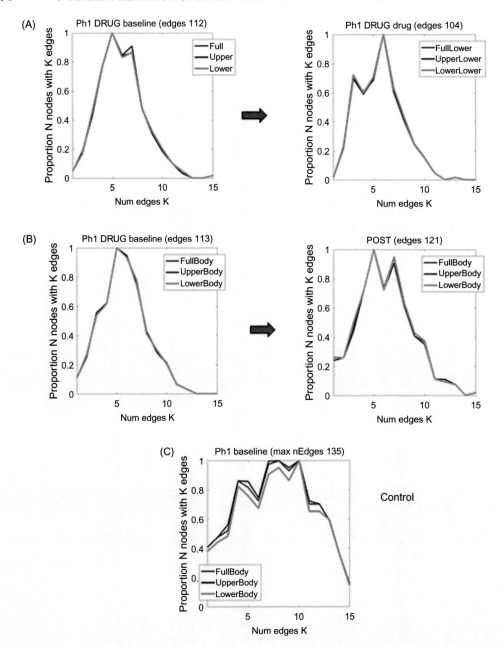

FIGURE 7.46 **Connectivity patterns change with the drug trial at different time scales.** (A) The 9-year-old male has an immediate change in the link distribution (edges of the network) right after taking the drug. (B) The change in the 14-year-old female is delayed to the post trial visit. (C) The patterns of the age-matched control.

somatic-motor patterns in very dramatic ways, shown in various parameters and visualization tools. Fig. 7.47 retakes the feet distributions and contrasts the baseline patterns with the maximal changes observed during the clinical trial. As with all other network and stochastic signatures, some of these changes were immediate for some of the children while other took longer to manifest. All PMS children who underwent this trial unanimously moved in the direction of the controls. Fig. 7.47A refers to the 10 children we assessed in the phase I of the clinical trial. One quit the trial after the baseline (the last one in the graph) so we only have the baseline values (open circles). The other 9 shifted the p-values of the Hartigan dip test of unimodality toward lower values and reached significance for evidence of nonunimodal distribution in one or both feet for several cases (notably 6 and 8 reached significance both feet, while 1 and 3 reached significance for one foot). To appreciate the departure from their baseline values towards the values that age- and sex-matched neurotypicals manifested, we plot the controls in Fig. 7.47B.

Other metrics also revealed the changes between the first and last visit (pre- and post-drug trial). For example, Fig. 7.48 shows the patterns of the angular acceleration and the shifts for the two groups of younger and older PMS participants. As before, we plot the estimated Gamma parameters in Fig. 7.48A with dashed lines denoting the stochastic shift, i.e., the change in PDF shape and dispersion. Fig. 7.48B shows the estimated Gamma moments (mean and variance) and their corresponding shifts. We plot the representative neurotypical male and female and the representative iASD as reference. Fig. 7.48C plots the older group with the shifts and the representative controls and iASD of matching ages and sex. The bottom panels of the figure contrast the controls on the Gamma parameter plane spanning a broader range of values of the shape (more symmetric ranges) and lower scale (dispersion) of the distributions. The other panels reveal the changes in PDF for the younger and older PMS. As before, I plot the ideal case to demonstrate its stochastic signatures as a reference.

INDIVIDUALIZED TRACKING OF TRIAL EFFECTS

These biometrics are best appreciated when we examine the evolution of each individual child (**Appendix** Figs. A7.1—A7.9) and the dramatic shifts in stochastic signatures that broadcast changes across the body, including changes in individual body regions. The body-maps we can draw from empirically estimated NSRs of various biosensors will help us understand the type of time-dependent somatic-motor feedback that the brain is likely receiving from the peripheral nervous systems. The estimation of the systems' ability to accurately predict the sensory consequences of impending actions will inform us of issues of controllability and autonomy. These two components can serve to advise us about the likelihood of the treatment to improve the person's quality of life.

By improving the motor feedback that the person's brain receives, it is possible to improve autonomy of the brain over the body. The continuous improvements in the signal's bandwidth and quality that the periphery echoes back to the brain can lead to and rise the person's comfort. Likewise, if the noise levels increase and the randomness worsens, the drug is most certainly not working properly. For example, the person may improve memory performance according to a questionnaire, but if the person cannot

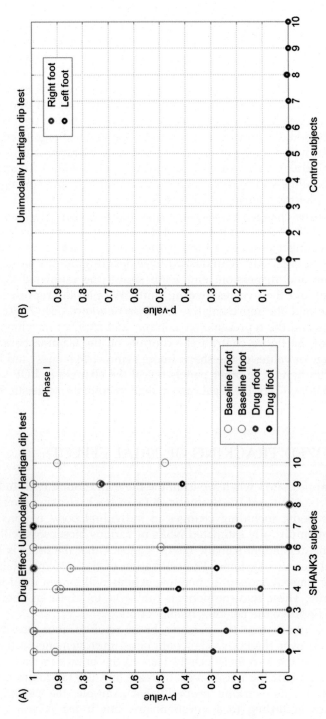

FIGURE 7.47 **The drug effect on the rotational angle patterns of the feet.** (A) The 10 participants in the phase I (participant 10 quit the trial but we do not know why). The open circles are the baseline measurements. The closed circles are the maximal change the child experienced during the trial. The control values are $p < 0.05$, the Hartigan dip test outcome when the distribution fails the unimodality test. All participants started with a unimodal distribution, with a *p*-value near 1, indicating that there was no second bump denoting the controlled turns. Yet all participants shifted these values. Participants 1–3 shifted them for one foot, but participants 6 and 8 shifted them for both feet. The patterns will be individually shown in the Appendix Figs, where we see the profound changes in the gait patterns these children experienced with this trial. (B) Age- and sex-matched controls show the normative data so we can appreciate the departure of the PMS participants from it.

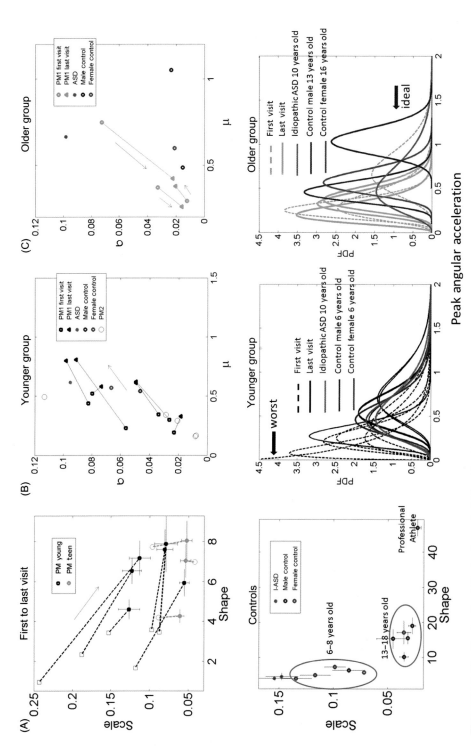

FIGURE 7.48 **Stochastic shifts in the probability distribution functions of bodily patterns or peak angular acceleration with the clinical trial.** (A) Nine participants at baseline (open marker) shifted the Gamma parameters with the drug (filled marker). Black are the PMS young (5–9 years old) while the green are the older PMS (12–16 years old). Bottom panel are controls of various ages to match the two groups (PMS-Y and PMS-O, with the additional patterns of the athlete (professional martial artist). The three iASD are also plotted for reference. They are the points with the PDFs with the highest dispersion (NSR) and the lowest shape values (highest skewness) despite their age (older than 12 years old). (B) The empirically estimated Gamma moments on the top panel for the PMS-Y and representative controls of similar age (one of each sex) plus the iASD for reference. Bottom panel are the PDFs corresponding to the PMS-Y. (C) The changes in PMS-O (top panel) with the corresponding PDFs (bottom panel). Controls of matching age are also plotted. Note the ideal PDF marked with an arrow.

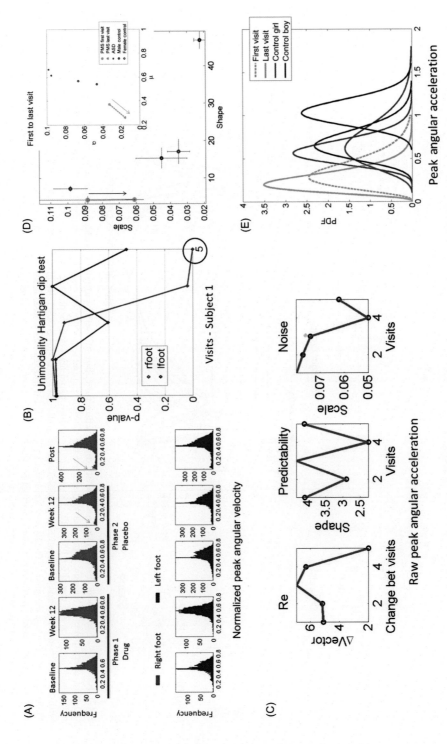

FIGURE A7.1 **Delayed changes in patterns with drug visit.** This subject is one of the two teenage females (16 years old). She started with the drug in the Phase one of the trial. (A) Marked changes in feet patterns appeared later in the placebo Phase of the trial upon the drug Phase had ended. A second bump grouping high speed rotations on average appeared in the frequency histogram of peak angular velocities. This histogram was unimodal during the baseline visit and also 12 weeks after the drug visit. This bump of faster turns became more pronounced in week 12 for the right foot (non-significant p-value) and turned significant $p < 0.01$ in the post-trial visit. By then the bump also appeared in the left foot histogram (although not reaching statistical significance) as shown in (B). The changes are quantified in the p-values obtained using the Hartigan's dip test of unimodality across the visits. Changes in the p-value quantify the significance of the failure of the distribution's unimodality. (C) The underlying angular acceleration paterns across the body also showed fluctuations between visits. In particular the fourth visit shows a drop in the noise 12 weeks upon the drug. The predictability also dropped, raising its value in the post-trial visit. (D) The drop in the noise (marked by arrow) can be appreciated in the change between the first and last visits, away from the 6-year-old control and toward the 16-year-old female control. The controls of comparable age however fall on the Gamma parameter plane far away from this participant with much lower level values of the NSR and higher values of the shape (towards the ideal athlete case). The inset shows an increase in the Gamma variance and the Gamma mean. The shift in the mean toward higher values accounts for the drop of the NSR (the scale) on the Gamma parameter plane. (E) The PDF graph shows that the change in the shape parameter from the first (dashed line) to the last (full line) visit are away from the patterns of the controls. The speed of the angular rotations on average turned faster, away from the lower rates of change values of the controls. Notice that the reduction in the NSR (the scale) in D was not due to a reduction on the variance (which increased) but rather attained by an increase in the value of the mean angular velocity, due to faster rotations (which bring the normalized peak angular velocity ratio to a lower value shifting it from right to left along the mean axis in the inset graph of panel D).

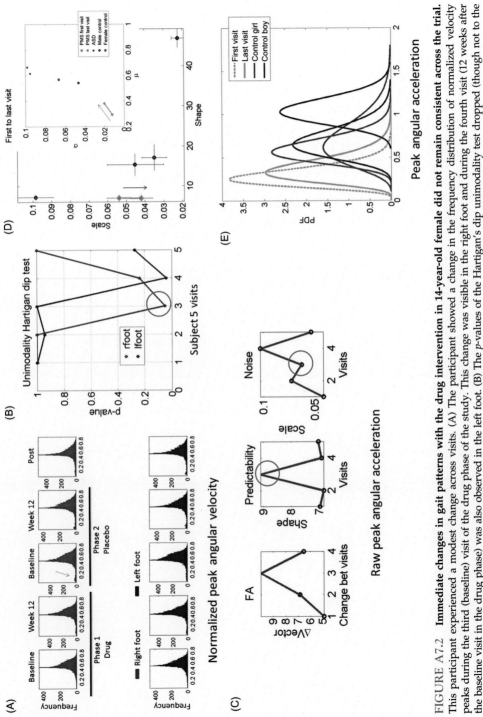

FIGURE A7.2 **Immediate changes in gait patterns with the drug intervention in 14-year-old female did not remain consistent across the trial.** This participant experienced a modest change across visits. (A) The participant showed a change in the frequency distribution of normalized velocity peaks during the third (baseline) visit of the drug phase of the study. This change was visible in the right foot and during the fourth visit (12 weeks after the baseline visit in the drug phase) was also observed in the left foot. (B) The *p*-values of the Hartigan's dip unimodality test dropped (though not to the significant levels) $p > 0.05$ suggesting a change in the shape of the unimodal distribution toward additional bumps. (C) An increase in predictability for the angular acceleration peaks underlying the changes in angular speed is visible in the third visit which also coincides with a drop in the noise. (D, E) The Gamma parameter plane of estimated shape and scale values for the normalized peak acceleration reflects the drop in the noise-to-signal level for this (female) participant towards those of the female control her age. This change is rather modest but brings the values of the normalized peak acceleration closer to typical ranges than those in the first visit. By the last visit the probability distribution function derived from the normalized peak angular accelerations across the body had shifted (dashed green to full green curve) toward PDFs of the 16-year-old female control (red).

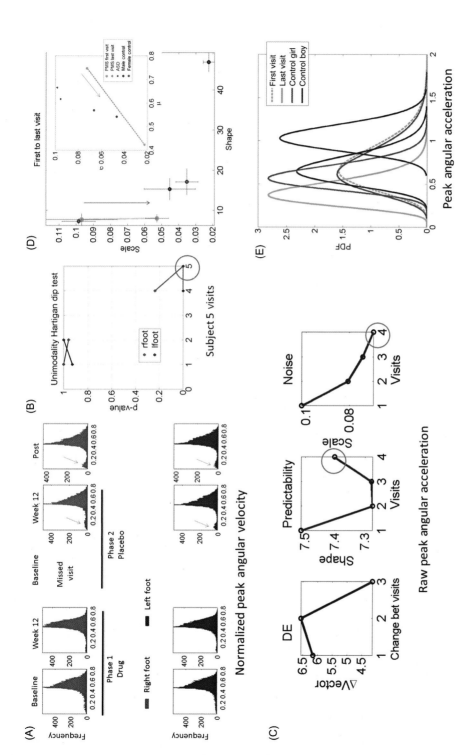

FIGURE A7.3 **Marked changes with drug intervention in 8-year-old male with PMS remain post-trial.** (A) Frequency histograms of the normalized peak angular velocity of the feet. As with all PMS participants, the initial patterns are unimodal. In the fourth visit (participant missed the third visit) of the drug Phase (12 weeks after the baseline-drug visit) this participant showed a well-defined bump of faster rotations of the left foot (marked by the arrow) and a modest bump in the patterns of the right foot as well. These bumps are indicative of an increase in the frequency of faster turns, which are present in typical controls. By the post-trial visit both bumps corresponding to faster and slower joint speed of the angular rotations of the feet were well defined in both feet. (B) These patterns in the frequency histograms of normalized peak angular velocity attained significance for the left foot in the third visit and for both feet in the post-trial visit, according to the drop in the p-value $\ll 0.01$ of the Hartigan's dip unimodality test. (C) The analyses of the peak angular acceleration underlying the peak angular velocity across visits show fluctuations across the two phases of the trial. A drop in the noise-to-signal (scale values) was quantified (circled) in the post-trial visit along with a modest change in the shape of the distribution. (D) The patterns comparing the first to the last (post-trial visit) are shown on the Gamma parameter plane corresponding to the estimated shape and scale parameters averaged across the 15 joints of the body. Noise-to-signal (scale) levels were comparable to those of a 6-year-old female control (red dot) in the first visit. They dropped to the levels of an 8-year-old male control (blue dot), but never reached the shape values of the control. The inset plots the estimated Gamma mean and variance values and the shift experienced from the first to the last visits. The arrow marks the direction of the shift along the mean and variance axes of this statistics parameter plane. The value of the mean peak angular acceleration changed towards lower values (away from the controls plotted in red (female) and blue (male)). The values of the variance moved towards lower values (closer to the controls and away from the values of the two participants with idiopathic ASD, plotted in magenta). The changes are also reflected in the PDF curves (E) where the shape of the distribution changed from the first visit (dotted line) to the last visit (full line).

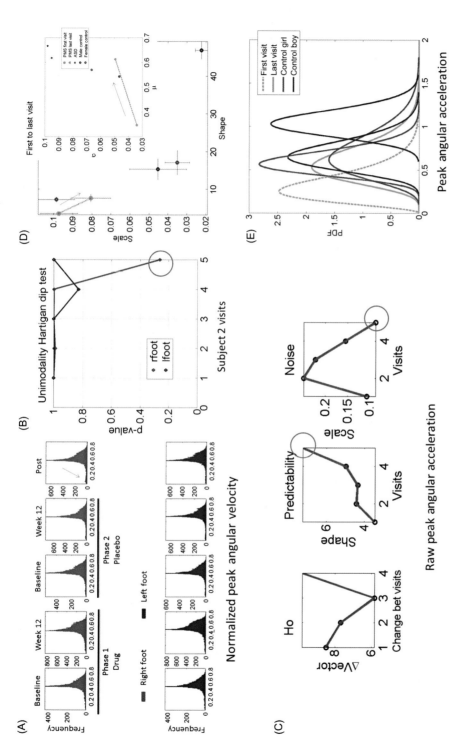

FIGURE A7.4 This participant (6-year-old female) showed improvements in the feet patterns (A) in the post-trial visit (a small bump appeared in the patterns of the right foot, marked by the arrow). (B) The course of the *p*-value from the Hartigan's dip test of unimodality shows the change (non-significant $p > 0.05$) but nonetheless affected by the drug. (C) The patterns of angular acceleration changed from visit to visit with an overall trend that increased the predictability and lowered the noise. By the post-trial visit these patterns showed a marked improvement relative to the first visit. The shape and scale values shifted toward the desirable typical regimes. (D) The signatures of normalized peak angular velocity shifted toward those of the female control of similar age and in the direction of the older controls as well. The noise-to-signal values dropped and the shape values increased in magnitude. This is best appreciated in the PDF plots (E) where the curve shifts towards that of the 6-year-old female control and away from the original curve at baseline.

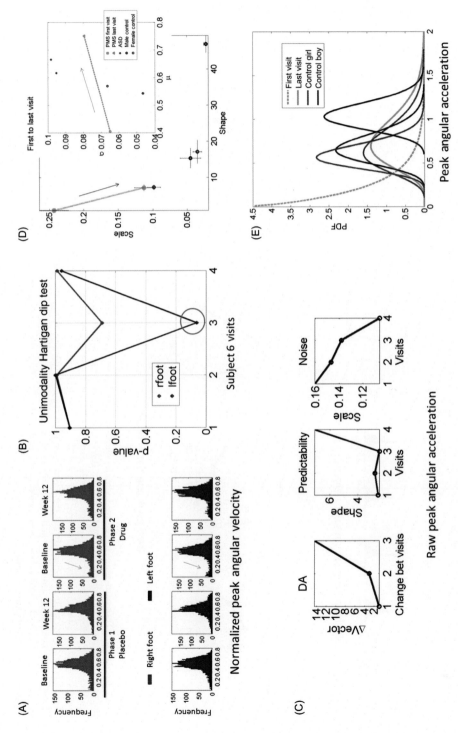

FIGURE A7.5 **Immediate and pronounced change with the drug in 7-year-old male with PMS was transient.** This participant experienced the largest statistical change of the group. His patterns shifted from exponential (random and noisy) to skewed distributions of normalized peak angular velocities. The latter are close to those of a representative typical (younger) child of comparable age in the control group. (A) The third visit corresponding to the baseline epoch of the drug phase of the trial showed a bump in the frequency histogram of the normalized peak velocity values of the feet, particularly of the left foot. (B) The p-value from the Hartigan's dip test of unimodality was not at the significant level in the third visit, yet the change was quite dramatic compared to the first visit. The p-value regressed closer to those of the first visits by the week 12 of the drug Phase, so the change was not retained. (C) The patterns of peak angular acceleration underlying the peak angular velocity also changed. Most notably was a shift in the estimated shape and scale parameters plotted on the Gamma parameter plane. Along the shape axes changes in the predictability of the angular acceleration were registered by the fourth visit. These were accompanied by a drop in the noise levels which decreased on average across the body. (D) The Gamma parameter plane shows the shift of the estimated shape and scale parameters corresponding to the fluctuations in angular acceleration across the body, taken from the first and the last visits. These estimated parameters and 95% confidence intervals are the average across the 15 joints registering motions. They shifted closer to those of the 6-year-old representative girl than to the patterns of the 8-year-old male control. The inset plots the estimated Gamma statistics which show that the drop in the Fano Factor or NSR (the Gamma scale parameter) was primarily due to the increase in the mean value of the peak acceleration. This child had a modest increase in the variance in his angular acceleration but a large change on the mean value. (E) This change is also appreciated in the PDF curves graphing the dramatic switch from exponential (the most random distribution) to skewed-log normal distribution (green dashed to full traces).

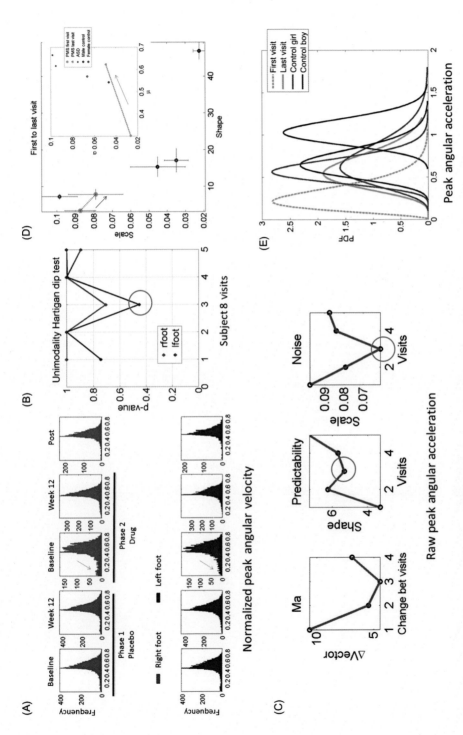

FIGURE A7.6 **Immediate changes with the drug intervention for 6-year-old female with PMS do not remain in the feet but overall improvements across the body are quantified** This participant experienced a large change in motion patterns that aligned the patterns closer to those of a control of similar age. (A) During the third visit (baseline drug) in the second phase, this participant developed a different shape of the frequency histogram of both feet. (B) This is reflected in the drop of the p-value (nonsignificant, $p > 0.05$) for the Hartigan's dip unimodality test. The rest of the trial and the post-trial visit returned to the original feet patterns. (C) The signatures of peak angular acceleration changed from visit to visit showing an overall increase in predictability and a decrease in the NSR. Visit 3 had the largest drop in noise which accompanied the change in the frequency distribution of the feet's normalized velocity peaks (the emergence of the bump of faster rotations). (D) The Gamma parameter plane from the normalized peak angular accelerations shows the drop in the noise (marked by the arrow) form the first to the last visit. An increase in predictability (shift right) is also shown toward the stochastic signatures of the typical controls. The inset shows the estimated Gamma statistical parameters. The large increase in the mean normalized peak angular acceleration accounts for the drop in the Gamma scale parameter (the Fano Factor computed using the variance over the mean ratio). Here the variance has a modest increase in value in relation to the change in the mean value. The trends of these patterns towards the typical control of the same age are also appreciated in the PDF curves in (E).

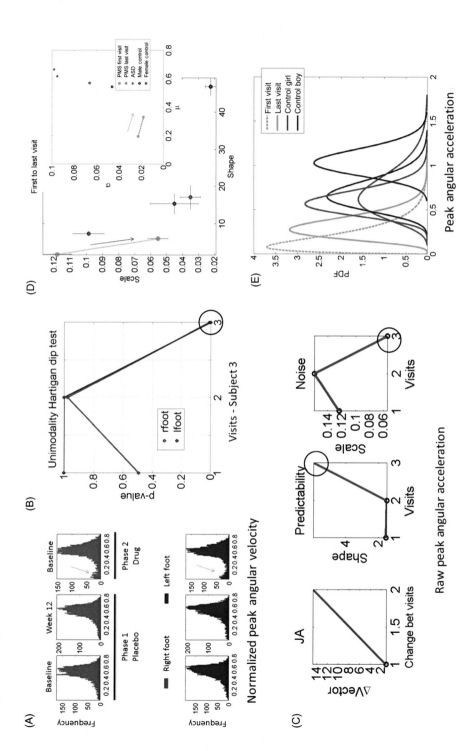

FIGURE A7.7 **Marked immediate changes with the drug intervention** (6-year-old female with PMS). (A) This participant stopped the visits at the baseline of the drug phase, but that visit showed a remarkable change in the feet patterns, transitioning from a completely unimodal frequency distribution to a distribution with a marked second bump of faster rotations. (B) The change was quantified by the Hartigan's dip test of unimodality which output a significant p-value <<0.01 rejecting the null hypothesis of unimodality. (C) The underlying patterns of angular acceleration across the body also showed marked changes in the magnitude of the 15-dimensional vector of joint angular accelerations between visits. The values of the shape parameter shifted towards higher magnitudes indicating higher predictability and the noise-to-signal levels (scale values) dropped indicating increase in reliability of this parameter. (D) The changes in the patterns of normalized angular acceleration across the body shifted away from baseline and towards the patterns of the typical controls. This is well appreciated in the PDF plots (E) where the curve shifted away from baseline and towards the shape of the curves in older controls.

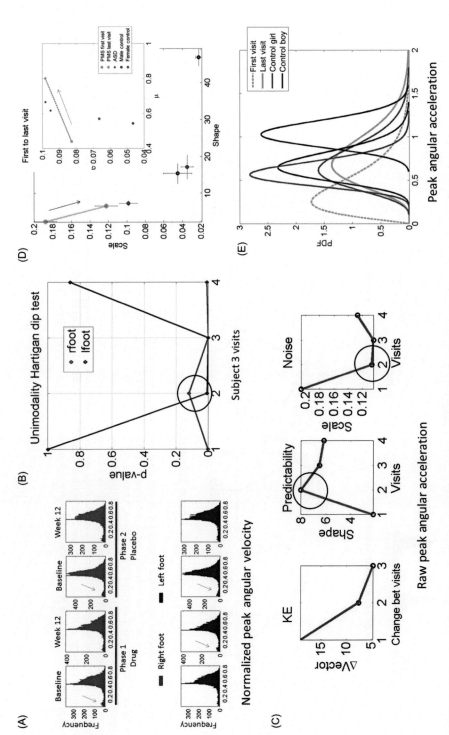

FIGURE A7.8 **Marked immediate changes with the drug intervention (6-year-old female with PMS).** (A) This participant started out with feet patterns that showed both faster and slower rotations for the right foot. Yet these patterns transferred to the left foot and remained significant in the right foot immediately after the first drug visit (baseline drug visit). (B) The change was significant for both feet during the third visit according to the outcome of the Hartigan's dip test of unimodality ($p \ll 0.01$). (C) Changes across visits were also registered by the angular accelerations across the body. The raise in the predictability index (shape of the distribution) is quantified by a drop in the NSR. (D) The shift on the Gamma parameter plane of the normalized peak angular acceleration indicated a drop in the NSR (Gamma scale parameter) and an increase in the value of the Gamma shape parameter. These changes shifted the signatures to the values of the typical control participants of comparable age (6-year-old control female) and in the general direction projecting towards the older teen-age controls as well. The inset figure shows the increase in variability (higher estimated Gamma variance) and the much higher increase in the value of the Gamma scale value. The latter accounted for the decreased in the Fano Factor (variance/mean) explaining the drop in the Gamma scale value. This longitudinal trend can also be appreciated in the PDF graph (E) where the curve corresponding to the last visit falls within those of the age-matched control.

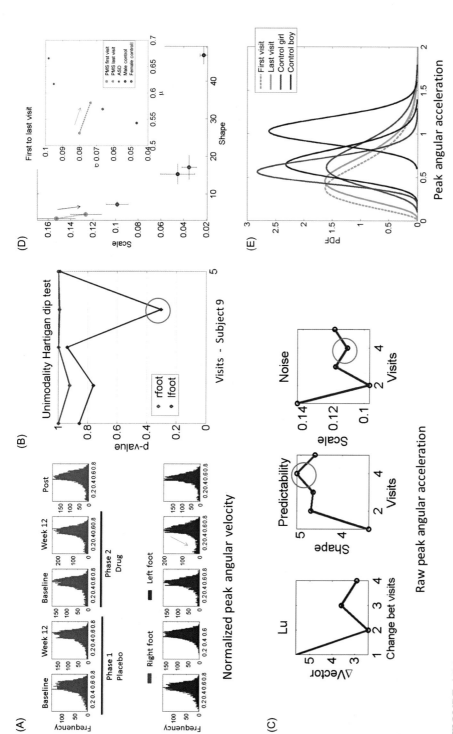

FIGURE A7.9 This participant experienced modest overall change in motion patterns. (A) During the fourth visit, in week 12 of the drug phase, the participant showed a nonsignificant change in the frequency of normalized angular velocity peaks corresponding to faster rotations (smaller values of the normalized parameter). (B) The quantification of the unimodality of the distribution of normalized peak angular velocities for each foot revealed a change (nonsignificant) for the left foot upon 12 weeks of the drug. (C) This change coincided with an overall increase in predictability and a drop in the noise for the peaks in the rates of change of the angular speed. The peak angular acceleration had a change in the shape of the distribution during the 4th visit that indicated more predictable motions with lower NSR than in the previous baseline-drug visit. (D) The Gamma parameter plane with points localizing the empirically estimated shape and scale parameters of the continuous Gamma family of probability distributions shows the drop in the scale (the NSR, or Fano Factor) from the first to the last visit. This change is shown in relation to the neurotypical controls. (E) The probability density functions empirically estimated from the data are also shown across the first and last visit in relation to the 2 controls of similar age (closer in the graph) and the 2 college-age young controls. This child had a more modest change than others in that age group.

behave autonomously (e.g., cannot navigate to get to work or to simply buy food, cannot sleep at night or if it loses mobility in general) and lives with a body in a dysregulated state, then the person's comfort decreases and with that the overall quality of life decreases. Biomarkers that do not alert us of normal vs. dysregulated states of the nervous systems are useless, because they cannot give us indication of change in factors that ultimately influence quality of life.

To define the "normal" state for a person, one must sample that person at its baseline state. Then, one can characterize departures from that state by longitudinally harnessing data from that person and empirically estimating the dynamics of the person's nervous systems without the treatment and with the treatment. As such, the biometrics we build to detect the risks and benefits of a given intervention therapy need to provide a personalized notion of *change* and rates of change in parameters of the nervous systems within contexts that indicate gain (or loss) in self-regulation, autonomy, agency, and self-control. An objective metric-scale of volitional control of the brain over the body, one derived using self-emerging signatures (rather than a priori imposed signatures) will provide useful information in a clinical trial. Everything else derived using subjective means will be plagued with opinions, confirmation bias, and conflict of interests. These may serve some hidden agenda of a researcher, clinician or person from industry; but it will be of very little help to the affected person.

We close this chapter with an invitation to combine forces and integrate multidisciplinary work from basic science, clinical practice and industry sectors to advance the analytical tools, clinical paradigms and technology that can help us help the nervous systems find the path of self-healing. Our lab has initiated a research program to that end (Fig. 7.49).

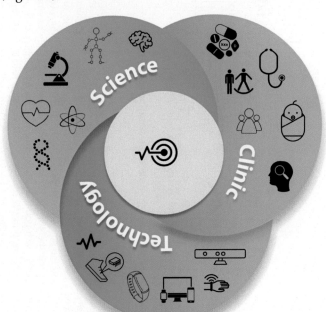

FIGURE 7.49 **Our logo designed by Caroline and her sister and brother in law after I suggested a logo that looked more like a recycling advert.** We decided to open a Research Topic in Frontiers to bring papers of different fields able to collaborate towards the common goal of improving the quality of life of people affected by neurological disorders. The aim? To improve sensory—motor feedback and help develop autonomy of the brain over the body. Technology, clinical, and scientific areas can actively collaborate towards this aim.

References

1. Easterling PE, Knox B. Greek Literature. 1st pbk. ed. The Cambridge History of Classical Literature. Cambridge, UK/New York: Cambridge University Press; 1989.
2. Cox RW. AFNI: software for analysis and visualization of functional magnetic resonance neuroimages. *Comput Biomed Res*. 1996;29(3):162−173.
3. Lord C, et al. Autism diagnostic observation schedule: a standardized observation of communicative and social behavior. *J Autism Dev Disord*. 1989;19(2):185−212.
4. Lord C, et al. The autism diagnostic observation schedule-generic: a standard measure of social and communication deficits associated with the spectrum of autism. *J Autism Dev Disord*. 2000;30(3):205−223.
5. American Psychiatric Association. and American Psychiatric Association. DSM-5 Task Force. *Diagnostic and Statistical Manual of Mental Disorders: DSM-5*. 5th ed. Washington, D.C.: American Psychiatric Association; 2013: xliv, pp 947.
6. Torres EB, Denisova K. Motor noise is rich signal in autism research and pharmacological treatments. *Sci Rep*. 2016;6:37422.
7. Jerrell JM. Pharmacotherapy in the community-based treatment of children with bipolar I disorder. *Hum Psychopharmacol*. 2008;23(1):53−59.
8. Loy, J.H., et al., Atypical antipsychotics for disruptive behaviour disorders in children and youths. Cochrane Database Syst Rev, 2012. 9: p. CD008559.
9. Chang KD. The use of atypical antipsychotics in pediatric bipolar disorder. *J Clin Psychiat*. 2008;69(Suppl 4):4−8.
10. Adler BA, et al. Drug-refractory aggression, self-injurious behavior, and severe tantrums in autism spectrum disorders: a chart review study. *Autism*. 2015;19(1):102−106.
11. Ho JG, et al. The effects of aripiprazole on electrocardiography in children with pervasive developmental disorders. *J Child Adolesc Psychopharmacol*. 2012;22(4):277−283.
12. McDougle CJ, et al. Atypical antipsychotics in children and adolescents with autistic and other pervasive developmental disorders. *J Clin Psychiatry*. 2008;69(Suppl 4):15−20.
13. McCracken JT, et al. Risperidone in children with autism and serious behavioral problems. *N Engl J Med*. 2002;347(5):314−321.
14. Schur SB, et al. Treatment recommendations for the use of antipsychotics for aggressive youth (TRAAY). Part I: a review. *J Am Acad Child Adolesc Psychiat*. 2003;42(2):132−144.
15. Sikich L, et al. A pilot study of risperidone, olanzapine, and haloperidol in psychotic youth: a double-blind, randomized, 8-week trial. *Neuropsychopharmacology*. 2004;29(1):133−145.
16. Hansel K. Rethinking insurance coverage of "experimental" Applied Behavioral Analysis therapy and its usefulness in combating autism spectrum disorder. *J Leg Med*. 2013;34(2):215−233.
17. Donnellan AM. *Progress Without Punishment: Effective Approaches for Learners with Behavior Problems. Special Education Series*. New York: Teachers College Press; 1988. xi, 168 p.
18. Donnellan AM, Leary MR. *Movement Differences and Diversity in Autism/Mental Retardation: Appreciating and Accommodating People with Communication And Behavior Challenges. Movin on Series*. Madison, WI: DRI Press; 1995:107.
19. Foxx RM, Mulick JA. *Controversial Therapies for Autism and Intellectual Disabilities: Fad, Fashion, and Science in Professional Practice*. 2nd ed. New York: Routledge, Taylor & Francis Group; 2016. xix, 568 pages.
20. American Academy of Pediatrics. Sensory integration therapies for children with developmental and behavioral disorders. *Pediatrics*. 2012;129(6):1186−1189.
21. Torres EB. Two classes of movements in motor control. *Exp Brain Res*. 2011;215(3-4):269−283.
22. Torres EB, et al. Neonatal diagnostics: toward dynamic growth charts of neuromotor control. *Front Pediatr*. 2016;4(121):1−15.
23. Torres EB, Heilman KM, Poizner H. Impaired endogenously evoked automated reaching in Parkinson's disease. *J Neurosci*. 2011;31(49):17848−17863.
24. American Psychiatric Association and American Psychiatric Association. Task Force on DSM-IV, *Diagnostic and Statistical Manual of Mental Disorders: DSM-IV*. 4th ed., Washington, DC: American Psychiatric Association; 1994: xxvii, pp 886.
25. Zito JM, et al. Psychotropic practice patterns for youth: a 10-year perspective. *Arch Pediatr Adolesc Med*. 2003;157(1):17−25.

26. Chai G, et al. Trends of outpatient prescription drug utilization in US children, 2002-2010. *Pediatrics*. 2012;130 (1):23−31.

27. Zhang T, et al. Prescription drug dispensing profiles for one million children: a population-based analysis. *Eur J Clin Pharmacol*. 2013;69(3):581−588.

28. Di Pietro N. *The Science and Ethics of Antipsychotic Use in Children*. Boston, MA: Elsevier; 2015.

29. Peng CK, et al. Quantification of scaling exponents and crossover phenomena in nonstationary heartbeat time series. *Chaos*. 1995;5(1):82−87.

30. Stanley HE, et al. Statistical physics and physiology: monofractal and multifractal approaches. *Physica A*. 1999;270(1-2):309−324.

31. Castiglioni P, et al. Local-scale analysis of cardiovascular signals by detrended fluctuations analysis: effects of posture and exercise. *Conf Proc IEEE Eng Med Biol Soc*. 2007;2007:5035−5038.

32. Kaipust JP, et al. Gait variability is altered in older adults when listening to auditory stimuli with differing temporal structures. *Ann Biomed Eng*. 2013;41(8):1595−1603.

33. Hausdorff JM, et al. Altered fractal dynamics of gait: reduced stride-interval correlations with aging and Huntington's disease. *J Appl Physiol (1985)*. 1997;82(1):262−269.

34. Qiu L, et al. Multifractals embedded in short time series: an unbiased estimation of probability moment. *Phys Rev E*. 2016;94(6-1):062201.

35. Terrier P. Fractal fluctuations in human walking: comparison between auditory and visually guided stepping. *Ann Biomed Eng*. 2016;44(9):2785−2793.

36. Botcharova M, et al. Resting state MEG oscillations show long-range temporal correlations of phase synchrony that break down during finger movement. *Front Physiol*. 2015;6:183.

37. Torres EB, et al. Autism: the micro-movement perspective. *Front Integr Neurosci*. 2013;7:32.

38. Denisova K, et al. Cortical interactions during the resolution of information processing demands in autism spectrum disorders. *Brain Behav*. 2017;7(2):e00596.

39. Denisova K, Zhao G. Inflexible neurobiological signatures precede atypical development in infants at high risk for autism. *Sci Rep*. 2017;7(1):11285.

40. Torres EB, Whyatt C. *Autism: The Movement Sensing Perspective*. Boca Raton: Taylor & Francis; 2018. p.

41. Damasio AR, Maurer RG. A neurological model for childhood autism. *Arch Neurol*. 1978;35(12):777−786.

42. Maurer RG, Damasio AR. Vestibular dysfunction in autistic children. *Dev Med Child Neurol*. 1979;21 (5):656−659.

43. Maurer RG, Damasio AR. Childhood autism from the point of view of behavioral neurology. *J Autism Dev Disord*. 1982;12(2):195−205.

44. Leary MR, Hoyle RH. *Handbook of Individual Differences in Social Behavior*. New York: Guilford Press; 2009:624. xv.

45. Hill DA, Leary MR. *Movement Disturbance: A Clue to Hidden Competencies In Persons Diagnosed with Autism and Other Developmental Disabilities*. "Movin' on" Beyond Facilitated Communication. Madison, Wis: DRI Press; 1993:31. viii.

46. Donnellan AM, Hill D, Leary MR. Rethinking autism: implications of sensory and movement differences for understanding and support. Frontiers in Integrative. *Neuroscience*. 2013;6(124):1−11.

47. Minderaa RB, et al. Snout and visual rooting reflexes in infantile autism. *J Autism Dev Disord*. 1985;15 (4):409−416.

48. Reed P. In: Mesmere BS, ed. The Return of the Reflex: Considerations of the Contribution of the Early Behaviorism to Understanding, Diagnosing, and Preventing Autism, *in* New Autism Research Developments. Hauppauge NY: Nova Science Publishers; 2007:19−24.

49. Teitelbaum P, et al. Infantile reflexes gone astray in autism. *J Develop. Learn Disord*. 2002;6:15.

50. Jansiewicz EM, et al. Motor signs distinguish children with high functioning autism and Asperger's syndrome from controls. *J Autism Dev Disord*. 2006;36(5):613−621.

51. Noterdaeme M, et al. Evaluation of neuromotor deficits in children with autism and children with a specific speech and language disorder. *Eur Child Adolesc Psychiat*. 2002;11(5):219−225.

52. Teitelbaum O, et al. Eshkol-Wachman movement notation in diagnosis: the early detection of Asperger's syndrome. *Proc Natl Acad Sci USA*. 2004;101(32):11909−11914.

53. Fournier KA, et al. Motor coordination in autism spectrum disorders: a synthesis and meta-analysis. *J Autism Dev Disord*. 2010;40(10):1227−1240.

54. Gowen E, Stanley J, Miall RC. Movement interference in autism-spectrum disorder. *Neuropsychologia*. 2008;46 (4):1060−1068.
55. Fournier KA, et al. Decreased static and dynamic postural control in children with autism spectrum disorders. *Gait Post*. 2010;32(1):6−9.
56. Minshew NJ, et al. Underdevelopment of the postural control system in autism. *Neurology*. 2004;63 (11):2056−2061.
57. Jones V, Prior M. Motor imitation abilities and neurological signs in autistic children. *J Autism Dev Disord*. 1985;15(1):37−46.
58. Mostofsky SH, et al. Developmental dyspraxia is not limited to imitation in children with autism spectrum disorders. *J Int Neuropsychol Soc*. 2006;12(3):314−326.
59. Rinehart NJ, et al. Movement preparation in high-functioning autism and Asperger disorder: a serial choice reaction time task involving motor reprogramming. *J Autism Dev Disord*. 2001;31(1):79−88.
60. Rogers SJ, et al. Imitation and pantomime in high-functioning adolescents with autism spectrum disorders. *Child Dev*. 1996;67(5):2060−2073.
61. Williams JH, et al. Imitation, mirror neurons and autism. *Neurosci Biobehav Rev*. 2001;25(4):287−295.
62. Haswell CC, et al. Representation of internal models of action in the autistic brain. *Nat Neurosci*. 2009;12 (8):970−972.
63. Gidley Larson JC, et al. Acquisition of internal models of motor tasks in children with autism. *Brain*. 2008;131 (Pt 11):2894−2903.
64. Izawa J, et al. Motor learning relies on integrated sensory inputs in ADHD, but over-selectively on proprioception in autism spectrum conditions. *Autism Res*. 2012;5(2):124−136.
65. Bernstein N. *The Co-ordination and Regulation of Movements*. Oxford: Oxford Press; 1967.
66. Thelen E. Grounded in the world: developmental origins of the embodied mind. *Infancy*. 2000;1(1):3−28.
67. Kalampratsidou, V. and E.B. Torres. *Outcome Measures of Deliberate and Spontaneous Motions. Third International Symposium on Movement and Computing, MOCO'16*. 2016. Thessaloniki, GA, Greece: ACM.
68. Powers SK, Howley ET. *Exercise Physiology : Theory and Application to Fitness and Performance*. 10th ed New York: NY: McGraw-Hill Education; 2018. p.
69. Cosgrove L, et al. Conflicts of interest and disclosure in the American Psychiatric Association's clinical practice guidelines. *Psychother Psychosom*. 2009;78(4):228−232.
70. Cosgrove L, Krimsky S. A comparison of DSM-IV and DSM-5 panel members' financial associations with industry: a pernicious problem persists. *PLoS Med*. 2012;9(3):e1001190.
71. Cosgrove L, et al. Financial ties between DSM-IV panel members and the pharmaceutical industry. *Psychother Psychosom*. 2006;75(3):154−160.
72. Cosgrove L, et al. Tripartite conflicts of interest and high stakes patent extensions in the DSM-5. *Psychother Psychosom*. 2014;83(2):106−113.
73. Cosgrove L, et al. From caveat emptor to caveat venditor: time to stop the influence of money on practice guideline development. *J Eval Clin Pract*. 2014;20(6):809−812.
74. Greenberg DS. Medicare overhaul wins congressional support. *Lancet*. 2003;362(9398):1816.
75. Torres EB, et al. Stochastic signatures of involuntary head micro-movements can be used to classify females of ABIDE into different subtypes of 3 neurodevelopmental disorders. *Front Integr Neurosci*. 2017;11(10):1−17.
76. Volkmar FR, Szatmari P, Sparrow SS. Sex differences in pervasive developmental disorders. *J Autism Dev Disord*. 1993;23(4):579−591.
77. Mandy W, et al. Sex differences in autism spectrum disorder: evidence from a large sample of children and adolescents. *J Autism Dev Disord*. 2012;42(7):1304−1313.
78. Torres EB, et al. Strategies to develop putative biomarkers to characterize the female phenotype with autism spectrum disorders. *J Neurophysiol*. 2013;110(7):1646−1662.
79. Takahashi T, et al. Enhanced brain signal variability in children with autism spectrum disorder during early childhood. *Hum Brain Map*. 2016;37(3):1038−1050.
80. Cheng W, et al. Autism: reduced connectivity between cortical areas involved in face expression, theory of mind, and the sense of self. *Brain*. 2015;138(Pt 5):1382−1393.
81. Falahpour M, et al. Underconnected, but not broken? Dynamic functional connectivity MRI shows underconnectivity in autism is linked to increased intra-individual variability across time. *Brain Connect*. 2016;6 (5):403−414.

82. Alaerts K, Swinnen SP, Wenderoth N. Sex differences in autism: a resting-state fMRI investigation of functional brain connectivity in males and females. *Soc Cogn Affect Neurosci*. 2016;11(6):1002−1016.

83. Supekar K, Menon V. Sex differences in structural organization of motor systems and their dissociable links with repetitive/restricted behaviors in children with autism. *Mol Autism*. 2015;6:50.

84. Schaer M, et al. Sex differences in cortical volume and gyrification in autism. *Mol Autism*. 2015;6:42.

85. Pantelis PC, Kennedy DP. Estimation of the prevalence of autism spectrum disorder in South Korea, revisited. *Autism*. 2016;20(5):517−527.

86. Rogers DM. *Motor Disorder in Psychiatry: Towards a Neurological Psychiatry*. Chichester, UK/New York: J. Wiley & Sons; 1992:159. viii.

87. Esler AN, et al. The autism diagnostic observation schedule, toddler module: standardized severity scores. *J Autism Dev Disord*. 2015;45(9):2704−2720.

88. Hus V, Lord C. The autism diagnostic observation schedule, module 4: revised algorithm and standardized severity scores. *J Autism Dev Disord*. 2014;44(8):1996−2012.

89. Hus V, Gotham K, Lord C. Standardizing ADOS domain scores: separating severity of social affect and restricted and repetitive behaviors. *J Autism Dev Disord*. 2014;44(10):2400−2412.

90. Gotham K, Pickles A, Lord C. Standardizing ADOS scores for a measure of severity in autism spectrum disorders. *J Autism Dev Disord*. 2009;39(5):693−705.

91. Gotham K, et al. The autism diagnostic observation schedule: revised algorithms for improved diagnostic validity. *J Autism Dev Disord*. 2007;37(4):613−627.

92. Mosconi MW, Sweeney JA. Sensorimotor dysfunctions as primary features of autism spectrum disorders. *Sci China Life Sci*. 2015;58(10):1016−1023.

93. Mosconi MW, et al. Feedforward and feedback motor control abnormalities implicate cerebellar dysfunctions in autism spectrum disorder. *J Neurosci*. 2015;35(5):2015−2025.

94. Brincker M, Torres EB. Noise from the periphery in autism. *Front Integr Neurosci*. 2013;7:34.

95. Torres EB, et al. Toward precision psychiatry: statistical platform for the personalized characterization of natural behaviors. *Front Neurol*. 2016;7:8.

96. Torres EB, Donnellan AM. Autism: The Movement Perspective. In: Torres EB, Donnellan AM, eds. *Front Integr Neurosci*. 2012:1−374.

97. Donnellan AM, Hill DA, Leary MR. Rethinking autism: implications of sensory and movement differences for understanding and support. *Front Integr Neurosci*. 2012;6:124.

98. Whyatt C, Craig C. Sensory-motor problems in autism. *Front Integr Neurosci*. 2013;7:51.

99. Whyatt CP, Craig CM. Motor skills in children aged 7-10 years, diagnosed with autism spectrum disorder. *J Autism Dev Disord*. 2012;42(9):1799−1809.

100. Lord, C., et al., *Autism Diagnostic Observation Schedule ADOS Manual*, 2012. Western Psychological Services (WPS), Torrance, CA,140 pages.

101. Whyatt, C. E.B. Torres. *The Social-Dance: Decomposing Naturalistic Dyadic Interaction Dynamics to the 'Micro-Level'. in Fourth International Symposium on Movement and Computing, MOCO'17*. 2017. London, UK: ACM.

102. Wilson HL, et al. Molecular characterisation of the 22q13 deletion syndrome supports the role of haploinsufficiency of SHANK3/PROSAP2 in the major neurological symptoms. *J Med Genet*. 2003;40(8):575−584.

103. Phelan MC, et al. 22q13 deletion syndrome. *Am J Med Genet*. 2001;101(2):91−99.

104. Durand CM, et al. Mutations in the gene encoding the synaptic scaffolding protein SHANK3 are associated with autism spectrum disorders. *Nat Genet*. 2007;39(1):25−27.

105. Moessner R, et al. Contribution of SHANK3 mutations to autism spectrum disorder. *Am J Hum Genet*. 2007;81(6):1289−1297.

106. Bonaglia MC, et al. Molecular mechanisms generating and stabilizing terminal 22q13 deletions in 44 subjects with Phelan/McDermid syndrome. *PLoS Genet*. 2011;7(7):e1002173.

107. Roussignol G, et al. Shank expression is sufficient to induce functional dendritic spine synapses in aspiny neurons. *J Neurosci*. 2005;25(14):3560−3570.

108. Von Holst E, Mittelstaedt H. The principle of reafference: interactions between the central nervous system and the peripheral organs. In: Dodwell PC, ed. *Perceptual Processing: Stimulus Equivalence and Pattern Recognition*. New York: Appleton-Century-Crofts; 1950:41−72.

109. Von Holst E. Relations between the central nervous system and the peripheral organs. *Br J. Anim Behav*. 1954;2(3):89−94.

110. Raab M, Boeckers TM, Neuhuber WL. Proline-rich synapse-associated protein-1 and 2 (ProSAP1/Shank2 and ProSAP2/Shank3)-scaffolding proteins are also present in postsynaptic specializations of the peripheral nervous system. *Neuroscience*. 2010;171(2):421−433.

111. Aldinger KA, et al. Cerebellar and posterior fossa malformations in patients with autism-associated chromosome 22q13 terminal deletion. *Am J Med Genet A*. 2013;161A(1):131−136.

112. Soorya L, et al. Prospective investigation of autism and genotype-phenotype correlations in 22q13 deletion syndrome and SHANK3 deficiency. *Mol Autism*. 2013;4(1):18.

113. Bisi MC, Stagni R. Development of gait motor control: what happens after a sudden increase in height during adolescence? *Biomed Eng Online*. 2016;15(1):47.

114. Sutherland DH, et al. The development of mature gait. *J Bone Joint Surg Am*. 1980;62(3):336−353.

115. Wu, D., E.B. Torres, and J.V. Jose. Peripheral micro-movements statistics leads to new biomarkers of autism severity and parental similarity. *The Annual Meeting of the Society for Neuroscience*. 2014. Washington DC.

116. Leblond CS, et al. Meta-analysis of SHANK mutations in autism spectrum disorders: a gradient of severity in cognitive impairments. *PLoS Genet*. 2014;10(9):e1004580.

117. Betancur C, Buxbaum JD. SHANK3 haploinsufficiency: a "common" but underdiagnosed highly penetrant monogenic cause of autism spectrum disorders. *Mol Autism*. 2013;4(1):17.

118. Battaglia A. Sensory impairment in mental retardation: a potential role for NGF. *Arch Ital Biol*. 2011;149 (2):193−203.

119. Gentili RJ, et al. Brain biomarkers of motor adaptation using phase synchronization. *Conf Proc IEEE Eng Med Biol Soc*. 2009;2009:5930−5933.

120. Aydore S, Pantazis D, Leahy RM. A note on the phase locking value and its properties. *Neuroimage*. 2013;74:231−244.

121. Sporns O. *Networks of the Brain*. Cambridge, MA: MIT Press; 2011. xi, 412 p., 8 p. of plates.

122. Sporns O. *Discovering the Human Connectome*. Cambridge, MA: MIT Press; 2012. p.

123. Mandelbrot B, Van Ness J. Fractional Brownian motions, fractional noises and applications. *SIAM Rev*. 1968;10(4):422−437.

124. Delignieres D, et al. Fractal analyses for 'short' time series: a re-assessment of classical methods. *J Math Psychol*. 2006;50:525−544.

125. Chen Z, et al. Effect of nonstationarities on detrended fluctuation analysis. *Phys Rev E Stat Nonlin Soft Matter Phys*. 2002;65(4 Pt 1):041107.

126. Peng CK, et al. Fractal mechanisms and heart rate dynamics. Long-range correlations and their breakdown with disease. *J Electrocardiol*. 1995;28(sssuppl):59−65.

127. Hurst HE. Long-term storage capacity of reservoirs. *Trans Am Soc Civil Eng*. 1951;116:770.

APPENDIX

The following document outlines sources for costs presented in the ADOS_Costs document as well as a description of how values were selected.

Autism Diagnostics by Site

This table collects different types of autism diagnostics used throughout the Autism Brain Imaging Data Exchange (ABIDE) databases (http://fcon_1000.projects.nitrc.org/ indi/abide/). This data was accessed via the NeuroImaging Informatics Tools and Resources Clearinghouse (NITRC) site on January 31, 2017 (http://www.nitrc.org/ include/about_us.php). Numbers reported in this document reflect data collected from ABIDE 1 and ABIDE 2 prior to this date. Please note that the two new sites as released by the ABIDE group on March 27, 2017 are not reflected in this table.

Subjects with the following score reports are included:

- ADI-R: The Autism Diagnostic Interview-Revised (Lord et al., 1994)
- ADOS: Autism Diagnostic Observation Schedule (Lord et al., 2000)
- ADOS-2: Autism Diagnostic Observation Schedule-2nd Edition (Lord et al., 2012)
- ADOS_Gotham: Autism Diagnostic Observation Schedule Revised Algorithm (Gotham et al., 2007) (https://www.ncbi.nlm.nih.gov/pubmed/17180459)
- ADOS_G: Autism Diagnostic Observation Schedule Generic (or ADOS-2 Module 4)

ADI-R

All cost estimates were obtained from the following site:
https://www.wpspublish.com/store/p/2645/adi-r-autism-diagnostic-interview-revised

ADOS

Booklet costs: https://www.wpspublish.com/store/p/2647/ados-autism-diagnostic-observation-schedule

No available prices were published for the ADOS additional resources. Costs were estimated using pricing for ADOS-2. https://www.wpspublish.com/store/p/2648/ados-2-autism-diagnostic-observation-schedule-second-edition

ADOS_Gotham

Booklet costs were estimated using the ADOS booklet costs: https://www.wpspublish.com/store/p/2647/ados-autism-diagnostic-observation-schedule

No available prices were published for the ADOS additional resources. Costs were estimated using pricing for ADOS-2. https://www.wpspublish.com/store/p/2648/ados-2-autism-diagnostic-observation-schedule-second-edition

ADOS_G

Booklet costs were estimated using the ADOS booklet costs: https://www.wpspublish.com/store/p/2647/ados-autism-diagnostic-observation-schedule

No available prices were published for the ADOS additional resources. Costs were estimated using pricing for ADOS-2. https://www.wpspublish.com/store/p/2648/ados-2-autism-diagnostic-observation-schedule-second-edition

ADOS-2

All cost estimates were obtained using publically provided cost information from the official ADOS-2 seller:
https://www.wpspublish.com/store/p/2648/ados-2-autism-diagnostic-observation-schedule-second-edition

Demographic Infor PMS participants and Controls (note de-identified letter and number codes are internal to our lab and has no relation to names or personal info)

Subject	Chronological Age (Years)	Sex	Subject Type
FINISHED TRIAL—PHASE I			
1 re	16.9	F	PMS—finished trial
2 fa	14.4	F	PMS—finished trial
3 de	9.2	M	PMS—finished trial
4 ho	6.4	F	PMS—finished trial
5 da	7.7	M	PMS—finished trial
6 ma	5.4	F	PMS—finished trial
7 ja	8.7	F	PMS—quit trial 3rd vis
8 ke	6.0	F	PMS—finished trial
9 lu	5.0	M	PMS—finished trial
NEUROTYPICAL CONTROLS			
17 −6 (sib)	5	F	Neurotypical (sib)
18 −6 yo	6	F	Neurotypical
19 −7 yo	7	F	Neurotypical
20 −8 yo	8	M	Neurotypical
21 −13 yo	13	M	Neurotypical
22 −7 yo (sib)	7	M	Neurotypical (sib)
23 − TO	16	F	Neurotypical
24 − UY	19	M	Neurotypical (College)
25 − CP	18	F	Neurotypical (College)
26 −ES	16	M	Neurotypical

Adding Dynamics to the Principle of Reafference: Recursive Stochastic Feedback Closed Control Loops to Evoke Autonomy

"That rug really tied the room together, did it not?" —**Walter Sobchak, The Big Lebowski**
(1998)

PART I EVERYTHING IS SOUND

When I was a postdoctoral fellow at the California Institute of Technology (CALTECH), for learning the recording techniques of brain electrophysiology I used to listen to the spike trains of the cortical neurons I was recording from the Posterior Parietal Cortex (the PPC) of rhesus monkeys. I recall being in awe the first few weeks I spent with Chris Buneo, the postdoctoral fellow that trained me, because he could know several properties of the neurons he was recording from, just by listening to the spikes. While searching for neurons with good reach-planning activity in the Parietal Reach Region of the PPC—a region that had been well characterized in Richard Andersen's lab[1–3]—he would say to me, "Here Liz, this neuron likes it up and to the right." That meant, relative to the monkey's hand, the neuron was maximally responsive (as it spiked a lot) when the animal was planning the hand reach toward a target in that region of space. Initially, I could not come to grips with such an ability. Yet, within weeks of learning the ropes of these electrophysiological recordings, I noticed that I too could begin to tell which region of space a given neuron liked best, in the "impending-reach" sense. The animal had not made a reaching motion yet, it was merely thinking ahead, so it was not exactly the execution of the reach that I was studying at that point. It was the intention to reach.

I put sensors on the body of the animals that I was recording cortical spiking activities from and registered their bodily motions continuously, so I could see what transpired before the animal made the reaches, that is, as these parietal neurons planned over a second ahead of the action, something about the arm and the hand motions to a visual target in the animal's peripersonal space. As it turned out, the biorhythms output by these sensors revealed that underlying the strong responses the cell had to the visual stimuli and to the plan locating where to reach a second later, there were many subtle postural readjustments prior to the reach. I was recording these micro-motions with biosensors that sampled at 120 Hz (Polhemus Fastrak, Colchester, VT), but I only had 4 of those located at the shoulder, upper arm, forearm, and at the hand. I could not know what other body parts were doing during the preparatory activity before the target was cued. I could, however, see what the arm was doing and recorded continuously those anticipatory postural changes. They had never before been studied in relation to PRR neurons. This was in contrast to studies involving cells in area 5, on the surface of the PPC, where a postural code had been reported.[4–7] Those experiments had probed two different postures that the experimenter passively changed with an apparatus. The experiment I was running involved the avoidance of obstacles in the path the hand would have to take straight to the target. As such, the curved trajectories of the hand evoked a total reconfiguration of the arm joints, and the paths in joint angle space were very different than those from any of the plain straight reaches the animals had performed. For plain reaches, that is, reaches straight to the targets located on the board that I was using, there was no need to change the initial body posture. We had never seen the PRR cells respond to such postural rearrangements during the baseline activity, even prior to the cue indicating the target for reach. I tried two situations, that is, letting the animals reconfigure the arm posture on their own and then passively changing the initial arm posture with the help of an apparatus we built, to closely match the posture that the sensors had systematically recorded during this anticipatory phase of the experiment. In this way, I could at least learn if the anticipatory postural rearrangement that the animals self-initiated had a similar neural code than those passively imposed. At some level they did. In both cases the cell had an overall qualitative response that was similar during both postural rearrangements, but the difference emerged during the execution of the actual reach.

Fig. 8.1 shows a posterior parietal cortex cell with tuning activity across the space we probed. These activities on the color-coded board correspond to the planning (so called memory) epoch prior to the reaching action. There one can see the schematic of the arm showing the anticipatory postural adjustment that we recorded from the arm sensors during the baseline epoch, 300Ms before the visual target was cued. Because the animal had seen the obstacle set up relative to the starting position of the hand (in the center of the vertically oriented board) and it could guess the impending path around this obstacle, the postural rearrangement was compliant with the required rotation of the wrist to avoid the obstacle on the way to potential targets on that side of the work space.[8] The activity of this neuron increased dramatically across the board, but it did much more so on the potential locations at the bottom of the board, switching the cell's original preferred direction (down and to the left) to the opposite site (down and to the right).

I have yet to complete the analyses of all that neural data I gathered for several years from two animals. I plan to do this during my upcoming sabbatical. But the remarkable

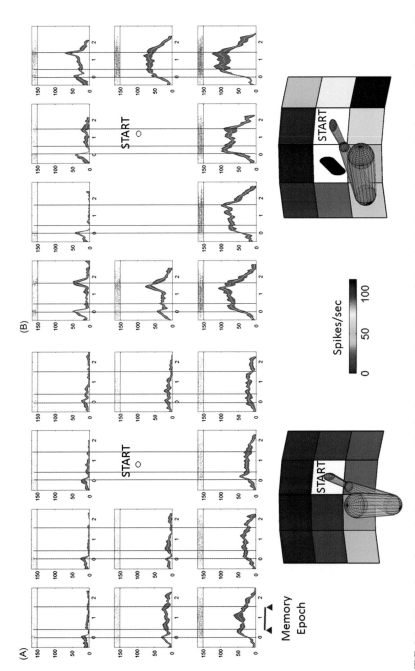

FIGURE 8.1 The spikes of a cortical neuron sampled in the context of a task consisting of pointing motions to visual targets in the animal's peripersonal space. (A) Direct reaches from the START location to 11 locations around the center. The points between the two horizontal lines in each panel represent spikes while the curves show the result of convolving the spikes with a Gaussian Kernel and averaging the trials aligned to the touch of the button at the START location. The vertical lines show the epochs of the experiment. This is the baseline activity from the animal waiting (300Ms), followed by the visual cue (300Ms). Then there is a variable period of up to a second with random variations in the time length of the period when the animal must plan the reach to the target that was cued in the visual-cue epoch. A light in the center of the board indicates the time to initiate the execution of the reach, which in this case was allotted enough time for the animal to do it at its own comfortable pace. The board with colors indicates the average firing rate of the memory period, with a preference for the location down and to the left of the board at 50 spikes per second. The arm posture was recorded with sensors at the shoulder, upperarm, forearm, and the hand. (B) Positioning an obstacle (OB) along the hand path that would otherwise reach the target along a near straight line changes the firing rates of the neuron. In this case, the presence of the OB increases the firing rates even at the baseline epoch. Then during the memory period, the cell's preferred location switches to the opposite side of the board (down and to the right). The postural configuration of the arm recorded by the sensors also changes, even before the visual cue is ON.

fact that I acquired a sense of the neuron's spatial tuning by combining the spatial maps I was plotting with the spike's sound maps across the space prompted me to think about movement planning and sound in very different ways than one would normally do. Essentially, I began to formulate in my mind a notion that every biophysical rhythm we record in electrophysiology as time series of fluctuations in a signal could be treated as a point process (much like we treated the neuronal spikes), and as such, converted to sound. I reasoned that after some time recording the movement activities of a person, and anchoring it to a repeatable behavioral landmark that we defined (as a point of reference), we could begin to identify periodic activities using the waveforms of the motions we were capturing. Fig. 8.2 provides an example of a simple pointing task where I aligned the trials

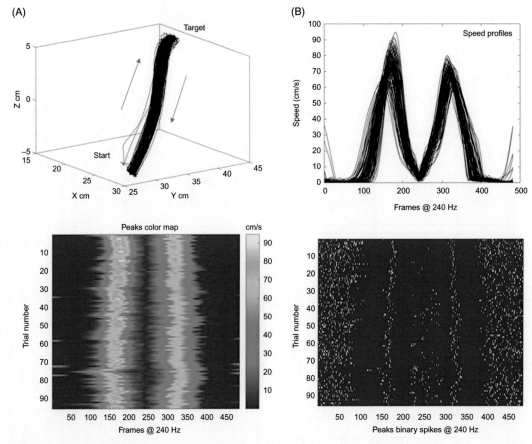

FIGURE 8.2   The spikes from our biorhythms. A pointing task can be examined from trial to trial as a periodic motion. (A) Three dimensional trajectories of the hand from a start location to a target location. (B) The speed profiles of the hand trajectories to the target and back to rest are aligned to the touch event and superimposed trial by trial. (C) Heat colormaps of the motions with color scale indicating the range of physical speeds (cm/s). Maxima are highlighted in yellow. (D) Spikes from the peaks (local and global maxima) of the speed temporal profiles reveal periodic patterns.

to the touch of the target recorded by a touch screen. By simply plotting these activities as heat maps and trial by trial extracting the peaks, I could see the self-emerging periodic patterns that resembled the neuronal spike code. This signal from the bodily motions, I thought, could be converted to sound and played back to the participant. But achieving that in real time proved to be a major feat.

It involved the synchronization of multiple biosensors to eliminate as much as possible temporal disparities in the acquisition of the signal. It also involved understanding differences in sampling resolution across different sensor types and different physical units (e.g., degrees *vs.* cm, or m/s^2 vs. μV, etc.) and varying internal delays of the equipment processing the data and performing the computations to reparameterize the signals from the nervous systems and feed them back to the end user of our interfaces. When one transitions from open loop settings, where the data are only examined a posteriori, and instead closes the loop in experiments that attempt to witness the immediate nervous systems reactions to external and internally sensed stimuli (including kinesthetic reafference from self-generated motions), everything becomes more complex. These technical issues at the front end of the data acquisition and its instantaneous playback were undertaken one by one by the lab members, while at the same time empirically informing us at each step, through experimentation, what the person's reaction was to all manipulations in the audio-visual domain.

For the initial open loop version of the pointing task, we placed the position sensor on the back of an iPad to record both the touch of the finger at the target and the physical position of the target on the iPad. I also placed sensors around the corners of the table where the iPad rested, to know the physical positions of the iPad in relation to the global frame of reference recording the motions of the person performing the task. Once I had all elements of the task expressed relative to the same coordinate system, I proceeded to record the motions of the participants, as they made pointing movements to a target displayed on the iPad. An important component of this experiment was that the flow of motion was self-controlled, meaning that the person moved at his/her own comfortable pace. This may seem like a trivial remark, but it is actually very important to point out this distinction from what the rest of the motor control field does.

Typically, experiments that involve pointing motions to a visual target preprogram the duration of the reach. This is a very hard constrain on the nervous system, and it is one that assumes once again an *one size fits all* model. Every person has anatomical variations that lead to slightly different dynamics and different internal transduction and transmission delays for sensory-motor signals. As such, the duration of the reach is not the same for everyone. It is not even the same for repetitions of the same trial by the same participant.

In experiments of motor control, this disparity is somewhat alleviated by pretraining the participant before the experiment begins so that the person attains the predefined duration of the trajectory. The participant makes many pointing motions to the visual target until it perfects the desired time duration of a trajectory. This may lead to some level of fatigue, which is never mentioned in the literature; but it is something we must seriously consider in our clinical cases. Some of the theoretical motivations underlying assumptions of a predefined time course of the motion were discussed in Chapter 7 within the context of trajectory formation models. The empirical data that I have collected and analyzed over the span of 20 years point to the need for different models and different experimental approaches altogether. Here, besides the clinical constraints we have when defining our experiments, we also strive for a new theoretical conceptualization of motor control: one that conceives the central and the peripheral nervous systems working in tandem, using a multilayered, multimodal, multifunctional approach. In this new conceptualization of the problems of motor control, we revisit the degree of freedom (DoF) problem and the variability problem that Nikolai Bernstein prosed in the beginnings of the 20th century.[9]

A WINK AND A BLINK CONTAIN A SPONTANEOUS SEGMENT— AND SO DOES A REACH, A WALK, AND EVERY OTHER COMPLEX BODILY MOVEMENT

I long ago decided to study all movements in natural behaviors, without any type of constraint, that is, as a person would normally move during activities of daily living. In the lab our strategy was to place as many sensors as we could afford on the body and simultaneously record the kinematics of the body in motion, the heartbeat, the brain waveforms, the muscles through electromyograpgy (EMG), and so forth. The idea was to catch phenomena invisible to the naked eye and happening so fast or being so subtle that we would not even know it existed. This process of data acquisition, of course, evolved as we acquired more sensors through grants and endowments to the lab, and as several companies donated their sensors to us for testing and evaluation. In the following experiments, I will gradually introduce different examples of sensors we have used and will build up my case, aiming toward a full grid of sensors recording in tandem most activities we can harness noninvasively from the nervous systems in action. The transition from an open-loop exploratory setting to a closed-loop one, where we actually intervene with feedback that we take from the end user of our interfaces, and re parameterize in a personalized manner to alter the person's actions was rather gradual because we explored many technical issues and many issues related to the psychophysics of the motor phenomena.

Returning to the case of pointing, while the sensor at the hand recorded the full trajectories and provided the information to obtain the endpoint variability relative to the target, we also explored other variables. From the hand motion trajectory, we could extract the velocity fields along the paths, and for each of those velocity vectors, compute their magnitude to obtain a temporal profile of the speed of the hand on its way to the target; but we could also continuously record the hand speed profile as the hand returned to rest. This simple motion, which happens without instruction and beneath the person's awareness, turned out to be critical later to help us design our interventions, to evoke agency in children with autism. As mentioned in Chapter 7, this movement segment had been discarded as a nuisance in most experiments of motor control, but it turned out to contain an important clue to help us help the child with autism discover self-control of the thoughts over the actions and acquire a sense of agency.

When from moment to moment the activity repeated some patterns, in due time the contiguity of this process would build and antecedent mark to help begin associating events in the past with present events such that an initially equally probable random event would eventually acquire biases and form probabilistic associations with other events, and as such, would begin to change the relevance of the phenomena to the system: that signal would rise over the background noise level. Importantly, because this is not a mere stimulus response association but rather a continuous random process whereby the re-entrant activity is being self-generated by the nervous system and the external stimuli may come from the external environment or from within, it requires the system to have an inherent ability to discern between the part that it has control over (internally generated) and the part that it does not. Such an ability is not obvious. I believe that its origins may be traced back to conception when the immune system of the mother ought to host the nascent immune system of the fetus. At some point a differentiation of the two systems must occur. That bridge between the nascent immune system and the nascent nervous system is one that we do not have a good understanding of in neuro-developmental disorders in general. Yet, building a basic mechanism to differentiate between self-generated flow and not self-generated flow of activities seems fundamental to the recursive model of multilayered, multimodal and multifunctional reafference that we have in mind.

In our setup, the surprise that we evoked, set the path for the child to self-discover the goal of a task without any instructions. This was similar to when we, as adults, discover what a problem is and redefine an entire line of inquiry. The aha moment is often preceded by an unexpected event that alerts us of a bifurcation point whereby other alternatives we would have never considered suddenly appear in front of us. The process of trial and error exploration the child initiated eventually led to the spontaneous discovery of the goal of the task ahead.

The child was rather surprised when *out of the blue*, a movie played at the computer he was facing. The change of his body position in reaction to the surprise was paired with setting the movie off. But his returning the hand to the same position as before made the movie play back again. Upon systematic verification that his hand could trigger the movie only if he placed it in a certain region of space, the child discovered that placing the hand in that region was the goal that fulfilled a purpose and gave him a reward. This reward, built from the bottom up, using the self-generated movements of the child, was important to evoke and sustain the motivation to carry on the trial and error. The setup had a goal (that the child had to self-discover). The goal led to the playing of the audio-visual media and that had two types of rewards: external, from the external audio-visual stimuli; and internal, from the sense of cause and effect that the child self-discovered. It is this latter one that interests me beyond the stimulus-response association the former can build. The internally triggered reward, recursively produced by the child's self-generated motions, leads to the acquisition of body-part ownership: "This is my hand; I am in control of it because when I move it, I cause this effect of playing the media." In simpler words, the neurotypical controls, one by one, expressed that sense of ownership and self-control. Although the nonverbal children we tested felt this empowerment, they could not articulate it with spoken words. They made gestures and facial expressions that revealed their excitement. The simple task we designed had all the elements we needed to evoke the sense of agency in the child. And it all started with the unexpected event: it all started with a surprise.

Surprises in our thought processes (i.e., unexpected departures from a seemingly logical train of thoughts) may help us think differently and redefine a given problem to advance a line of inquiry. In other words, creative thinking cannot be directly instructed. It has this inherent property of self-emergence that is often preceded by unexpected outcomes.

When we examined the fluctuations in the child's hand signal, we were able to see how the contextual variability of the spontaneously retracting motions proved to upset the baseline variability the system came with to the context of a pointing action. Consequently, the moment by moment spontaneous variations—traditionally treated as superfluous motor noise,[10] contributed nontrivially to the inference of systematic changes in the outcome. These systematic consequences of the child's action were eventually conducive of transitions from spontaneous (equally probable) random noise to well-structured signal. We saw this type of stochastic phase transition in our longitudinal research with neonates in Chapter 1. We saw indeed the emergence of predictive codes from the self-sampling of self-generated motions.[11] This alerted us that a necessary condition for the emergence of agency in a system is to have the sensitivity to context dependent variations, that is, these variations need to be such that the system *takes note* of them at some point, and their accumulation of evidence leads to the understanding of whether the variations were consequential of self-generated actions (from within the system), or came from the outside. Internally, self-generated variations are those through which the system can discover ownership and aim to control. I think that external variations are more difficult to control, unless anchored to an internal source conducive of cause and effect pairings. This sort of hunch (call it an exploratory hypothesis) was something I set to pursue in more detail.

To that end, I used our empirically informed model of movement classes, paired with the notions of surprise, internally generated reward and recurrent kinesthetic reafference.[12] Under this umbrella of multifunctional reafferent levels of control, we could distinguish such contextual variations in the deliberate *vs.* the spontaneous (consequential) motions that co-existed in the athletes' boxing routines, or in the tennis players' serve. The class of spontaneous goal-less movements, which are hidden to the conscious experience of the person—as they transpire largely beneath awareness, proved to be more sensitive to contextual dynamics than the deliberate ones. The deliberate ones were performed under conscious voluntary control. As such, I figured through empirical exploration that the spontaneous motions contributed differently to the adaptability of the system and to the generalization of one stable (context free) solution to turn adjustable to another (new) situation. But in a subtler way, the retraction motions prompted me to think of them as the ones to be modified with minimal reactive resistance from the controller systems in the brain. This was so because manipulating these motions largely *under the conscious radar* posed no resistance to the self-discovery of agency across all 25 nonverbal children with autism spectrum disorder (ASD) that we tested.[13] It posed no resistance either in the 10 neurotypical age- and sex-matched controls that we tested. That work gave us a new sense for what we needed to do if we wanted to steer the autistic system toward neurotypical regimes of anticipatory performance, leading to a predictive code in the motor signal that was serving as kinesthetic reafference.

My friend Maria Brincker, a Professor of Philosophy at the University of Massachusetts, Boston, calls these movements the "listening ones." We met at a talk I gave at the CUNY graduate center in NYC. There we initiated our fruitful collaboration leading to deep discussions on the possible roles of these movements to scaffold the emergence of social exchange in neurodevelopment.[14,15]

In the context of sports (see also Chapter 5), where I discovered these supplemental motions, spontaneously co-occurring with the goal-directed ones, I could take random trials of the participant's performance (while repeating a boxing routine) and blindly predict without error at all, whether that trial came from a spontaneous or a deliberate segment of the routine. Fig. 8.3A shows the traces of the speed profiles that I derived from the hand trajectories in **3B**. They show that the segments intended to punch the opponent were impervious to the speed with which such motions were performed. The speed type was called at random, so when I gathered all the data and grouped it by speed type, I could see that the retracting trajectories (shown here for the hook [H2] and upper cut [U2] segments) changed the geometry of the path. In marked contrast, those deliberately staged toward the opponent remained invariant to speed. Their variations served as input to a linear classifier trained to decode the segment type for a trial that was not present in the training set (using a leave one out decoder).[16] The important outcome of this classification exercise was that the off-diagonal boxes in the confusion matrix of Fig. 8.3C (for the full group of 10 participants) and those of Fig. 8.3D (for the worst individual performance, i.e., the one with the highest number of false-positives) had no confusion at all about the predicted and real goal-directed and the predicted and real spontaneous segments. Within each segment type, the worst performance would confuse the deliberate Jab with the Cross, or with the Hook, or with the Uppercut, etc. at 50–60% level (chance was 1/8, 12.5%). And the group performance was at the 80–90% level for the deliberate segments. In the case of spontaneous segments, the performance for the group was 100% correct (no confusion) and 70–80% for the worst performance of the classifier. Yet, the classifier never confused a goal-directed with a spontaneous segment. This implies that their curvature-based variability (i.e., the fluctuations in the variability of the hand's curvature along the geometric path) that I used to train the classifier was such, that a system using it could blindly tell, given a trial, whether that trial came from a deliberate segment forward to the target, or from a spontaneous retraction toward the body. The analyses yielded similar results for the speed-dependent variations, suggesting that more than one motion parameter could serve the purpose of automatically classifying movement classes based on the moment by moment fluctuations in motor performance.

The spontaneous movements, invisible to the naked eye, are the ones Maria and I decided to call the *listening movements* after that discussion from my talk at CUNY Graduate Center. In any given situation they support the visible (deliberate) segments; but one of the distinguishing features these consequential movement segments have is that they are malleable, as they funnel out the contextual variations and can be readily turned into deliberate acts. For example, if you were to complement a tennis player on his *perfect* serve, very likely you will transiently upset the serve. He would be trying to deliberately attend to an automated motion of his wrist that would otherwise occur spontaneously beneath his full awareness. This property of the spontaneous motions, that is, to easily move in and out of the awareness realm, is a fundamental departure from reflexes. Reflexes inevitably react to actions in highly predictable ways, unlike the highly adaptive and flexible spontaneous movements that can react differently as a function of context.

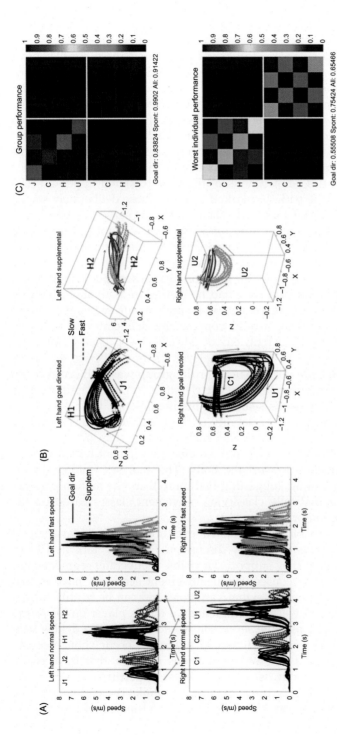

FIGURE 8.3 Blind prediction of deliberate *vs.* spontaneous movement types. (A) Speed profiles from different movement classes in boxing routines (Jab, Cross, Hook, Upper cut) co-exist in fluid natural motions. Forward segments (denoted with 1 and plotted in black) co-exist with backward segments (denoted with 2 and colored in red for slow speeds and green for fast speeds). (B) Corresponding deliberate hand trajectories for the forward Jab (J1) and forward Hook (H1), and for the forward Cross (C1) and forward Uppercut (U1), are speed invariant. Their geometric path is conserved despite changes in dynamics. Colored trajectories for the spontaneous movements (retracting the hand toward the body after the punch) do change the geometric shape of the hand trajectory under different speeds. (C) Leave-one-out linear classifier does not confuse deliberate and spontaneous classes of movements, even in the worst performance of the decoder. (D) The predicted and real deliberate movement segments are not confused with the spontaneous ones.

Because of all these features of the spontaneous segments of natural behaviors, I targeted them as the part of the biorhythmic bodily activities that we needed to use in ASD, to evoke agency and autonomy: (1) they are by default beneath awareness; (2) they pose minimal reaction at a conscious level; (3) they are highly malleable as they can be easily brought up to awareness and turned into deliberate states; (4) they are highly sensitive to contextual variations and when let alone to freely vary, they self-organize and provide structure to inform the voluntary system on the readiness to turn the motion into an automatic, yet controllable one.

In this sense, the spontaneous segments of our behaviors know how to "listen" to the variations that a new context or situation elicits. They are an example of the context sensitive constraints that philosopher Alicia Juarrero refers to when distinguishing these constraints from constraints that are context free.[17] Within the research program of motor control that I have been building over the past 20 years, the classes of constraints that Juarrero's work refers to (i.e. context-free vs. context-dependent), map well onto the types of variabilities my empirical work has uncovered in these movement classes.[12] Namely, those that are extremely sensitive to the movement dynamics (context sensitive) and those which are invariant to it (context free) in the precise sense that the geometry of their movements' paths remains unperturbed by the variations the dynamics elicit in the trial to trial fluctuations of the kinematic parameters.[12,18–21]

The variability of the motor trajectories they describe automatically give away the movement class that a system is likely outputting. As such, if we listened to their biorhythms, we could tell what type of context the person was likely experiencing (new or habitual) when the action was being performed. In the context of sports, we could tell whether we were in the presence of an expert or a novice (Fig. 8.4); but we could also tell if the person was moving in complete darkness, or with the lights of the room ON.[12,22] Typically, these motions do listen to the context, are highly sensitive to the need for adaptation, and funnel out in their variability the process of learning that ensues in a training session, or simply the process that ensues while the person experiences a habitual motion in a new situation.

In autism this was different (Fig. 8.5) because the types of variations we encountered were well characterized by the Exponential distribution (see also Fig. 8.4A). Unlike the neurotypicals in (5 A and 5B), the decoder could not distinguish variations in peak curvature (5 C) or peak speed (5D) to blindly predict if the trial was coming from a deliberate or from a spontaneous segment of the action. From moment to moment, the "good noise" from contextual variations could not distinguish when a motion was intended from when it was spontaneously co-occurring. In other words, the autistic system had a hard time separating a blink from a wink; but the spontaneous variability was still highly malleable.[13,23]

Our uncovering of the autistic systems' capacity to process information in the *here and now* gave us a sense that the real-time processing of reparameterized, self-generated motor output was possible. So long as we could dampen the noise and bring the instantaneous stochastic signatures to a more symmetric distribution, we could effectively intervene. Within the time scale of minutes, we witnessed changes in the stochastic parameters of the speed-dependent motions of the children in the spectrum.[13,23] Because of this, we set to build the closed loop interface whereby we could recursively play back as sound, in real

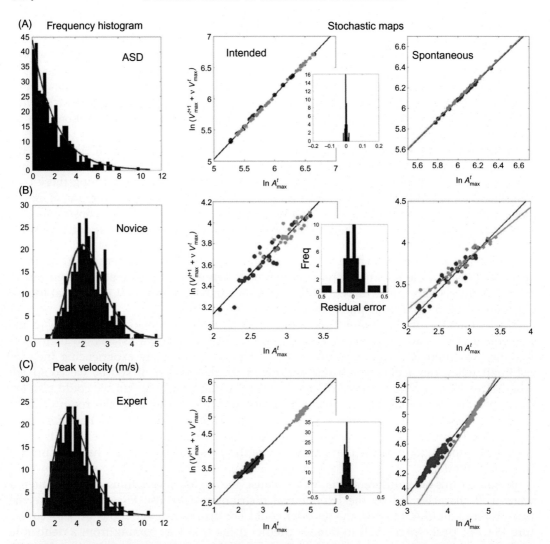

FIGURE 8.4 Comparison of patterns of variability extracted from the peak velocities of the hand trajectories from a Jab motion. Frequency histograms of the magnitude of the velocity vectors along the velocity flow are well fit by the Exponential distribution in ASD, and the Gamma distributions in the neurotypical participants (a naïve [B] and an expert participant [C]) The stochastic maps are computed using a rule explained in[22] whereby the speed maximum and the maximum acceleration in a previous trial can be used to predict the speed maximum in the subsequent trial. Importantly, the rule distinguishes the trial speed (which was instructed at random by a computer program), and the spontaneous variations typically separate the speed levels in the retractions the participants performed largely beneath awareness. These distinctions and predictions are different from those of neurotypical controls in the case of the autistic participant.

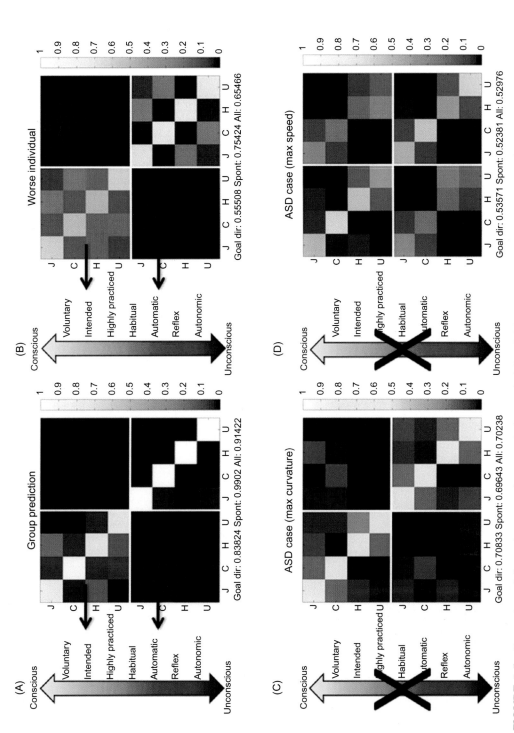

FIGURE 8.5 Comparison of the performance of neurotypical controls (A) and (B) in Fig. 8.3 and the participant with autism (C) and (D). In the case of autism, the curvature and the speed parameters confuse deliberate and spontaneous variations from trial to trial. As such, blindly predicting if a trial belongs in one movement class or another is more error prone in autistic participants than doing so in controls.

time (i.e. in the here and now), the very biorhythms the system self-produced. We just needed to use the "back door" of motions beneath awareness to that end.

LET'S PLAY BACK THE SOUNDS WE MAKE WHEN WE MOVE

Several hunches made me think of sound as a good avenue to channel these different types of variabilities and explore the different types of constraints that Alicia Juarrero refers to in her fascinating work. One was transient abilities I found myself acquiring when listening to the cortical spikes and pairing them with spatial locations. I could very fast learn which location the cortical cell preferred using the auditory inputs from the spike trains. The other was the slow rate of response to sound that I found myself having when I came out of anesthesia after a minor surgical procedure I underwent. The world experienced through sound—the first sense that came back to consciousness for me, seemed to have slowed down considerably. Yet I could process sound somehow before processing vision or any other bodily sensation. I could already hear and decode speech, even though it was in a disembodied and blind way. These personal experiences suggested to me that sound was a good channel to explore the spontaneous movements beneath awareness and blend sound into their biorhythms, somehow.

As with the cortical spikes, the spikes I found in the peaks of the temporal profiles of the motions' speed waveforms in Fig. 8.2 resembled the ebb and flow of soundwaves (Fig. 8.6A). These spikes' representation I used in the studies of cortical neurons'

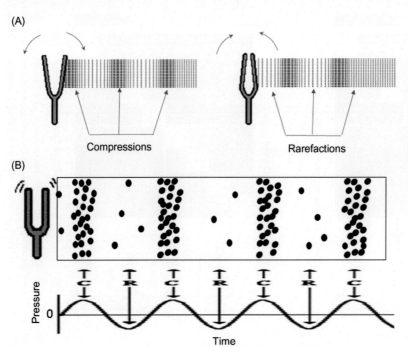

FIGURE 8.6 Analogy of waveforms from biorhythms and waveforms from sound representation to represent the spikes extracted from biorhythms as sound. (A) Tuning fork motions evoking compressions and rarefractions in alternating patterns. (B) Particle representation and sinusoidal waveform representing the sound traveling in a medium.

electrophysiology could also be used in experiments that involved pointing behaviors under different cognitive loads. We used these studies to explore biorhythms activity across different layers of the nervous systems. As with the neuronal spikes, we could turn the bodily biorhythms' spikes into sound. Recall here that I had converted the peaks of the waveforms of these bodily biorhythms into spike trains normalized between 0 and 1. Di Wu, a PhD student of Professor Jorge V. Jose, our collaborator from Theoretical Physics at Indiana University, Bloomington, had pointed me to the normalization procedure that I ended up adopting to later build the micro-movements. She went on and further developed the idea of spikes from behavior with a more detailed model that took care of several important issues related to sampling frequency across multiple sensors. Her work will be very important soon, but because she thinks *out of the box*, it will be a while before others in the field that she is trying to alert of these issues, will begin to catch up.[24,25]

Sound waves in this context of biorhythmic spikes can be thought of as repeating patterns of high *vs.* low areas of pressure that move through a medium. Such fluctuations in pressure can be detected by our ears or by instrumentation. They tend to occur at periodic and regular time intervals such that a plot of pressure versus time would appear as a sinusoidal curve (Fig. 8.6B). In this curve, the peak points correspond to compressions; the low points correspond to rarefactions; and the "zero points" correspond to the pressure that the air would have if there were no disturbance moving through it.[26]

Diagrams in the form of a sinusoidal waveform over time commonly depict the correspondence between the longitudinal time-profile of a sound wave in air and the fluctuations in pressure with respect to a fixed detector location. Fig. 8.6 show such diagrams for a vibrating tuning fork producing a combination of low and high pressure, as particles in the air move with the forward and backward motions of the fork tines. In particle physics, one would think of the high concentration of particles (compressions) and low concentration of particles (rarefactions) as they group or as they spread in the air respectively.[26] To represent the alternating nature of the phenomena, we use a sinusoidal waveform as a schematic of how the longitudinal wave would propagate in time along some distance (the wavelength). Clearly, this is just a representation, but it provides the notion that we can readily relate this representation of sound to the waveform of the biorhythms that we can harness from multiple layers of the nervous systems, using contemporary biosensors. For example, when we prepare a cup of coffee in any given morning, one occurrence of the full sequence of motions by itself will not tell us anything periodic about it. But if we record this routine every day for a long time, and anchor the waveforms of the hands' movement trajectories to a given event (e.g. the event of turning the stove on), we may begin to see patterns—much as we saw them emerge in the pointing task of Fig. 8.2 when we aligned each trial to the touch of the screen, the critical point when the finger attained the target location.

The question we had was whether contextual variations, even the slightest ones, could be distinguished in the spike trains representation of movements that were biomechanically identical but performed slightly different in different situations. If we were to convert them to sound and play them back, could we hear those subtle differences?

Jihye Ryu, a third-year PhD student in my lab, carried out a series of experiments during her Master's Thesis, whereby she recorded variations of the simple Pointing task of Fig. 8.2. These variations consisted of adding the estimation of the time duration of a tone as the person pointed to a touch screen and touched the target, to indicate the guess on how long the tone was; or simply pointed and touched the target while counting backward. In each variation of the pointing task, there was a slightly different stochastic response that significantly changed the spikes capturing the moment by moment fluctuations in the amplitude of the peaks from the hand's linear speed. The hands' angular acceleration reflected these changes and enabled us to measure departures from the baseline stochastic signatures, when there was no cognitive load.

Most surprisingly yet, these slight changes in the cognitive load significantly shifted the timing of the spikes of the heartbeat from moment to moment. The underlying biomechanical pointing motion was identical in all variations of the cognitive load, but the subtle changes they evoked in the nervous systems' responses were quantifiable and revealed different patterns. Fig. 8.7A shows these differences for the hand speed spikes and **7B** shows them for those derived from the heart signals.

We also depict these changes on the space of Gamma moments, as vectors reflecting the shifts in the empirically estimated moments of the continuous Gamma family of probability distribution functions best characterizing different spike types from different parameters and movement classes of the pointing loop. Fig. 8.8A comes from forward reaches as characterized by spikes reflecting the fluctuations in amplitude of the angular acceleration peaks. Fig. 8.8B measures the stochastic shifts of the time intervals between the spikes derived from the angular acceleration peaks. Fig. 8.8C reflects the stochastic shifts from the spike trains derived from the fluctuations in the inter-heartbeat-interval timings. One of the features of this work is that the micro-movements data type allows us to plot all these different signals using a standardized unitless real valued scale varying between 0 and 1. Yet, we preserve the ranges of physical parameters (**8AB** deg/s^2; **8 C** Ms) and plot these along a color-scale gradient for each variation of the task (e.g., in **8AB**).

The work with neurotypical controls gave us a sense for what to expect in the periodicity of the spikes derived from the peaks of these multimodal signals' time-series, harnessed from different layers of the nervous systems. So, we recorded these motions from children with autism, to quantify the departure from the neurotypical cases. The differences were striking (shown in Fig. 8.9) for a representative child in relation to the age- and sex-matched neurotypical control that we presented in Fig. 8.2; here depicted again for comparison on the left-hand side of the figure).

These disparity in biorhythmic patterns between the system with a diagnosis of autism and the neurotypical controls prompted us to develop new ways to intervene using the biorhythms across all layers of the nervous systems (not just those from the hand). Because we conceptualized movement as sound and provided audio-visual feedback triggered spontaneously (without instructions) by the child's self-generated movements, we were able to transition the motion's stochastic signatures from equally probable random spontaneous noise to regimes of well-structured signals leading to goal-directed behavior. We discussed these experiments in Chapter 2. They resulted in the child's self-discovery of agency over his/her self-initiated motions. Because the setup evoked the self-discovery of cause and effect in systematic ways that were eventually predictable with high certainty, we decided to scale this idea a notch and in so doing, fix some caveats.

Despite its success in evoking agency, the work had several shortcomings. A major one was that we did not parameterize the audio-visual media we used to engage the child's

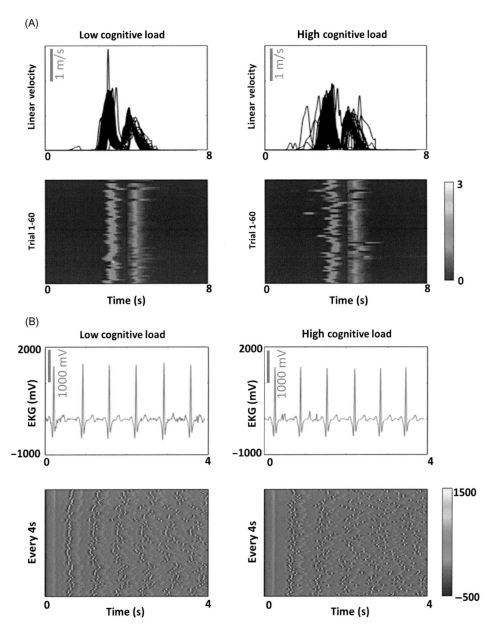

FIGURE 8.7 Visualizing the effects of cognitive loads on the biorhythms of hand motions (speed-peaks) (A) and heartbeat (based on the R-peaks inter time intervals) (B).

nervous systems. We did not record the heartbeat either, so the multilayered, multimodal approach we later officially adopted in the lab was only theoretical in nature at the time the original work to evoke agency was done. We really wanted to do this since 2009 when I started developing these ideas under my NSF-funded program to study social

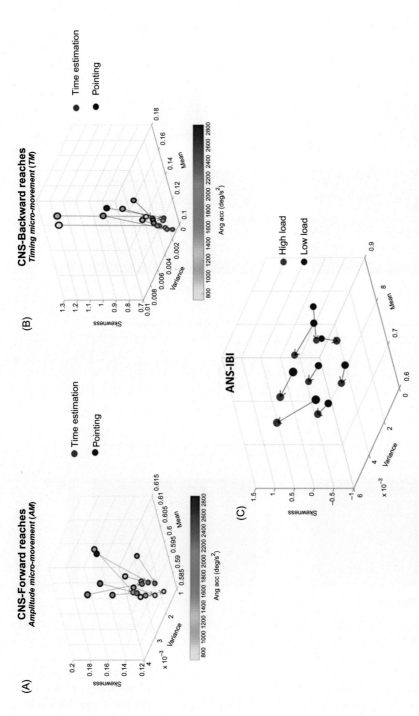

FIGURE 8.8 Personalized stochastic profiles: Visualizing the effects of cognitive loads on the stochastic signatures of different biorhythms and movement classes. (A) Speed peak amplitude micro-movements of the forward pointing motions shift the stochastic signatures with cognitive loads. (B) Spontaneous segments retracting the hand in the backward reaches shift the stochastic signatures of inter-peak-time intervals. (C) Shifts in stochastic signatures of inter beat interval times from the heart's R-peaks. All plots represent the empirically estimated Gamma moments for each participant: the mean along the X-axis, the variance along the Y-axis, the skewness along the Z-axis, and the kurtosis represented by the size of the marker. In (A) and (B) the color bar represents the median value of the physical range of values for each participant across all participants (deg/s^2) while the values of the Gamma-space axes are along unitless scales from the micro-movements derived from the raw angular acceleration data.

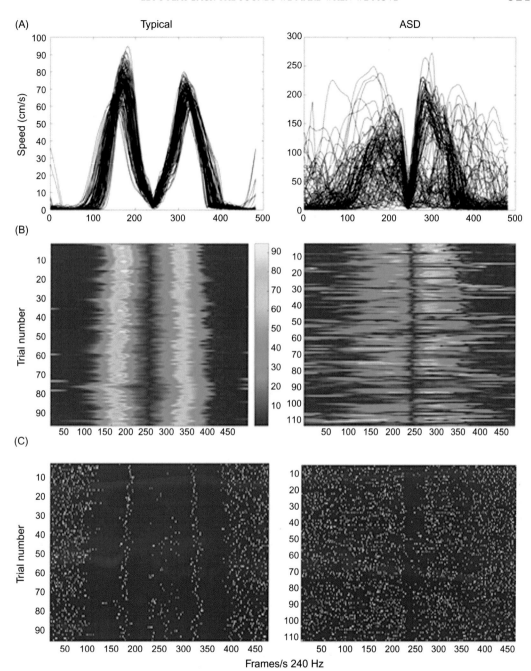

FIGURE 8.9 Comparison between the neurotypical case and the age- and sex-match autistic case. Biorhythms derived from the hand-pointing motions are random and noisy in excess.

interactions under a new theoretical construct. I had, however, no instrumentation to do it. Years later, we acquired the instruments and developed the means in the lab to synchronize all these disparate signals from different biosensors. Then, we initiated a path of exploration and new avenues of inquiry. We could finally turn the theoretical ideas into a physical reality for empirical examination.

Ji Ryu and Joe Vero built a set up to collect in tandem 32 channels using EEG sensors, the ECG signal using a wireless heartrate monitoring system, and the motor signal from the kinematics of 12 IMUs using a grid across the body. They used the Lab Stream Layer (LSL) open access platform and requested the APIs from several companies whose sensors we had acquired. This enabled Joe to write the proper code and synchronize the waveforms from these various sensors. All companies cooperated and through a collective effort the lab and the LSL community helped us gain the ability to harness these signals in tandem. Joe placed his code in GitHub for others to use, and now we have ways to continue to collaborate on this open access community project.

Some biosensors are already multimodal and synchronously collect signals at the same sampling rate from various body parts. Yet, there is no system that allows harnessing EEG and bodily motions in tandem at the same rate. The Neuroelectrics company, with headquarters in Barcelona and an office in Boston, did have a wireless EEG system with an accelerometer embedded in the head cap, which we could use to help remove motor artifacts from the EEG waves. This IMU sensor "speaks" the same language as our other IMUs in the bodily grid, so we started synching the two signals in that way. The heads of Neuroelectrics, Ana Maiques and Giulio Ruffini, were very kind throughout the process of getting these instruments to work in ways we could integrate with the other instruments. I had met them personally at a meeting in Boston, at the headquarters of the Nancy Lurie Marks Family Foundation Board, when the Foundation awarded me the Career Development Award to build these interactive interfaces.

Their Enobio (500 Hz, Neuroelectrics) wireless EEG system also comes with a great interface that allows us to track the quality of the recordings in real time. One of the channels in the EEG can be used to output the heart-related signal in the form of an ECG. And because the company is really in tune with researchers' needs (as they are researchers themselves), they maintain a Wiki page and have uploaded all the necessary software toolkits and APIs to the LSL community sites. The synchronization of the various bodily sensors with the EEG signal from the Enobio system was then feasible.

As discussed in Chapter 5, Joe figured it all out and made it possible for the lab to move into a different era of multimodal signal streaming and real-time computing. I designed a new type of parameterization involving weighted directed graphs to represent connected nodes in a network. Adapting tools from connectivity analyses to these brain-body or body-body networks, we could then track the dynamics of synergistic activities self-emerging within these dynamically coupled networks and extend them to social networks involving bodies in motion.

The nodes of these wireless bio-sensing electrodes and IMUs were conceived as a dynamically evolving brain-body network with self-emergent behavior (a dynamic complex system) (Fig. 8.10). A challenge here was to find a unified sampling rate and try to set a common frequency across all sensors. This is an open problem in other fields, but for now, we resorted to up-sampling using interpolation and filters, and bringing all

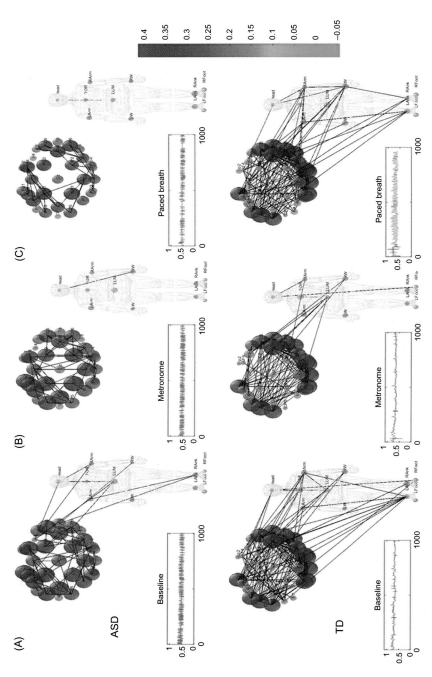

FIGURE 8.10 Brain-Body coupled network dynamics and corresponding heart IBI traces for each of the three conditions. Top row from an autistic individual and bottom row from a neurotypical control. (A) Natural walking; (B) Walking with a metronome in the background (no instruction given to observe if the biorhythms spontaneously entrain, or not); (C) Walking with a metronome in the background while breathing to the metronome's beat upon instruction (gestural instructions in the cases of nonverbal individuals). Data from two representative participants.

waveforms to a common number of points. This still does not alleviate the drawback of different frequencies in the original instrumentation, so we used the rates of change in the fluctuations in amplitude of the peaks to be able to align fluctuations in one domain over fluctuations in another domain (e.g., brain node and body node) while sampling both at 500 Hz. We then examined peak activities in the frequency domain.

As explained in Chapter 6, we used cross-coherence analyses to establish pairwise coincidences in both directions (i.e., EEG sensor [i] paired with body sensor [j] in the i→j direction; and the same in the j→i direction). We computed the cross-coherence pairing both ways and examined the peaks in each direction along with the shifts in phase and the frequency corresponding to the maximum across of all peaks for a given paired comparison.

This method will have to be improved for a variety of reasons beyond the scope of this book. However, I have teamed up with Professor George V Moustakides of Computer Science and Electrical Engineering to develop new methods that deal with temporal disparities using sensors' fusion schemes;[27] tackle disparities in frequency using multirate processing schemes,[28,29] and modeling of spatial and temporal features across sensors harnessing activities from different parts of the nervous systems. We have a full plan of work with very well-defined goals for the immediate future.

Once the temporal and rate disparities are resolved, we are ready to move on to new statistical methods of analyses. For the time being, I am working with the micromovements of each sensor type, thus using the rates of change of fluctuations in the amplitude of the peaks, normalized to a unitless real value in the 0−1 scale, so we can integrate different statistical features across different brain-body structures. Moustakides and I submitted several proposals already to the NSF and the McKnight Foundation, but either one of those will be hard to get. In the meantime, we are not waiting around for funding; we are quite busy already working out these wrinkles of these challenging problems.

Under the current schemes and acknowledging the caveats of the methods, we set to then track CNS-PNS interactions using tasks that can probe nervous systems' functions and elicit activities that we can easily distinguish from one situation to the next. One of these tasks involves three conditions that we explained in Chapter 6. These were registering activities with the multimodal grid, while the person automatically walked, as it normally would; while the person walked to the beat of a metronome set up in the background without any notification or instruction; and while the person walked and was instructed to deliberately breathe to the rhythm of the metronome. When we examined the connectivity patterns of the network as self-emerging synergies manifested from reciprocally connected body nodes and clusters of EEG-leads' activities, they differed dramatically between each person with autism that we tested and the corresponding age- and sex-matched neurotypical control.

Patterns from representative individuals are depicted in Fig. 8.10 for the ASD (top) and neurotypical control (bottom) using methods of statistical inference and kinematics analyses developed in the lab. Ji is expanding her work to the Fragile X population that Caroline is studying in the dyadic settings. We will be able to address familial links across generations of affected child and premutation carriers in the family.

The surprising result from these analyses was that in the autistic cases, the heart biorhythms did not spontaneously entrain with the metronome, nor they did with the deliberate breathing. The insets of Fig. 8.10 show traces of the heart inter beat interval (IBI) timings and their less periodic features in the ASD participant. This result, which we have also found in other participants of this highly heterogeneous disorder, was very worrisome and prompted us to think about ways to bypass this problem and perhaps use the "back door" of spontaneous motions—those between blinks or winks, between touching a target and resting the hand, between punching an opponent and retracting the hand back, because as we have seen, these consequential movements transpire largely beneath mental awareness and their variability funnels out the context-sensitive constraints of a given situation.

EXPLORING SOUND PREFERENCES: ENTRAINMENT BENEATH AWARENESS

Under mental awareness that evokes emotions, we listen to music and let our body's rhythms entrain to the rhythms of the music. Vilelmini Kalampratsiduo, (Vilmi) our PhD student from CS and a classical ballet dancer (featured on our cover), is deeply interested in developing a research program at the intersection between technology and the performing arts. When she heard us talking about converting all biorhythms to sound, she became an integral enabling part of this enterprise. Her PhD thesis work then took an interesting turn and her co-adaptive/interactive avatar interface began to lead the lab toward a completely new era of real-time stochastic recursive feedback control: that is, feedback control using kinesthetic reafference from self-generated bodily motions.

I had watched a TED talk by astronomer Wanda Diaz Merced on her sonification of the graphs she used to be able to employ to visualize her data, before she turned blind. By replacing vision with audio to quantify her star-related data, she could continue to do science despite her lack of sight. I immediately connected her talk with the spike-audio representation I used to study neurons in the PPC of the monkey's brain and contacted her by email to try and get help to sonify our biorhythms. She responded very enthusiastically willing to help us out. However, as it turned out, Wanda is from Puerto Rico. After the 2017 hurricane Irma that devastated the island, we have lost contact. We know she is okay, but feel that perhaps our collaboration ought to wait.

To pursue our ideas, we reconnected with Professor Steve Kemper of our George Mason School of Music. I had been in touch with him long ago to collaborate on a project involving the sonification of the heart signals. I wanted to convert the heart signals to music, because at the same CUNY Graduate Center talk that I met Maria there was a member of the audience (David Sassian) that alerted me of a very interesting attempt by a jazz musician to convert heart rhythms and pitches, caused by muscles and valves' movements, into music.

David brought me all the way to Queens, NYC to the home of Mr. Milford Graves, the jazz musician featured on a New York Times piece. I remember vividly how Mr. Graves received us with open arms and showed us his laboratory in the basement of his home. (http://query. nytimes.com/gst/fullpage.html?res = 9507EFD9103CF93AA35752C1A9629C8B63)

Corey Kilgannon published the NY Times piece on November 9th, 2004, but my visit to Mr. Graves' home was in 2012 after my talk at CUNY (hosted by Philosopher David

Rosenthal), where I met this fellow aficionado musician who alerted me of Mr. Graves' work. The reporter wrote from the interview with Graves:

> "A lot of it was like free jazz," Mr. Graves said one day last week in his basement. "There were rhythms I had only heard in Cuban and Nigerian music." He demonstrated by thumping a steady bum-BUM rhythm on a conga with his right hand, while delivering with his left a series of unconnected rhythms on an hourglass-shaped talking drum.

The Cuban music connection opened our conversation. We talked about my homeland and the African heritage we are so proud of in Cuba. Then we moved to medicine and the healing powers of music. I told him of my desire to help the children with autism better connect with their own bodies in motion, so they could gain volition and develop agency. It was an amazing experience to see this musician's creativity at work in that magic basement full of interesting musical instruments from all over the world and computers of all kinds "cooking" music, literally from the heart.

Back at Rutgers, Steve Kemper and I tried to start a collaborative project. Although we had a concrete project in mind, where he would help us sonify all the bodily biorhythms, we both got swallowed up by our tenure-track commitments' madness. We did not pursue this collaboration further until Vilmi implemented the audio-visual co-adaptive interactive interface for her PhD thesis. This project alerted us of the need for real time computations, because we wanted to play back the sounds from the biorhythms embedded in some music of the child's preference.

We had managed to simultaneously harness the biorhythms of the EEG, ECG, and body while the person walked around and listened to different types of music, or simply had the eyes closed. We had also combined these audio signals with an avatar. The interface projected on a large TV screen the bodily motions of the end-user through this avatar, in real time, but Vilmi wanted to improve the avatar's visualization and continue to provide feedback to the end-user through her visually driven stimuli (the one that we described in Chapter 6).

At this point, we decided to first explore sound preferences in open loop mode (i.e., analyzing the data post hoc) to evaluate entrainment of the biorhythms with the music. The idea was to be able to identify the music that helped the bodily biorhythms gain structure and form synergies. This time around though, Steve pointed us to various links that explained how to process the music and how to extract the rhythms, the pitch, and other features from the waveforms.

One of these open access software packages was the MIR Matlab toolbox for music processing. We used its routines, and under our micro-movements' umbrella we converted all biorhythms to audio, got their envelope, and converted them to sound. We could then play them back as sound and Steve could help us turn them into music, so we could play them back to the person, hidden in the rhythms of various musical pieces. We noted that the micro-movements conserved the structure of the original music. They were a sort of compressed version of the original piece, which made sense, since they were extracted from the peaks of the piece's original waveform, and as such, reflected their key fluctuations.

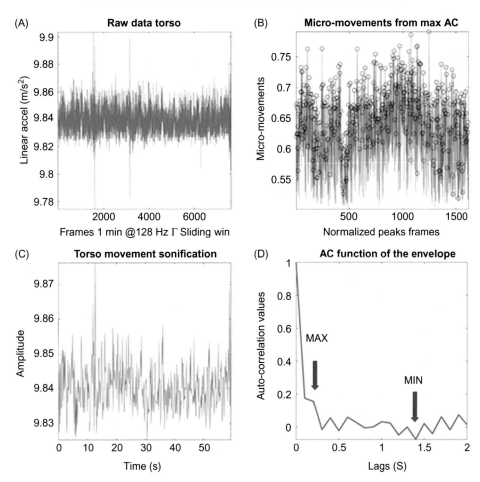

FIGURE 8.11 Pipeline of data processing to sonify the biorhythms and eventually play them back. (A) Sample trace from biosensor (IMU at 128 Hz). (B) Micro-movements of the peak accelerations that maximally deviate from the overall mean and normalized 0−1 real range. (C) Envelope of the micro-movements obtained using the MIR toolbox of Matlab. (D) Autocorrelation function with maximum and minimum values at the indicated lags.

We then computed the autocorrelation of each biorhythmic sonified channel (each body part and heart, leaving the EEG separate for later). The pipeline for data analyses is shown in Fig. 8.11.

We had a set of 15 Polhemus Liberty sensors sampling at 240 Hz and IMUs from the APDM sampling at 128 Hz. The latter has the advantage that can be taken to the home to monitor sleep (in addition to other activities of daily life). Fig. 8.11 illustrates the order of steps for an IMU sensor located on the torso area. The first step (**11 A**) is to gather the data minute by minute, as we explained using a ½ minute sliding window with an overlapping size drawn at random from the Gamma distribution. This ensures that the blocks we use to construct the stochastic trajectory of empirically estimated Gamma parameters are not from identically independently distributed points. We then determine the peaks of the time series in each block and normalize them to build the micro-movements series (**11B**). These are sonified and the envelope plotted in **11 C**. We then compute the auto-correlation function and determine the maximum and minimum for each block sequence. The maxima and minima are saved to compute the corresponding Gamma parameters from the original micro-movements in step **11B**. Then we plot the point at the corresponding lag (the indexes we saved when obtaining the maximum and minimum). For example, for the **11D** case, the maximum occurs at 0.2 seconds while the minimum occurs at 1.4 seconds; but these change from body part to body part and from minute to minute. The time window to select the size of the block depends on the sampling resolution of the biosensors and the confidence intervals we set for the estimation procedure. Here we set 95% confidence intervals for the estimation of the Gamma parameters using MLE and used 128 Hz for the IMUs and 480 Hz for the Phase Space. As such, in one case the window is 1 minute long and in the other 1 second long. This is the caveat we mentioned in relation to disparate sampling resolutions across biosensors in the market. We are working toward a unifying rate of change based code, but for the time being we try different block sizes and sliding window sizes to find values that conserve the overall qualitative nature of the results. But we do alert of this caveat up front, so we know the problem and aim at solving it. When in doubt, make sure the confidence levels for empirical estimation are high.

Staking up the auto-correlation waveforms from all body parts gives us the auto-correlation matrix, which we can visualize as a heat colormap. For one condition and one representative neurotypical participant we show it in Fig. 8.12, where we also plot some of

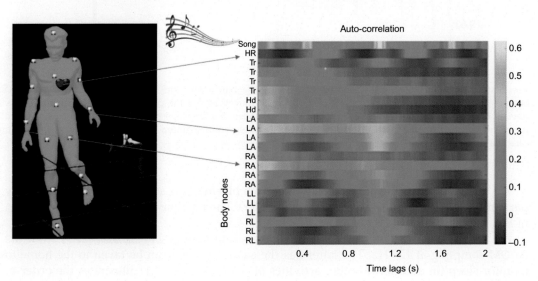

FIGURE 8.12 Sample of sensor placements to build a grid across the body and auto-correlation matrix built by stacking up all the traces from the music, the heart and all body parts recorded with a Phase space impulse system at 480 Hz.

the sensor placement locations. Here we used the Phase Space (480 Hz, San Leandro, CA) and had 22 LEDs across the body. The figure shows a subset of the locations.

We were able to begin the process of ascertaining the signatures of variability for each individual node along the continuum of the session, as the participant moved through all the different conditions (no music eyes open, no music eyes closed, music with eyes open, and then with eyes closed, using different melodies and then using the favorite song of the person). These conditions were played as the participant naturally walked around the room. Upon extracting the micro-movements of all biorhythms and examining them as audio signals, we could unveil those nodes for which the past events and the future events were maximally auto-correlated *vs.* those which were minimally auto-correlated.

The auto-correlation matrix of Fig. 8.12 uses blue shades to represent the low auto-correlation values and orange to yellow gradients to represent high auto-correlation values. The top row is the auto-correlation profile of the song, the second row is from the heartbeat, and the rest are from the body locations, including the head. In this case we had additional (redundant) locations where we placed LEDs because camera systems suffer from the problem of occlusions and although we have a grid of 8 cameras and the Phase Space has the unique feature in the market of motion caption that the active LEDs are always captured, we did not want to risk losing any data due to occlusions or due to the person walking near the edge of the motion caption field. One of the conditions was with closed eyes, so the possibility that the person would walk too close to the edge was real. In any case, we did not want unnecessary data loss and sampled as many locations across the body as we could afford.

Using our home-made visualization tools that Vilmi implemented into an avatar-driven interface, and the analytic power of the Gamma process for the spike trains, we were able to estimate the stochastic trajectories of each node. As such, we began the process of exploring which music was the most likely to physically entrain the bodily biorhythms.

The signal underlying this inquiry was the linear speed, obtainable from the positional trajectories. For each condition lasting 5 seconds at 480 Hz, we could obtain thousands of peaks so that using, for example, one-second-long blocks with overlapping sliding windows of up to $1/2$ second, we sampled from the Gamma distribution at random and built our data sets to input to a Gamma process. In this way, we relaxed the identically independent distributed (I.I.D.) assumption that we had made before. Now the spikes were obtained from nonidentically distributed events and they were not assumed to be independent. This was more in tune with how the nervous systems' signals may be generated, as they co-exist with features that are at times inter-related and entrained, and at times independent.

The different contexts where the participant walked naturally without music, then with various types of music (rhythmic, romantic, classical, and while listening to the favorite song they told us they liked) elicited different variations in the biorhythms' peaks. They changed the variability of the spike trains of fluctuations in the amplitudes of the speed peaks. As such, the auto-correlation values separated processes that were maximally auto-correlated in time *vs.* those that were more independent. Some of these responses from different contexts are plotted in Fig. 8.13 for a neurotypical control and an adolescent with a diagnosis of Asperger's syndrome (AS). The questions we had were whether music could modulate the biorhythms, and whether some music could do so the best possible way,

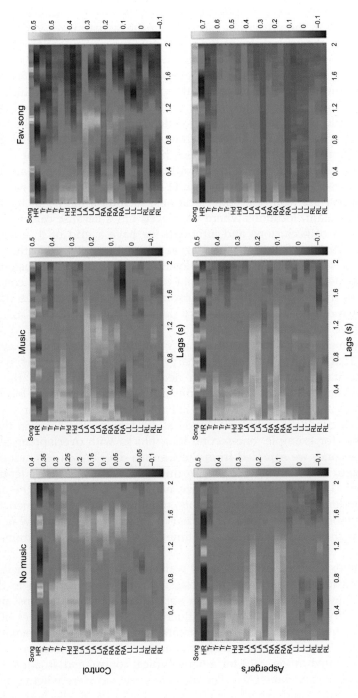

FIGURE 8.13 Sample auto-correlation matrices as in Fig. 8.12, for three different naturalistic walking contexts: no music, music, and favorite song. The two individuals are a neurotypical male and a male with AS diagnosis.

that is, by allowing us to distinguish different types of dispersion and shapes of the probability distributions we estimated for the activities corresponding to maximal and minimal auto-correlation.

To address those questions, we examined the micro-movements of the linear speed of the participants' body parts that corresponded to the lag for which the autocorrelation was maximal (or minimal). In other words, the Gamma parameter plane that we had been using to study the stochastic processes and the evolution of their trajectories over time could now have another dimension (the time lags) to add dynamics to the probability landscape. For the quadrants that the median lines delineated, that is, spanned by the median shape and the median scale values, we could examine the auto-correlation values of each node micro-movement trajectory. As such, we could see the points on the Gamma plane for which the auto-correlation was maximal at the Right Lower Quadrant (the RLQ). This is the median-delineated quadrant where the shape values are high, denoting a skewed to symmetric distribution with lower dispersion than those in the Left Upper Quadrant (the LUQ). The LUQ is the median delineated quadrant for which the dispersion was higher and the shape values lower, that is, tending toward the Exponential distribution. Fig. 8.14 shows in schematic form the various planes of the Gamma parameter plane lifted across the various time lags that we used to compute the critical points of the auto-correlation function for each node micro-movements' trajectory.

In our conceptualization of this graph, we thought of the *dynamic memory* of the system for processes where events in the past maximally correlated with events in the present along the RLQ *vs.* those minimally auto-correlated, and as such, more independent from the past events. We also examined the opposite pattern, namely the RLQ minimally auto-correlated points and the LUQ maximally auto-correlated points.

This way of examining the data was more informative than the Gamma plane alone, because it introduced the notion of time lags and posed the question of how these scatters of point evolved across the probability landscape with a given context. It enabled us to examine the changing dynamics of the context-sensitive constraints and context-independent constraints' variability. Fig. 8.15 shows the case of Vilmi's experiment for the participants in Fig. 8.13. Indeed, the favorite song of each participant resulted in a different pattern of the bodily probability distributions. We examined the four cases mentioned above and could readily see the separation of each scatter: the RLQ with maximal auto-correlation (red diamonds) are at the bottom of the graph, while the RLQ with minimal auto-correlation (red squares) are scattered across the various lags and much sparser in the neurotypical than in the AS participant.

The LUQ counterparts also differ between participants, suggesting different bodily patterns rather distinguishable with the music type. This gave us the sense that we could use these analyses to track musical preferences and distinguish different bodily activity across different contexts, as shown for the same representative participants of Fig. 8.16.

The results from these experiments encouraged Vilmi to work out the technical kinks to get real time sonification of the biorhythms, so we could play them back to the person, blended with the favorite music. The idea here was to do this at the unconscious level, by embedding the beat we extracted from the micro-movements into the favorite music and see which synergies developed and which patterns of entrainment worked best (i.e., evolved toward lowest dispersion and most symmetric PDFs).

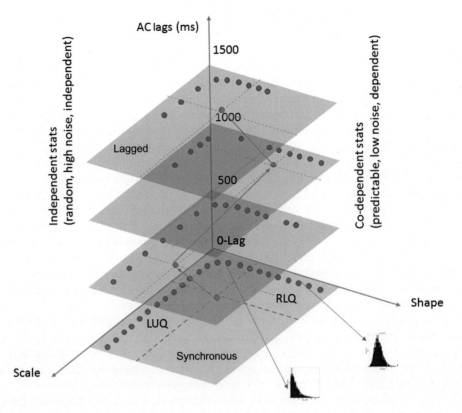

FIGURE 8.14   Schematics of the lifting of the Gamma parameter plane and the dynamically changing median delineated quadrants. At each plane obtained for each lag of the critical points of the auto-correlation function, the median shape and the median scale delineate the RLQ with limiting Gaussian distribution and low dispersion (scale value) and the LUQ, with limiting Exponential distribution and high-scale value.

My ongoing hypothesis for autism is that we may be able to regularize the biorhythms of the child by using the mom's heart's signals and embedding them in the child's favorite music. This guess is grounded on the fact that the mom's heartbeat is the first autonomic source of sound the nascent nervous system of the fetus is exposed to. Perhaps the music will be most effective for the autistic child if it is played in combination with the heartbeat of the mom. The metronome did not spontaneously entrain the heartbeat in the autistic child. The biorhythms did not entrain either with the deliberate breathing patterns, that is, when the child was instructed (through gestures) to breathe at the pace of the metronome. Music by itself was, however, more engaging to the heart rhythms than to the rest of the body.

Our game plan now is to assess the mom—child dyad first during a regular interaction; then assess the dyad with music; and finally with the reparameterization of the music, feeding the music back as kinesthetic reafference, containing this time the rhythms of the mom's heartbeat blended into the music. This recursive scheme whereby we create

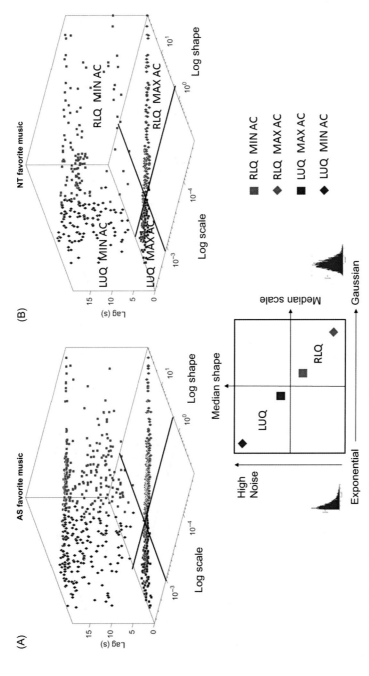

FIGURE 8.15 The favorite song condition gives rise to different patterns for the participants in Fig. 8.13. Schematics of the empirically determined stochastic regimes from the bodily biorhythms (drawn from the linear speed parameter).

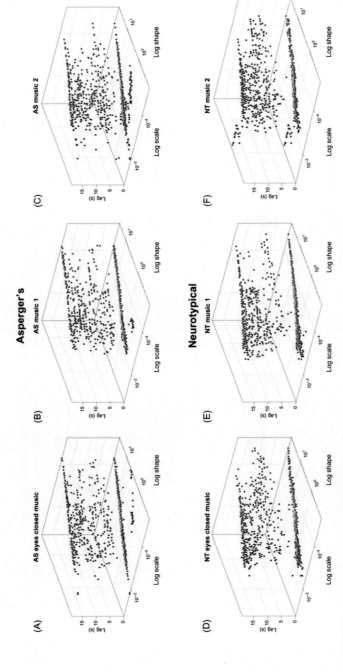

FIGURE 8.16 Different contextual variations in stochastic patterns: Personalized profiling.

feedback loops between the mother and the child, using as re-entrant input the child's own generated biorhythms and those of the mom, may help us create rapport across the biorhythms and gain our first formal entry point into the autistic system to encourage social exchange.

Thus far, Vilmi has managed to musicalize the heartbeat in real time. Now we need to blend it with the music and play it back as sound. Steve is a composer and is helping us with the musical aspects of the problem. In piloting sessions with neurotypical controls, we have begun to test the reactions the biorhythms show to their own real-time sonification by using different music played through earphones. It is really exciting that as the person hears the music through the earphones, we hear the change in the sound the biorhythms produce at our end (through the speakers). As we transition the person through different music types, the sonified heart rhythms also change. Sometimes the change is slight, and other times, they are more pronounced. We think that we can do this now though the outcomes of the program that we are gradually building will be material for a different book.

OBJECTIVE OUTCOME MEASURES OF TREATMENTS: DRUG TRIAL REVISITED

In Chapter 7 we described the outcome of a clinical trial that we measured for a clinical research lab at Mount Sinai. The trial involved children with Phelan McDermid Syndrome and a drug trial of insulin like growth hormone factor (IGF1). Here we walk the user through one example of a child for whom we applied these biometrics and used the graphs as visualization tools to show us the impact of the drug on the motor patterns.

These probability maps (each point is an empirically estimated Gamma PDF) are plotted on a linear scale. If we were to use a logarithmic scale, the scatters would be along a plane and partitioned into four main areas of interest (as in Figs. 8.15–8.16) where we fused both cases (max RLQ with min RLQ and min LUQ with max LUQ). Plotting them separately enables us to see the interactions between the scatters during each condition in one experimental session. An example is shown in Fig. 8.17 for a neurotypical participant matched in age and sex with the participant undergoing the drug trial in Fig. 8.18.

In the present case, we obtain the longitudinal evolution of the drug trial across 5 visits (Fig. 8.18). Initially the scatters are very sparse. The top panels are the points with maximal RLQ auto-correlation (the past events highly correlated with the present events, and the present events highly correlated with the future events) with distributions tending to the symmetric Gaussian and the lowest dispersion values. The points with minimal LUQ make up the other scatter (very sparse in Fig. 8.18 for the visits prior to the drug). Those are points with minimal auto-correlation values, implying more independence from past events and tending to the Exponential distribution with higher dispersion. The bottom panels are the opposite for RLQ and LUQ, and as the top panel, the evolution of the scatters shows a change toward the patterns of the neurotypical control female of Fig. 8.17.

The overall map of the scatter (where each point represents a PDF from the micro-movements of a node) evolved dramatically with the drug trial. The significance of these changes in the contextual variability sensitive to the drug/placebo stages can be

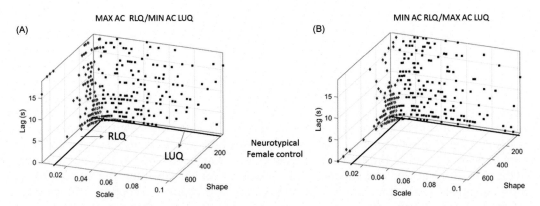

FIGURE 8.17   Neurotypical female control to provide normative data and help examine departures from normality in a clinical drug trial.

appreciated in the visualization of the patterns of variability across the body pre- and post-trial. These are plotted in Fig. 8.19 for this participant in relation to the control.

This longitudinal evolution of the variability patterns and the inferences we can make using auto-correlation minima and maxima across the bodily biorhythms provided proof of concept to use these methods in the longitudinal tracking of treatment outcomes during therapeutic interventions. More precisely, as we discovered through the NSF I-Corps program that sensory-motor-based occupational therapy had no insurance coverage in the US, because of the subjective methods in use; we set to try our objective biometrics to quantify progress during such OT interventions. To that end, we received funding from the Henry Wallace Family Foundation and the Colorado STAR center helped us recruit participating families for our study. Dr. Lucy Miller, the Director of the center had been instrumental in developing the inventories to create the sensory-processing-disorders (SPD) assessment tests. As with the autism diagnostic observation schedule (ADOS) task to diagnose ASD, here we adapted the SPD inventories to objectivize the diagnosis and measure the therapist-child dyad during therapeutic intervention.

OBJECTIVE OUTCOME MEASURES OF TREATMENTS: SENSORY-BASED OT

One of the lessons I have learned over the years is that each pathology of the nervous system has a group of people that specializes on one aspect of the problem and to claim that turf, and financially own it, sacrifices the bigger picture. There are accreditation programs that confer some certification to the person who will diagnose and treat the disorder at a cost that is not in any way related to the outcome of the treatment: that is, whether the person improves or not, the earnings of the therapist remain unchanged. There is no inherent motivation on the part of the clinician to improve the methods toward an

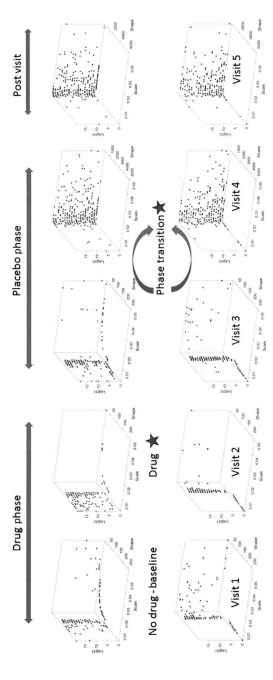

FIGURE 8.18 Case study from a drug trial tracked longitudinally across several months.[30] This is a 14-year-old female participant with Phelan McDermid syndrome. The drug started in the indicated phase. The first few visits showed very large departure from neurotypical control in Fig. 8.17. Upon the drug, by visit 4, already in the placebo phase of the trial, there is a change in stochastic signature and a shift in the probability landscape (phase transition type of change) that brought the patterns close to the neurotypical. This pattern remained post trial in a visit when the participant returned to the lab for testing.

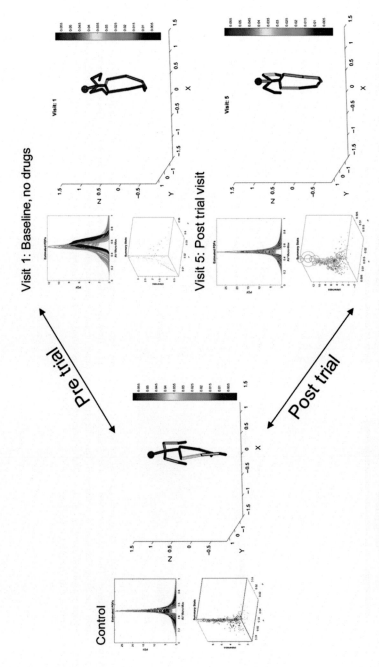

FIGURE 8.19 Visualization of the pre and posttrial patterns of Gamma moments across all body parts (based on the angular velocity) and color coding the avatar by levels of variability across the body in relation to the control patterns.

objective scale. Generally, some organization makes the diagnoses and treatments official by building a Board that legitimizes the whole enterprise and dictates the *"science"* of it all.

In autism, we saw the ADOS claiming that the affected children had no sensory-motor issues. The test presumes the condition to be exclusively about cognitive and social issues (as if sensory motor phenomena could be stripped away from the behavior they observe and use as guidance to score the child). Sadly, every piece of scientific work is forced to use this test as the validating criterion for the study. Without the ADOS scores no paper will be published in peer-reviewed journals.

The DSM also claims that motor issues are irrelevant because they may be the byproduct of medication, and the children with a nascent nervous system do have issues with the psychotropic medications (but the APA will allow and encourage the prescription of these medications nonetheless).

The medications they prescribe, however, never came from a proper clinical trial in Pediatrics. In fact, the NIH cannot afford such trials. It has no budget to that end. Consequently, the whole operation is under the control of Big Pharma and the FDA. This is a tad problematic—I thought when I found out; because science being the one entity that could validate such trials does not have (1) proper methodologies to do so and (2) does not have a budget to be doing this in a conflict-of-interest free fashion. Moreover, the scientific community at large has no access to data from clinical trials; so there is no way to verify, validate, or blindly reproduce the work of the small interest groups that run the trials. These groups are mostly soft-money researchers, who do not have a steady income from a home University. Their income is paid by grants they secure from the NIH and from private Pharmaceutical companies; but even the NIH grants are funneling the funds from Pharma to the clinical settings in ways that are not entirely transparent to the public or to the scientific community. These researchers who run the trials for Big Pharma and discuss their outcomes within closed doors among themselves are of a different category than the rest of us because they work under different financial constraints than we do. Indeed, I learned all of it (and more) rather recently. It was shocking to see that science has no agency of its own to monitor clinical trials in a neutral manner.

These groups of scientists that hold stocks in and are paid consultation fees by the very pharmaceutical company they run the trials for have full green light to do what they do without supervision from neutral observers outside their circles. More important yet, the skills set of these fellows is far from ideal to do anything computational or mathematical of their own. As such, they mostly outsource their data analyses to some handful of companies in industry. They get back numbers and percentages from instrumentations and softwares that do it all in a highly black box manner. That whole operation is called *science* in such circles and the results from these black-box analyses are cited as the hard-core scientific evidence favoring the approval of the drug. I, as many others out there still do, thought that the word *science* had a different meaning altogether. Today I know it has many shades and is in fact in ***high jeopardy***.

Some of the information about these issues came from detective work the lab did at the library, some came from my direct involvement with such groups in an attempt to try and make the clinical trials data open access (to no avail), and some came from personal interviews at the NSF I-Corps, when Jill Nguyen, my doctorate student back in 2015, and

I went around interviewing over 120 stakeholders in the ecosystem of autism. The sad news was that most families affected by autism cannot afford the types of therapeutic interventions that seem to alleviate the sensory-motor problems of children. The children are over medicated with these awful drugs that alter their neurodevelopment in uncertain ways. While the medical industry prescribes psychotropic meds (for which no adequate clinical trials exist for adults, let alone for children) and the psychological testing industry diagnoses and prescribes ABA. This form of behavioral therapy aimed at making the child look socially appropriate is the only insurance-covered behavioral therapy. It has no outcome measures either, but it has claimed scientific evidence from observation as an "evidence-based" practice. The lack of objective evidence, however, has begun to question the validity of it all, and sadly, that therapy too will no longer be covered by insurance unless objective outcome measures provide true scientific evidence beyond opinion.

This is tragic because institutionalized therapies and diagnostic systems such as the ABA and the ADOS do have at their disposal a superb infrastructure to develop knowledge and accreditation. However, there is no neuroscientific grounding in what they do and no proper statistical approaches aiming at personalized profiling in either one of them.

Amid these issues, and as I learned to navigate the autism ecosystem to discover problems and address them in the lab, I met Dr. Lucy Miller. Lucy was the last student of Dr. Anna Jean Ayres (1920–1988) who advocated for people with special needs and is known for her definition of sensory integration and learning disorders. She wrote the Sensory Integration Tests and revised them into the Sensory Integration and Praxis Tests in 1989. She developed a theoretical account of these problems using the science of her times as the foundation to support her ideas.

Several of these tests were later expanded to other disorders, notably attention deficit hyperactivity disorder (ADHD), but sensory integration and sensory processing disorders met with resistance by the practitioners in Pediatrics and Psychology owing to the lack of objective means to measure outcomes or to even diagnose. Lucy had spent 15 years developing and perfecting observational inventories and sensory-based therapeutic interventions when we met. I offered to measure the children at her STAR center of Colorado, but we had no resources to do so in the lab at the time we met.

The meeting took place at a conference I attended in NYC in 2014. We were both invited speakers and because I had to run right after my talk to the Annual Meeting of the Society for Neuroscience in Washington DC, I had no time to further speak with Lucy. We met at 6:00 AM before her talk and discussed possible ways forward. One of these plans involved applying for a grant to the Henry Wallace Family Foundation and proposing a project to derive objective biometrics that would serve as outcome measures of the therapy to treat sensory processing disorder (SPD). This was the term used to augment and help operationalize sensory integration disorder, so it could also be treated.

At the STAR center a group of scientists had been funded by the Wallace Family to do research that disambiguated and demonstrated the existence of this disorder. One of the problems I saw with the whole strategy was the insistence to remain separate from other neurodevelopmental disorders like ASD and ADHD, as if that was at all possible. All these disorders have underlying sensory motor issues. Any attempts to try and stratify groups with one disorder and not the other had failed, not only because of the pervasiveness of sensory-motor processing and integration issues across all neuro-developmental disorders, but also because the observational methods to track the continuous flow of behaviors leave out important information that can contribute to the definition of the problem and to its treatment as well.

My movement sensing approach, grounded on the classical principle of reafference, but now extended to the spontaneous movements, taking a multilayered, multimodal and multifunctional approach to the real-time analyses and tracking of behavior has a fair chance to profile each child in a personalized manner, and as such, provide the outcome measures of the treatment that they needed. I submitted the proposal and it got funded, so we set up the study at the STAR center that Lucy directs.

The research involving such therapeutic interventions is a bit tricky, because one does not want to interfere with the hour-based therapy the parent is paying for. The tasks to track the progress must be brief and naturally integrated into the therapy session. Further, the biosensors ought to be unobtrusive and provide high sampling rates and synchronous options to register data in tandem from both the therapist and the child. Part of the intervention success can be tracked according to the evolution of rapport between the therapist and the child over time.

Our biometrics of dyadic exchange are designed to precisely track synergies across the two bodied in motion, in naturalistic settings. Here we are still tracking 10 children with SPD and 10 age- and sex-matched controls in tandem with the therapist as a dyad, across 15 sessions. Although the work is still in progress, we already have our first results and they are encouraging.

To track the progress of the therapies we performed a small set of tasks that involved (1) pointing; (2) walking; and (3) resting state. These are shown in Fig. 8.20. The latter we called the statue game, so the children were willing to do it pre and post session, together with the therapist in a playful way.

These tasks are performed within a period of 5–10 minutes, 5 times during the intervention (every three sessions) to track the trends in performance along the multifunctional levels of control in the nervous systems. Pointing probes both voluntary control and spontaneous retracting motions. In Fig. 8.20A we show the child pointing forward to the wall and then in Fig. 8.20B we show the retracting motion, which has an initial segment which is spontaneously returning from the wall, and then a second segment that is directed to a goal location on the head, the nose.

Walking probes automatic control. Resting under the instruction to remain still (like a statue) probes the presence of involuntary motions (small tremor like fluctuations and ticks) as a measure of volitional control to remain motionless at will.

The tasks help us delineate interactions between the child and therapist according to external stimuli (from visually guided tasks), vestibular stimuli (from balance-related task components), internal stimuli (from postural-related task components), and planning (from components involving praxis, hand-use in grasping tasks and following instructions, etc.). Since these tasks are performed naturally, the video tapes tell us the order in which they were administered. The order is tailored to each child and may differ across sessions. An example of a graph summarizing the probability map the child and therapist spanned over the course of 15 sessions is shown in Fig. 8.21. Here the color gradient denotes the different sources of stimuli and the small markers are from the child while the large markers are from the therapist (the same therapist throughout all sessions).

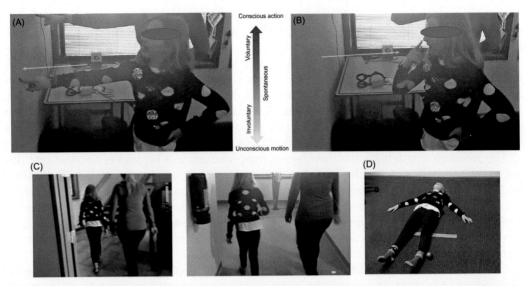

FIGURE 8.20   Tracking sensory-based occupational therapy at the STAR center of Colorado. Wearable biosensors (IMUs) across the body of the child and therapist collect data while the child performs a set of three tasks to probe (1) Voluntary goal directed motions in (A) using a pointing forward task and automatic (spontaneous) retractions in (B) using the immediate motions after touching the wall to touch the nose. (2) Automatic level of control during natural walk across the therapy center in (C). (3) The statue task requires the child to remain still (D) to measure the volitional control of the brain over the body and the presence of involuntary motions and ticks.

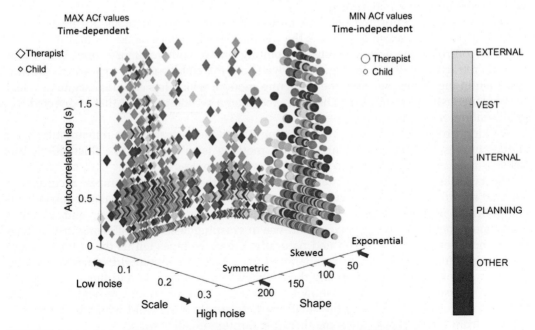

FIGURE 8.21   The probability map of the dyad interacting over the span of 15 sessions.

Fig. 8.22 shows the evolution of the graph per session, for one child and his therapist summarizing the landscape of estimated PDFs across 12 sessions. As before, the small markers denote the child while the larger markers denote the therapist. The color gradient denotes the type of task component the marker corresponds to and the axes denote the types of PDFs that the therapy sampled. Note that the uniformly sampled space of probabilities is about 500Ms along the Z-axis of time lags. The areas of high noise and low-shape values plot the points corresponding to the minimum auto-correlation values (circles). The areas of low noise and high-shape values plot the points of maximal auto-correlation values (diamonds).

When the therapy started, the spaces representing processes, whereby past events are maximally correlated with future events and those in which past and future events are minimally correlated, were disjointed and sparse. The therapist was primarily in the independent region (to the right of the map) and the child in the region to the left, indicating points of maximal auto-correlation, mostly alone. By the third visit, there is a distribution of the points representing the child across the space for the first 500Ms lag, and the scatter evolves such that by the last four visits there is more covering and more uniformity in the distribution of these PDFs for the child and the therapist. Fig. 8.23 shows the full map for the body parts (see legend) color coded by body part and sensor location. As before, the larger markers represent the therapist's body parts, and the smaller markers represent the child's body parts. The unfolding of these over time is shown across 12 sessions in Fig. 8.24. These correspond to those shown in Fig. 8.22. There we can visualize the body parts that contribute to the coverage of the space of probabilities, as the therapist and the child build good rapport. The trunk is particularly involved in this evolution of the scatter toward the coverage of the space. A summary of the trends of the PDFs can be appreciated in Fig. 8.25 for each visit. Two main families are distinguishable. One that is more symmetric with lower dispersion (high-shape values and low-scale values corresponding to the maximally auto-correlated nodes). The other is more skewed, toward the Exponential range and with high dispersion (small values of the shape and large values of the scale corresponding to the minimally auto-correlated bodily nodes).

To track all body parts of the child as he performed the three tasks of Fig. 8.20, we show the evolution of a ratio we computed in Fig. 8.26. This figure shows the value of the ratio for each node of the grid of biosensors. This ratio quantifies the number of time-independent body parts (defined as those which minimally auto-correlate past events with future events) over the time-dependent ones (which maximally auto-correlate the past and future events). If the quantity is above one, it means that there are more body parts acting independently than body parts maximally auto-correlated. The color shows the evolution per visit. We can see that generally the thorax and lumbar areas (the core of the child's body) act independently, whereas the wrists have a ratio below 1, indicating the prevalence of maximally auto-correlated states. The ankles start out with ratio above 1 but gradually move toward a more auto-correlated prevalence, suggesting more synergistic activities likely higher coordination of the legs and hands. This result is congruent with the dyadic analyses whereby the trunk areas built the bridge between these types of processes, that is, connecting the initially disjointed regions of the probability map. Finally, using the similar idea for the ratio, we plot in Fig. 8.27 the evolution of each body part across the visits. There we can see that the lumbar area remains above 1 (independent)

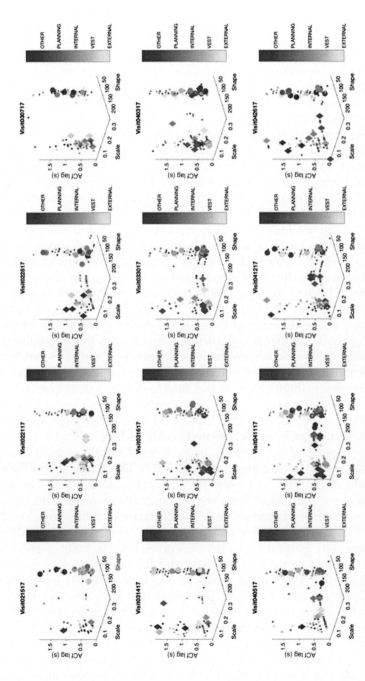

FIGURE 8.22 Longitudinal break down of the probability map as the coverage of the space evolved with the therapy. Regions that were initially disjointed began to merge, thus indicating the dyadic reciprocity across the space of probabilities. Color bars indicate the sensory-based type of task.

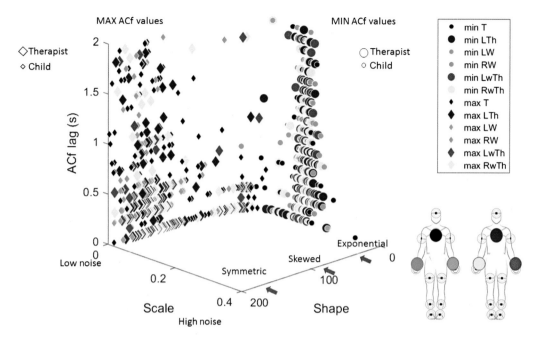

FIGURE 8.23   The probability map of Fig. 8.21 expressed as a function of body parts color-coded for the child and the therapist.

while the thorax oscillates initially but turn above 1 after the 7th visit. This quantifies a higher frequency of this node in the minimally auto-correlated state, suggesting more independence, in contrast to the hands, which remain mostly below value of 1. This suggests for the hands a prevalence of maximally auto-correlated states. The legs oscillate between values of the ratio below and above 1. This suggest a bodily hierarchy of probabilistic processes whereby the core is more independent with the legs keeping a balance between independent and co-dependent activity from events in the past, and the hands with a prevalence of co-dependence from past events. The maximal auto-correlation at the hands indicates their extensive deliberate use in the tasks the child must coordinate with the therapist at all times. When the child is walking or pointing there is rapport with the therapist that builds gradually and is captured by the longitudinal trends of the probability spaces we uncovered, coded by sensory input in Fig. 8.22 and by body parts in Fig. 8.24.

These biometrics can be used to track the dyadic exchange between the child and the therapist as it evolves longitudinally, and as it unfolds in each session. Besides serving as outcome measures of therapy effectiveness, these biometrics also inform on the relationship that the therapist's body builds with the child's body over time.

Our way of measuring the outcomes of a therapy is a radical departure from the notion that the patients' nervous systems respond to treatment in isolation and should be tracked in isolation. In any situation involving a child, there will likely be an adult involved with

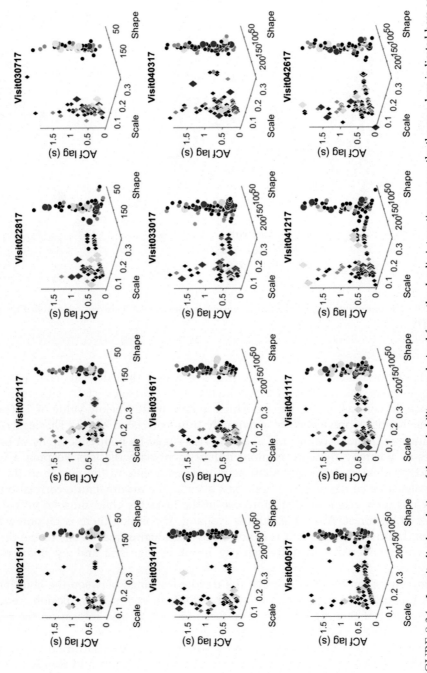

FIGURE 8.24 Longitudinal evolution of the probability map derived from the dyadic interactions across the therapy days indicating to body nodes participation in each region.

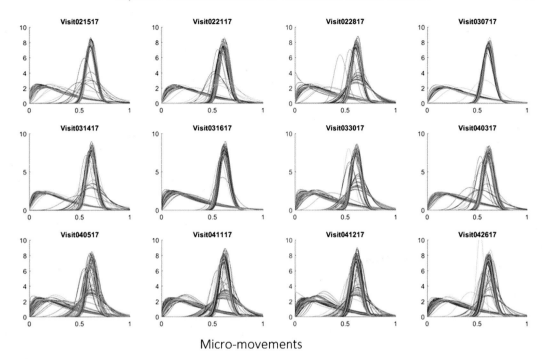

Micro-movements

FIGURE 8.25   Longitudinal changes in the PDFs of the maximal auto-correlation RLQ regions and the minimal autocorrelation LUQ regions. Activity drawn from the linear acceleration amplitude micro-movements.

the child, who will tend to take the lead in the interaction to guide the child's actions. Yet, if the adult leads all the time, the opportunities for the child to self-discover agency in social contexts will be impeded. In this case we examined, the interaction was well balanced and led the child's body to develop coordinated motions with an anchor at the core of the body (the trunk and lumbar areas). This development was gradual and took place within a true dyadic exchange.

Dyadic exchange may take place in situations at a clinic (with the clinician or therapist), at the school (with the teacher) and at the home (with the parent). In all instances, understanding the dyadic interaction as a basic unit of social exchange will be important. One of the applications of this idea for dyadic exchange is in Bernstein's DoF problem that we discussed in Chapter 4. I submit the idea that entraining the DoF from one's body with those from the other person's body in the dyad mitigates Bernstein's redundancy problem of finding a consistent path to a set of goals in external space. One solution to this problem that we discussed in Chapter 4 was the locally linear isometric embedding (LLIE) equation (1), which partitions the DoF variability into the rank and the null subspaces of the linear transformation the equation resolves locally. While revisiting the model in light of the variability and the 2 autocorrelation-gamma spaces that we defined in Fig. 8.15, we found a correspondence between the DoF partition and the movement classes that points to the statistical interpretation of the previous DoF analyses we performed of the forward

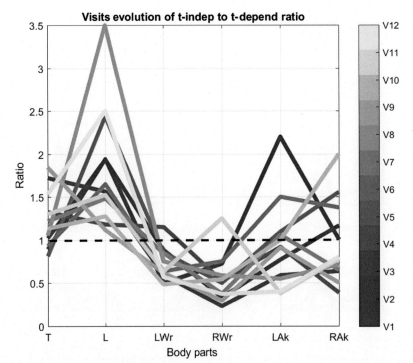

FIGURE 8.26 Evolution of the ratio of independent over co-dependent variations color-coded by the visit order by body part. Values above 1 imply a prevalence of processes whereby events in the past are minimally auto-correlated with future events. Values below 1 imply a prevalence of processes whereby events in the past are maximally auto-correlated with future events.

and backward pointing task. This is expressed in Fig. 8.28. The figure in panel A highlights the solution of the problem of finding a local inverse map at a one to one correspondence between the goal space and the internal configuration space of joint angles (in this case). The local solution depicted in panel **28B** and panel **28 C** shows the two subspaces where one produces motion toward the target (**28 C**), thus reducing recursively the hand-target distance, and the other which is redundant to the task and produces self-motions (**28B**) without advancing the hand to the target.

The space in **28 C** spans the task relevant dimensions of the internal configuration space. The space in **28B** spans the task incidental dimensions of the configuration space. On the Gamma parameter plane, the variabilities corresponding to these two subspaces have very distinct signatures for movements that are deliberate (e.g., the motions of the hand from resting state to the target) and movements that are spontaneous (e.g., the retracting motions from the target to the resting hand position).

When examining neurotypical adults, we found that the LUQ space is in correspondence with the task incidental dimensions (the null space) of both the deliberate and spontaneous motions (Fig. 8.28D). The RLQ space is in correspondence with the task relevant dimensions (the rank) of both the deliberate and the spontaneous segments. In other words, the variability of the forward reaches deliberately aimed at the target, tend to the Gaussian ranges of the Gamma shape parameter (i.e., most symmetric case in panel D) and have the lowest dispersion (lowest Gamma scale value). In the RLQ we have the maximal auto-correlation values and the minimal auto-correlation values, which map to

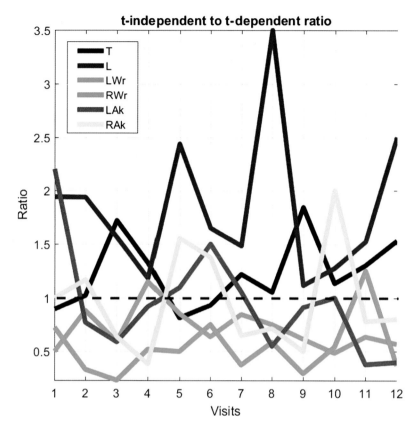

FIGURE 8.27 Plot of evolution over visits of the ratio in Fig. 8.26 computed by body part. Similar explanation. Different style of visualization of the same data.

forward (deliberate) and backward (spontaneous) segment in the task relevant rank space. Here mapping our metrics to Alicia Juarrero's notion of constraints[17], the forward movement variability is dynamics invariant (a form of context free constraint); but the backward movement variability is dynamics dependent (a form of context sensitive constraint *beneath awareness*). The DoF in the self-motion, task incidental LUQ also provide insights into the types of variability we empirically observe. These have a "here and now" random feature for the spontaneous retractions (backward null space) well characterized by the Exponential distribution; but a deliberate component of the LUQ brings awareness in the self-motion manifold, also a context sensitive constraint *above awareness*. Table 8.1 depicts different types of variability that we have empirically identified and modeled with our geometric equation (1) of Chapter 4.

When the set of goals in a social context involves synergizing with the body in motion of an external agent that the actor is trying to communicate with for social exchange, one of the underlying issues is the distribution of DoF across the spaces that separate DoF in above mentioned rank and null subspaces. The recruitment and release of these DoF will depend on the levels of entrainment these biorhythms have, as that shapes the synergies that self-emerge during the exchange. Our research program aims at creating the conditions to scaffold the ability of the bodily biorhythms of the child with a NND to entrain

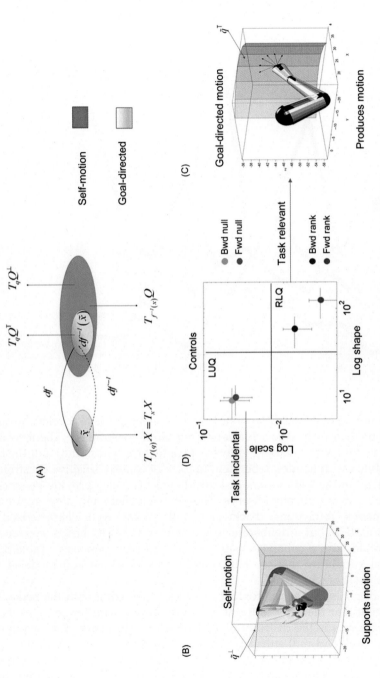

FIGURE 8.28 Rethinking Bernstein's DoF problem in terms of the types of variability we have empirically quantified for each of the rank and null spaces and each of the deliberate and spontaneous movement classes.

TABLE 8.1 Suggested empirically informed parallels between human behavior and Juarrero's model of context sensitive and context free constraints

LUQ BWD NULL (Spontaneous)	RLQ BWD RANK (Spontaneous)
Beneath Awareness	Beneath Awareness
Min AC (Future—Past Events Independence)	Min AC (Future—Past Events Independence)
Beneath Awareness: Here and Now	***Beneath Awareness***: Context Sensitive
LUQ FWD NULL (Deliberate)	RLQ (FWD RANK (Deliberate)
Above Awareness	Above Awareness
Max AC (Future—Past Events Independence)	Max AC (Future—Past Events Dependence)
Above Awareness: Context Sensitive	*Above Awareness*: Context Free

with the external rhythms, including those of the social agents in different contexts. We believe that taking this research path and utilizing these objective biometrics will lead us to the quantification of the person's autonomy of the brain over the body as an important index to express overall changes in the person's quality of life.

CLOSING REMARKS

In this chapter, we have offered new objective methods to track dyadic social exchange over time and to provide a sense of treatment effectiveness at the personalized level and at the level of dyadic social interactions. At the basic science level, the biometrics we introduce have implications for the study of neuro-developmental disorders in general. At the societal level, they invite legislators to rethink the issue of insurance coverage for sensory-motor-based OT, as these activities are now quantifiable with simple means that include the use of commercially available instrumentation with good sampling rates and synchronously registered multimodal signals.

Our experimental paradigms define simple tasks, that can be easily integrated into the therapy session, and probe multilayered nervous systems functioning. These can be adopted as standardized tasks to perform at home on a regular basis to help profile in a personalized manner the child's neurodevelopment. These methods are a step forward toward a type of personalized medicine that has the potential to help develop social exchange. They can indeed help us define quality of life for the individual receiving the treatments and interacting with its social medium with agency.

References

1. Andersen RA, et al. Posterior parietal areas specialized for eye movements (LIP) and reach (PRR) using a common coordinate frame. *Novartis Found Symp.* 1998;218:109—122. Discussion122-8, 171-5.
2. Batista AP, et al. Reach plans in eye-centered coordinates. *Science.* 1999;285(5425):257—260.
3. Buneo CA, et al. Direct visuomotor transformations for reaching. *Nature.* 2002;416(6881):632—636.

4. Kalaska JF, Caminiti R, Georgopoulos AP. Cortical mechanisms related to the direction of two-dimensional arm movements: relations in parietal area 5 and comparison with motor cortex. *Exp Brain Res.* 1983;51 (2):247–260.
5. Georgopoulos AP, Caminiti R, Kalaska JF. Static spatial effects in motor cortex and area 5: quantitative relations in a two-dimensional space. *Exp Brain Res.* 1984;54(3):446–454.
6. Scott SH, Sergio LE, Kalaska JF. Reaching movements with similar hand paths but different arm orientations. II. Activity of individual cells in dorsal premotor cortex and parietal area 5. *J Neurophysiol.* 1997;78 (5):2413–2426.
7. Cui H, Andersen RA. Different representations of potential and selected motor plans by distinct parietal areas. *J Neurosci.* 2011;31(49):18130–18136.
8. Torres EB, et al. Neural correlates of learning and trajectory planning in the posterior parietal cortex. *Front Integr Neurosci.* 2013;7:39.
9. Bernshteĭn NA. The co-ordination and regulation of movements. 1st English ed. Oxford, New York: Pergamon Press; 1967:196. xii.
10. Faisal AA, Selen LP, Wolpert DM. Noise in the nervous system. *Nat Rev Neurosci.* 2008;9(4):292–303.
11. Torres EB, et al. Neonatal diagnostics: toward dynamic growth charts of neuromotor control. *Front in Pediatr.* 2016;4(121):1–15.
12. Torres EB. Two classes of movements in motor control. *Exp Brain Res.* 2011;215(3-4):269–283.
13. Torres EB, Yanovich P, Metaxas DN. Give spontaneity and self-discovery a chance in ASD: spontaneous peripheral limb variability as a proxy to evoke centrally driven intentional acts. *Front Integr Neurosci.* 2013;7:46.
14. Brincker M, Torres EB. Why study movement variability in autism? In: Torres EB, Whyatt CP, eds. *Autism: the movement sensing approach.* US: CRC Press-Taylor and Francis; 2017:1–37.
15. Brincker M, Torres EB. Noise from the periphery in autism. *Front Integr Neurosci.* 2013;7:34.
16. Theodoridis S, Theodoridis S. *Introduction to pattern recognition: a MATLAB approach.* Burlington, MA: Academic Press; 2010:219. x.
17. Juarrero A. What does the closure of context-sensitive constraints mean for determinism, autonomy, self-determination, and agency?. *Prog Biophys Mol Biol.* 2015;119(3):510–521.
18. Torres EB. New symmetry of intended curved reaches. *Behav Brain Funct.* 2010;6:21.
19. Torres E, Andersen R. Space-time separation during obstacle-avoidance learning in monkeys. *J Neurophysiol.* 2006;96(5):2613–2632.
20. Torres EB, Zipser D. Simultaneous control of hand displacements and rotations in orientation-matching experiments. *J Appl Physiol (1985).* 2004;96(5):1978–1987.
21. Torres EB, Zipser D. Reaching to grasp with a multi-jointed arm. I. Computational model. *J Neurophysiol.* 2002;88(5):2355–2367.
22. Torres EB. Signatures of movement variability anticipate hand speed according to levels of intent. *Behav Brain Funct.* 2013;9:10.
23. Torres EB, et al. Autism: the micro-movement perspective. *Front Integr Neurosci.* 2013;7:32.
24. Wu D, et al. *Computational Psychiatry modelling leads to an empirically derived biomarker in ASD clicnial trial.* Annual meeting of the society for *Neuroscience.* Washington, DC; 2017.
25. Wu D, Torres EB, Jose JV. Micromovements: the s-Spikes as a way to "zoom in" the motor trajectories of natural goal-directed behaviors. In: Torres EB, Whyatt C, eds. *Autism: the movement sensing approach.* Boca Raton, London, New York: CRC Press; 2017:217–223.
26. Parker BR. *Good vibrations: the physics of music.* Baltimore: Johns Hopkins University Press; 2009:274. vi.
27. Cristian F. Probabilistic clock syncrhonization. *Distrib Comput.* 1989;3(3):146–158.
28. Mitra SK. *Digital signal processing: a computer-based approach.* 4th ed. New York, NY: McGraw-Hill; 2011. xx, 940 p.
29. Moustakides GV. *Basic techniques in digital signal processing.* Greece: Editions Tziola; 2003.
30. Kolevzon A, et al. A pilot controlled trial of insulin-like growth factor-1 in children with Phelan-McDermid syndrome. *Mol Autism.* 2014;5(1):54.

Index

Note: Page numbers followed by "*f*" and "*t*" refer to figures and tables, respectively.

A

ABA. *See* Applied behavioral analysis (ABA)
ABIDE. *See* Autism Brain Imaging Data Exchange (ABIDE)
"ABSTRACT" condition, 81, 83–85
Acceleration, 12, 318
Accelerometers, 360
Action–Perception Psychology program, 145
Actual social dyadic exchange, 107
Adaptive learning, 295
Adaptive motor learning, 162
  consequent motor performance upon cognitive variations, 163*f*
Adaptive process, 295, 315, 322
ADHD. *See* Attention deficit hyperactivity disorder (ADHD)
Adjacency matrix, 344
Adolescents, 169
ADOS. *See* Autism Diagnosis Observational Schedule (ADOS)
ADOS-generic (ADOS-G), 448, 450–452, 451*f*, 452*f*
  test, 380–381
  versions, 381
"Affordances", 199–200
AFNI. *See* Analysis of Functional NeuroImages (AFNI)
Age-dependent biometrics, 292–293
Age-dependent shifts
  age-dependent shifts in stochastic signatures, 435*f*
  fundamentally different dynamics of spontaneous involuntary activity, 436*f*
  in probability distribution functions, 433–439
AHA! moment, 126–135
  anchors of etiology and longitudinal dynamic diagnoses, 133–135
  drawing board, 131–133
  empirically driven ideas on biorhythms' statistics, 128–131
AIMS. *See* Alberta Infant Motor Scores (AIMS)
Alberta Infant Motor Scores (AIMS), 31–32
Alonso, Alicia, 325, 326*f*
American Pediatrics Academy, 400–401

American Psychiatric Association (APA), 399–400
American Recovery and Reinvestment Act, 112–113
American Recovery Reconstruction Act (ARRA), 329
  NSF grant, 329
"*Analisis Matematico*", 272
Analysis of Functional NeuroImages (AFNI), 396
Anatomical increment
  differences in rates, 31
  differences with absolute anatomical values, 32*f*
Anchors of etiology, 133–135
Angular speeds (AS), 410
Angular velocity vector, 246–247
Animal models of disorder, 401
ANS. *See* Autonomic nervous system (ANS)
APA. *See* American Psychiatric Association (APA)
Applied behavioral analysis (ABA), 57, 155, 333, 400–401
Arm-hand system, 220
Arm-hand systems, 196, 232–234
ARRA. *See* American Recovery Reconstruction Act (ARRA)
AS. *See* Angular speeds (AS); Asperger's syndrome (AS)
ASD. *See* Autism spectrum disorders (ASD)
Asperger's syndrome (AS), 440, 529–531
Attention deficit hyperactivity disorder (ADHD), 540
Audio-visual media, 518–522
Audio-visual reafferent activity, 289–290
Auditory brainstem responses, 366–368
Auditory processing, 366
Autism. *See* Autism spectrum disorders (ASD)
Autism Brain Imaging Data Exchange (ABIDE), 381
  ABIDE I, 396
  ABIDE II, 396
  databases, 396
  demographic records, 429–430
  demographics, 421
  repositories, 422
  sets, 406
Autism Diagnosis Observational Schedule (ADOS), 12, 399–400, 439, 536
  ADOS-2, 381, 448, 450, 451*f*, 452*f*

Autism Diagnosis Observational Schedule (ADOS)
(*Continued*)
  ADOS-based scoring systems, 450–452
  cost of, 388
  dyadic exchange, 380–381
  test, 388
Autism spectrum disorders (ASD), 12, 56–57, 62–63,
    86, 97–98, 112, 170*f*, 179*f*, 195, 331, 337–339, 377,
    379, 381–382, 396, 446–448, 510, 539
  AHA! moment, 126–135
  autistic somatic-motor phenotype, 168–173
  behavior, 135–140
  brain, 441
  excess noise in, 425–427
  group, 446–448
  involuntary head motion in FMRI to faulty
      diagnosis, 396–399
  measurement, 111–118
    set-up for experiments, 114*f*
    tracking complex boxing motions, 113*f*
  motor research with children, 150–162
  one plain random process *vs.* many running in
      tandem within interconnected systems, 118–126
  open access to rescue, 176–180, 180*f*
  parameterizing changes in nervous systems,
      142–145
  physiological signals underlying cognitive decisions,
      145–150
  science meets technology transfer and
      commercialization, 140–142
  scientific community and clinicians, 173–174
  sensing through movement, 162–168
  sensory-motor plasticity assessment, 372–373
  sensory-motor systems, 177
  speed-dependent stochastic signatures of head
      micro-movements, 426*f*, 427*f*
Autistic somatic-motor phenotype, 168–173
  data-driven separation, 175*f*
  differentiating females and males, 174*f*
  stochastic evolution of decision movement latency,
      173*f*
  stochastic trajectories of participants with
      shifts, 172*f*
Autistic system, 401
Auto-correlation
  matrix, 528–529, 530*f*
  values, 548–549
Automatic
  movement segments of PD patients, 243–248
  separation between forward and backwards
      segments, 73
Autonomic nervous system (ANS), 137–139, 162, 258,
    356–357, 366–368

Autonomic/autonomy, 513
  autonomous control, 292–293
    of muscle synergies, 315
  biorhythmical motions, 135–137
  biorhythms, 372
  from bottom-up process, 92–100
  and enteric systems, 354–355
  system, 135, 136*f*
  systems, 378–379
Avatars dyadic interactions with children, 377–379

B
Ballet partnering, 339–353
  activity profiled during dance, 347*f*
  bottom–up approaches, 350*f*
  building avatar for visualization purposes, 341*f*
  complex positional trajectories from LEDs, 342*f*
  dancers in T-pose for calibration, 340*f*
  local curvature data type, 343*f*
  network configurations, 349*f*
  network connectivity metric, 348*f*
  new data type to quantify complex trajectories, 343*f*
  outputs from system aid visualization of complex
      dances and precise behavioral modeling, 341*f*
  pipeline of data analyses and signal processing, 345*f*
  representation of peripheral network, 345*f*
  stochastic signatures, 353*f*
  top–down metric of "togetherness", 352*f*
  top–down tracking, 347*f*
  visualization tool to tracking dynamics of each
      network and coupled network, 346*f*
Ballet to personalized precision psychiatry, 325–326
Baltimore Longitudinal Study of Aging, 184
Baseline condition, 354–355
Basic pointing task, 143, 145–152, 147*f*, 148*f*
BBAI tool. *See* Brain body avatar interface tool (BBAI
    tool)
BCBA. *See* Board Certified Behavioral Analyst (BCBA)
BCI. *See* Brain–computer interface (BCI)
Behavior(al), 135–140
  analysis, 401
  analysts, 139
  inventories, 176–177
  methods, 401
  in PM knowledge network, 339
  therapies, 56–57, 377
Bending metric, 342–343
Bernstein's DoF problem, 211, 547–548, 550*f*
  in boxing routines, 305–317
    angular speed temporal profiles, 309*f*
    de-adaptation from new dynamics, 313–317
    full routine breakdown to upper limbs'
        motions, 306*f*

goal identification, 310*f*
mapping from Torres-Zipser PDE, 311*f*
path conservation from DoF decomposition, 312*f*
sample hand kinematics of complex sequences, 316*f*
Toy model of the two link arm motions, 308*f*
Uri Yarmush performing J-C-H-U, 314*f*
model, 201, 201*f*
Binary PLV matrices, 288–289
Biomarkers, 381
Biometrics, 20, 322, 545
for clinical trials, 454–455
ADOS, 439
age-dependent shifts in probability distribution functions, 433–439
from deliberate autonomy to measure of quality of life, 393–396
DSM, 403–415
dual diagnosis and treatments in neurodevelopmental disorders, 399–403
excess noise in autism, 425–427
females to males ratio of ASD, 440–454
hope keepers of SMIL, 392, 393*f*
increment data in neurodevelopment, 429*f*
incremental IQ scores, 432*f*
individualized tracking of trial effects, 479–494
involuntary head motion in FMRI to faulty diagnosis of ASD, 396–399
longitudinal tracking of somatic-motor change, 472–479
Pandora's box, 391–392, 415–425
SHANK3 deletion syndrome, 456–457
significance and potential consequences of results, 428–433
in sports research, 295
Biorhythms, 283, 328–329, 334–335, 516, 525, 532
parameters, 91–92
statistics, 128–131
comparison of current statistical model, 129*f*
shifts in stochastic signatures of speed amplitude and timing, 132*f*
Biosensors, 282
"Black box" approach, 285–286
Board Certified Behavioral Analyst (BCBA), 57, 377
Bodily biorhythms, 54–55, 79–80
Bodily network of connections, 362
Bootstrapping method, 406–408, 411*f*
Bottom up connectivity, 344–346
Bottom–up processes, 347–348, 364
autonomy from, 92–100
Boxing routines, Bernstein's DoF problem in, 305–317
Bradykinetic movements, 178
Brain, 15

brain–body
loops, 62–63
network dynamics, 522–524, 523*f*
togetherness, 369
EEG waveforms, 284–285
to heart, 164–168
network, 362
signal, 285–286
Brain body avatar interface tool (BBAI tool), 106*f*
Brain–computer interface (BCI), 120
Brainless body approaches, 364–365
BrainNet Viewer software tool, 288–289
Brownies to PDE, 183–185
Bernstein DoF problem, 185*f*

C
Caballero, Carla, 406, 410
California Institute of Technology (CALTECH), 211–212, 221–231, 503
CAR. *See* Clinically at risk (CAR)
Cartesian configuration space, 266
CDC. *See* Center for Disease Control (CDC)
CDFs. *See* Cumulative distribution functions (CDFs)
Center for Disease Control (CDC), 19
CDC–WHO data, 26
Central nervous system (CNS), 1–2, 125, 188–189, 354–355, 366–368, 428
Central system, 135, 136*f*
Cerebro-cortical biorhythms, 372
Children, 169
dyadic interactions with avatars, 377–379
motor research with, 150–162
sensory-motor, 154–155
CIs. *See* Confidence intervals (CIs)
Clinical trial adventure, 454
Clinically at risk (CAR), 28
Closed-loop co-adaptive interfaces, 372
computer-body interface, 372–373
Closing/closed feedback loops
in parametric form, 372–377
co-adaptive body-computer interface, 374*f*
playing biorhythms back to participants, 375*f*
between PNS and CNS
first-order stochastic rule, 42–43
physical growth and neurodevelopment in neonates, 18–40
searching for volition in comma state, 1–18
Clustering coefficient, 472
CNS. *See* Central nervous system (CNS)
Co-adaptive interface, 91–92
dynamic interface, 331
visuo–motor interface for ASD, 131

Cognitive loads, 162
    induces adaptive motor learning, 157−161, 159f
    quantifying stochastic shifts between fluctuations,
        160f
Cognitive neuroscience, 404
Cognitive psychology, traditional theories of, 87−88
Cognitive science
    PHD in, 186−190
    program of UCSD, 183−184
Cognitive-motor task, 440
    designing, 145−150
        basic pointing task output from hand movement
            trajectories, 148f
        experimental paradigm to study natural
            movements, 147f
Coherence network, 366
Color bar, 351−353
Color-coding motions, 33
Comma state, searching for volition in
    revisiting principle of reafference, 17−18
    sample wearables and output traces, 2f
    serendipitous encounters, 1−3
    from spontaneous random noise to well-structured
        signals, 3−17
Commercialization, 140−142
Comorbidities, 56, 333
Complex bodily movement, 508−516
Computational simulation, 58
Computer programming, 279−280
Confidence intervals (CIs), 128
Connectivity
    index, 472
    metrics, 380
Conscious control, 144
Conscious motor signature, 154
Conscious volition, 191
Contemporary biosensors, 517
Continuous dynamics, 144
Continuous feedback, 239−240
Continuous flow of physiological parameters, 166−167
Continuous motion, 135−137
Continuous physiological signals, 9−11
Continuous random process, 508−509
Continuous time-cost functional, 186
Control(s), 28
    autonomous, 292−293
    conscious, 144
    strategy, 315
Coordinate charts and inner product, 259
Coordinate functions, 262−263
Coordinate system, 259
Coordinate transformations, 263
    matrix, 206
        numerical estimation of positive definite, 266−267
    and new distance metrics, 266

Coping biological system, 455
Cost surfaces, 197−198
Coupled biorhythms, dance of, 339−353
Coupled brain−body synergies, 362
Coupled dynamics, 107, 335−337, 362, 372
    network, 362
Coupled network, 350
    of connections, 362
Coupled PNS−CNS modules, 369
Critical baseline measurement, 457−469
    bar plots of proportion of count of actual angular
        velocity peaks, 466f, 467f
    different sources of biophysical data, 460f
    examples of atypical stochastic signatures, 470f
    experimental set-up and sample parameters, 459f
    frequency histograms of normalized peak angular
        velocity for footsteps, 463f
    raw data from angular speed capturing joint
        rotations, 462f
    stochastic signatures of micro-movements, 468f
Cross-coherence analyses, 362
Cuban Math-CS program, 278
Cubic spline interpolation, 361−362
Cumulative distribution functions (CDFs), 143, 469
CUNY Graduate Center, 511

D
Dancer
    complex positional trajectories from LEDs, 342f
    dyadic network, 369
    identify synergies, 348−350
    individual, 344
    rhythms, 344−346
    in T-Pose for calibration, 339−340, 340f
    tracking of multiple biorhythms, 328−329
    visualization tool to track dynamics, 346f
Dancing ensemble, 327
Dark-matter, 335−337
"Dark Matter of Science", 327, 327f, 331
DARPA. See Defense Advanced Projects Agency
        (DARPA)
Data
    acquisition, 508
    analyses, 327−328
    data-driven approach, 31−38
    inclusive of younger infants, 26
    processing pipeline, 527, 527f
DDDC. See Douglass Developmental Disability Center
        (DDDC)
De-adaptation from new dynamics, 313−317
Decision-making, 143, 145−146, 152, 171
    embodied, 74f, 80f
    motor behavior, 157
    paradigm to study, 78−83
Defense Advanced Projects Agency (DARPA), 4

Deficit models of condition, 55
Degree distribution graph, 287–288
Degrees of freedom (DoF), 115, 183, 305–306, 507–508
    decomposition, recruitment, and release, 209–211, 211*f*
    geometrization of DOF problem, 197–200
    in self-motion, 548–549
Deliberate autonomy, 137–139, 144
    to measure of quality of life, 393–396
Deliberate motions, 307, 309, 315
Deliberate processes, 125–126, 162, 165*f*, 166*f*, 355, 379
Deliberate segments, 298
Derivative data. *See* Increment data
Detrended fluctuation analyses (DFA), 406, 409*f*
Developmental Douglass Disability Center, 373–375
DFA. *See* Detrended fluctuation analyses (DFA)
Diagnostics Statistical Manual (DSM), 133, 403–415, 539
    ASD outliers on psychotropic meds, 416*f*
    automatic classification of ASD *vs.* TD participants, 414*f*
    DFA, 409*f*
    DSM-5, 399–400
    DSM-based criteria, 448
    excess noise accumulation, 407*f*
    makers, 438
    mean of alpha values, 413*f*
    quantifying outlier's patterns, 417*f*
    relationship between α values and types of noise, 408*f*
    spontaneous involuntary head motions, 405*f*
    systematic increase of dispersion and skewness, 417*f*
Discrete clinical scores, 9–11
Discrete observational clinical inventories, 365
Discrete trials training (DTT), 155
"Disembodied actions", 55
Disembodied brain approaches, 364–365
Dispersion, 128–130
Dissecting simulation theory, 60–65
Distributed intelligence hypothesis, 356–357
DoF. *See* Degrees of freedom (DoF)
Dorsal visual stream, 133
Double blind cross over, 455
Douglass Developmental Disability Center (DDDC), 112, 168–169
Drug trial, 535, 537*f*
    revisited, 535–536
DSM. *See* Diagnostics Statistical Manual (DSM)
DTT. *See* Discrete trials training (DTT)
dx^{MODEL} vector, 245
Dyadic behavior, 332
Dyadic exchange, 339, 547–548
Dyadic interactions, 339

biometrics, 375–376
between children and avatars, 377–379
co-adaptive closed loop interface for stochastic-based feedback control, 379*f*
Dynamic(s), 188–189
    coupled brain–body–heart networks, 368
    diagnosis, 134, 365
    memory, 531
    sensory-augmentation, 248

E

Early intervention program (EIP), 20, 377–378, 388, 400–401
ECG. *See* Electrocardiography (ECG)
EDA. *See* Electrodermal activity (EDA)
EEG. *See* Electroencephalography (EEG)
EIP. *See* Early intervention program (EIP)
Electrocardiography (ECG), 354–355, 360–361
    heart data, 361–362
    sensors, 522, 524
    signals and track, 285–286
Electrodermal activity (EDA), 150
Electroencephalography (EEG), 142–143, 150, 283, 357, 360–361, 375–376
    biorhythms of, 526
    noninvasive techniques, 52
    sensors, 522, 524
    signals, 357–358
    visualization tools to study EEG activity, 287*f*
    wireless EEG system, 522
Electromyography (EMG), 54–55, 142, 508
EMG. *See* Electromyography (EMG)
End effector, 307
English as Second Language (ESL), 274
Enobio accelerometer, 362
Enobio wireless EEG device, 360–361
Enteric nervous system (ENS), 137–139
Entrainment beneath awareness, 525–535
Error, 266–267, 319
    error correction codes, 60
ESL. *See* English as Second Language (ESL)
"*Estaba escrito*", 276
Euclidean geodesic path, 207–208
Euler-Rodrigues angle-vector, 310–311
Euler-Rodrigues parameter, 243–244
Excess noise in autism, 425–427
Excess skewness, 430
Exogenous sources, 122–124
Experimental paradigm, 237–242, 238*f*
    hand trajectories to target and back to posture, 241*f*
    speed profiles for patient's trajectories, 242*f*
    susceptibility of τ to lesion in left PPC manifests, 242*f*

Exponential distribution, 115, 119, 120*f*, 126–128, 297, 318, 513, 514*f*, 548–549
External observer, 101–103, 193–195, 194*f*
External sensory source, 201, 232–234
External vantage point of observer, 332
Externally reinforced process, 169
Extraneous noise, self-generated kinesthetic noise distinguishing from, 91–92
Eye Link Eye tracker system, 291

F
F5 neurons, 48–49
Fancy camera-based method, 402
Fano factor, 38, 253, 254*f*, 490*f*
FAO. *See* Food and Agriculture Organization (FAO)
Fast dynamic coupled exchange, 328–329
Fatigue, 317, 507–508
Females to males ratio of ASD, 440–454
    ADOS-2 and ADOS-G scores, 451*f*
    age corrected ADOS-2 scores, 449*f*
    data-driven approach for cluster detection, 446*f*, 447*f*
    differences in relative physical head excursions, 445*f*
    involuntary head excursions, 444*f*
    movement decisions, 442*f*
    simple pointing behavior, 441*f*
First-order stochastic rule, 42–43
Fitbits, 450
Food and Agriculture Organization (FAO), 18–19, 28
Frequency matrix, 344, 362
Frequency-domains, 327–328
Functional magnetic resonance imaging (fMRI), 52, 53*f*
    involuntary head motion in, 396–399
    signals, 28–29
Fundamental biorhythms of heart, 335, 375–376

G
Gait biometrics, clinical trial tracking using, 457–472.
    See also Biometrics for clinical trials
    critical baseline measurement, 457–469
    visualization tools, 469–472
"GAIT" condition, 81, 83–85
Gamma distributions, 126, 529
Gamma LUQ. *See* Gamma plane left upper quadrant (Gamma LUQ)
Gamma moments, 288–289, 479, 518, 538*f*
Gamma parameter plane, 70, 128, 133, 142, 156, 531, 532*f*, 548
Gamma parameters, 256, 302–303, 370–372, 408–410, 415, 430, 464, 527–528
Gamma PDFs. *See* Gamma probability distribution functions (Gamma PDFs)
Gamma plane left upper quadrant (Gamma LUQ), 33
Gamma plane right lower quadrant (Gamma RLQ), 33
Gamma plane visualization tool, 255

Gamma probability distribution functions (Gamma PDFs), 117, 253, 370–372, 464–469
Gamma probability distributions, 117, 302
Gamma process, 9, 126, 127*f*, 168, 253, 319, 351–353, 361, 529
Gamma RLQ. *See* Gamma plane right lower quadrant (Gamma RLQ)
Gamma signatures, 134, 172*f*
Garbage of Cognitive Neuroscience, 396, 397*f*
Gauss Map and differential, 260–261
Gaussian distribution, 20–22, 24, 36, 118*f*, 131
Gaussian ruler, 146–149
Geodesics, 203, 210*f*
    curvature, 231
    geodesic-generating PDE invariants, 219–221
Geometric approach to movement modeling, 196
Geometric invariants of unconstrained actions, 231–235
    area-perimeter symmetry, 232*f*
    geometric nature of τ and map across reachable board space, 236*f*
Geometric solutions, 201–206
    locally linear isometric embedding, 205*f*
    parameterization of posture space, 202*f*
Geometric symmetry, 239, 240*f*
Geometrization of DOF problem, 197–200
    geometrization of dynamics problem, 200*f*
    static perception *vs.* perception for action, 199*f*
Geometry, 199–200
Gestural micromovements, 48, 55
Global solution, 209
Gold standard to diagnose autism, 380–381, 439
Gradient, 203, 204*f*
Graph theory, 104–105, 380
Growth charts, 19, 22, 406–408

H
Habilitate motor control, 142
Hand motion trajectory, 508
Hand path in space, 190
Hand signatures, 82–83
Hand trajectories of typical child, 152, 153*f*
"HAND" condition, 81, 83–85
Hartigan dip test of unimodality, 464, 465*f*, 475–479
Health mobile concept, 154
Heart IBI signal, 357–358, 360, 523*f*
High-risk (HR), 38, 40*f*
Home-made visualization tools, 529
HR. *See* High-risk (HR)
Human biorhythms, 333–334
Human mirror neuron systems, 52–59, 53*f*
    mirror neuron systems without reafferent codes, 54–59
Human observation-based methods, 450

I

iASD. *See* Idiopathic autism spectrum disorder (iASD)
IBI. *See* Interbeat interval (IBI)
I-Corps. *See* Innovation Corps (I-Corps)
IDD events. *See* Independent identically distributed events (IDD events)
Ideal expert zone of RLQ, 302–303
Ideal population, 5, 22
Idiopathic autism spectrum disorder (iASD), 457, 461–469, 470f
IEP. *See* Individualized educational program (IEP)
IGERT. *See* Integrative Graduate Education and Research Traineeship (IGERT)
IGF-I. *See* Insulin like growth hormone factor I (IGF-I)
IID assumption. *See* Independent and identically distributed assumption (IID assumption)
Imitation, 48
"Impending-reach" sense, 503
IMU. *See* Inertial measurement unit (IMU)
Increment data, 399
Independent and identically distributed assumption (IID assumption), 9–11, 64f, 529
Independent identically distributed events (IDD events), 14f
Individualized educational program (IEP), 377–378, 400–401
Individualized tracking of trial effects, 479–494.
    See also Longitudinal tracking of somatic-motor change
    delayed changes in patterns with drug visit, 482f
    immediate and pronounced change with drug, 488f
    immediate changes
        with drug intervention, 490f
        in gait patterns, 484f
    improvements in feet patterns, 487f
    marked changes with drug intervention, 485f
    marked immediate changes with drug intervention, 491f, 492f
    modest overall change in motion patterns, 493f
    research program, 494f
Inertial measurement unit (IMU), 283, 360–361, 363f, 522–524
    Opal IMUs, 361–362
    sensor, 522, 527–528
Inevitable autonomy process, 125–126, 137–139, 144
Inevitable process, 125–126, 162, 165f, 356–357, 379
Infant data, 28
Innovation Corps (I-Corps), 2–3
Institutional Review Board (IRB), 421
"Instrumental conditioning". *See* "Operant conditioning"
Insulin like growth hormone factor I (IGF-I), 454, 535
Insurance-covered behavioral therapy, 539–540

Integrative Graduate Education and Research Traineeship (IGERT), 66, 67f, 71, 104
Intended motions, 302
Interbeat interval (IBI), 164, 356–357, 524–525
Interdisciplinary research, 337–339, 403
Internal configuration space, 186, 548
Internal nervous system, 193–195
Internal observer, 194–195
Internal sensory source, 201
Internal vantage point, 332
    external vantage point *vs.*, 194f
    of persons nervous systems, 335–337
Internally reinforced process, 169
Interpolation techniques, 342–343
Intrinsic space, 188
Invariants of geodesic-generating PDE, 219–221
Inverse Jacobian, 265
Invisible spontaneous movements, 154, 158
Involuntary head excursions, 404, 444f
Involuntary head motions, 28–29
    in FMRI to faulty diagnosis of ASD, 396–399
IRB. *See* Institutional Review Board (IRB)
Iterative process, 208–209

J

Jab-Cross-Hook-Uppercut routine (J-C-H-U routine), 295, 298
Jacobian coordinate-transformation matrix, 264
Jacobian matrix, 245–246
Jacobian transformation matrix, 201–203
Joint angle path in arm space, 190
Journal of Neuroscience, The, 421–424
Juarrero, Alicia, 513

K

Kalampratsiduo, Vilelmini, 329, 330f, 525
Kinematics, 188–189
    analyses of hand trajectories, 223
    criteria, 152, 188
    distinguishing deliberateness from spontaneity in signals, 12–15
Kinesthetic feedback, 48, 83, 162, 375–376
Kinesthetic reafference, 46–48, 86–88, 124, 186, 195
"Kinesthetic self" notion, 89
Kolevzon, Alexander, 454
Kolmogorov–Smirnov test, 143, 448, 469

L

Lab streaming layer (LSL), 283, 357, 360, 396, 522
Lateral intra parietal region (LIP region), 223
Lead-lag phase information, 344
Learning disorders, 540

Learning to detect expertise in sports
  Bernstein's DoF problem in boxing routines,
    305–317
  motor learning measurement in sports, 293–297
  sensing future speed in different contexts, 317–322
  spotting genius, 280–293
    Jillian Nguyen, 291–293
    Joe Vero, 281–283, 281f
    Neha Tadimeti, 283–290
    Ushma Majmudar, 283, 285f
  uncertain times, 271–280
    place values Math, 272f
    school of math and computer science at UH, 273f
    surviving immigration, 279–280
"Leave-one-out" decoding algorithm, 84
LED. See Light emitting diode (LED)
Left upper quadrant (LUQ), 12–14, 255, 353–354, 464,
    531
  counterparts, 531
  Gamma plane, 15
Light emitting diode (LED), 239, 339–340, 342–343
Linear map, 260
Linear speeds (LS), 410, 412f
  amplitude, 84
  maxima, 156
LIP region. See Lateral intra parietal region (LIP
    region)
Listening
  motions, 191
  movements, 158
Living nervous system, 118–119
LLE algorithm. See Local linear embedding algorithm
    (LLE algorithm)
LLIE. See Locally linear isometric embedding (LLIE)
Local curvature metrics, 342–343
Local linear embedding algorithm (LLE algorithm), 212
Local solution, 209
Locally linear isometric embedding (LLIE), 547–548
  of X into Q, 206–209
    DoF decomposition, 207f
    simulated tasks of reach family, 208f
Log-normal distribution, 115, 116f, 117–118
Log–log Gamma parameter plane, 77, 443–445
Longitudinal dynamic diagnoses, 133–135
Longitudinal tracking of somatic-motor change,
    472–479. See also Individualized tracking of trial
    effects
  connectivity patterns change, 478f
  drug effect on rotational angle patterns of feet, 480f
  network activity patterns, 477f
  stochastic shifts in probability distribution functions,
    481f
Low-functioning ASD, 94

Lowest Gamma scale value, 548–549
LS. See Linear speeds (LS)
LSL. See Lab streaming layer (LSL)
LUQ. See Left upper quadrant (LUQ)

M

MABC. See Movement assessment battery for children
    (MABC)
Magic spot, 378, 379f
Male ruler, 446–448
Malnutrition, 18–19
MARC. See Minority Access to Research Careers (MARC)
Match to Sample task (MTS), 156
  task, 150–151, 151f
  paradigm, 440
"Matematica-Cibernetica" program, 272
Mathematical Analysis. See "Analisis Matematico"
Mathematical modeling, 58
MATLAB language, 282, 406
Maturation process, 156
Maximum likelihood estimation (MLE), 76, 143
Mean curvature, 261
Media
  storm, 331
  types, 372
Meds (MEDS), 410
Melissa case, 14–15
Melissa's subtle movements, 9
Memory, 504
  memory-less exponential, 12
  photographic, 115
  retinal, 178
Mental decisions, physical movement kinematics as
    window into, 90–91
Mental endoafference, 122
Mental function, 249–250
Mental intent, bridge between physical volition and,
    100–104
Mental rehearsal, 327, 330
Metronome, 358
  brought spontaneous periodicity to signal, 358
  condition, 354–355
  sound, 366
m-Health. See Mobile Health (m-Health)
Micro-movements, 5, 362
  data, 415
    type, 9
  extraction from kinematics data, 10f
  parameterization, 285–286, 428
  perspective, 9–12
  trajectories, 344
  waveforms, 9

Mind blindness, 56, 333
Minority Access to Research Careers (MARC), 276–277
MIR Matlab toolbox for music processing, 526
Mirror-neuron systems theory (MNST), 46–50
  bridge between mental intent and physical volition,
    100–104
  habilitation and enhancement of volition, 104–107
  human mirror neuron systems, 52–59, 53f
  mirror neurons, 46f
    system, 47f
  moving own body with agency, 66–83
    automatic decomposition, 69f
    different paradigm to study decision-making,
      78–83
    embodied decision-making, 74f
    experimental epochs for decision-making
      paradigm, 72f
    locations of sensors, 77f
    mental and physical decisions, 78f
    participants performed different movement types,
      69f
    spontaneous hand retractions, 75f
    types of avatar, 71f
  neurodevelopmental disorders, 91
  new paradigm for, 49f
  physical movement kinematics as window into
    mental decisions, 90–91
  predicting task from performance variability, 83–86
  rethinking simulation theory, 59–65
  self-discovering cause and effect, 92–100
  self-supervision, 91–92
  small pedagogical parenthesis, 89–90
Mirror(ing), 45
  feedback, 317
  neurons, 45, 46f
    aspects, 50
    system, 47f, 54–59
  problem, 46–48
  property, 45
Mistry, Sejal, 406–408
MLE. *See* Maximum likelihood estimation (MLE)
MNST. *See* Mirror-neuron systems theory (MNST)
Mobile Health (m-Health), 2–3
Modularity, 351
  index, 472
  modularity-based coding of synergies, 351–353
Modules, 344, 350–351
Moment-by-moment fluctuations, 156, 157f
Motion, 166–167, 508
Motion Monitor software, 295
  graphs, 306–307
  of InnSports, 70
  package, 113–114

Motor
  control
    to classroom in do-it-yourself setting, 89–90
    experiments, 507–508
    system, 214–215
    formulation, 199–200
    learning measurement in sports, 293–297
      Uri Yarmush, 295–297, 314f
    phenomena, 508
    research with children, 150–162
      adaptive motor learning, 162
      cognitive loads induces adaptive motor learning,
        157–161
Motor control experiment designing using sports,
    297–305
  analytical methods, 301f
  different effects of speed changes, 300f
  movement trajectory decomposition, 299f
  statistics of normalized maximum speed labeling
    subjects, 304f
  velocity-dependent parameters reveal learning, 303f
Movement assessment battery for children (MABC),
  382
Movement modeling, geometric approach to, 196
Movement variability, 115
MTS. *See* Match to Sample task (MTS)
Multidimensional anthropomorphic robotic arm, 228
Multilayered biologically sound approach, 333
Multilayered nervous systems, 327–328, 551
Multimodal signals, 518, 551
Multiple biosensors synchronization, 507
Multiplicative random process, 117
Multirate processing schemes, 524

N
National Database of Autism Research (NDAR), 422
National Institute of General Medical Sciences
  (NIGMS), 278
National Institute of Health (NIH), 278
National Science Foundation (NSF), 2–3
National Science Foundation Innovative Corps (NSF I-
  Corps), 56–57, 419
  program, 536
Natural behaviors, 508
Natural gradient learning, 212
Natural movements, 361
NCE. *See* No-cost extension (NCE)
NCM. *See* Neural motor control (NCM)
NDA. *See* Non-Disclosure Agreement (NDA)
NDAR. *See* National Database of Autism Research
  (NDAR)
Neocortex, 188

Neocortical regions, 122
Neonatal intensive care unit (NICU), 1
Neonates, 2−3
  physical growth and neurodevelopment
    data-driven approach, 31−38
    detective work, 18−27
    pre-labeled data, 31
    scale-free relation, 38−40
    USC baby data, 28−31
Nervous system, 119, 135−137, 137f, 138f, 235, 248, 280−281, 360, 518
  disorders, 337
  parameterizing changes in, 142−145
Network
  behavior, 364
  connectivity analyses, 340
  graph, 366
Neural code, search for evidence within, 221−231
Neural control of movements, 332
Neural motor control (NCM), 45, 190−193
Neuroanatomical principles, 380
Neuroanatomy, 377
Neurodevelopment in neonates
  data-driven approach, 31−38
  detective work, 18−27
    contrasting linear vs. nonlinear change, 22f
    evolution of empirically estimated PDFs, 27f
    irregular growth profile of preterm babies, 25f
    non-Gaussian nature of growth data, 24f
    nonlinear and stochastic nature, 23f
    poised for maturation, 21f
  pre-labeled data, 31
  scale-free relation, 38−40
  USC baby data, 28−31
Neurodevelopmental disorders, 54, 91, 295−296, 332
  dual diagnosis and recommended treatments in, 399−403
    psychotropic medications, 400t
Neurological conditions, 337
Neurological disorders, 332
Neuromotor control, 36, 327−328
  problem architecture, 5
  taxonomy, 6f
Neurons, 226
Neurophysiological principles, 380
Neurophysiology, 377
  of nervous systems, 176−177
  for systems neuroscience, 55
Neuropsychiatric
  conditions, 337
  disorders, 332
Neuroscience, 139, 222

Neurotypical (NT), 358
  controls, 120−124, 173, 509
    age-matched, 443, 464−469
    comparison of performance, 515f
    sex-matched, 464−469
  female control, 535, 536f
  regimes, 353−354
Neutral objective observer, 339
  ADOS
    automatically tracking physiological somatic-motor signatures, 387f
    dyadic exchange, 380−381
    issues, 381−382
    performance, 386f
  autism research, 329
  ballet
    partnering, 339−353
    to personalized precision psychiatry, 325−326
  closing feedback loops in parametric form, 372−377
  Development Career Award, 332
  dyadic interactions between children and avatars, 377−379
  dynamic diagnostics and outcome measures, 353−372
  knowledge network gap, 332−339
  multifaceted student, 330f
  network connectivity metrics and stochastic analyses, 383f
  NLMF Foundation, 331−332
  pipeline of data analysis, signal processing, 384f
  sample network states, 385f
  thoughts on social mental spaces, 327−329
Neutral observer, 328−329
New Jersey Governor's Council, 417−418
Newark-Rutgers University Brain Imaging Center (RUBIC), 434
Nexus 10 software Biotrace, 360−361
Nguyen, Jillian, 291−293, 292f
  representative hand trajectories, 293f
  variability patterns of hand trajectories speed, 294f
NIC Neuroelectrics, 360−361
NICU. See Neonatal intensive care unit (NICU)
NIGMS. See National Institute of General Medical Sciences (NIGMS)
NIH. See National Institute of Health (NIH)
NLMF Foundation (autism), 331−332
No-cost extension (NCE), 329
"Noise from periphery", 162
Noise signatures, 76
Noise-cancellation
  techniques, 126
  therapy, 357−358

Noise-dampening techniques, 372
Noise-to-signal ratio (NSR), 5, 36, 38, 253, 254f, 415
Noise-to-signal transitions
  frequency, 40f
  tracking rate of adaptive change through, 39f
NoMEDS. *See* Not on meds (NoMEDS)
Non-Disclosure Agreement (NDA), 422
Non-Gaussian processes of complex system
  scale-free relation between, 38–40
Non-linear complex dynamics, 305
Nonlinear processes of complex system
  scale-free relation between, 38–40
Normalization, 126
  types, 130t, 150
Not on meds (NoMEDS), 410
NSF. *See* National Science Foundation (NSF)
NSF I-Corps. *See* National Science Foundation
    Innovative Corps (NSF I-Corps)
NSR. *See* Noise-to-signal ratio (NSR)
NT. *See* Neurotypical (NT)
Numerical estimation of positive definite coordinate
    transformation matrix, 266–267

O

OB. *See* Obstacle (OB)
Objective analyses, 139
Objective biometrics, 396, 402, 540, 549–551
Objective quantification, 58
Obstacle (OB), 222, 233f
Occupational and Physical Therapists (OT/PT), 161
Occupational therapists/therapies (OTs), 2–3, 56–57,
    292–293, 377–378
Office of Technology Transfer and Commercialization
    (OTC), 141
One plain random process *vs.* many running in
    tandem within interconnected systems, 118–126
"One-size-fits-all" approach, 20, 62, 339, 446–448
  statistical model, 5, 333–334
Opal IMUs, 360–362
Open access brain imaging data banks, 441
Open loop mode, 399, 526
Operant conditioning, 57
Optimal control
  framework, 187
  theory, 186
Ordinal clinical scores. *See* Discrete clinical scores
OT/PT. *See* Occupational and Physical Therapists
    (OT/PT)
OTC. *See* Office of Technology Transfer and
    Commercialization (OTC)
OTs. *See* Occupational therapists/therapies (OTs)
Overt motion patterns, 382–384

P

Paced-breathing condition, 354–355
Pandora's box, 391–392, 415–425, 418f
PAR. *See* Partially at risk (PAR)
Parallel distributing processing (PDP), 212
Parameterizing changes in nervous systems in real
    time, 142–145
Parametric form
  closing feedback loops in, 372–377
  co-adaptive body-computer interface, 374f
  playing biorhythms back to participants, 375f
Parietal reach region (PRR), 221, 224f, 225f, 227f, 230f
Parietal–occipital pre-frontal networks, 120–122
Parkinson's disease (PD), 115, 381–382
  automatic movement segments of PD patients,
    243–248
Parsing DOF reveals excessing deliberateness, 243–248
  forward reach goals, 243
  goals for reaching back to sensed posture, 243–248
  implementation of model using empirical data, 244f
Partially at risk (PAR), 38
PD. *See* Parkinson's disease (PD)
PDFs. *See* Probability distribution functions (PDFs)
PDP. *See* Parallel distributing processing (PDP)
Performance IQ (PIQ), 430
  of TD controls, 433
Performance variability
  blind trial classification of mental and physical
      variability, 85f
  predicting task from, 83–86
Peripheral intelligent nodes, 137–139
Peripheral nervous systems (PNS), 3, 125, 143–144,
    188–189, 366–368
Personalized precision psychiatry, ballet to, 325–326
Personalized somatic-motor driven criteria, 250–251
Phase locking value metrics (PLV metrics), 285–286,
    288–289, 469–470
Phase Space software, 329–330, 339–340
PHD thesis in cognitive science, 186–190, 186f
Phelan–McDermid syndrome
    (PMS), 438, 454–457
  tracking pharmacodynamics in IGF-1 trial of, 476f
Phylogenetically orderly taxonomy
  acquiring data from nervous systems, 354f
  additional tools to summarizing coupled PNS–CNS
      modules, 369
  autonomic signals, 358f
  dynamic diagnostics and outcome measures,
    353–372
  dynamic Gamma parameter plane, 355f
  experimental paradigm and instrumentation, 356f
  Gamma process, 359f

Phylogenetically orderly taxonomy (*Continued*)
  pipeline of analyses for coupled brain—body
        dynamics, 363*f*
  reciprocally connected network, 366—368
  statistics profile, 370—372, 371*f*
  top—down *vs.* bottom—up approaches, 365*f*
  weighted directed graphs for connectivity network
        analyses, 364*f*
Physical deafferentation, 122
Physical endoafference, 122
Physical growth in neonates
  data-driven approach, 31—38
  detective work, 18—27
    contrasting linear *vs.* nonlinear change, 22*f*
    evolution of empirically estimated PDFs, 27*f*
    irregular growth profile of preterm babies, 25*f*
    non-Gaussian nature of growth data, 24*f*
    nonlinear and stochastic nature, 23*f*
    poised for maturation, 21*f*
  pre-labeled data, 31
  scale-free relation, 38—40
  USC baby data, 28—31
Physical kinesthetic reafference, 119
Physical measurement, 58
Physical movement kinematics as window into mental
        decisions, 90—91
Physical volition, bridge between mental intent and,
        100—104
Physically disembodied brain, 122
Physiological biorhythms, 428
Physiological rhythms, 9
Physiological signals underlying cognitive decisions,
        145—150
  designing cognitive-motor tasks, 145—150
PIQ. *See* Performance IQ (PIQ)
Plasticity of autistic sensory-motor signatures, 102*f*, 104
PLV metrics. *See* Phase locking value metrics
        (PLV metrics)
PM. *See* Precision Medicine (PM)
PM knowledge network, 335—337, 339
PMS. *See* Phelan—McDermid syndrome (PMS)
PMS-O, 464
PMS-Y, 464
PNS. *See* Peripheral nervous systems (PNS)
Pointing
  behavior, 152—153
  trajectories of child, 156
Polhemus Liberty system, 68—70, 164
Polyfit, 443—445
Polyval, 443—445
Positive scalar function, 203
Post-synaptic density (PSD), 456
Post-Traumatic Stressed Syndrome, 58
Posterior Parietal Cortex (PPC), 115—117, 286, 503—504
Postural motor behavior, 230—231

Postures, 185, 187
  of arm, 186
  experimental paradigm to study coding of, 224*f*
  map, 191
PPC. *See* Posterior Parietal Cortex (PPC)
Pre-labeled data, 31
Precision Medicine (PM), 2—3, 332, 335
Precision Psychiatry, 332, 335, 339
Probability distribution functions (PDFs), 9—11, 62, 64*f*,
        117, 335, 518
  age-dependent shifts in, 433—439
  gamma, 117, 253
Probability distributions, 302, 305, 431, 529—531
Probability map, 535, 541, 545*f*, 546*f*
PRR. *See* Parietal reach region (PRR)
PSD. *See* Post-synaptic density (PSD)
Pseudo-science, 333
Psychiatrists, 176, 258
Psychological methods, 403
Psychologists, 176, 448
PTNs. *See* Pyramidal tract neurons (PTNs)
PubMed peer-reviewed papers, 424
Pullback action, 204
Push forward action, 206
Pyramidal tract neurons (PTNs), 50
  with mirror neuron activity, 51*f*
Python, 282, 406

Q
22q13 deletion syndrome, 454—457
Quality of life, from deliberate autonomy to measure
        of, 393—396

R
Random process(es), 117—119
RDoC. *See* Research Domain Criteria (RDoC)
Reaction time (RT), 72—73, 139—140
Reafference principle, 122
  adding dynamics to
    exploring sound preferences, 525—535
    let's play back sounds, 516—525
    objective outcome measures of treatments,
        535—551
    pipeline of data processing, 527*f*
    remarks, 551
    sound, 503—508
    tracking sensory-based occupational therapy, 542*f*
    wink and blink contain spontaneous segment,
        508—516
  classical, 541
  revisiting, 17—18
Reafferent codes, mirror neurons system without,
        54—59
Real-time biofeedback, 378

Reciprocally connected network, 366–368
  connectivity metrics to quantify top–down and
      bottom–up activities, 367f
  dissimilar reciprocal brain–body connections, 368f
Reflexes, 511
Regions of interest (RoI), 364, 369
Rehabilitation path, volition to, 15–17
Representational deafferentation, 130–131
Research Domain Criteria (RDoC), 250
"Reshape" behavior, 162
Rethinking Diagnoses and Treatments of Disorders
  ADOS
    automatically tracking physiological somatic-
        motor signatures, 387f
    dyadic exchange, 380–381
    issues, 381–382
    performance, 386f
  autism research, 329
  ballet partnering, 339–353
  ballet to personalized precision psychiatry, 325–326
  closing feedback loops in parametric form, 372–377
  Development Career Award, 332
  dyadic interactions between children and avatars,
      377–379
  dynamic diagnostics and outcome measures,
      353–372
  knowledge network gap, 332–339
    age-dependent evolution of stochastic signatures,
        336f
    need to improving science, 334f
    twenty years of autism research and trends, 338f
  multifaceted student, 330f
  network connectivity metrics and stochastic
      analyses, 383f
  NLMF Foundation, 331–332
  sample network states, 385f
  thoughts on social mental spaces, 327–329
Retracting motion, 171, 302, 541
Right lower quadrant (RLQ), 12–14, 255, 353–354, 464,
    531
RLQ. *See* Right lower quadrant (RLQ)
Robotic arm-brain interface, 228
RoI. *See* Regions of interest (RoI)
RT. *See* Reaction time (RT)
RUBIC. *See* Newark-Rutgers University Brain Imaging
    Center (RUBIC)
Rutgers University Center for Cognitive Science
    (Ruccs), 50–51

S

San José State University (SJSU), 274
Scalar function, 203
Scale-free relation, 38–40

Scaling factor, 252
Schizophrenia case, 112–113
  coordinate charts and inner product, 259
  coordinate transformations and new distance
      metrics, 266
  DOF
    decomposition, recruitment, and release, 209–211
    parsing DOF reveals excess deliberateness,
        243–248
    problem geometrization, 197–200
  empirical evidence for speed invariance, 211–218
  experimental paradigm, 237–242
  Gauss Map and differential, 260–261
  generating unique geometric solutions, 201–206
  geometric
    approach to movement modeling, 196
    invariants of unconstrained actions, 231–235
  invariants of geodesic-generating PDE, 219–221
  locally linear isometric embedding of X into Q,
      206–209
  neural motor control, 190–193
  numerical estimation of positive definite coordinate
      transformation matrix, 266–267
  patients, 248–258
    forward and back pointing task, 250f
    geometric analyses of three dimensional scatter,
        252f
    statistics of individualized local-linear
        decomposition, 257f
    stochastic signatures of DoF locally linear
        decomposition, 255f
  PHD in cognitive science, 186–190
  predictive reach, 183–185
  in search for evidence within neural
      code, 221–231
  switching
    research paths for clinical applications, 235–237
    between tasks, 263–265
  testing geodesic property and curvature measures
      with model, 261–262
  two vantage points, 193–195
Science meets technology transfer and
    commercialization, 140–142
Self-corrective processes, 125
Self-discovering cause and effect, 92–100
Self-emergence, 509–510
Self-generated kinesthetic noise distinguishing from
    extraneous noise, 91–92
Self-generated motions, 15, 33
Self-motion manifold, 205
Self-noise, 83
Self-supervision, 91–92, 356
Sensed posture, goals for reaching back to, 243–248

Sensed posture, goals for reaching back to (*Continued*)
   modulation of DoF in task-relevant, 247*f*
Sensing
   future speed in different contexts, 317–322
   through movement, 162–168
      from brain to heart, 164–168
Sensors' fusion schemes, 524
Sensory, 187
   augmentation, 235
   integration, 540
   processing function, 188–189
   sensory-based OT, 536–551, 542*f*
   sensory-motor
      approaches, 57
      augmentation, 396
      biomarkers, 322
      consequences, 57
      control, 112
      feedback loops, 119
      issues, 56–57
      issues in autism, 176
      patterns, 2–3
      phenomena, 178, 332
      substitution, 396
      systems, 313
      transformations, 192
   sensory-space dimension, 192
   sensory-substitution, 235
   space, 199–200
Sensory Motor Integration Lab (SMIL), 392
   and friends at total solar eclipse, 394*f*
   group, 280–281
   hope keepers, 392, 393*f*
Sensory processing disorder (SPD), 417–418, 536, 540
SHANK-related disruptions, 456–457
SHANK1 gene, 456
SHANK2 gene, 456
SHANK3 deletion syndrome. *See* Phelan–McDermid
   syndrome (PMS)
SHANK3 gene, 456
Shifts in PDF shape and dispersion with aging, 131
Shortest distance paths, 197–198, 351
"Significant hypothesis testing" paradigm, 173–174
Silber, Helbert, 276, 277*f*
Similar brain–body system, 61–62, 107
Simulation, 317
Simulation theory
   dissecting, 60–65
   rethinking, 59–65
Single Gamma PDF, 126
Singular value decomposition (SVD), 209
Sinusoidal waveform, 517
SJSU. *See* San José State University (SJSU)
Skewness, 131
Smart and connected health, 3

Smart watches, 450
SMIL. *See* Sensory Motor Integration Lab (SMIL)
Smooth transitions, 51–52
Social dance
   form, 380–381
   type, 52
Social dyad, 52, 337, 454
Social interactions, 48, 50
Social mental spaces, thoughts on, 327–329
   dark matter of science, 327*f*
   poised for accelerated change in medical research
      and patient care, 328*f*
Somatic system, 135, 136*f*
Somatic-motor
   autonomy, 119
   biorhythms, 372
   features, 448
   parameters, 186
   peripheral networks, 354–355
   physiology, 339
Sound, 503–508
   exploring sound preferences, 525–535
   waves, 517
Space, geometrization of DOF problem to move along
   shortest paths in, 197–200
SPD. *See* Sensory processing disorder (SPD)
Speech therapies, 56–57, 377–378
Speed, 146–149
   speed-independence of deliberate motions, 309
Speed invariance, 231, 235*f*
   empirical evidence for, 211–218
      empirical data from human subjects, 216*f*
      empirically registered speed invariance, 219*f*
      experimental setup and simulations, 214*f*
      model simulations and empirical data, 217*f*
SPIBA. *See* Statistical platform for individualized
   behavioral analyses (SPIBA)
Spline-based techniques, 342–343
Spontaneous goal-less movements, 510
Spontaneous media triggering, 98
Spontaneous motions, 315, 322, 511, 524–525
Spontaneous movements, 152–153, 511
"Spontaneous" movement segments, 154
Spontaneous process, 125–126, 162, 165*f*, 167*f*, 355, 366,
   379
Spontaneous random
   motions, 191
   noise to well-structured signals, 3–17
      comma patient, 4*f*
      data waste with current statistical assumptions, 7*f*
      distinguishing deliberateness from spontaneity,
         12–15
      integrated micro-movements, 11*f*
      integrated micro-movements waveform, 16*f*
      micro-movements perspective, 9–12

SPIBA, 8f
 volition, 17f
 from volition to path of rehabilitation, 15—17
Spontaneous retracting motions, 541
Spontaneous segment
 J2-C2-H2-U2, 298
 wink and blink containing, 508—516
Spontaneously self-discovered process, 373
Sports
 motor control experiment designing using, 297—305
 motor learning measurement in, 293—297
Spotting genius, 280—293
Springer Nature Scientific Reports, 423
State dependent force field, 220
State-of-the-art motion caption systems, 329
Stationary transitions, Gamma plane quadrant, 38
Statistical platform for individualized behavioral
  analyses (SPIBA), 2—3, 8f, 28—29
 analytical framework, 396—397
 framework, 9
 for IDD events, 14f
 methods, 34—36
 technology, 421—422
Statue game, 541
Stochastic process, 353—354
Stochastic signatures, 62, 68—70, 81, 82f, 91—92, 99,
  105f, 333—334, 351—353
"Straight lines", 197—198, 203—204
Straight pointing motor behavior, 157
Studies of Psychological and Cognitive Sciences, 78
Sub-networks togetherness
 coupled PNS—CNS modules, 369
 modularity patterns of brain, 369f
 quantification of top—down activity
   patterns, 370f
 visualization tools, 371f
Subconscious motor signature, 154
Superfluous motor noise, 510
Suppression mirror-neurons, 50
Surface cortical map, 120—122
Surviving immigration, 279—280
 faculty recognition, 280f
SVD. See Singular value decomposition (SVD)
Switching
 research paths for clinical applications, 235—237
 between tasks, 263—265
Synchronization, 522
Synergies, 351—353

T
Tadimeti, Neha, 283—290, 290f
 examples of visualization tools, 288f
 tracking stochastic signatures of learning, 289f
 visualization tools to study EEG activity, 287f

Task-relevant DoF, 247—248, 311, 313
TD. See Typically developing (TD)
Temporal disparities, 524
Temporal dynamics of motions, 76—77
Testing geodesic property and curvature measures
  with model, 261—262
Theoretical Gaussian distribution, 146—149
Theoretical geometric model, 331
Theory of Mind (ToM), 55
Therapeutic interventions, 541
Time
 geometrization of DOF problem to move along
   shortest paths, 197—200
 time-dependent forces, 199, 220
 time-domains, 327—328
 time-independent formulation, 219—220
 time-normalized curve, 312
 time-varying forces, 220
Togetherness, 364—365
 metric, 351
 top—down metric, 352f
ToM. See Theory of Mind (ToM)
Top down connectivity, 344—346
Top—down approach, 339, 344—346
Top—down processes, 364
Torres-Zipser PDE approach, 309
Track connectivity, 344—346
Tracked multisensory inputs, 375—376
Tracking of multiple biorhythms, 328—329
Transduction
 delay, 232—234
 transduction-transmission delays, 188—189
Transformation matrix, 266
Transient phenomenon, 220
Transmission delay, 232—234
Treatments, objective outcome measures of, 535—551
Trial and error process, 18, 376, 399, 509
Two-dimensional "toy model", 197—198
Typically developing (TD), 38, 97—98
 brain, 441
 children, 162, 372—373
 individuals, 28—29
 infants, 86

U
UH. See University of Havana (UH)
Unconstrained arm movements, 190
Unfolding DoF decomposition, 312
Unfolding stochastic signatures computation, 353—354
University of Havana (UH), 271—272, 272f
University of Southern California (USC), 2—3
 baby data, 28—31
Unprompted media triggering. See Spontaneous media
  triggering

USC. *See* University of Southern California (USC)
Ushma Majmudar, 283, 285*f*

V

Valdés, José Antonio, 274, 275*f*
Variability, 128–130, 218, 234, 511
Variable rates of change in shift, 131
Vector transformation action, 206
Velocity, 318
Velocity-dependent data. *See* Increment data
Ventral visual stream, 133
Veridical signatures, 372
Vero, Joe, 281–283, 281*f*
Virtual deafferentation, 122, 124
Virtual region of interest (vRoI), 94, 97–99, 372–373
Virtually embodied, 122
Visual illusion, 291
    stimulus, 283
Visual Science Society (VSS), 133
Visualization tool, 351–353, 364–365, 368, 469–472
    network of 14 joints dynamically unfolding in time, 473*f*
    visualization of dynamic states of peripheral network connectivity, 471*f*
Visuomotor systems, 199
Vocalization, habilitation and enhancement of volition to evoking, 104–107
Volition(al), 3, 17*f*
    control, 395
    habilitation and enhancement to evoking vocalization, 104–107
    searching for volition in comma state
        revisiting principle of reafference, 17–18
        sample wearables and output traces, 2*f*

serendipitous encounters, 1–3
from spontaneous random noise to well-structured signals, 3–17
Voluntary control, 541
vRoI. *See* Virtual region of interest (vRoI)
VSS. *See* Visual Science Society (VSS)

W

Walking experiment, 357–358
Wearable biosensing instrumentation registering signals, 88
Wearable biosensors, 284*f*
Wearable sensor, 2–3, 34*f*
Weight, 18–19
    actual data from weight parameters, 24
Weighted directed graph, 344
Western Psychological Services (WPS), 439
WHO. *See* World Health Organization (WHO)
Wild card, 298
Wireless
    bio-sensing electrodes, 522–524
    Nexus-10 device, 360–361
    technology, 360
Works and Days, The, 391
World Health Organization (WHO), 19
WPS. *See* Western Psychological Services (WPS)

Y

Yarmush, Uri, 295–297
Young adults, 169–171

Z

"Zeus eyes", 420, 424

Printed in the United States
By Bookmasters